ATOMIC PHYSICS 15

ATOMIC PHYSICS 15

FIFTEENTH INTERNATIONAL
CONFERENCE ON ATOMIC PHYSICS,
ZEEMAN-EFFECT CENTENARY

AMSTERDAM, THE NETHERLANDS,
5–9 AUGUST 1996

EDITORS:

H. B. VAN LINDEN VAN DEN HEUVELL
J. T. M. WALRAVEN
M. W. REYNOLDS
UNIVERSITY OF AMSTERDAM,
THE NETHERLANDS

World Scientific
Singapore • New Jersey • London • Hong Kong

Published by
World Scientific Publishing Co. Pte. Ltd.
P O Box 128, Farrer Road, Singapore 912805
USA office: Suite 1B, 1060 Main Street, River Edge, NJ 07661
UK office: 57 Shelton Street, Covent Garden, London WC2H 9HE

British Library Cataloguing-in-Publication Data
A catalogue record for this book is available from the British Library.

ATOMIC PHYSICS 15

Copyright © 1997 by World Scientific Publishing Co. Pte. Ltd.

All rights reserved. This book, or parts thereof, may not be reproduced in any form or by any means, electronic or mechanical, including photocopying, recording or any information storage and retrieval system now known or to be invented, without written permission from the Publisher.

For photocopying of material in this volume, please pay a copying fee through the Copyright Clearance Center, Inc., 222 Rosewood Drive, Danvers, MA 01923, USA. In this case permission to photocopy is not required from the publisher.

ISBN 981-02-3186-5

Printed in Singapore.

Contents

Preface ix

Generation of a "Schrödinger cat" of radiation and observation of
its decoherence 1
    S. Haroche, M. Brune, J.M. Raimond, E. Hagley, C. Wunderlich, A. Maali,
    J. Dreger, X. Maitre

Synthesis of entangled states and quantum computing 16
    J.I. Cirac, S. Gardiner, T. Pellizzari, J. Poyatos, P. Zoller

Entangled states of atomic ions for quantum metrology and computation 31
    D.J. Wineland, C. Monroe, D.M. Meekhof, B.E. King, D. Leibfried,
    W.M. Itano, J.C. Bergquist, D. Berkeland, J.J. Bollinger, J. Miller

Entanglement and indistinguishability: Coherence experiments with photon
pairs and triplets 47
    A. Zeilinger

Atom optics as a testing ground for quantum chaos 62
    C.F. Bharucha, J.C. Robinson, F.L. Moore, K.W. Madison, S.R. Wilkinson,
    B. Sundaram, M.G. Raizen

Coherent ultra-bright XUV lasers and harmonics 82
    J.S. Wark, K. Burnett, D. Chambers, M.H. Key, S.G. Preston, A. Sanpera,
    C.G. Smith, J.B. Watson, M. Zepf, J. Zhang, C.L.S. Lewis, A.G. McPhee,
    A.E.Dangor, P. Lee, A. Dyson, G.J. Pert, P.B. Holden, J.A. Ploues,
    G. J. Tallents, L. Dwivedi, M. Holden, P.A. Norreys, D. Neely, P. Fews,
    S. Moustazis, M. Bakarezos, P. Loukakos

Hollow atoms 97
    R. Morgenstern

Interdisciplinary experiments with polarized noble gases 113
    E.W. Otten

The creation and study of Bose–Einstein condensation in a cold
alkali vapor 132
    C.E. Wieman, E.A. Cornell, D. Jin, J. Ensher, M. Matthews, C. Myatt,
    E. Burt, R. Ghrist

Macroscopic quantum phenomena in trapped Bose-condensed gases 145
    Yu. Kagan, E.L. Surkov, G.V. Shlyapnikov

Doppler-free spectroscopy of trapped atomic hydrogen 158
    T.C. Killian, D.G. Fried, C.L. Cesar, A.D. Polcyn, T.J. Greytak,
    D. Kleppner

QED and the ground state of helium 169
    W. Hogervorst, K.S.E. Eikema, W. Vassen, W. Ubachs

Towards coherent atomic samples using laser cooling 180
    J. Dalibard

Bose–Einstein condensation of a weakly-interacting gas 192
    C.G. Townsend, N.J. van Druten, M.R. Andrews, D.S. Durfee,
    D.M. Kurn, M.-O. Mewes, W. Ketterle

Zeeman and his contemporaries: Dutch physics around 1900 212
    A.J. Kox

Zeeman's great discovery 221
    P.F.A. Klinkenberg

The Zeeman effect: A tool for atom manipulation 237
    C. Cohen-Tannoudji

The Zeeman effect a century later: New insights into classical physics 253
    J.B. Delos

QED effects in few-electron high-Z systems 271
    I. Lindgren, H. Persson, S. Salomonson, P. Sunnergren

Lamb shift experiments on high-Z one- and two-electron systems 289
    Th. Stöhlker

Fundamental constants of nature 305
    L.B. Okun

Response of atoms in photonic lattices 313
    D. van Coevorden, R. Sprik, P. de Vries, A. Lagendijk

Hydrogen-like systems and quantum electrodynamics 328
    M.G. Boshier

New experiments with atomic lattices bound by light 343
    A. Görlitz, M. Weidemüller, T. Hänsch, A. Hemmerich

Bloch oscillations of atoms in an optical potential 358
    E. Peik, M.B. Dahan, I. Bouchoule, Y. Castin, Ch. Salomon

Quantum decoherence and inertial sensing with atom interferometers 374
    D.E. Pritchard, M.S. Chapman, T.D. Hammond, D.A. Kokorowski,
    A. Lenef, R.A. Rubenstein, E.T. Smith, J. Schmiedmayer

Quantum effects in He clusters 391
    J. Harms, M. Hartmann, W. Schöllkopf, J.P. Toennies, A.F. Vilesov

Atoms in super-intense radiation fields 409
    H.G. Muller

Wave packet dynamics of excited atomic electrons in intense laser fields 422
    K.C. Kulander, K.J. Schafer

Nonlinear laser-electron scattering 431
   D.D. Meyerhofer

Comparing the antiproton and proton and progress toward
cold antihydrogen 446
   G. Gabrielse, D.S. Hall, A. Khabbaz, T. Roach, P. Yesley, C. Heimann,
   H. Kalinowsky, W. Jhe, B. Brown

Author Index 460

Preface

Atomic Physics 15 extends the series of books containing the invited papers presented at each International Conference on Atomic Physics (ICAP). The ICAP meetings, which are held every two years, provide the atomic physics community an opportunity to review problems of current interest and to consider future directions in the field. The fifteenth such meeting, which also celebrated the centenary of the discovery of the Zeeman effect, took place at the University of Amsterdam, The Netherlands, from 5-9 August 1996.

The program for this conference consisted of 31 invited papers and 247 contributed papers. The invited papers were presented in plenary lectures and all speakers have provided a written account of their contribution for this book. Unfortunately, Professor Okun was unable to come to Amsterdam to present his lecture. We are glad that the written version of his lecture could be included. The contributed papers were presented as posters; the abstracts were printed separately in a booklet that was provided to all conference attendees.

The initiative to organize ICAP-15 in Amsterdam, because of the celebration of 100 years of the Zeeman effect, was taken by the chairman of the conference, Professor P. Klinkenberg. Four speakers were invited to enlighten various aspects of the history and present role of the Zeeman effect in atomic physics. Evidently, many others did the same, intentionally or not. Think, for instance, of the Bose-Einstein-condensation experiments – this was the first ICAP where these experiments were discussed – as an illustration that the Zeeman effect is still prominently present in contemporary atomic physics.

The program of speakers was made by J.T.M. Walraven, H.B. van Linden van den Heuvell (both of the Van der Waals-Zeeman Institute of the University of Amsterdam, as well as of the FOM Institute for Atomic and Molecular Physics) and by W. Hogervorst (Free University). This was done in close collaboration with the Program Committee: A. Aspect, K. Burnett, J. Dalibard, G. Drake, T.F. Gallagher, M. Gavrila, T.W. Hansch, D. Kleppner, and D.J. Wineland. The evolution of the conference program started with the advice of the International Advisory Committee: E. Arimondo, V.I. Balykin, S. Chu, C. Cohen-Tannoudji, G. Drake, E.N. Fortson, T.W. Hansch, S. Haroche, V.W. Hughes, D. Kleppner, R.R. Lewis, I. Lindgren, H. Narumi, E. Otten, P.G.H. Sanders, F. Shimizu, B. Sonttag, S. Svanberg, H. Takuma, H. Walther, C.E. Wieman, D.J. Wineland, and J.P. Woerdman.

The actual organization of the conference was handled very enthusiastically by the conference secretary Friedje Witzenhausen and by the local committee consisting of A. Donszelmann, J.E. Hansen, T.W. Hijmans, J. Klinkenberg, and the editors of this book.

We are grateful for the sponsorship of the Genootschap ter bevordering van Natuur- Genees- en Heelkunde, the International Union of Pure and Applied Physics (IUPAP), the Nederlandse Organisatie voor Wetenschappelijk Onderzoek (NWO), the

Stichting Physica, the Stichting Pieter Zeeman Fonds, the Stichting voor Fundamenteel Onderzoek der Materie (FOM), Coherent Nederland, Delta Electronica bv, Elsevier Science, Institute of Physics Publishing, Newport bv, Princeton Instruments, Spectra Physics bv, Springer Verlag GmbH & Co KG, Technolas Umwelt und Industrieanalytik GmbH, the European Commission, the city of Amsterdam, and the University of Amsterdam.

Amsterdam, October 1996

H.B. van Linden van den Heuvell
J.T.M. Walraven
M.W. Reynolds

GENERATION OF A "SCHRÖDINGER CAT" OF RADIATION AND OBSERVATION OF ITS DECOHERENCE

S. HAROCHE, M.BRUNE, J.M. RAIMOND, E. HAGLEY, C. WUNDERLICH,
A.MAALI, J. DREYER and X. MAITRE

Département de Physique de l'Ecole Normale Supérieure
Laboratoire Kastler Brossel[a]
24 rue Lhomond 75231, Paris Cedex 05, France

We report the generation of a mesoscopic superposition of quantum states involving radiation fields with classically distinct phases – a "Schrödinger cat" like system. The experiment uses circular Rydberg atoms interacting one at a time with a few photon coherent field trapped in a high Q microwave superconducting cavity. The mesoscopic superposition is the equivalent of an "atom +measuring apparatus" system in which the meter is pointing simultaneously towards two different directions. By performing a two–atom correlation experiment, we have monitored the progressive decoherence transforming this superposition into a statistical mixture. This experiment provides a direct insight into a process at the heart of quantum measurement. Future developments are briefly discussed.

1 "Schrödinger Cats" and the Measurement Process in Quantum Theory.

The properties of mesoscopic superpositions of quantum states and the study of their decoherence is a fundamental topic in quantum physics. Over the last fifteen years, a lot of attention has been devoted to various theoretical approaches to this subject[1]. Recently, the development of new experimental techniques has opened this field to experimental observations. We report here the results of a cavity QED experiment performed at ENS in this domain[2]. We have generated a mesoscopic superposition of radiation states with different phases trapped in a high Q microwave cavity and observed the progressive quantum decoherence of this superposition. Before describing this experiment, it is useful to recall the general context of these studies.

The logic of the quantum world is quite at odds with our classical intuition. A particle may be simultaneously at different places and behave as a wave subject to interference phenomena. Two particles flying apart may get entangled together and exhibit striking non local properties[3]. It is very hard to understand these quantum effects because our intuition is based on the observation of nature at a macroscopic level, where all these puzzling phenomena are washed out. We have thus developed an intuition of the world where particle interferences have no real place. One of the main difficulties of the quantum theory is precisely to define the boundary between the micro– and the macro–world and to understand how the quantum effects progressively vanish when the size of the system under investigation is increased.

Schrödinger has illustrated this difficulty in his famous "cat" metaphor[4] which describes an hypothetical procedure to reproduce at the macroscopic scale a microscopic superposition of quantum states. A simple version of this metaphor involves

---

[a]Laboratoire de l'Ecole Normale Supérieure, associé au CNRS (URA 18) et à l'Université Pierre et Marie Curie

a cat trapped in a box with a single excited atom. The photon emitted by this atom is used to trigger a contraption which will kill the cat. Since there is at any given time an amplitude that the photon has been emitted and an amplitude that the atom is still excited, does it mean that the cat is, before being observed (i.e.before the box is opened), in a linear superposition of the dead and alive states?

One can give to this question a very simple and operational answer: if the cat were in a superposition state, it should be possible to design an experiment sensitive to the interference between the "cat alive" and "dead cat" states. In other words, the physics of the cat in the box should be different from the classical situation where one knows that the cat is dead with a given probability or alive with the complementary probability, without interference between these two possibilities. For a system as complex as a cat, which clearly cannot be described by a wave function, the outcome of any experiment of this kind is quite obvious and it is of course impossible that any interference effect could be observed. The issue is precisely to understand why the coherence of a macroscopic object is not observable and, subsidiarily, to put an upper limit to the size of feasible "Schrödinger cat"–like systems.

This discussion is important because it is directly connected to the theory of measurement in quantum physics [5]. In short, one may say that the cat is a macroscopic meter used to detect the photon. The "dead cat" state corresponds to the position of the meter when the photon has been emitted, while the "cat alive" state corresponds to the position of the meter when the photon is not yet there. One could as well replace the cat by a meter with two positions and rephrase the discussion in similar terms. The real question is "why do we never observe interferences between the states of a macroscopic meter?" or "why does a meter evolve instantaneously into one or the other position?". Stated in this way, the discussion is obviously related to the problem of the wave function collapse in quantum mechanics.

The measurement paradox has been discussed at length during the last six decades. Among the various explanations given for the non–observation of quantum coherences at the macroscopic level, the decoherence theories appear as the most convincing. They invokes the irreversible damping of the macroscopic coherences, due to the unavoidable coupling of the system to its large size environment [1]. This perturbation transforms the superposition involving different meter states into a statistical mixture described by a diagonal density matrix, for which all the information contents can be analyzed in classical terms. The nature of the relaxation process defines the basis in which this density matrix is diagonal, thus lifting the quantum ambiguity of the "microscopic object +meter" superposition state, which could be expressed in many arbitrary bases. An important parameter in these discussions is the size of the apparatus, in short the size of the meter. Since the quantum coherence does exist at the microscale of small systems and disappears at the large scale of macroscopic objects, there should be an intermediate "mesoscopic" scale at which the decoherence process should be slow enough to be experimentally accessible.

Theoretical models of decoherence in a quantum measurement have been considered. The simplest consists in coupling a spin–like two–level atom (the micro–system) to a quantized harmonic oscillator (the "meter"). An oscillator in a coherent

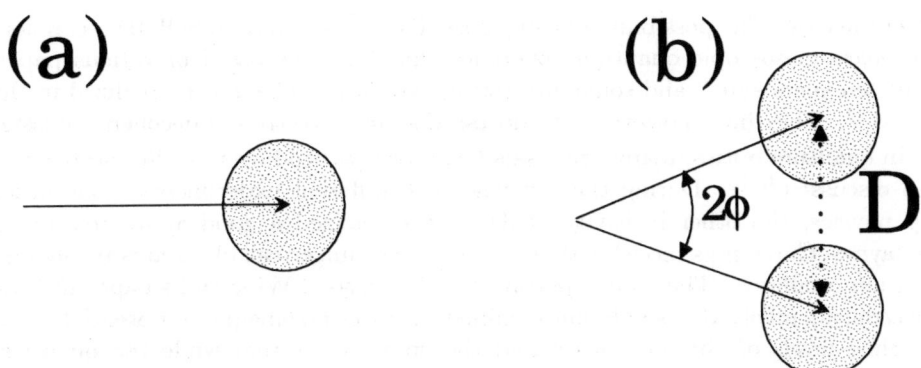

Figure 1: (a): Pictorial representation in phase space of a coherent state of a quantum oscillator. (b) the two components separated by a distance D of a "Schrödinger cat" corresponding to Eq. (1).

state [6] is indeed defined by a c–number α, represented by a vector in phase-space ($|\alpha| = \sqrt{n}$ where n is the mean number of oscillator quanta). Quantum fluctuations make the tip of this vector uncertain, with a circular gaussian distribution of radius unity (Fig. 1(a)). The interaction between the atom and the oscillator is described by the Jaynes–Cummings Hamitonian [7]. Consider the ideal measurement where the interaction entangles the phase of the oscillator ($\pm\Phi$) to the state of the atom (e or g), leading to the combined state:

$$|\Psi\rangle = \frac{1}{\sqrt{2}} \left(|e, \alpha e^{i\Phi}\rangle + |g, \alpha e^{-i\Phi}\rangle \right) \tag{1}$$

When the distance $D = 2\sqrt{n}\sin\Phi$ between the meter states is larger than 2, we will say that a "Schrödinger cat" state is obtained.

The environment is represented in this model by a reservoir of "bath oscillators" slowly dissipating the energy of the meter in a characteristic time T_r. This simple model lends itself to a complete analytical calculation of the meter relaxation [8,9]. It is found that the transformation of $|\Psi\rangle$ into a statistical mixture of the form

$$\frac{1}{2} \left(|e, \alpha e^{i\Phi}\rangle\langle e, \alpha e^{i\Phi}| + |g, \alpha e^{-i\Phi}\rangle\langle g, \alpha e^{-i\Phi}| \right) \tag{2}$$

occurs within a time scale $2T_r/D^2$, provided that $D \gg 1$. This result illustrates the basic feature of the quantum to classical transition. Mesoscopic superpositions made of a few quanta are expected to decohere in a finite time interval shorter than T_r, while macroscopic ones ($n \gg 1$) decohere quasi instantaneously and cannot be observed in practice.

2 From "Gedanken" to Real "Schrödinger Cats": the Cavity QED Route.

Up to recently, the discussions of Schrödinger cat situations belonged to the realm of "Gedankenexperiments". Owing to many technological progresses, this is no longer the case. The possibility of using Josephson junctions and SQUID technology to prepare mesoscopic quantum coherences has been discussed in various papers about ten years ago [10] and some interesting experiments have been realized in this context [11]. They have however not addressed so far directly the decoherence issue.

In quantum optics, many proposals to prepare Schrödinger cat like states have been discussed [12,13]. During the last year, two different experiments – one in ion trap physics, the other in cavity QED – have realized a good approximation of the Jaynes–Cummings model and generated Schrödinger cat like states involving a quantum oscillator. The trap experiment [14] (see also D.Wineland's paper in these proceedings) involved a single ion confined in an electromagnetic potential. Two hyperfine levels of the ion constituted the spin–like system while the quantized vibration of the ion in the trap modeled the oscillator. The decoherence of the vibration state superposition has however not yet been studied in this experiment.

In cavity QED experiments, atoms are coupled one by one to a single mode of the radiation field in a high Q cavity, which realizes the quantum oscillator. Many interesting studies have been performed with this system, both in the optical and in the microwave domains [15]. The possibilty of realizing Schrödinger cat like states of the field and exploring their quantum decoherence with Rydberg atoms coupled to superconducting microwave cavities has been discussed [9,16,17,18]. We have now realized an experiment of this kind [2] and we report here its results, which constitute the first observation of a phenomenon which lies at the heart of the quantum measurement process.

The mesoscopic state of the field is generated by sending a single Rubidium atom, prepared in a superposition of two circular Rydberg states e and g across a high Q microwave cavity storing a small coherent field $|\alpha\rangle$. The coupling between the atom and the cavity is defined by the coupling frequency Ω corresponding to the rate at which the two systems exchange a single photon, when they are at resonance [18,19]. For the Schrödinger cat experiment, the $e \to g$ atomic transition and the cavity frequency are slightly tuned off resonance (detuning δ), so that the atom and the field cannot exchange energy, but only undergo $1/\delta$ dispersive frequency shifts [12,16]. This shift corresponds to a single atom index effect with a quite unusual order of magnitude. The atom–field coupling during time t produces an atomic level dependent dephasing of the field. The index effect of an atom in e is opposite to the one produced by an atom in g. The interaction with the atom in a linear superposition of e and g thus generates an entangled state given (for $\Omega/\delta \ll 1$) by Eq. (1) with $\Phi = \Omega^2 t/\delta$ [18].

The states e and g are circular Rydberg levels with principal quantum numbers 51 and 50 (transition frequency $\nu_0 = 51.099$ GHz). These states correspond to the maximum possible value of the valence electron angular momentum, whose orbital is a thin circular torus around the atom's core [20]. They have a very long radiative life time (30 ms) and a very strong coupling to radiation, making them ideal tools for this experiment [18]. The cavity C is a Fabry-Pérot resonator with its axis normal to the atomic trajectory. It is made of two superconducting niobium mirrors (mirror distance 2.7 cm, mode waist $w_0 = 6$ mm). The coupling $\Omega/2\pi$, easily calculated from the size of the circular atom orbit and the volume of the cavity mode, is 25 kHz [19]. The cavity Q factor is 5.1 10^7 (photon life time $T_r = 160$ μs). The cavity is

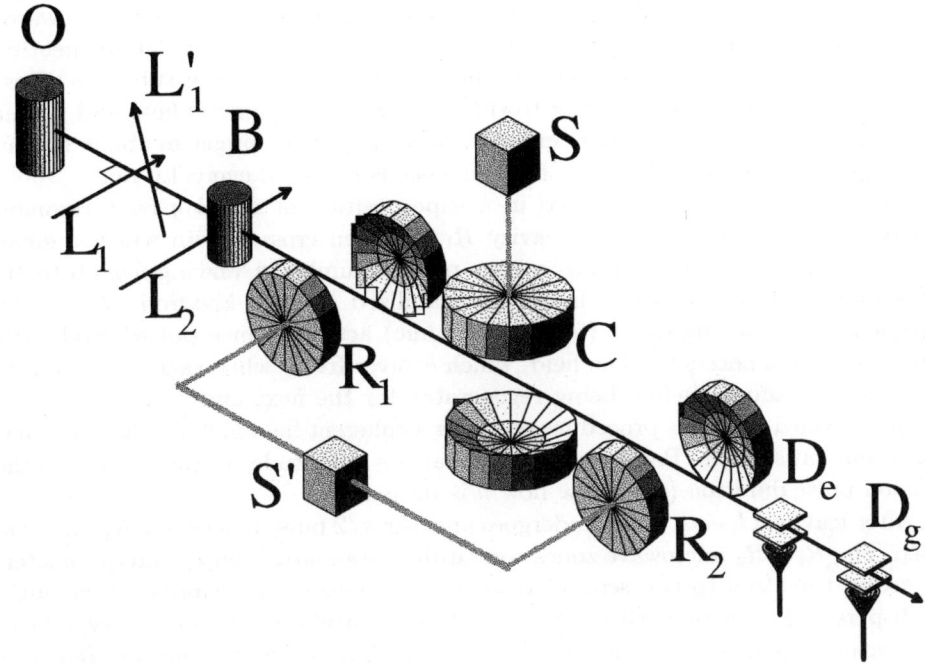

Figure 2: Sketch of the experimental set-up

tuned by adjusting the mirror separation, varying $\delta/2\pi$ between 70 and 700 kHz.

During its flight across the cavity, each atom experiences a time varying coupling with the cavity field, corresponding to the exploration of the transverse gaussian profile of the cavity mode. The atom–field interaction is adiabatically switched on and off, so that real transitions between e and g are avoided, even when the detuning $\delta/2\pi$ is as small as 70 kHz (i.e. less than 3Ω). This point, predicted by numerical simulations, was checked experimentally before we performed the Schrödinger cat experiment. When analyzing the atom–field interaction, it is convenient to replace the time varying coupling by a square pulse, with an effective interaction time defined as $t = (\sqrt{\pi/2})w_0/v$ where v is the atom's velocity. This effective interaction time is set to 19 μs by selecting atoms with $v = 400$ m/s. The dephasing angle Φ, inversely proportional to δ, is thus varied between 0.1 and 1 radian, an unusually large single atom index effect.

We now give more details about the set-up and the experimental procedure. The apparatus is sketched in Figure 2. It is cooled to 0.6 K by a He4-He3 cryostat making thermal radiation negligible (mean blackbody photon number in C: 0.05). The Rb atoms effusing from the oven O are pumped out of the $F = 3$ ground hyperfine level by a diode laser L_1 and optically repumped by a diode laser L'_1 oriented at 58^0 relative to the atomic beam. With a proper tuning of L'_1 in the Doppler profile, only atoms at 400 ± 6 m/s are prepared in $F = 3$. The atoms are then excited into the circular state e in box B. This process consists of a three step laser excitation from $F = 3$ (lasers L_2), followed by an adiabatic transfer towards the circular level involving radiofrequency transitions feeding angular momentum into the atom [21]. These transitions are induced between Rydberg sublevels whose degeneracy is removed by applying to the atom a static magnetic field and a time varying electric field pulse. Each "circularisation sequence" prepares about 1 atom in average in a 2 μs time interval. The sequence is repeated every 1.5 ms.

Each circular atom is prepared in a superposition of e and g by a resonant microwave $\pi/2$ pulse in a low Q cavity R_1. It then crosses C in which a small coherent field is injected, with an average photon number n varying from 0 to 10. The source S of this field is a frequency stabilized X–band klystron followed by a frequency multiplying diode (fourth harmonic) activated in a pulsed mode just before each atom enters C. The field, which evolves freely while each atom crosses C, relaxes to vacuum before being regenerated for the next atom ($T_r \ll 1.5$ ms). We have verified that this procedure prepares a coherent field in C by checking that its photon statistics are Poissonian [19] and that it's amplitude is proportional to the injection pulse duration (see below how n is measured).

After leaving C, each atom undergoes another $\pi/2$ pulse in a cavity R_2 identical to R_1. The $R_1 - R_2$ microwave zones constitute a standard Ramsey interferometer, subjecting the atom to two separated oscillatory pulses, one before and one after the dispersive interaction with C [22]. The time separation between the two zones for an atom moving at 400 m/s is $T = 230$ μs. The Ramsey zones are fed by a c.w. source S' whose frequency ν is swept across the atomic transition frequency ν_0. The atoms are finally counted in e and g by two field ionization detectors (D_e, D_g, detection efficiency 20%). With 50 000 events recorded in 10 min, the probability $P_g^{(1)}(\nu)$ to find the atom in g as a function of ν is finally reconstructed.

3 Preparation of a "Schrödinger Cat": a Complementarity Experiment.

When C is empty and far off resonance with the atomic transition, its effect on each atom may be neglected and this experiment then merely records the Ramsey fringes produced in the $e \rightarrow g$ transition probability by a sequence of two pulses acting on the atom in R_1 and R_2. Figure 3(a) shows the obtained fringe signal, when C is detuned by $\delta/2\pi = 712$ kHz. It is very instructive, in the context of this experiment, to view these fringes as a quantum interference effect. The $e \rightarrow g$ transition can occur either in R_1 or in R_2, corresponding to two paths in which the atom crosses C in g or in e. These paths are of course indistinguishable, leading, in the final transition rate, to an interference term between the corresponding probability amplitudes. The phase difference between these amplitudes is $2\pi(\nu-\nu_0)T$, so that $P_g^{(1)}(\nu)$ oscillates with the period $1/T = 4.2$ kHz. The fringe contrast, ideally 100%, is reduced to about 60% by various spurious effects (static and microwave fields inhomogeneities between R_1 and R_2 over the 0.7 mm atomic beam diameter, finite atomic life time, atom count noise).

Figure 3(b), 3(c) and 3(d) show the fringes for $\delta/2\pi = 712, 347$ and 104 kHz when there is a field in C, containing on average 9.5 photons ($|\alpha| = 3.1$). We notice that when δ is reduced the contrast of the fringes decreases and their phase is shifted. The fringe contrast and shift are plotted versus Φ in Figs. 4(a) and (b). Points

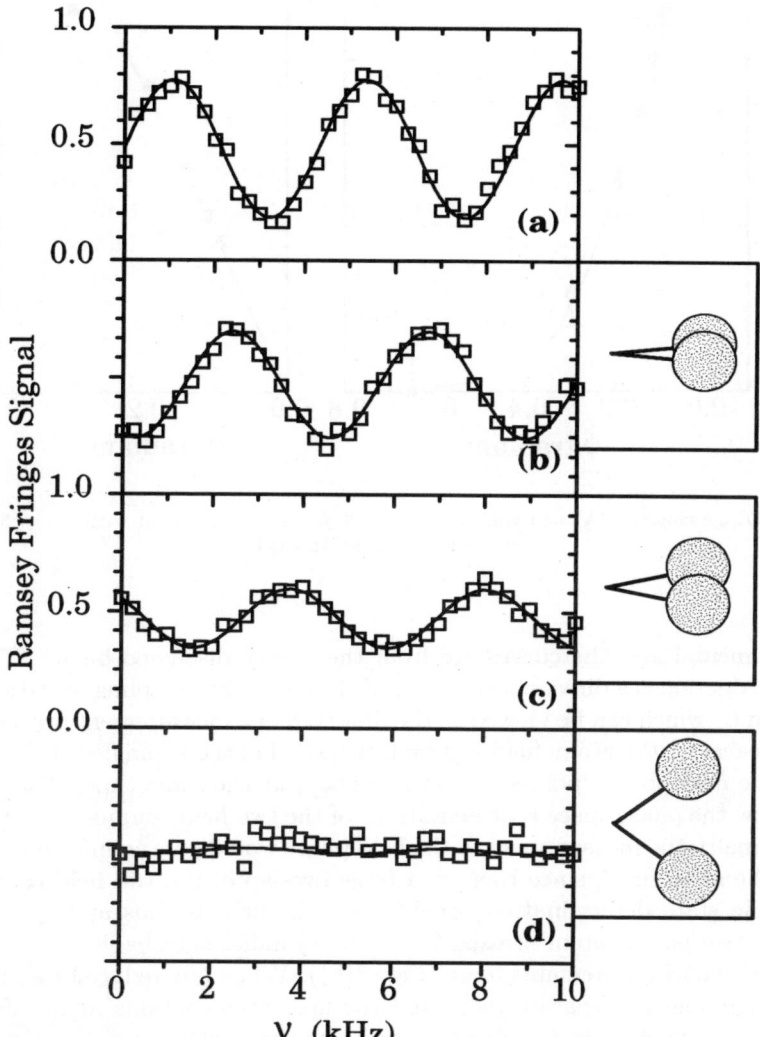

Figure 3: $P_g^{(1)}(\nu)$ signal exhibiting Ramsey fringes: (a) C empty, $\delta/2\pi = 712$ kHz; (b) to (d) C stores a coherent field with $n = 9.5$ photons on average ($|\alpha| = 3.1$), $\delta/2\pi$=721, 347 and 104 kHz respectively. Points are experimental and curves are sinusoidal fits. Inserts show the phase space representation of the field components left in C.

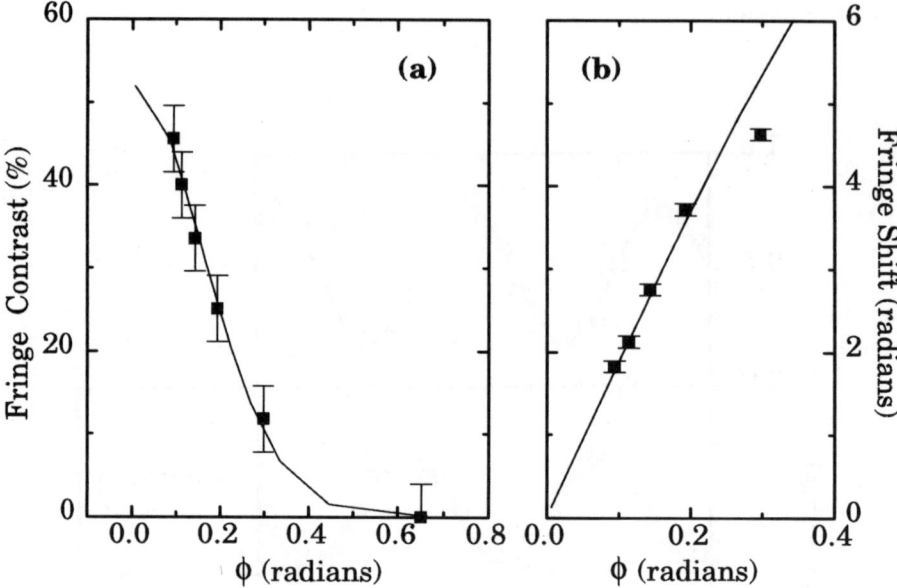

Figure 4: Fringe contrast (a) and shift (b) versus Φ for a coherent field with $|\alpha| = 3.1$ (points: experiments; line: theory).

are experimental and the curves are from the theory discussed below. The fringe contrast reduction is a direct consequence of the dispersive coupling of the atom with the field in C, which can be viewed as the first step in a measurement process. When an atom leaves C, the atom–field system is prepared in the entangled state of Eq.(1), so that the field phase "points" towards e *and* g at the same time. The inserts in Fig. 3 show the phase space representations of the two field components. When δ is large (Φ small), the measurement of the field phase would give no information on the state of the atom in C, since there is a large overlap of the two field components. The atomic state determination provided by the field is thus ambiguous, which leaves the two paths (atom crosses C in g or e) indistinguishable. Therefore, the contrast of the fringes remains large (Fig. 3(b)). When δ is reduced (Φ increased), the field components separate more and the field thus contains more information about the atomic state in C. The two paths become partially distinguishable and the fringe contrast decreases (Fig. 3(c)). It vanishes when the field components do not overlap at all (Fig. 3(d)).

We recognize in the above discussion the ingredients of a "which path" experiment illustrating the basic aspect of complementarity [23,24,25,26]. If we can tell in which level (e or g) the atom crossed C, then the interference between the two probability amplitudes vanishes. It does not matter that the field is actually not measured. The mere fact that the atom leaves in C an information which *could* be read out is enough to destroy the interference. A simple calculation confirms the results of this discussion and shows that the fringe contrast is reduced by a factor equal to the modulus of the overlap integral between the two field components:

$$\langle \alpha e^{i\Phi} | \alpha e^{-i\Phi} \rangle = \exp(-D^2/2) \exp(in \sin 2\Phi) \tag{3}$$

The line in Fig. 4(a), which represents the variations versus Φ of $\exp(-D^2/2)$ is in very good agreement with the measured points. The same analyzis shows that the phase of the fringes (represented versus Φ in Fig. 4(b)) is shifted by an angle equal to the phase of the same overlap integral, namely $n \sin 2\Phi$. The fringe phase shift is proportional to n, a result which we had already demonstrated in a previous experiment [22], and the photon number is thus determined from the slope of the fringe phase variation. The line on Fig. 4(b) corresponds to the best fitted value $n = 9.5 \pm 0.2$.

4 Observing the Decoherence of the "Schrödinger Cat" by a Two–Atom Correlation Measurement.

The Ramsey fringe complementarity experiment demonstrates the separation of the field into two components with classically distinguishable phases, but does not prove the quantum coherence between them. In order to demonstrate this coherence, and to study the ensuing decoherence of the system, we have performed a second experiment in which we record the correlations between two atoms crossing the cavity with a variable time interval between them.

This experiment follows closely a proposal described in a recent paper [9]. Its principle is simple: a first atom creates in C a superposition state involving two separated field components (Fig. 5(a)). A second "probe" atom crosses C with the same velocity after a short delay τ and dephases again the field by an angle $\pm \Phi$ (Fig. 5(b)). The two field components then turn into three, with phases $\pm 2\Phi$ and zero. The $\pm 2\Phi$ components are separated by the distance $D' = 2\sqrt{n} \sin(2\Phi)$ and do not overlap, unless Φ is close to $\pi/2$, a situation we will exclude in this discussion. The "zero phase" component however is obtained via two different paths since the first atom may have crossed C in e and the second in g or vice versa. In both cases, the second atom "undoes" the phase shift produced by the first one, leading to a partial recombination of the state components around the initial phase. Since the atomic states are mixed again in R_2 after C, the paths corresponding to the e, g and g, e combinations are indistinguishable in the end. As a result, there is an interference term in the joint probabilities $P_{ee}^{(2)}$, $P_{eg}^{(2)}$, $P_{ge}^{(2)}$ and $P_{gg}^{(2)}$ of detecting any of the four possible two atom configurations.

It is convenient to consider the combination η of conditional probabilities:

$$\eta = \frac{P_{ee}^{(2)}}{P_{eg}^{(2)} + P_{ee}^{(2)}} - \frac{P_{ge}^{(2)}}{P_{ge}^{(2)} + P_{gg}^{(2)}} \tag{4}$$

which is obviously equal to zero if there are no correlations between the two atoms. This correlation signal can be calculated analytically, taking into account the field relaxation during the time interval τ between the two atoms. The calculation follows closely the method described in [9]. We will not give here the rather cumbersome expression of η which depends only of the dimensionless parameters $n, \Phi, (\nu - \nu_0)T$

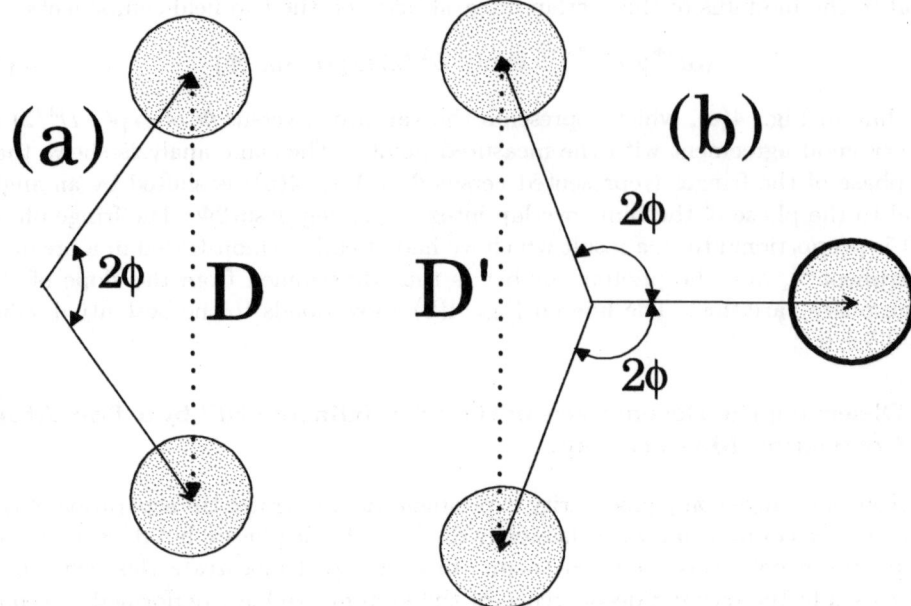

Figure 5: (a) Pictorial representation in phase space of the two phase components created by the first atom. (b) the three phase components after interaction with the second atom. Two indistinguishable quantum paths lead to the zero phase component (horizontal arrow).

and τ/T_r. For the following discussion, it will suffice to say that η can be expressed as the sum $\eta_0(\tau) + \eta_m(\nu,\tau)$ of a ν–independent term $\eta_0(\tau)$ plus a ν–modulated term $\eta_m(\nu,\tau)$. The modulated part $\eta_m(\nu,\tau)$ has a zero ν–average and a negligible amplitude provided that the conditions $D, D' \gg 1$ are satisfied. These conditions mean simply that Φ departs enough from 0 and $\pi/2$ so that the components of the Schrödinger cat state left by the first atom in the cavity do not overlap ($D > 1$) and the $\pm 2\Phi$ components generated by the probe atom are also non overlapping ($D' > 1$). When these conditions are met, $\eta_0(\tau)$ is expected to be equal to 0.5 for $\tau = 0$ and to decay to zero with increasing τ. The initial 0.5 value is entirely due to the two atom interference described above and the decay of η_0 is thus a direct indication of the washing out during time τ of the quantum coherence of the Schrödinger cat produced by the first atom in C. When η_0 has decayed to zero, this "cat" has evolved into a fully incoherent statistical mixture.

In order to measure η, the Rydberg state preparation pulse is replaced by a pair of pulses separated by τ, varied from 30 to 250 μs. The sequence is, as before, repeated every 1.5 ms and statistics on double detection events are accumulated. Due to the low atomic flux, the atom pair rate is ten time smaller than the single atom count rate. For each delay τ, 15000 coincidences are detected in about 2 hours. Figure 6 (a) shows η versus ν for $n = 3.3$, $\delta/2\pi = 70$ kHz ($\phi = 0.98$ rd) and $\tau = 40$ μs. As predicted, a correlation signal with no statistically significant ν–dependence is observed in this case. A ν–averaged value η_0 of 0.11±0.01 is found for this τ value.

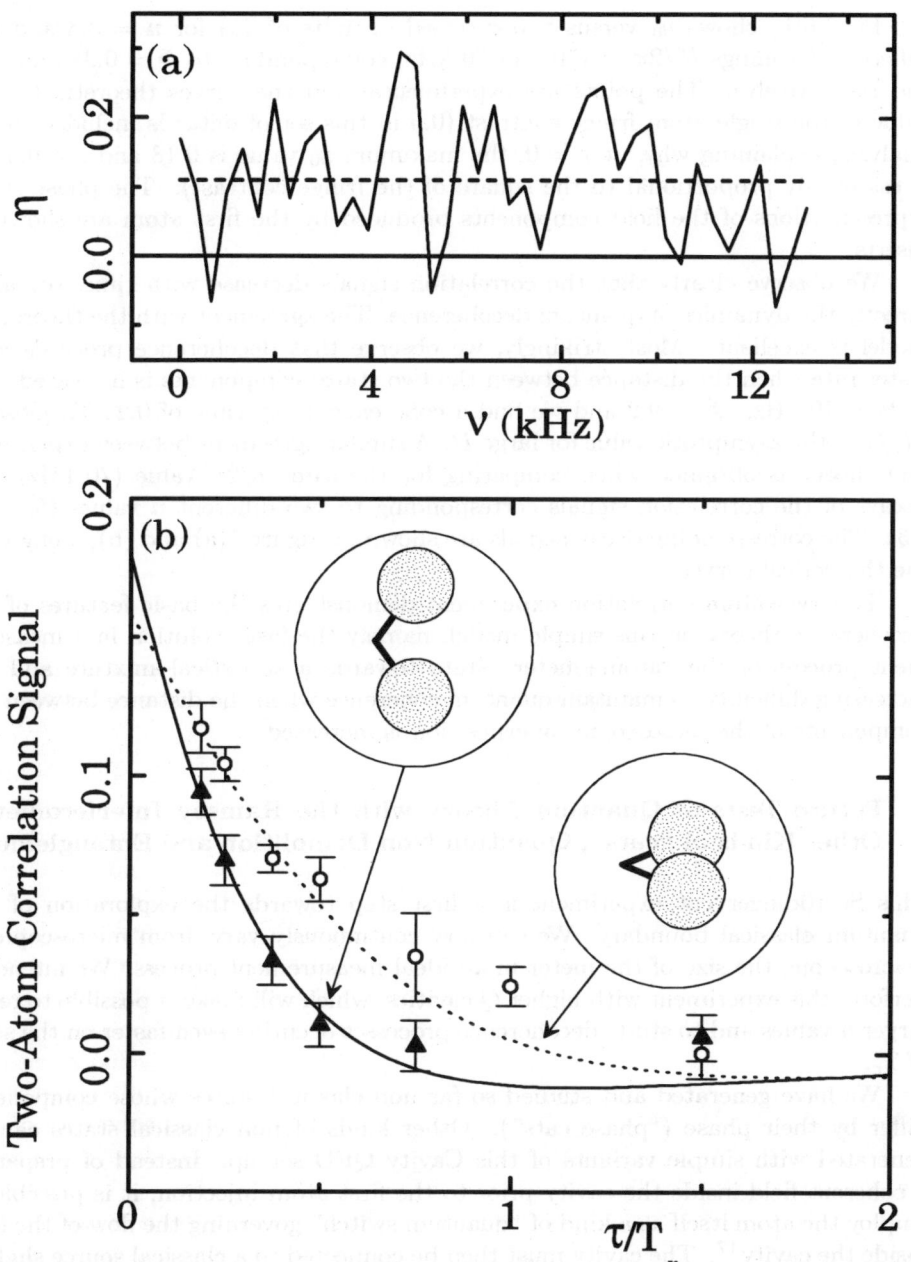

Figure 6: (a) Two-atom correlation signal η versus ν for $n = 3.3$, $\delta/2\pi = 70$ kHz and $\tau = 40$ μs. (b) ν–independent term η_0 versus τ/T_r for $\delta/2\pi = 70$ kHz (triangles) and 170 kHz (circles). Dashed and solid lines are theoretical. Inserts: representation of the field components left by the first atom.

Fig. 6(b) shows η_0 versus τ (expressed in units of T_r) for $n = 3.3$ and two different detunings ($\delta/2\pi = 170$ and 70 kHz, corresponding to $\Phi = 0.98$ and 0.40 rad respectively). The points are experimental and the curves theoretical. The value of the single atom fringe contrast (0.6 in this set of data) is included in the analyzis, explaining why, at $\tau = 0$, the maximum η_0 value is 0.18 and not 0.5 (η_0 is essentially proportional to the square of the fringe contrast). The phase space representations of the field components produced by the first atom are shown in inserts.

We observe clearly that the correlation signals decrease with time, revealing directly the dynamics of quantum decoherence. The agreement with the theoretical model is excellent. Most strikingly, we observe that decoherence proceeds at a faster rate when the distance between the two states components is increased. For $\delta/2\pi = 70$ kHz, $D^2 = 9.2$ and we find a coherence decay time of 0.24 T_r, close to $2T_r/D^2$, the asymptotic value for large D. A similar agreement between experiment and theory is obtained when comparing for the same $\delta/2\pi$ value (70 kHz), the decays of the correlation signals corresponding to two different n values (5.1 and 3.3). The corresponding decay signals are shown in Figure 7(a) and (b), along with the theoretical curves.

This two–atom correlation experiment demonstrates the basic features of the decoherence theory on this simple model, namely the fast evolution in a measurement process of the "atom+meter" state towards a statistical mixture and the increasing difficulty to maintain quantum coherence when the distance between the components of the mesoscopic superposition is increased.

5 Future Tests of Quantum Theory with the Ramsey Interferometer: Other Kinds of "cats", Quantum Non Demolition and Entanglement.

This Schrödinger cat experiment is a first step towards the exploration of the quantum–classical boundary. We can now continuously vary, from microscopic to macroscopic, the size of the meter in an ideal measurement process. We intend to perform this experiment with higher Q cavities, which will make it possible to reach larger n values and to study decoherence processes occuring even faster on the scale of T_r.

We have generated and studied so far non classical states whose components differ by their phase ("phase cats"). Other kinds of non classical states can be generated with simple variants of this Cavity QED set up. Instead of preparing a coherent field inside the cavity prior to the first atom injection, it is possible to employ the atom itself as a kind of "quantum switch" governing the flow of the field inside the cavity[17]. The cavity must then be connected to a classical source slightly detuned, so that, in the absence of an atom, it cannot feed any field inside C. The atomic parameters are then adjusted so that an atom crossing C in level e provides exactly the mode frequency shift required to tune it into resonance with the source. The atom in level g leaves the cavity and the source detuned. We take again here advantage of the single atom index effect, the atom behaving as a kind of dispersive "plunger" tuning C in and out of resonance with the source. Such a device will allow us to prepare "amplitude cat states" of the form $|\Psi\rangle = 1/\sqrt{2}(|e,\alpha\rangle + |g,0\rangle)$,

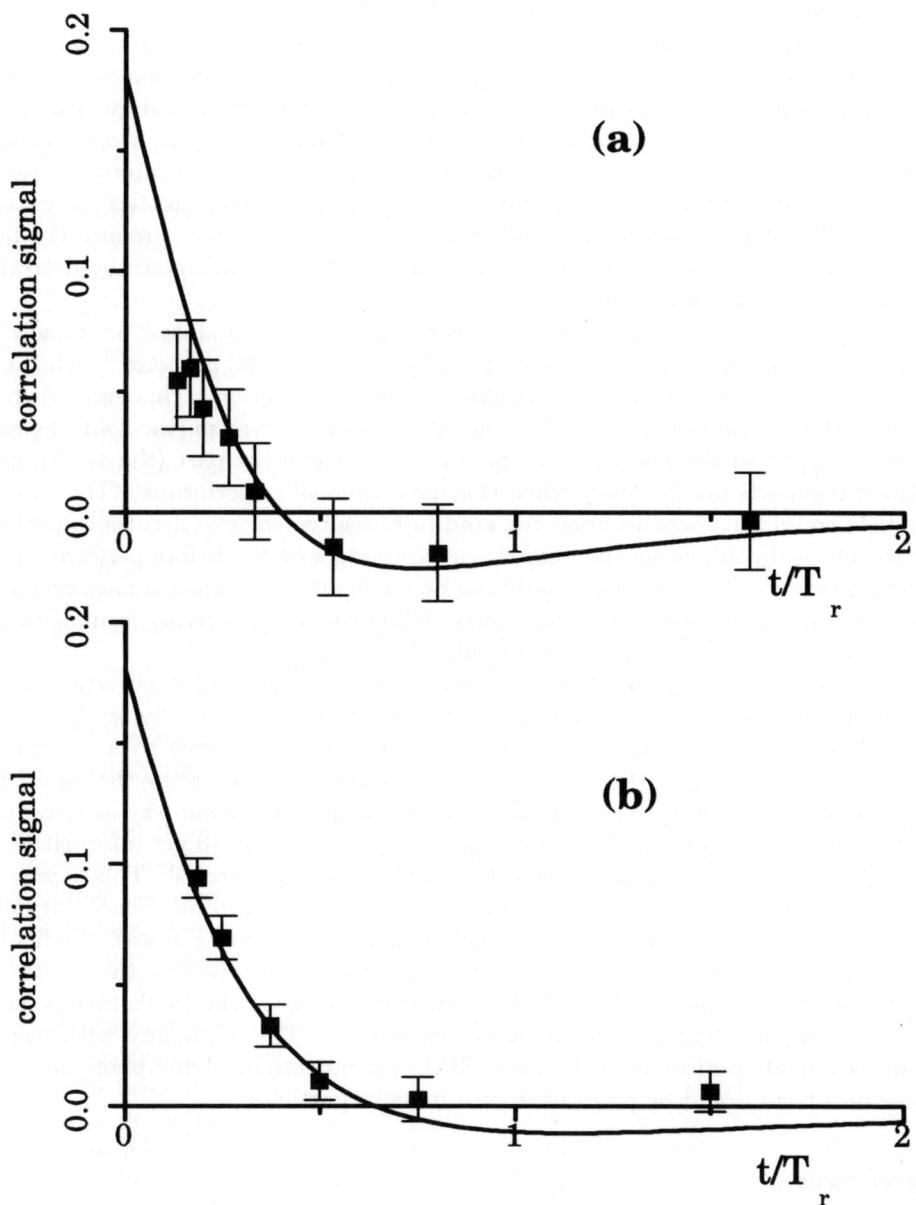

Figure 7: Two–atom correlation signal η_0 versus τ/T_r for $\delta/2\pi = 70$ kHz. Average photon number $n = 5.1$ (a) and 3.3 (b).

whose coherence will also be tested by sending a second atom across the system and measuring the conditional probability that both atoms end up being detected in the same or in different quantum states [9].

The Ramsey interferometer used in these experiments is very versatile and opens the way to other tests of the quantum theory. Let us first notice that this set up lends itself to quantum nondemolition measurements of field quanta in the cavity [16,27]. In fact, as noticed above, the determination of the fringe phase is a measurement of n. Since the interaction of the field and the atoms is purely dispersive, without photon emission or absorption, the method satisfies the criteria of a non demolition measurement. We plan to use this method to reduce the field in the cavity to a Fock state and to monitor continuously this state by detecting atoms crossing the cavity one by one.

In fact, the processes leading to the generation of the "cat states" are closely related to those involved in the Einstein–Podolski–Rosen (EPR) paradox [28]. The atom and the field in the cavity are entangled by their interaction. This entanglement survives the system separation. The final state of one subsystem (the field) depends upon the result of the measurement performed on the other part (the atom), even if these two parts are far apart when this measurement is performed. The state of the field could thus depend upon the kind of measurement one decides to perform on the atom (by adjusting the microwave parameters in R_2, before performing the field ionization). This decision could even be made after the systems have ceased to interact (one can change these parameters while the atom is flying from C to R_2, thus realizing a "delayed choice" experiment).

There is however a practical difference with the usual EPR situation which involves some correlation between the states of spin like particles flying apart from each other. Here, we correlate a spin like particle (the atom) to an harmonic oscillator like system (the field). Moreover, we know how to measure the "spin" (by the field ionization detection), but we do not have any convenient way of measuring directly the field stored in C. In fact, the only practical way to get information on the field is to couple it to atoms which are subsequently detected. This is why we have been naturally led to two-atom correlation detection in our Schrödinger cat experiment. These studies could be easily generalized to more atoms, the field in the cavity behaving as a "correlator" for a train of atoms crossing the cavity one at a time. Such experiments could be turned into tests of the Bell's inequalities [3] or their generalization to three or more particles [29]. The originality with respect to other already performed or planned EPR experiments involving photons is that these new tests would be performed with massive particles.

References

1. W.H. Zurek, *Physics Today* **44**, 10 p.36 (1991); W.H. Zurek, *Phys. Rev. D* **24**, 1516 (1981) and **26**, 1862 (1982); A.O. Caldeira and A.J. Leggett *Physica A*, **121**, 587 (1983); E. Joos and H.D. Zeh, *Z. Phys. B* **59**, 223 (1985); R. Omnès, *The Interpretation of Quantum Mechanics*, Princeton University Press (1994).
2. M. Brune, E. Hagley, J. Dreyer, X. Maître, A. Maali, C. Wunderlich, J.M. Raimond and S. Haroche, Phys. Rev. Lett. submitted.

3. J.S. Bell, *Physics*, **1**, 195 (1964).
4. E. Schrödinger, *Naturwissenschaften* **23**, 807, 823, 844 (1935). Reprinted in english in [5].
5. J.A. Wheeler and W.H. Zurek, *Quantum theory of measurement*, Princeton University Press (1983).
6. R.J. Glauber, *Phys. Rev.*, **131**, 2766 (1963).
7. E.T. Jaynes and F.W. Cummings, *Proc. IEEE*, **51**, 89 (1963).
8. D.F. Waals and G.J. Milburn, *Phys. Rev. A*, **31**, 2403 (1985).
9. L. Davidovich, M. Brune, J.M. Raimond and S. Haroche *Phys. Rev. A*, **53**, 1295 (1996).
10. A.J. Leggett in *Chances and Matter, Les Houches Summer School, session XLVI*, J. Souletie, J. Vannimenus and R. Stora eds., North Holland (1987).
11. J.M. Martinis, M.H. Devoré and J. Clarke, *Phys. Rev. B*, **35**, 4682 (1987); R. Rouse, Si Yuan Han and J.E. Lukens, *Phys. Rev. Lett.*, **75**, 1614 (1995).
12. C.M. Savage, S.L. Braunstein and D.F. Walls, *Opt. Lett.*, **15**, 628 (1990).
13. B. Yurke and D. Stoler *Phys. Rev. Lett.*, **57**, 13 (1986); B. Yurke, W. Schleich and D.F. Walls, *Phys. Rev. A*, **42**, 1703 (1990); G. Milburn, *ibid.*, **33**, 674 (1986); V. Buzek, H. Moya-Cessa, P.L. Knight and S.D.L. Phoenix, *Phys. Rev. A*, **45**, 8190 (1992).
14. C. Monroe, D.M. Meekhof, B.E. King and D.J. Wineland *Science*, **272**, 1131 (1996).
15. *Cavity Quantum Electrodynamics, Advances in Atomic Molecular and Optical Physics, Supplement 2*, P. Berman ed., Academic Press (1994).
16. M. Brune, S. Haroche, J.M. Raimond, L. Davidovich and N. Zagury, *Phys. Rev. A*, **45**, 5193 (1992).
17. L. Davidovich, A. Maali, M. Brune, J.M. Raimond and S. Haroche, *Phys. Rev. Lett.*, **71**, 2360 (1993).
18. S. Haroche and J.M. Raimond, in [15], p. 123.
19. M. Brune, F. Schmidt-Kaler, A. Maali, J. Dreyer, E. Hagley, J.M. Raimond and S. Haroche, *Phys. Rev. Lett.*, **76**, 1800 (1996).
20. R.G. Hulet and D. Kleppner *Phys. Rev. Lett.*, **51**, 1430 (1983).
21. P. Nussenzveig, F. Bernardot, M. Brune, J. Hare, J.M. Raimond, S. Haroche and W. Gawlik, *Phys. Rev. A* **48**, 3991 (1993).
22. M. Brune, P. Nussenzveig, F. Schmidt-Kaler, F. Bernardot, A. Maali, J.M. Raimond and S. Haroche, *Phys. Rev. Lett*, **72**, 3339 (1994).
23. M.O. Scully, B.G. Englert and H. Walther *Nature*, **351**, 111 (1991).
24. T. Pfau, S. Spälter, Ch. Kurtsiefer, C.R. Ekstrom and J. Mlynek *Phys. Rev. Lett.*, **73**, 1223 (1994).
25. M.S. Chapman, T.D. Hammond, A. Lenef, J. Schmiedmayer, R.A. Rubenstein, E. Smith and D.E. Pritchard, *Phys. Rev. Lett.*, **75**, 3783 (1995).
26. S. Haroche, M. Brune and J.M. Raimond, *Appl. Phys. B*, **54**, 355 (1992).
27. M. Brune, S. Haroche, V. Lefèvre, J.M. Raimond and N. Zagury, *Phys. Rev. Lett.*, **65**, 976 (1990).
28. A. Einstein, B. Podolski and N. Rosen, *Phys. Rev.*, **47**, 777 (1935).
29. D.M. Greenberger, M.A. Horne and A. Zeilinger, *Am. Journ. of Phys.*, **58**, 1131 (1990).

Synthesis of Entangled States and Quantum Computing

J.I. Cirac,[†] S. Gardiner,[*] T. Pellizzari,[*] J. Poyatos[*] and P. Zoller[*]

[†] *Departamento de Fisica, Universidad de Castilla–La Mancha, 13071 Ciudad Real, Spain*

[*] *Institut für Theoretische Physik, Universiät Innsbruck, Technikerstrasse 25, A–6020 Innsbruck, Austria.*

We summarize recent theoretical work and proposals to control entanglement in atomic and quantum optical systems.

1 Introduction

One of the most intriguing features of Quantum Mechanics is the existence of entangled states. These are superposition states of composite systems, i.e. states which cannot be expressed as a product of states of the individual system. The simplest examples involve entangled states of two spin-$\frac{1}{2}$ particles which provide the basis for tests of violation of Bell's inequalities, teleportation and quantum cryptography.[1,2,3,4,5] Entangled states of three spin-$\frac{1}{2}$ particles (such as the so–called GHZ states) predict violation of locality without any need of inequalities.[1] In addition, the possibility of entangling more particles is the basis of quantum computing.[2,6] Quantum computing in general is based on the obervation that using the superposition and interference principles of quantum mechanics might speed up certain computations dramatically. The potentially useful algorithms for quantum computers so far include prime factorization[7], and simulation of certain quantum mechanical systems.[8] Furthermore, the possibility to entangle atoms promises a novel atomic spectroscopy with resolution better than the standard quantum limit.[9,10] Thus, the preparation and controlled manipulation of entangled states does not only serve as a testing ground for basic concepts of Quantum Mechanics, but has the potential for applications that might be useful in the future.

Given these results, a natural question is whether the generation and manipulation of entangled states can ever be achieved reliably in the laboratory. We will summarize here recent theoretical work and proposals in atomic physics and quantum optics with the goal to take first steps in this direction (see also the contributions by S. Haroche[12], D. Wineland[11] and A.Zeilinger[13] in these proceedings). Most of the recent work has focused on the realization of the universal quantum gate which is the basic building block to achieve two-particle entanglement[14,15,16,17,18,19,20,21] (see also 22,23). The general entangled states can be synthesized as a sequence of one and two-bit operations (quantum network).[6] The questions to be discussed below are:

(i) We need to identify a physical mechanism to entangle the states of the atoms (or photons) to perform the two-bit quantum gates, and to combine these operations into a quantum network.

(ii) We wish to characterize the input / output characteristics of this two-bit quantum gate to evaluate implementations of quantum gates in the laboratory, as well as a theoretical tool to compare the expected performance of specific quantum computer model systems.

(iii) In practice, the central obstacle is the fragility of macroscopic spin-1/2 superpositions with respect to decoherence by coupling to an environment. It is thus of crucial importance to understand and study the effects of decoherence.

(iv) To (actively) suppress environmental effects is essential to create entangled states in mesoscopic systems. This question is closely related to the problem of error correction in quantum computers.

2 Quantum Computing

In the following we find it convenient to phrase the manipulation of N–particle entangled states adopting the language of quantum computing. For details on quantum computing and quantum computing algorithms (in particular Shor's factorization algorithm[7]) we refer to the excellent reviews.[2,6] A quantum computer (qc) can be thought of as N spin-1/2 systems with levels $|0\rangle$ and $|1\rangle$, representing the quantum bit or qubit. The most general state of the qc is an entangled state

$$|\psi\rangle = \sum_{x=0}^{2^N-1} c_x |x\rangle \equiv \sum_{\underline{x}=\{0,1\}^N} c_{\underline{x}} |\underline{x}\rangle$$

of quantum registers $|\underline{x}\rangle = |x_{N-1}\rangle_{N-1} \ldots |x_0\rangle_0 \in \mathcal{H}(N-1)_2 \otimes \ldots \otimes \mathcal{H}(0)_2$ with $x = \sum_{n=0}^{N-1} x_n 2^n$ the binary decomposition of x. Quantum computations correspond to processes $|\psi_{\text{in}}\rangle \to |\psi_{\text{out}}\rangle = \hat{U}|\psi_{\text{in}}\rangle$ where a given input state is mapped to an output state by a unitary transformation $\hat{U}$. This can be carried out as a sequence of elementary steps (quantum gates) involving operations on a few qubits. In particular, a two-bit gate operation corresponds to conditional dynamics[15] represented by the unitary operator,

$$\hat{U} = |0\rangle_{11}\langle 0| \otimes 1_2 + |1\rangle_{11}\langle 1| \otimes \hat{U}_2$$

where we apply the unitary operation $\hat{U}_2$ to the target bit if the control bit is in state $|1\rangle$. It has been shown [16] that any operation can be decomposed into universal two-bit gates, for example a controlled–NOT gates between two qubits, and rotations on a single qubit. A controlled–NOT is defined by $\hat{C}_{12}$: $|\epsilon_1\rangle|\epsilon_2\rangle \to |\epsilon_1\rangle|\epsilon_1 \oplus \epsilon_2\rangle$ with $\epsilon_{1,2} = 0, 1$, and $\oplus$ denotes addition modulo 2. The question is, therefore, how to implement two-bit entanglement operations on systems of atoms and photons in the laboratory.

3 Quantum Optical Implementations of the 2-Bit Quantum Gate

Before discussing specific realizations of quantum gate implementations, we will summarize the basic requirements for our physical systems. First, the system must provide reliable storage of the quantum bits. This requires that the qubits do not undergo decoherence for the time of the experiment. Second, we need a physical interaction to perform one-bit and two–bit entanglement operation of the universal two-bit quantum gate. The key idea in the proposals discussed below is to provide a coupling between the qubits via auxiliary degrees of freedom which acts as a "bus for quantum information". Again we require small damping and quantum noise. Third, we need a physical mechanism to perform state measurements with high efficiency. This is necessary to read the output after the quantum computation. In addition error correction schemes rely require state measurements during the computation to apply operations conditioned on the outcome of these measurements

Due to experimental progress in recent years, the requirement that the coherent interactions are much larger than the dissipative rates can be fulfilled in some quantum optical and atomic systems, in particular in ion traps[21,24,25,26] (for a review see 27) and in cavity quantum electrodynamics (CQED).[22,23,28] Below we will review the essential physical ideas to implement two-bit quantum gates in these systems.

3.1 Ion Trap

Laser cooled trapped ions are a unique experimental system: unwanted dissipation can be made negligible for very long times, much longer than typical times in which an experiment takes place. Furthermore, arbitrary quantum states of the ion's motion can be synthesized and coherently manipulated using laser radiation[24,25,30,31,32,33,34,35,36] (for a review see 37). In addition, the state of motion can be completely determined in the sense of tomographic measurements.[26,38,15,41] In a series of remarkable experiments, Wineland and collaborators have generated a variety of non–classical states of ion motion [24] and a "Schrödinger cat state".[25]

At present the most realistic proposal for a quantum computer model is based on a string of cold ions interacting with laser light and moving in a linear trap.[17] The basic elements of the computer (i.e., the qubits) are the ions themselves. The two states of the n-th qubit are identified with two of the internal states of the corresponding ion. In this system independent manipulation of each individual qubit is accomplished by directing different laser beams to each of the ions. The two-bit quantum gate can be implemented by exciting the collective quantized motion of the ions with lasers. The coupling of the motion of the ions is provided by the Coulomb repulsion which is much stronger than any other interaction for typical separations between the ions of a few optical wavelengths. The confinement of the motion along X, Y and Z directions can be described by an (anisotropic) harmonic potential of frequencies $\nu_x \ll \nu_y, \nu_z$. The ions have been previously laser cooled so that they undergo very small oscillations around the equilibrium position. In this case, the motion of the ions is described in terms of normal modes. Furthermore, it is assumed that sideband cooling has left all the normal modes in their corresponding (quantum) ground states. For this to be possible, one has to assume that the Lamb–Dicke limit holds for all the modes, i.e. the oscillation amplitudes are much smaller than the laser wavelength.[27]

Single qubit gates can be performed by acting with a resonant laser beam on the corresponding ion. Two–qubit gates are implemented by entangling two ions through exchange of a phonon with a laser tuned to the lower motional sideband. The scheme is outlined in Fig. 1(a). A universal two–bit quantum gate between ions a and b can be carried out in three steps[17]: (i) A π laser pulse swaps the qubit of the ion a to the center-of-mass mode;. (ii) A conditional sign change is introduced through an auxiliary state $|e'\rangle$ with the help of a 2π pulse on ion b. (iii) The qubit of ion a is restored by inverting step (i). This corresponds to the sequence

$$
\begin{aligned}
C_{ab} : |\varepsilon_1\rangle_a |\varepsilon_2\rangle_b |0\rangle_p &\stackrel{(i)}{\to} |0\rangle_a |\varepsilon_2\rangle_b |\varepsilon_1\rangle_p \\
&\stackrel{(ii)}{\to} (-1)^{\varepsilon_1 \varepsilon_2} |0\rangle_a |\varepsilon_2\rangle_b |\varepsilon_1\rangle_p \\
&\stackrel{(iii)}{\to} (-1)^{\varepsilon_1 \varepsilon_2} |\varepsilon_1\rangle_a |\varepsilon_2\rangle_b |0\rangle_p
\end{aligned}
\quad (1)
$$

where $|\varepsilon_{1,2} = 0, 1\rangle_{a,b}$ represent the state of the ion, and $|n\rangle_p$ refers to phonons in the center-of-mass motion. The two key elements behind the above implementation of quantum gates are (i) a nonlocal entanglement between individual qubits is achieved by transferring the internal atomic coherence to and from the CM motion shared by all the ions ; and (ii) as an intermediate step we "hide atomic amplitudes" corresponding to the qubits in a third internal atomic level level, and induce 2π–rotations via this state to selectively change the sign of atomic amplitudes.

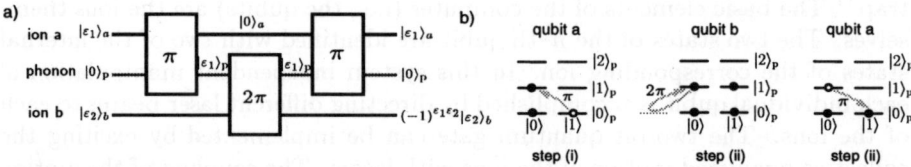

Figure 1: Implementation of the universal two-bit gate $C_{ab} : |\varepsilon_1\rangle_a |\varepsilon_2\rangle_b \to (-1)^{\varepsilon_1 \varepsilon_2} |\varepsilon_1\rangle_a |\varepsilon_2\rangle_b$ in an ion trap qc: a) schematic diagram, b) ion operations steps (i) to (iii) according to Equ. (1).

In a remarkable experiment a two-bit quantum gate based on a single trapped ion has been demonstrated at NIST Boulder,[21] and we refer to the contribution by Wineland *et al.* in the present proceedings for a summary of this work.[11]

Decoherence[29] in an ion trap is due to spontaneous decay of the internal atomic states, and damping and heating of the motion of the ion. Application of stored ions in ultrahigh precision spectroscopy, and time and frequency standards[9,27] shows that this decoherence times can be quite long, longer than the time required to perform many operations. Spontaneous emission is suppressed using metastable or Raman transitions.[21] Collisions with background atoms can be avoided at sufficiently low pressures for very long times, and other couplings that affect the moving charges can be made sufficiently small.[9] Furthermore, the final readout of the quantum register (state measurement of the individual qubits) at the end of the computation can be accomplished using the quantum jumps technique with essentially unit efficiency.[21] On the other hand the number of gate operations required to perform even simple computational tasks is significant[42,43], and we should not expect that the present model systems can be scaled up to useful computing machines in the near future.[43,44]

3.2 Cavity QED

Cavity QED both in the microwave[12,23,28] and optical domain[22] are candidates of systems to implement quantum gates. Specific proposals been made.[15,16,19,20,45]

We have proposed a model system based on optical Cavity QED. In this scheme N atoms representing the qubits communicate via their interaction with a single quantized mode of a high-Q optical cavity.[20] It is assumed that the atoms are fixed inside the cavity at distances apart much larger than the wavelength of the cavity mode and interacting individually with laser beams, which allows for sequences of operations between any two qubits and thus the implementation of a whole quantum network. The qubits are stored in Zeeman ground state levels of the trapped atoms. Quantum gates are implemented by coupling atoms to individual lasers and entangling them by exchange of a cavity photon. Specifically, it can be shown that the Zeeman coherence between ground state levels of two atoms can be transferred between two atoms by employing adiabatic passage via a dark state of the two–atom + cavity system. This is the basis of the quantum gate proposed by Pellizzari et al..[20] Sources of decoherence are thus spontaneous emission from the excited states of the atoms, and cavity decay during the gate operation. An important features of this scheme is that, although the lasers and cavity are tuned on resonance, the excited states are (in principle) never populated and thus the transfer is immune to spontaneous emission. Furthermore, the interaction can be resonant throughout the transfer, and thus there will be cavity decay can only occur during the gate operation.

Haroche and coworkers have performed a series of fundamental Cavity QED experiments on topics closely related to quantum gate experiments: for a review we refer to Haroche et al. in the present proceedings.[12]

A quantum logic gate based on single polarized photons as carrier of the quantum information (qubits) has been proposed by Kimble an coworkers,[22] and first experimental steps towards the realization of a universal quantum gate have been reported. Specifically, these authors have demonstrated conditional dynamics at the single photon level between two fields of different frequencies in an optical resonator. The physical mechanism was based on the circular birefringence of an atom strongly coupled to the resonator to rotate the linear light polarization of a transmitted probe beam. The phase shift between the circular polarization states $\sigma^{\pm}$ is conditioned upon the intensity of the pump via a Kerr type nonlinearity. Experimentally a polarization conditional phase shift of 16 degree per intracavity photon field was obtained. At present the experiment employs coherent input fields, while demonstration of a quantum gate requires single photon sources.

4 Tomography of a Quantum Gate

Both from an experimental and theoretical point of view it is essential to characterize the input / output (transfer) characteristics of a quantum gate in the sense of a "black box"[22] (for general references on quantum *state* tomography see[38,15,40,41]). In general, a given quantum dynamics $\mathcal{E}$ transforms input states ρ_{in} into output states ρ_{out}, i.e. $\hat{\rho}_{\text{in}} \longrightarrow \hat{\rho}_{\text{out}} = \mathcal{E}[\hat{\rho}_{\text{in}}]$ with $\mathcal{E}$ a linear mapping. The aim is to characterize the process $\mathcal{E}$, by a sequence of measurements in such a way that it is possible to predict what the output state will be for any input state. In particular, we outline a procedure for characterizing a two-qubit gate.[46] This can be implemented using only (i) product states as inputs, and (ii) single qubit measurements on the outputs This avoids utilizing any interaction (entanglement) between the qubits which would be required to prepare Bell state inputs and perform Bell measurements: otherwise the decoherence and errors induced by the measurement itself would distort the characterization of $\mathcal{E}$.

A general scheme to characterize quantum dynamics has been developed.[46] Let us assume that the system is initially prepared in the pure state $|\Psi_{in}\rangle = \sum_{i=0}^{N} c_i |i\rangle$, and will denote by $|E\rangle$ the initial state of the environment (extension to mixed input states is straightforward). Thus, the initial state of the system-plus-environment is transformed according to $|i\rangle|E\rangle \xrightarrow{\mathcal{E}} \sum_{j=0}^{M} |j\rangle|E_j^i\rangle$ where the states $|E_j^i\rangle$ are unnormalized states of the environment. Tracing over the environment degrees of freedom we get the reduced system density operator,

$$\hat{\rho}_{out} = \sum_{i,i'=0}^{N} c_i [c_{i'}]^* \hat{R}_{i'i} \equiv \sum_{i,i'=0}^{N} c_i [c_{i'}]^* \left[\sum_{j,j'=0}^{M} (E_{j'}^{i'}|E_j^i) |j\rangle\langle j'| \right]$$

Knowledge of the "process operators" $\hat{R}_{i'i}$ (which are independent of the inputs) is sufficient to predict the final density operator for any input state and, the problem of fully characterizing the physical process $\mathcal{E}$ is reduced to finding the $\hat{R}_{i'i}$. Suppose now we prepare the system intially in $(N+1)^2$ different initial (pure) input states, and in each of these cases we obtain a corresponding output density matrix ρ_{out} (which can be measured by standard tomographic techniques). This gives a set of linear equations relating the $\hat{R}_{i'i}$'s to the ρ_{out}'s. It can be shown that with an appropriate choice of input states this set equations can always be inverted.

In the case of a universal two-qubit gate the system is composed of two two-level subsystems 1 and 2 with levels $|0\rangle_{1,2}$ and $|1\rangle_{1,2}$ each. A basis for the initial states $|i\rangle = |i_1\rangle_1 |i_2\rangle_1$ (with $i = 2i_1 + i_2$, and $i_1, i_2 = 0, 1$). The quantum tomography of the output states can be carried out following the lines proposed by Wootters.[47] One writes the output operators as $\hat{\rho}_{out} = \sum \lambda_{q_1 q_2} \hat{\sigma}_{q_1}^1 \otimes \hat{\sigma}_{q_2}^2$. By measuring the observables $\hat{\sigma}_{q_1}^1 \otimes \hat{\sigma}_{q_2}^2$, one can determine the coefficients λ_{q_1,q_2}. An example of 16 possible product input states are states of the form with $|\psi_a\rangle_1 |\psi_b\rangle_2$ ($a, b = 1, \ldots, 4$) with $|\psi_1\rangle = |0\rangle$, $|\psi_3\rangle \propto |0\rangle + |1\rangle$, $|\psi_2\rangle = |1\rangle$, $|\psi_4\rangle \propto |0\rangle + i|1\rangle$. On the basis of this tomographic measurement of the gate Ref.[46] introduces *three global parameters* to characterize the action of the quantum gate: first, a "Gate Fidelity" ($\mathcal{F}$), that expresses how close the experimental gate is to the ideal one; second, the "Gate Purity" ($\mathcal{P}$), that reflects

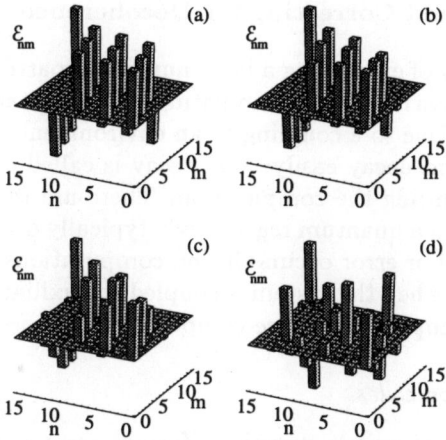

Figure 2: Quantum transfer matrix $\mathcal{E}_{n,m}$ in arbitrary units for a two–qubit gate in the ion trap quantum computer: (a) ideal gate; (b,c,d) numerical simulation with the following parameters: Lamb–Dicke parameters $\eta_{cm} = \eta_r = 0.5$, detuning from resonance $\Delta_1 = \Delta_2 = -\nu$, and Rabi frequencies $\Omega_1 = \Omega_2 = 0.1, 0.2, 0.5$. [46]

to what extent the gate experiences decoherence; and third, the "Quantum Degree of the Gate" ($\mathcal{Q}$), a measure of the possibility to produce (quantum) entangled states out of initial product states.

Simulations illustrating this procedure are illustrated in Fig. 2 which plots the transfer matrix $\mathcal{E}$.[46] These results obtained by this tomographic technique for the quantum gate of Equ. (1) on the basis of a realistic model of a two ions moving in a trap. The corresponding Schrödinger equation was numerically integrated for this system we have simulated the measurement of the operators $\hat{R}_{i'i}$. For Rabi frequencies Ω much smaller than the trap frequency ν (see Fig. 1a) this evolution approximates the ideal gate (1). [17] For finite Rabi frequencies, however, the result will not be ideal (to demonstrate the effect this assumes the unfavorable case of a running light wave). After the gate operation, some population will remain in the phonon modes, which would lead to decoherence. Moreover, there may remain some population in the auxiliary state (step (ii) in Equ.(1). Fig. 2 compares the ideal case of a perfect gate [Fig. 2(a)] with the simulation results with realistic parameters [Figs. 2(b,c,d)]. We have plotted the matrix elements of the 16 operators $\hat{R}_{i'i}$ (for details see[46]), and the parameters are chosen close to those in planned experiments. As it is shown, for moderately small Rabi frequencies the simulated results almost coincide with the ideal one.

5 Decoherence and Correcting for Decoherence

The central difficulty of entangling a large number of particles is decoherence.[29] Quantum computation involves manipulating quantum states that are in coherent superpositions. Due to a coupling to an environment, these superpositions tend to be fragile and decay easily: this decay is called decoherence and corresponds to errors during the computation. There are two kinds of errors. A superposition state in a quantum register will typically decay; this is a memory error. A second kind or error occurs during computations, for example during the gate operations, when the system is coupled to auxiliary degrees of freedom which undergoes damping; these are computational errors.

5.1 Error correction codes

During the last year various ingenious formal error correction codes have been developed to protect quantum superposition and entanglement (memory errors).[48] Shor[49] has proposed a scheme based on classical schemes using *redundancy*. The idea is to store quantum information not in a single qubit $|0\rangle$, $|1\rangle$ but in a logical qubit $|0\rangle_L$, $|1\rangle_L$ represented by an entanglement of nine. This scheme allows one to correct any error incurred by any one of the nine qubits. Steane[50], and Calderbank and Shor[51] have outlined a different scheme which uses only seven bits and demonstrated that this is the least required for the strategies inspired by classical coding theory. Finally, Laflamme et al.[52] have pointed out a true quantum code which distributes the quantum information over five qubits, the minimal number required for this task [48]. All error correction schemes have to be complemented with an appropriate encoding circuit which takes the initial superposition and the extra qubits in $|0\rangle$ to the encoded state. In addition one needs a decoding circuit, and reading the extra qubits at the decoder's output one learns which one of the possible alternatives (no error plus all possible 1-bit errors) was realized. The original unknown superposition state can then be restored by an appropriate unitary transformation. These schemes are based on the assumption of a perfect physical implementation of quantum circuits. A a sideremark we note that very recently an extension of the above described schemes has been proposed by Shor that allows for quantum error correction during computations with erroneous quantum circuits [53].

In quantum optical systems the coupling to an environment is typically described as Markovian dynamics, and decay and quantum noise modelled as quantum jumps of a system wave function.[55] Mabuchi and Zoller[56] have discussed the possibility of *inverting quantum jumps*, assuming that they are observed by continuously monitoring the system decay channels, to conserve

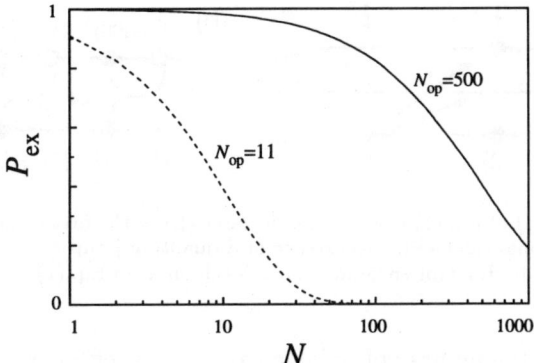

Figure 3: Probability of measuring the correct result P_{ex} after the application of N 2-bit quantum gates $C_{ab}^L : |\varepsilon_1\rangle_a^L |\varepsilon_2\rangle_b^L \to (-1)^{\varepsilon_1 \varepsilon_2} |\varepsilon_1\rangle_a^L |\varepsilon_2\rangle_b^L$ to the initial state $\propto |0\rangle_a^L |0\rangle_b^L + |1\rangle_a^L |1\rangle_b^L$. Dashed line: quantum gate without error correction. Solid line: present error correcting scheme. The parameters have been chosen such that the probability for an error within a single gate C_{ab}^L without error correction is $p = 0.09$.

an unknown superposition state.

5.2 Error correction in the ion trap qc

The error correction schemes discussed above have focused on preserving a given entangled state. We have outlined a method to correct for effects of decoherence in the dynamical process of preparation and modification of entangled states (gate errors).[57] The proposed scheme is a first-order error correction that allows to effectively square the number of gate operations relative to the uncorrected case. The motivation is that in quantum optical systems, entangled states are achieved by coupling qubits to another degree of freedom which in turn undergoes decoherence by coupling to a heat bath. For example, in the ion trap QC[17], the qubits can be stored in long–lived atomic ground states[21] with decoherence time $\simeq 1000$s.[9] Two bit quantum gates are implemented by coupling the ions to the collective center-of-mass motion in the trap, which decoheres in a time $\simeq 1$ms.[21] Thus, at least in present experiments, gate errors predominate.

Here we summarize some the key ideas, and illustrate how the correction scheme can be implemented in an ion trap QC we assume as our model that the phonons of the center-of-mass mode are coupled to a zero temperature reservoir. The essential results are summarized in Fig. 3, where we plot the fidelity for successful operation as a function of the number of applied two bit

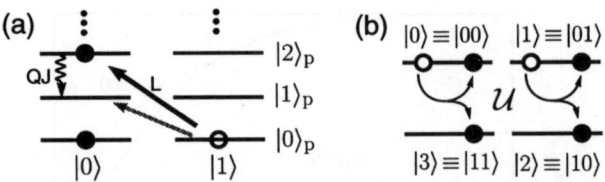

Figure 4: (A) Step (i) of Eq. (1) that is tune the laser (L) to the first (dim arrow) or second (black arrow) motional sideband. Occurrence of a quantum jump is indicated by QJ. (B) Redundant encoding in a 4–level ion as in Eq. (2).

gates. Note that the number of reliable gates $N_{\rm op}$ is effectively squared.

The error-corrected quantum gate is based on the following three elements (see Fig. 4):

(i) *Redundant encoding in logical qubits:* At the beginning of the gate operation between the ions a and b, we encode each of the logical qubits in two physical qubits $a_{1,2}$ and $b_{1,2}$, respectively,

$$|\varepsilon\rangle_x \stackrel{\mathcal{U}}{\to} |\varepsilon\rangle_x^L \propto |\varepsilon\rangle_{x_1}|0\rangle_{x_2} + |1-\varepsilon\rangle_{x_1}|1\rangle_{x_2} \qquad (2)$$

with $\varepsilon = 0, 1$ and $x = a, b$, and after the gate we decode. These *two physical qubits* are stored in a *single four-level ion* $|0\rangle_x \equiv |0\rangle_{x_1}|0\rangle_{x_2} \ldots$ [see Fig. 4b]. The unitary transformation $\mathcal{U}$ thus requires only a *single-ion* operation. In addition, these "qubits" can be manipulated independently with *single-ion* operations (laser pulses). Allowed computational inputs are states of the form

$$|\Psi(0)\rangle = \sum_{\varepsilon_1,\varepsilon_2=0,1} |\varepsilon_1\rangle_a^L |\varepsilon_2\rangle_b^L |0\rangle_{\rm p} \chi_{\varepsilon_1,\varepsilon_2} \qquad (3)$$

where the χ's denote coefficients in the superposition.

(ii) *Gate operation for logical qubits:* Our aim is to perform the universal gate $C_{ab}^L : |\varepsilon_1\rangle_a^L |\varepsilon_2\rangle_b^L \to (-1)^{\varepsilon_1 \varepsilon_2} |\varepsilon_1\rangle_a^L |\varepsilon_2\rangle_b^L$ between the logical qubits a and b stored in the physical qubits $a_{1,2}$ and $b_{1,2}$, respectively. Using Eq. (2) we decompose $C_{a,b}^L = C_{a_2,b_1} C_{a_1,b_2} C_{a_2,b_2} C_{a_1,b_1}$ (see Fig. 3). Each of these four subgates are now performed in the same way as in the uncorrected case Eq. (1).

Each of the four subgates C_{a_i,b_j} acts only on two physical qubits at a time. As an example we consider the operation C_{a_1,b_1}. The state of the QC Eq. (3) before the subgate can be rearranged as

$$|\Psi(0)\rangle = \sum_{\varepsilon_1,\varepsilon_2=0,1} |\varepsilon_1\rangle_{a_1}|\varepsilon_2\rangle_{b_1}|0\rangle_{\rm p}|R_{\varepsilon_1,\varepsilon_2}\rangle, \qquad (4)$$

with

$$|R_{\varepsilon_1,\varepsilon_2}\rangle = |0\rangle_{a_2}|0\rangle_{b_2}\chi_{\varepsilon_1,\varepsilon_2} + |0\rangle_{a_2}|1\rangle_{b_2}\chi_{\varepsilon_1,1-\varepsilon_2} \qquad (5)$$
$$+ |1\rangle_{a_2}|0\rangle_{b_2}\chi_{1-\varepsilon_1,\varepsilon_2} + |1\rangle_{a_2}|1\rangle_{b_2}\chi_{1-\varepsilon_1,1-\varepsilon_2}.$$

The subgate C_{a_1,b_1} operates on the first three kets of Eq. (4) only. So, if something goes wrong, we have a "backup" in $|R_{\varepsilon_1,\varepsilon_2}\rangle$.

(iii) We perform the gate operation (1) with *encoding on the second motional sideband involving two phonons*. Presence of one phonon after the gate operation indicates that there was an error, i.e. a decay $|2\rangle_p \to |0\rangle_p$ happened some time during the gate operation. Thus, after each gate operation a measurement is performed to detect the presence or absence of phonons. This measurement requires an extra ion (the *right light ion* in Fig. 5).

Thus we have the following steps (compare Fig. 5)): First, a phonon decay during one of the subgates will transform the state of the QC into a state with one phonon $|1\rangle_p$. The state after the jump $|2\rangle_p \to |0\rangle_p$ will remain unaffected until the end of the subgate (independent of the jump time). Thus the occurrence of a jump can be detected with the an extra ion (measurement $\mathcal{M}_1$ in Fig. 4). Second, if a jump was detected after one of the subgates, we wish to recover the state before this particular operation. This information can be "restored from the backup" (5). This involves single bit operations together with a measurement $\mathcal{M}_2$ (Fig. 4).

A remarkable feature of the proposal is that error testing measurements are performed *after* the gate operation to correct errors which accumulated *during* the computation. The overhead required by the scheme is a rather moderate: each logical qubit is encoded into two qubits that are stored in the same (four–level) ion. This has the advantage that one-qubit gates (for which decoherence is assumed to be negligible) are the same with or without the redundant encoding. On the other hand, implementation of the two–qubit gate requires an overhead of four two–qubit subgates. A proof-of-principle experiment to demonstrate the possibility of correcting effects of decoherence in dissipative quantum dynamics could be performed with three trapped ions which seems to be attainable with present technology.[21]

6 Conclusion

During the last two years novel ideas have been developed and first steps have been taken to realize two-bit entanglement operations in the laboratory. Currently several experiments are underway to build small working prototypes of "quantum computer" models. Even if small model systems can successfully

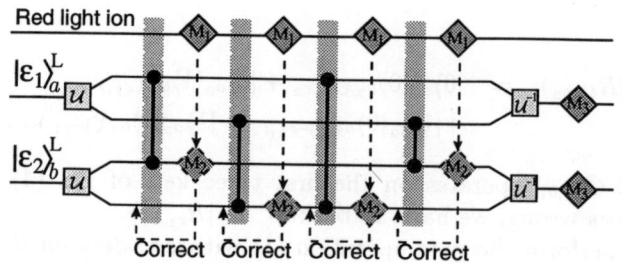

Figure 5: Logical two–qubit gate with error correction (see text).

built, scaling these up to computers that are large enough to yield useful computations will present formidable difficulties. On the other hand, in the not too distant future small model systems involving only a few atoms (or photons) will be the basis of experimental studies of fundamental aspects of quantum mechanics and improved resolution in atomic spectroscopy. An essential aspect of these investigations will be a systematic study of decoherence phenomena.

Acknowledgments

This work was supported in part by the Austrian Science Foundation, the "Aycciones Integradas" between Austria and Spain, and the European TMR network under contract ERB4061PL95–1412.

References

1. J. S. Bell, Physics **1**, 195 (1964); D. M. Greenberger *et al.*, Am J. Phys. **58**, 1131 (1990); N.D. Mermin, Am. J. Phys. **58**, 8 (1990).
2. D. P. DiVincenzo, Science **270**, 255 (1995).
3. C. H. Bennett, Phys. Today, Volume 24 (October 1995).
4. C.H. Bennett, G. Brassard, C. Crepeau, R. Josza, A. Peres, and W. K. Wootters, Phys. Rev. Lett. **70**, 1895 (1993).
5. C. H. Bennett, G. Brassard, S. Popescu, B. Schumacher, J.A. Smolin, W.K. Wootters, Phys. Rev. Lett. **76**, 722 (1996).
6. A. Ekert, Proc. 14th ICAP, ed. D. Wineland *et al* (AIP Press, 1995), p. 450; A. Ekert and R. Josza, Reviews of Modern Physics (in print).
7. P.W. Shor, in *Proceedings of the 35th Annual Symposium on the Foundations of Computer Science*, S. Goldwassser ed. (IEEE Computer Society Press, Los Alamitos, Ca, 1994), p. 124.
8. S. Lloyd, preprint.

9. D.J. Wineland, J.J. Bollinger, W.M. Itano, F.L. Moore, and D. Heinzen, Phys. Rev. A **46**, R6797 (1992).
10. D. J. Wineland, J.J. Bollinger, W. M. Itano, and D.J. Heinzen, Phys. Rev. A **50**, 67 (1994).
11. D. Wineland *et al.*, these proceedings.
12. S. Haroche *et al.*, these proceedings.
13. A. Zeilinger *et al*, these procedings.
14. S. Lloyd, Science **261**, 1569 (1993).
15. A. Barenco, D. Deutsch, and R. Josza, Phys. Rev. Lett. **74**, 4083 (1995).
16. T. Sleator and H. Weinfurter, Phys. Rev. Lett. **74**, 4087 (1995).
17. J.I. Cirac and P. Zoller, Phys. Rev. Lett. **74**, 4091 (1995).
18. D. P. DiVincenzo, Phys. Rev. A **51**, 1015 (1995).
19. P. Domokos, J.M. Raimond, M. Brune, and S. Haroche, Phys. Rev. A **52**, 3554 (1995).
20. T. Pellizzari, S. A. Gardiner, J. I. Cirac, and P. Zoller , Phys. Rev. Lett. **75**, 3788 (1995).
21. C. Monroe, D.M. Meekhof, B.E. King, W.M. Itano, and D.J. Wineland Phys. Rev. Lett. **75**, 4714 (1995).
22. Q. A. Turchette *et al.*, Phys. Rev. Lett. **75**, 4710 (1995);
23. M. Brune, P. Nussenzveig, F. Schmidt-Kaler, F Bernardot, A. Maali, J.M. Raimond, and S. Haroche, Phys. Rev. Lett. **72**, 3339 (1994).
24. D.M. Meekhof, C. Monroe, B.E. King, W.M. Itano, and D.J. Wineland , Phys. Rev. Lett. **76**, 1796 (1996).
25. C. Monroe, D. M. Meekhof, B. E. King, and D. J. Wineland, Science **272**, 1131 (1996).
26. D. Leibfried, D.M. Meekhof, B.E. King, C. Monroe, W.M. Itano, and D.J. Wineland, preprint.
27. R. Blatt, Proc. 14th ICAP, ed. D. Wineland *et al.* (AIP Press, 1995), p. 219.
28. M. Brune, F. Schmidt-Kaler, A. Maali, J. Dreyer, E. Hagley, J. M. Raimond, and S. Haroche, Phys. Rev. Lett. **76**, 1800 (1996).
29. R. Landauer, Philos. Trans. R. Soc. London A **353**,367 (1995); W. G. Unruh, Phys. Rev. A **51**, 992 (1995).
30. C.A. Blockley, D.F. Walls, and H. Risken, Europhys.. Lett. **17**, 509 (1992).
31. J. I. Cirac, R. Blatt, A. S. Parkins, and P. Zoller, Phys. Rev. Lett. **70**, 762 (1993).
32. J. I. Cirac, A. S. Parkins, R. Blatt, and P. Zoller, Phys. Rev. Lett. **70**, 556 (1993).

33. J. I. Cirac, R. Blatt, A. S. Parkins and P. Zoller, Phys. Rev. A **49**, 1202 (1994).
34. R. Blatt, J.I. Cirac, and P. Zoller, Phys. Rev. A **52**, 518 (1995).
35. S. Wallentowitz and W.Vogel, I. Siemers, and P.E. Toschek, Phys. Rev. A **54**, 943 (1996).
36. R. L. de Matos Filho and W. Vogel Phys. Rev. Lett. **76**, 4520 (1996).
37. For a review see J.I. Cirac, A. S. Parkins, R. Blatt and P. Zoller, Adv. in At. Mol. Phys., in press.
38. S. Wallentowitz and W. Vogel, Phys. Rev. Lett. **75**, 2932 (1995).
39. P. J. Bardroff, E. Mayr, and W. P. Schleich, Phys. Rev. A, **51** 4963 (1995).
40. C.D'Helon and G.J. Milburn, preprint.
41. J. F. Poyatos, R. Walser, J. I. Cirac, P. Zoller, Phys. Rev. A, **53**, R1966 (1996).
42. V. Vedral, A. Barenco, and A. Ekert, Phys. Rev. A **54**, 147 (1996).
43. R. J. Hughes, D.F.V. James, E.H. Knill, R. Laflamme and A.G. Petschek, preprint quant-ph/9604026, available on the Los Alamos National Laboratory preprint archive http://xxx.lanl.gov.
44. M. Plenio and P. Knight, Phys. Rev. A **53**, 2986 (1996).
45. J. I. Cirac and P. Zoller, Phys. Rev. A **50**, R2799 (1994).
46. J. Poyatos, J. I. Cirac and P. Zoller, submitted for publication.
47. W.K. Wootters, Ann. Phys. (N.Y.) **176**, 1 (1987).
48. E. Knill and R. Laflamme, preprint quant-ph/9600034 on http://xxx.lanl.gov.
49. P.W. Shor, Phys. Rev. A **52**, R2493 (1995)
50. A. Steane, Proc. Roy. Soc. London A, in press.
51. A. R. Calderbank and P. W. Shor, preprint quant-ph/9512032 on http://xxx.lanl.gov.
52. R. Laflamme, C. Miquel, J. P. Paz, and W. H. Zurek, Phys. Rev. Lett. **77**, 198 (1996).
53. Shor P. W. 1996, preprint quant-ph/9605011 on http://xxx.lanl.gov.
54. C.W. Gardiner, *Quantum Noise* (Springer, Berlin, 1991).
55. H. J. Carmichael, in *An Open Systems Approach to Quantum Optics*, Lecture Notes in Physics, m18 (Springer, Berlin, 1993); J. Dalibard *et al.*, Phys. Rev. Lett. **68**, 580 (1992); C. W. Gardiner, A. S. Parkins, and P. Zoller, Phys. Rev. A **46**, 4363 (1992).
56. H. Mabuchi and P. Zoller, Phys. Rev. Lett. **76**, 3108 (1996).
57. J.I. Cirac, T. Pellizzari, and P. Zoller, to be published in Science.

ENTANGLED STATES OF ATOMIC IONS
FOR QUANTUM METROLOGY AND COMPUTATION[†]

D.J. WINELAND, C. MONROE, D.M. MEEKHOF, B.E. KING, D. LEIBFRIED,
W.M. ITANO, J.C. BERGQUIST, D. BERKELAND, J.J. BOLLINGER, J. MILLER
Ion Storage Group, Time and Frequency Division, NIST, Boulder, CO, 80303, USA

A single trapped ^{9}Be$^+$ ion is used to investigate Jaynes-Cummings dynamics for a two-level atomic system coupled to harmonic atomic motion. We create and investigate nonclassical states of motion including "Schrödinger-cat" states. A fundamental quantum logic gate is realized using the quantized motion and internal states as qubits. We explore some of the applications for, and problems in realizing, quantum logic based on multiple trapped ions.

1 Introduction

Currently, a major theme in atomic physics is coherent control of quantum states. This theme is manifested in a number of topics such as atom interferometry, atom optics, the atom laser, Bose-Einstein condensation, cavity-QED, electromagnetic-induced transparency, lasing without inversion, quantum computation, quantum cryptography, quantum-state engineering, squeezed states, and wavepacket dynamics. A number of these topics are the subjects of other presentations at this meeting.

In this paper we report related trapped-ion research on (1) the study of Jaynes-Cummings dynamics for a two-level atomic system coupled to harmonic atomic motion, (2) the study of quantum mechanical measurement problems such as the generation of Schrödinger-cat-like superposition states and their relation to various decoherence phenomena, and (3) coherent quantum logic for the investigation of scaling in a quantum computer and for preparation of entangled states useful for spectroscopy.

2 Entanglement

An entangled quantum state is one where the wave function of the overall system cannot be written as a product of the wave functions of the subsystems. In this case, a measurement on one of the subsystems will affect the state of the other subsystems. For example, consider a two-level atom bound in a 1-D harmonic well. Suppose we can create the state

$$\Psi = \frac{1}{\sqrt{2}}(|\downarrow\rangle|n\rangle + e^{i\phi}|\uparrow\rangle|n'\rangle), \qquad (1)$$

where the kets $|\downarrow\rangle$ and $|\uparrow\rangle$ denote the two internal eigenstates of the atom (here, we use the spin-½ analog to a two-level system: $\sigma_z|\uparrow\rangle = +|\uparrow\rangle$, etc.), the second ket denotes a harmonic oscillator eigenstate $|n\rangle$, and ϕ is a (controlled) phase factor. If we measure the motional eigenstate of the atom and find it to be in level n, then it must also be found in the $\downarrow$ internal state if we measure σ_z. Similarly, if we find the atom in the n' motional state, it must be found in the ↑ internal state. Such correlations are at the heart of the "EPR" experiments[1]. Another state we will consider below is the state for N two-level atoms

$$\Psi = \frac{1}{\sqrt{2}}(|\downarrow\rangle_1|\downarrow\rangle_2...|\downarrow\rangle_N + e^{i\phi}|\uparrow\rangle_1|\uparrow\rangle_2...|\uparrow\rangle_N), \qquad (2)$$

where the subscript i (= 1, 2, ..., N) denotes the ith atom. This state is "maximally entangled" in the sense that a measurement of σ_z on any atom automatically determines the value of σ_z of all other atoms.

3   Jaynes-Cummings-type coupling between internal and motional states

To achieve entanglement from an initially nonentangled system, we need to provide a coupling between subsystems so that the state of one subsystem affects the dynamics of another. Coupling between spins or two-level atoms can, in principle, be achieved through a dipole-dipole interaction (like the hyperfine coupling between electron and proton in the hydrogen atom). In a system of trapped neutral atoms, dipole-dipole coupling may be difficult to control to the desired level; for trapped ions the Coulomb repulsion inhibits strong dipole coupling between ions. However, in the case of trapped ions, the motion can be strongly coupled to the internal levels with the application of inhomogeneous (classical) electromagnetic fields. For example, we consider an atom confined in a 1-D harmonic potential. The atom's dipole moment μ is assumed to couple to an electric field E(x,t) through the Hamiltonian

$$H_I = -\mu E(x,t) = -\mu \left[E(x=0,t) + \frac{\partial E}{\partial x}x + \frac{1}{2}\frac{\partial^2 E}{\partial x^2}x^2 + \cdots \right]. \qquad (3)$$

We have $\mu \propto \sigma_+ + \sigma_-$, where σ_+ and σ_- are the raising and lowering operators for the internal levels (in the spin-½ analog). In Eq. (3), the position x is an operator which we write as x = $x_o(a + a^\dagger)$, where a and $a^\dagger$ are the usual harmonic oscillator lowering and raising operators, and x_o is the rms spread of the n=0 zero-point state of motion. As a simple example, suppose the field is static and the motional oscillation frequency ω of the atom is equal to the resonance frequency ω_o of the internal state transition. In its reference frame, the atom experiences an oscillating field due to the motion through the inhomogeneous field. Since $\omega = \omega_o$, this field resonantly drives transitions between the internal states. If the extent of the atom's motion is small enough that we need only consider the first two terms in Eq. (3), H_I can be approximated as $H_{JCM} = \hbar\Omega(\sigma_+ a + \sigma_- a^\dagger)$ (in the interaction frame and using the rotating wave approximation) where Ω is a proportionality constant. This Hamiltonian is also obtained if E is sinusoidally time varying (frequency ω_L) and we satisfy the resonance condition $\omega_L + \omega = \omega_o$. This type of coupling was used to couple the spin and cyclotron motion in the classic electron g - 2 experiments of Dehmelt and coworkers[2]. Formally it is equivalent to the Jaynes-Cummings Hamiltonian of cavity-QED[3,4] which describes the coupling between a two-level atom and a single mode of the radiation field. This analogy has been pointed out in various papers[5-8]; for a review, see Ref. 9 and further references in Ref. 8.

3.1 Realization a Jaynes-Cummings-type coupling for a trapped $^9Be^+$ ion

To controllably manipulate the internal and vibrational levels of the ion, we must (1) initialize the ion in a well defined internal and motional state and (2) make the vibrational level spacing (trap frequency) much larger than any internal or motional relaxation rates. To accomplish this, we have built an rf (Paul) ion trap which confines a single $^9Be^+$ ion with pseudopotential harmonic trap frequencies of $(\omega_x, \omega_y, \omega_z)/2\pi \approx (11, 19, 29)$ MHz along the three principal axes of the trap[10].

The energy-level structure of $^9Be^+$ is summarized in Fig. 1. Because the ion is harmonically bound, the internal $^9Be^+$ electronic states must include the ladder of external harmonic oscillator levels of energy $E_n = \hbar\omega(n+\frac{1}{2})$, where we have considered only the x-dimension of the oscillator ($\omega \equiv \omega_x$) and its associated quantum number $n \equiv n_x \in (0, 1, 2, ...)$. The two internal levels of interest are the $^2S_{1/2}$ ground state hyperfine levels $|F=2, m_F=2\rangle$ (denoted by $|\downarrow\rangle$) and $|F=1, m_F=1\rangle$ (denoted by $|\uparrow\rangle$), which are separated in frequency by $\omega_0/2\pi \approx 1.25$ GHz. The other Zeeman levels are resolved from the $|\downarrow\rangle$ and $|\uparrow\rangle$ states by the application of a ≈ 0.2 mT magnetic field[8,11].

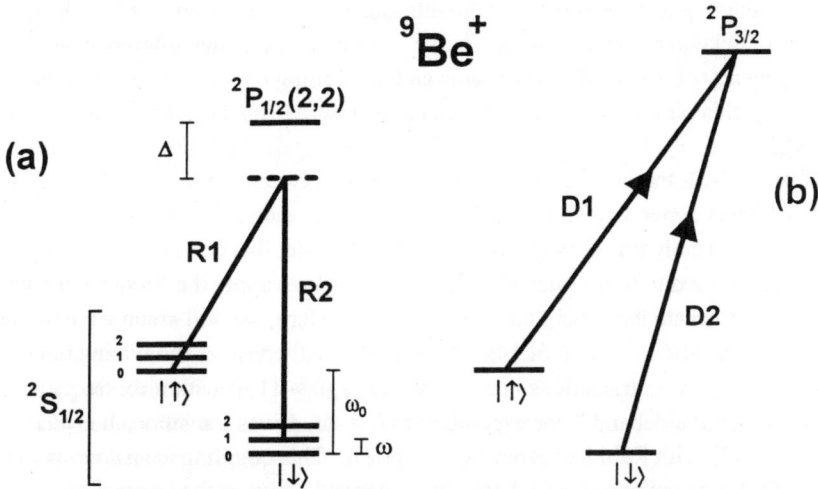

Fig. 1. (a) Electronic (internal) and motional (external) energy levels (not to scale) of the trapped $^9Be^+$ ion, coupled by indicated laser beams R1 and R2. The difference frequency of the Raman beams R1 and R2 is set near $\omega_0/2\pi \approx$ 1.250 GHz, providing a two-photon Raman coupling between the $^2S_{1/2}(F=2, m_F=2)$ and $^2S_{1/2}(F=1, m_F=1)$ hyperfine ground states (denoted by $|\downarrow\rangle$ and $|\uparrow\rangle$ respectively). The motional energy levels are depicted by a ladder of vibrational states separated in frequency by the trap frequency $\omega/2\pi \approx 11$ MHz. The Raman beams are detuned $\Delta/2\pi \approx -12$ GHz from the $^2P_{1/2}(F=2, m_F=2)$ excited state. As shown, the Raman beams are tuned to the red sideband. (b) Detection of the internal state is accomplished by illuminating the ion with σ^+-polarized "detection" beam D2, which drives the cycling $^2S_{1/2}(F=2, m_F=2) \to {}^2P_{3/2}(F=3, m_F=3)$ transition, and observing the scattered fluorescence. The vibrational structure is omitted from (b) since it is not resolved. Beam D1, also σ^+ polarized, provides spontaneous recycling from the $|\uparrow\rangle$ to $|\downarrow\rangle$ state.

Strong field gradients can be obtained with laser fields (e^{ikx} factor). In our experiment, the field corresponding to that in Eq. (3) is provided by two laser fields which drive stimulated-Raman transitions between the levels of interest (R1 and R2 of Fig. 1a). (The use of stimulated-Raman transitions has some technical advantages, but is formally equivalent to driving a narrow single-photon transition.) Two-photon stimulated Raman transitions between the $|\downarrow\rangle$ and $|\uparrow\rangle$ states can be driven by tuning the difference frequency of R1 and R2 to be near ω_o. The two Raman beams ($\lambda \approx 313$ nm) are generated from a single laser source and acousto-optic modulator, allowing excellent stability of their relative frequency and phase. Both beams are detuned $\Delta/2\pi \approx 12$ GHz from the excited $^2P_{1/2}$ electronic state (radiative linewidth $\gamma/2\pi \approx 19.4$ MHz), and the polarizations are set to couple through the $^2P_{1/2}(F=2, m_F=2)$ level (the next nearest levels are the $^2P_{3/2}$ states which are over 200 GHz away and can be neglected). Because $\Delta \gg \gamma$, the excited ^2P state can be adiabatically eliminated in a theoretical description, resulting in a coupling between the two ground states which exhibits a linewidth inversely proportional to the interaction time. When R1 and R2 are applied to the ion with wavevector difference $\delta\vec{k} = \vec{k}_1 - \vec{k}_2$ along the x-direction, the effective coupling Hamiltonian in the rotating-wave approximation is given by

$$H_I = g\left(\sigma_+ e^{i\eta(a^\dagger+a)-i\delta t} + \sigma_- e^{-i\eta(a^\dagger+a)+i\delta t}\right). \quad (4)$$

The coupling strength g depends on Δ and the intensity of the laser beams, $\eta = |\delta\vec{k}|x_0 \approx 0.2$ is the Lamb-Dicke parameter, $x_0 = (\hbar/2m\omega)^{1/2} \approx 7$ nm, and δ is the difference between the relative frequency of the two Raman beams and ω_o. Setting $\delta\vec{k}$ to be parallel to the x-axis of the trap, yields almost no coupling between the internal states and motion in the y- and z-directions.

If $\delta = \omega(n'-n)$, transitions are resonantly driven between the levels $|\downarrow,n\rangle$ and $|\uparrow,n'\rangle$ at a rate $\Omega_{n,n'}$ which is dependent on n and n'[8]. Starting from the $|\downarrow\rangle|n\rangle$ state, application of a Rabi -π pulse coherently transfers the ion to the $|\uparrow\rangle|n'\rangle$ state; this corresponds to applying the Raman beams for a duration τ such that $\Omega_{n,n'}\tau = \pi/2$. If we apply the Raman beams for half of this time, we create the entangled state of Eq. (1). Here, we will assume the ion is confined in the Lamb-Dicke limit ($|\delta\vec{k}|<x^2>^{1/2} \ll 1$) and will consider three transitions. The carrier, at $\delta = 0$, drives transitions between states $|\downarrow,n\rangle \leftrightarrow |\uparrow,n\rangle$ with Rabi frequency $\Omega_{n,n} = g$. The "first red sideband," corresponding to $\delta = -\omega$, drives transitions between states $|\downarrow,n\rangle \leftrightarrow |\uparrow,n-1\rangle$ with Rabi frequency $\Omega_{n,n-1} = g\eta\sqrt{n}$. This coupling is analogous to the case in cavity-QED[4] where energy is coherently exchanged between the internal and external degrees of freedom. The "first blue sideband," at $\delta = +\omega$, drives transitions between states $|\downarrow,n\rangle \leftrightarrow |\uparrow,n+1\rangle$ with Rabi frequency $\Omega_{n,n+1} = g\eta(n+1)^{1/2}$.

Preparation of the $|\downarrow\rangle|n=0\rangle$ state is accomplished by first Doppler cooling the ion to $\langle n\rangle \approx 1$, followed by sideband laser cooling using stimulated Raman transitions[11]. For sideband laser cooling, π pulses on the first red sideband ($|\downarrow\rangle|n\rangle \to |\uparrow\rangle|n-1\rangle$) are alternated with repumping cycles using nearly resonant radiation (Fig. 1b) - which results (most probably) in transitions $|\uparrow\rangle|n\rangle \to |\downarrow\rangle|n\rangle$. These steps are repeated (typically 5 times) until the ion resides in the $|\downarrow\rangle|0\rangle$ state with high probability (> 0.9).

From the $|\downarrow\rangle|0\rangle$ state, we are able to coherently create states of the form $|\downarrow\rangle\Psi(x)$, where the motional state $\Psi(x) = \Sigma_n C_n e^{-in\omega t}|n\rangle$ and the C_n are complex. We can analyze the motional state created as follows: The Raman beams are pulsed on for a time τ and the probability $P_\downarrow(\tau)$ that the ion is in the $|\downarrow\rangle$ internal state is measured. The experiment is repeated for a range of τ values. When the Raman beams are tuned to the first blue sideband, the expected signal is

$$P_\downarrow(\tau) = \frac{1}{2}\left(1 + \sum_{n=0}^{\infty} P_n \cos(2\Omega_{n,n+1}\tau) e^{-\gamma_n \tau}\right), \qquad (5)$$

where $P_n \equiv |C_n|^2$ is the probability of finding the ion in state n and γ_n are experimentally determined decay constants. The internal state $|\downarrow\rangle$ is detected by applying near-resonant σ^+-polarized laser radiation (beam D2, Fig. 1b) between the $|\downarrow\rangle$ and $^2P_{3/2}(F=3, m_F=3)$ energy levels. Because this is a cycling transition, detection efficiency is near unity[8,11]. The measured signal $P_\downarrow(\tau)$ can be inverted (Fourier cosine transform), allowing the extraction of the probability distribution of vibrational state occupation P_n. This signal does not show the phase coherences (phase factors of the C_n), which must be verified separately[8,12]. The most complete characterization is achieved with a state reconstruction technique[13].

3.2   Creation of Coherent and Schrödinger-Cat states

We have created and analyzed thermal, Fock, squeezed, coherent, Schrödinger-cat states, and superpositions of Fock states[8,12,13]; here we briefly describe the creation and measurements of coherent and Schrödinger-cat states. A coherent state of motion

$$\Psi(x) = |\alpha\rangle \equiv \exp(-|\alpha|^2/2) \sum_{n=0}^{\infty} \frac{\alpha^n}{\sqrt{n!}} |n\rangle, \qquad (6)$$

corresponds to a displaced zero-point wave-packet oscillating in the potential well with amplitude $2|\alpha|x_0$. From Eq. (5), $P_\downarrow(\tau)$ for a coherent state will undergo quantum collapses and revivals[14]. These revivals are a purely quantum effect due to the discrete energy levels and the narrow distribution of states[4,14].

We have produced coherent states of ion motion from the $|\downarrow\rangle|0\rangle$ state by applying either a resonant (frequency ω) classical driving field or a "moving standing wave" of laser radiation which resonantly drives the ion motion through the dipole force[8,12]. In Fig. 2, we show a measurement of $P_\downarrow(\tau)$ after creation of a coherent state of motion, exhibiting the expected collapse and revival signature. (For comparison, see the cavity-QED experiment of Ref. 4.) This data is fitted to Eq. (5) assuming a Poissonian distribution, allowing only $\langle n \rangle$ to vary. The inset shows the results of a separate analysis, which yield the probabilities of the Fock-state components, extracted by applying a Fourier cosine transform to $P_\downarrow(\tau)$ at the known frequencies as described above. These amplitudes display the expected

Poissonian dependence on n.

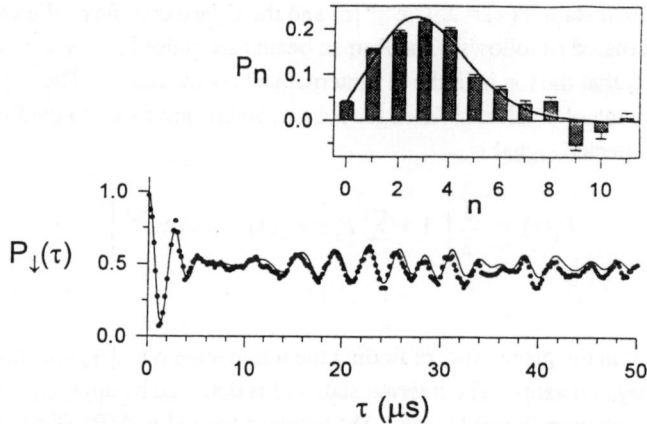

Fig. 2. $P_\downarrow(\tau)$ for a coherent state driven by the first blue sideband interaction, showing collapse and revival behavior. The data are fitted to a coherent state distribution, yielding $\langle n \rangle = 3.1(1)$. The inset shows the results of inverting the time-domain data by employing a Fourier cosine transform at the known Rabi frequencies $\Omega_{n,n+1}$, fitted to a Poissonian distribution, yielding $\langle n \rangle = 2.9(1)$. Each data point represents an average of ≈ 4000 measurements, or 1 s of integration.

A Schrödinger-cat state is a coherent superposition of classical-like states. In Schrödinger's original thought experiment[15], he described how one could, in principle, entangle a superposition state of an atom with a macroscopic-scale superposition of a live and dead cat. In our experiment[12], we construct an analogous state, on a smaller scale, with a single atom. We create the state

$$\Psi = \frac{1}{\sqrt{2}}(|\downarrow\rangle|\alpha_1\rangle + e^{i\phi}|\uparrow\rangle|\alpha_2\rangle), \qquad (7)$$

where $|\alpha_1\rangle$ and $|\alpha_2\rangle$ are coherent motional states and ϕ is a (controlled) phase factor. The coherent states of the superposition are spatially separated by mesoscopic distances much greater than the size of the atom wavepacket which has a spread equal to x_0.

Analysis of this state is interesting from the point of view of the "quantum measurement problem," an issue that has been debated since the inception of quantum theory by Einstein, Bohr, and others, and continues today[16]. One practical approach toward resolving this controversy is the introduction of quantum decoherence, or the environmentally induced reduction of quantum superpositions into classical statistical mixtures[17]. Decoherence provides a way to quantify the elusive boundary between classical and quantum worlds, and almost always precludes the existence of macroscopic Schrödinger-cat states, except at extremely short time scales. On the other hand, the creation of mesoscopic Schrödinger-cat states like that of Eq. (7) may allow controlled studies of quantum decoherence and the quantum-classical boundary. This problem is directly relevant to quantum computation.

In our experiment, we create a Schrödinger-cat state of the single-ion ^{9}Be$^+$ harmonic oscillator (Eq. (7)) with a sequence of laser pulses[12]. First, we create a state of the form $(|\downarrow\rangle + e^{i\xi}|\uparrow\rangle)|n=0\rangle/\sqrt{2}$ with a $\pi/2$ pulse on the Raman carrier transition (Sec. 3.1). To spatially separate the $|\downarrow\rangle$ and $|\uparrow\rangle$ components of the wave function, we apply a coherent excitation with an optical dipole force which, because of the polarization of the beams used to create the force, selectively excites the motion of only the $|\uparrow\rangle$ state. We then swap the $|\downarrow\rangle$ and $|\uparrow\rangle$ states with a π-carrier pulse and reapply the dipole force with a different phase to create the state of Eq. (7). In principle, if we could make $|\alpha_{1,2}|$ large enough, we could design a detector which could directly detect the (distinguishable) position of the particle and correlate it with a spin measurement[18]. Instead, to analyze this state in our experiment, we apply an additional laser pulse to couple the internal states, and we measure the resulting interference of the distinct wavepackets. With this interferometer, we can establish the correlations inherent in Eq. (7), the separation of the wavepackets, and the phase coherence ϕ between components of the wavefunction. These experiments are described in Ref. 12. The interference signal should be very sensitive to decoherence. As the separation $|\alpha_1 - \alpha_2|$ is made larger, decoherence is expected to exponentially degrade the fringe contrast[4,17].

We remark that other experiments generate Schrödinger-cats in the same sense as in our experiment. Examples are atom interferometers[19,20], and superpositions of electron wavepackets in atoms[21] (also, see additional citations in Ref. 12). However, as opposed to these experiments, the harmonic oscillator cat states of Eq. (7) do not disperse in time. This lack of dispersion provides a simple visualization of the "cat" (e.g., a marble rolling back and forth in a bowl which can be simultaneously at opposite extremes of motion) and should allow controlled studies of decoherence models.

4 Quantum Logic

Interest in quantum computation in the atomic physics community was stimulated, in part, by a talk given by Artur Ekert at the last ICAP meeting[22]. Subsequently, Ignacio Cirac and Peter Zoller[23,24] proposed an attractive scheme for a quantum computer which would use a string of ions in a linear trap as "qubits." This proposal has stimulated experimental efforts in several laboratories including those at Innsbruck, Los Alamos National Laboratory, IBM, and NIST.

Each qubit in a quantum computer could be implemented by a two-level atomic system; for the ith qubit, we label these states $|\downarrow\rangle_i$ and $|\uparrow\rangle_i$ as above. In general, any quantum computation can be comprised of a series of single-bit rotations and two-bit "controlled-NOT" (CN) logic operations[22,25]. We are interested in implementing these two operations in a system of ^{9}Be$^+$ ions. Single-bit rotations are straightforward and correspond to driving Raman carrier transitions (Sec. 3.1) for a controlled time. Such rotations have been achieved in many previous experiments. Next, we describe the demonstration of a nontrivial CN logic gate with a single ^{9}Be$^+$ ion[26].

4.1   "Conditional dynamics" and a single-ion controlled-not logic gate

The key to making a quantum logic gate is to provide conditional dynamics; that is, we desire to perform on one physical subsystem a unitary transformation which is conditioned upon the quantum state of another subsystem[22]. In the context of cavity QED, the required conditional dynamics at the quantum level has recently been demonstrated[27,28]. For trapped ions, conditional dynamics at the quantum level has been demonstrated in verifications of zero-point laser cooling[11,29]. Recently, we demonstrated a CN logic gate with the ability to prepare arbitrary input states (the "keyboard").

A two-bit quantum CN operation provides the transformation:

$$|\epsilon_1\rangle|\epsilon_2\rangle \rightarrow |\epsilon_1\rangle|\epsilon_1 \oplus \epsilon_2\rangle, \qquad (8)$$

where $\epsilon_1, \epsilon_2 \in \{0,1\}$ and $\oplus$ is addition modulo 2. The (implicit) phase factor in the transformation is equal to 1. In this expression ϵ_1 is the called the control bit and ϵ_2 is the target bit. If $\epsilon_1 = 0$, the target bit remains unchanged; if $\epsilon_1 = 1$, the target bit flips. In the single-ion experiment of Ref. 26, the control bit is the quantized state of one mode of the ion's motion. If the motional state is $|n=0\rangle$, it is taken to be a $|\epsilon_1=0\rangle$ state; if the motional state is $|n=1\rangle$, it is taken to be a $|\epsilon_1=1\rangle$ state. The target states are two ground-hyperfine states of the ion, the $|\downarrow\rangle$ and $|\uparrow\rangle$ states of Sec. 3.1 with the identification here $|\downarrow\rangle \leftrightarrow |\epsilon_2=0\rangle$ and $|\uparrow\rangle \leftrightarrow |\epsilon_2=1\rangle$. Following the notation of Sec. 3.1, the CN operation is realized by applying three Raman laser pulses in succession:

(1a) A "π/2-pulse" is applied on the spin carrier transition. For a certain choice of initial phase, this corresponds to the operator $V^{\frac{1}{2}}(\pi/2)$ of Ref. 23.

(1b) A 2π-pulse is applied on the first blue sideband transition between levels $|\uparrow\rangle$ and an auxiliary level $|aux\rangle$ in the ion (the $|F=2, M_F=0\rangle$ level in $^9Be^+$). This operator is analogous to the operator $U_n^{2,1}$ of Ref. 23. This operation provides the conditional dynamics for the controlled-not operation in that it changes the sign of the $|\uparrow\rangle|n=1\rangle$ component of the wavefunction but leaves the sign of the $|\uparrow\rangle|n=0\rangle$ component of the wavefunction unchanged.

(1c) A π/2-pulse is applied to the spin carrier transition with a 180° phase shift relative to step (1a). This corresponds to the operator $V^{\frac{1}{2}}(-\pi/2)$ of Ref. 23.

Steps (1a) and (1c) can be regarded as two resonant pulses (of opposite phase) in the Ramsey separated-field method of spectroscopy. We can see that if step (b) is active (thereby changing the sign of the $|\uparrow\rangle|n=1\rangle$ component of the wave function) then a spin flip is produced by the Ramsey fields. If step (1b) is inactive, the net effect of the Ramsey fields is to leave the spin state unchanged. This CN operation can be incorporated to provide an overall CN operation between two ions in an ensemble of N ions if we choose the ion oscillator mode to be the center-of-mass (COM) mode of the ensemble. Specifically, to realize a controlled-not $C_{m,k}$ between two ions (m = control bit, k = target bit), we first assume the COM is prepared in the zero-point state. The initial state of the system is therefore given by

$$\Psi = \left(\sum_{i=1}^{N} \sum_{M_i = \downarrow, \uparrow} C_{M_1, M_2, \ldots M_N} |M_1\rangle_1 |M_2\rangle_2 \cdots |M_N\rangle_N \right) |0\rangle . \tag{9}$$

$C_{m,k}$ can be accomplished with the following steps:

(2a) Apply a π-pulse on the red sideband of ion m (the assumption is that ions can be addressed separately[23]). This accomplishes the mapping $(\alpha|\downarrow\rangle_m + \beta|\uparrow\rangle_m)|0\rangle \rightarrow |\downarrow\rangle_m(\alpha|0\rangle - e^{i\phi}\beta|1\rangle)$, and corresponds to the operator $U_m^{1,0}$ of Ref. 23. We note that in our experiments, we prepare the state $(\alpha|\downarrow\rangle + \beta|\uparrow\rangle)|0\rangle$ using the carrier transition (Sec. 3.1). We can then implement the mapping $(\alpha|\downarrow\rangle + \beta|\uparrow\rangle)|0\rangle \rightarrow |\downarrow\rangle_m(\alpha|0\rangle - e^{i\phi}\beta|1\rangle)$. This is the "keyboard" operation for preparation of arbitrary motional input states for the CN gate of steps 1a - 1c above. Analogous mapping of internal state superpositions to motional state superpositions were demonstrated in Ref. 26.

(2b) Apply the CN operation (steps 1a - 1c above) between the COM motion and ion k.

(2c) Repeat step (2a).

Overall, $C_{m,k}$ provides the mappings $|\downarrow\rangle_m|\downarrow\rangle_k|0\rangle \rightarrow |\downarrow\rangle_m|\downarrow\rangle_k|0\rangle$, $|\downarrow\rangle_m|\uparrow\rangle_k|0\rangle \rightarrow |\downarrow\rangle_m|\uparrow\rangle_k|0\rangle$, $|\uparrow\rangle_m|\downarrow\rangle_k|0\rangle \rightarrow |\uparrow\rangle_m|\uparrow\rangle_k|0\rangle$, and $|\uparrow\rangle_m|\uparrow\rangle_k|0\rangle \rightarrow |\uparrow\rangle_m|\downarrow\rangle_k|0\rangle$ which is the desired logic of Eq. (8). Effectively, $C_{m,k}$ works by mapping the internal state of ion m onto the COM motion, performing a CN between the motion and ion n, and then mapping the COM state back onto ion m. The resulting CN between ions m and k is not really different from the CN described by Cirac and Zoller[7], because the operations $V^k(\theta)$ and $U_m^{1,0}$ commute.

4.2 Quantum Registers and Schrödinger Cats

The state represented by Eq. (9) is of the same form as that of Eq. (7). Both involve entangled superpositions and both are subject to the destructive effects of decoherence. Creation of Schrödinger-cats like Eq. (7) is particularly relevant to the ion-based quantum computer because the primary source of decoherence will probably be due to decoherence of the $|n=0,1\rangle$ motional states during the logic operations.

5 Potential for, and Problems with, Trapped-Ion Quantum Logic

Quantum computation has received a great deal of attention recently because of the algorithm proposed by Peter Shor for efficient factorization[30]. This has important implications for public-key data encryption where the security of these systems is due to the inability to efficiently factorize large numbers. To accomplish quantum factorization is extremely formidable with any technology; however, other applications of quantum logic may be more tractable.

5.1 Positive Aspects of Trapped-Ion Quantum Logic

Internal state decoherence can be relatively small in experiments on trapped ions. The ions' energy level structure is, of course, perturbed at some level by electric and magnetic fields. However, energy level shifts caused by electric fields (Stark shifts) tend to be quite small and, in many cases, level shifts due to magnetic fields can be controlled well enough. This is evident from trapped-ion atomic clock experiments where linewidths smaller than 0.001 Hz have been achieved[31,32], indicating internal state coherence times exceeding 10 min.

The required laser cooling to $|n=0\rangle$ has been demonstrated[11,29] for single ions. A string of laser-cooled ions (Fig. 3), which could be used as a quantum register, has been achieved in a linear ion trap[33,34], but an immediate future task will be to achieve zero-point cooling (for at least the COM mode) on an ensemble of ions. For a computation performed on an ensemble of ions in a trap, this need not be done extremely well. All we require is that the cooling be sufficient that the ion's COM mode is predominantly in the n=0 state, so the "correct" answer to a computation is obtained most of the time. Similarly, although nearly unit detection efficiency has been achieved with trapped ions[11,35], the basic requirement is that the noise in the "readout" of the quantum register should minimize the number of times the calculation is repeated.

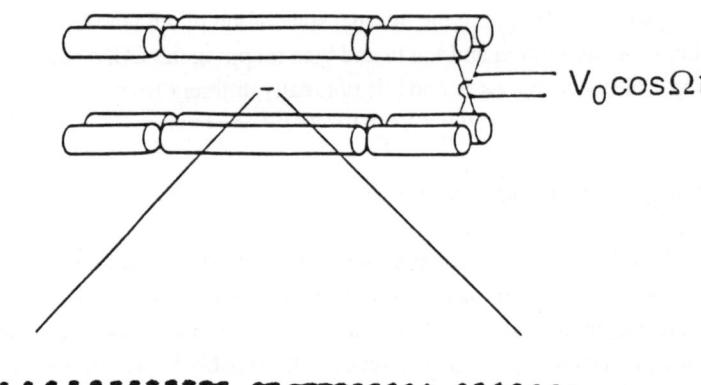

Fig. 3. The upper part of the figure shows a schematic diagram of the electrode configuration for a linear Paul-RF trap (rod spacing ≈ 1 mm)[34]. The lower part of the figure shows an image of a string of ^{199}Hg$^+$ ions, illuminated with 194 nm radiation, taken with a UV-sensitive, photon counting imaging tube. The spacing between adjacent ions is approximately 10 μm. The "gaps" in the string are occupied by impurity ions, most likely other isotopes of Hg$^+$, which do not fluoresce because the frequencies of their resonant transitions do not coincide with those of the 194 nm $^2S_{\frac{1}{2}} \rightarrow {}^2P_{\frac{1}{2}}$ transition of ^{199}Hg$^+$.

Estimates of motional decoherence times from the $|n=0,1\rangle$ states, due to fundamental causes, should be very long (more than 100 s)[36]. However, these predictions have not been realized experimentally and the causes of the observed decoherence[11,29] must still be found.

5.2 Problems and Possible Solutions

Many problems may conspire to prevent large-scale quantum computation; some of the problems relevant to trapped ions are briefly mentioned here. More complete analyses are given elsewhere,[37,38].

Motional decoherence can arise from fluctuating trap fields and radiative coupling to the environment. The integrity of the trap electrode structure is expected to play an important role, particularly if the number of ion qubits becomes large. Lithographic techniques for constructing electrodes[39] may be useful to insure accurate dimensional tolerances. With these techniques, it may also be possible to incorporate accurate (Josephson) voltage standards and (superconducting) flux magnetometers into the structure.

Laser power fluctuations will affect the fidelity of the rotations and logic gates (for example, $\pi/2$ rotations become $\pi/2 \pm \epsilon$ rotations where ϵ is unknown). Although the effects depend on the form of the noise and on the computational algorithm, a conservative estimate is to assume that phase errors accumulate linearly with the number of elemental operations. A computation requiring 10^6 elemental operations would therefore require an intensity stability of one part in 10^6 over the time of the computation. With current algorithms, factorization of large numbers will require even more elemental operations, so extreme laser stabilization will be required.

We have noted some of the advantages of using stimulated Raman transitions in quantum logic[26]. One apparent disadvantage is that, since the Raman beams are detuned from a virtual optical level, the energy levels are shifted by AC-Stark effects [see for example Ref. 36]. These effects are absent in single-photon transitions as assumed in Ref. 23. Therefore, if stimulated-Raman transitions are used and if the laser intensities fluctuate, additional Stark-induced phase errors will accumulate. However, we can show[38] that these errors are of the same order as those incurred from the angular errors of the preceding paragraph, if the two Raman laser intensities are approximately equal.

The scheme of Cirac and Zoller[23] assumes that the laser beams can separately address individual ion qubits. This necessitates a tradeoff between two factors. We desire that the ions be well separated spatially to allow a focused laser beam to address only one ion at a time. However, we also desire to spectrally isolate individual modes of the ion motion, to insure the fidelity of the logic. The closer in frequency the "contaminating" transitions (from coupling to other motional modes) are, the slower the logic speed must be to obtain isolation. To give an idea of the problem, we note that the separation of two $^9Be^+$ ions confined in a harmonic well of frequency $v \equiv \omega/2\pi$ is given by $d = 9.21 v^{-2/3}$ where d is in micrometers and v in megahertz. For longer strings, the spacing of the central ions becomes closer[40]. Although the focusing predicted with the use of Gaussian beam formulas implies the required isolation could be obtained for v up to a few MHz, in practice, stray light intensity will undoubtedly be a problem. With the use of stimulated-Raman transitions [26,36], one solution for this problem is to take advantage of the inherent AC Stark shifts. The basic idea is that the (resonant) Rabi frequencies g_1, g_2 of the two Raman beams are made substantially different, say $g_1 \gg g_2$. The transition frequency for the selected qubit is therefore shifted from the frequency of adjacent ions so that the adjacent ions are relatively unaffected. Unfortunately, the sensitivity to intensity fluctuations also becomes worse by the ratio g_1/g_2[38].

A large scale computation will require a large qubit register. This makes it extremely hard to isolate unwanted motional mode transitions from the desired one[38]. As noted in Ref. 23, the desired COM trap-axis mode frequency in a linear trap is smaller than other trap axis modes and can therefore be relatively well isolated spectrally. However, as the number of ions in the trap increases, the radial mode frequencies will tend to overlap the COM mode frequency. Also, multi-mode excitations may become a problem when the difference frequency of the modes is close to the COM mode frequency. Therefore, a multiplexing scheme for ion qubit registers seems desirable; we discuss one possibility below.

5.3   A 1- or 2-qubit ion accumulator

One possibility for multiplexing in a trapped-ion quantum computer is to perform all logic in minimal accumulators which hold one or two ions at a time[38]. Ions would be shuffled around in a "super-register" and into and out of the accumulators which are well shielded from the other ions. The shuffling could be accomplished with interconnected linear traps with segmented electrodes; this appears possible with the use of lithographic techniques. Single-bit rotations on the mth ion would be accomplished by moving that ion into an accumulator. Logic operations between ions m and k would be accomplished by first moving these ions into an accumulator. An accumulator would hold a second species of ion (say Mg^+) which could be used to provide laser cooling to the $|n=0\rangle$ level (of the mode used for the gate) if necessary. Therefore, for logic operations, an accumulator would hold two computational ions and the auxiliary ion. This scheme should make it easier to select ions with laser beams because it should be straightforward to address one ion while nulling the laser intensity on the other ion, even with very high trap frequencies. The very small number of logic ions in an accumulator (1 or 2) would make extraneous mode coupling much easier to avoid. The main problem appears to be that computational speed is reduced because of the time required to shuffle ions in and out of the accumulator and provide laser cooling with the auxiliary ion, if required. However, energy shifts of the ion's internal structure, due to the electric fields required to move the ion, need not be severe . For example, to move a $^9Be^+$ from rest to a location 1 cm away (and back to rest) in 1 μs would require a field of less than 50 V/cm. Electric fields of this order should give negligible phase shifts in qubits based on hyperfine structure[41]. The phase shift caused by time dilation would be less than 1 μrad.

5.4   Perspective on Ion Quantum Computation

To be useful for factorization, a quantum computer must be able to factorize a 200 digit decimal number. This will require a few thousand ions and perhaps 10^9 elementary operations[22]. Given the current state-of-the art (one ion and about 10 operations), we should therefore be skeptical. Decoherence will be most decisive in determining the fate of quantum computation. Already, decoherence from spontaneous emission appears to limit the number of operations possible[42,43]. The experiments can be expected to improve dramatically, but we must hope for more efficient algorithms or ways to patch them (such as error correction schemes[24]) before large scale factorization is possible.

Any quantum system that might be contemplated in quantum computation must be reproducible, stable, and well isolated from the environment. Quantum dots have the potential advantage of large scale integration using microfabrication; however at the present time, they suffer from lack of precise reproducibility and excessive decoherence. Trapped ions are reproducible and relatively immune to environmental perturbations - this is the reason they are candidates for advanced frequency standards[44]. In principle, high information density could be achieved by scaling down the electrodes; however, we must then worry about excessive environmental coupling such as magnetic field perturbations caused by impurities and/or currents in the (nearby) trap electrodes[38]. Electric field perturbations will also become important. Therefore, in terms of scale, the trapped ion system may be close to optimum.

Finally, factorization, discrete logs, and certain other mathematical computations appear to be the hardest problems that quantum logic might be applied to. One of the applications for quantum computation that Richard Feynman originally had in mind was to simulate quantum mechanical calculations[45]. This idea is being explored again with new possibilities in mind[46]. Below, we consider an application to atomic measurement.

6 Quantum Logic Applied to Spectroscopy

We conclude by discussing a possible application of quantum logic in the realm of atomic physics. This application has the advantage of being useful with a relatively small number of ions and logic operations.

Entangled atomic states can improve the quantum-limited signal-to-noise ratio in spectroscopy[6,36,47]. In spectroscopy experiments on N atoms, in which changes in atomic populations are detected, we can view the problem in the following way using the spin-½ analogy for two-level atoms. We assume spectroscopy is performed by applying (classical) fields of frequency ω_R for a time T_R according to the Ramsey method of separated fields[48]. After applying these fields, we measure the final state populations. For example, we might measure the operator $\tilde{N}_-$ corresponding to the number of atoms in the $|\downarrow\rangle$ state. In the spin-½ analog, this is equivalent to measuring the operator J_z, since $\tilde{N}_- = J\tilde{I} - J_z$ where $\tilde{I}$ is the identity operator.

If all technical sources of noise are eliminated, the signal-to-noise ratio (for repeated measurements) is fundamentally limited by the quantum fluctuations in the number of atoms which are observed to be in the $|\downarrow\rangle$ state. These fluctuations can be called quantum "projection" noise[49]. If spectroscopy is performed on N initially uncorrelated atoms (e.g., $\Psi(t=0) = \Pi_i |\downarrow\rangle_i$), the imprecision in a determination of the frequency of the transition is limited by projection noise to $(\Delta\omega)_{meas.} = 1/(NT_R\tau)^{½}$ where $\tau \gg T_R$ is the total averaging time. If the atoms can be initially prepared in entangled states, it is possible to achieve $(\Delta\omega)_{meas.} < 1/(NT_R\tau)^{½}$. Initial theoretical investigations[6,36] examined the use of correlated states which could achieve $(\Delta\omega)_{meas.} < 1/(NT_R\tau)^{½}$ when the population (J_z) was measured. More recent theoretical investigations[47] consider the initial state to be one where, after the first Ramsey pulse, the internal state is the maximally entangled state of Eq. (2).

After applying the Ramsey fields, we measure the operator $\tilde{O} = \Pi_i \sigma_{zi}$ instead of J_z (or $\tilde{N}$). For unit detection efficiency, we can achieve $(\Delta\omega)_{meas.} = 1/(N^2 T_R \tau)^{1/2}$ which is the maximum signal-to-noise ratio possible. For an atomic clock where T_R is fixed by other constraints, this means that the time required to reach a certain measurement precision (stability) is reduced by a factor of N relative to the uncorrelated-atom case. In terms of quantum computation, this amounts to a computation of the function $\cos(N(\omega - \omega_o)T)$. Of course, this computation has special significance for the measurement of ω_o (an intrinsic computer parameter) but otherwise is much better suited for a classical computer! See Ref. 50 for related work.

Cirac and Zoller[23] have outlined a scheme for producing the state in Eq.(2) using quantum logic gates. Using the notation of Sec. 3.1, we would first apply a $\pi/2$ rotation to ion 1 to create the state $\Psi = 2^{-1/2}(|\downarrow\rangle_1 + e^{i\phi}|\uparrow\rangle_1)|\downarrow\rangle_2|\downarrow\rangle_3...|\downarrow\rangle_N$. We then apply the CN gate of Eq. (8) sequentially between ion 1 and ions 2 through N to achieve the state of Eq. (2). An alternative method for generating this state, without the need of addressing individual ions is described in Ref. 47.

Acknowledgments

We gratefully acknowledge the support of the National Security Agency, the US Office of Naval Research, and the US Army Research Office. We thank P. Huang, M. Lombardi, C. Wood, and M. Young for helpful comments on the manuscript.

References

† Contribution of NIST; not subject to US copyright.
1. A. Einstein, B. Podolsky, N. Rosen, *Phys. Rev.* 47, 777 (1935).
2. H. Dehmelt, *Science* 247, 539 (1990).
3. *Cavity Quantum Electrodynamics*, ed. by P.R. Berman (Academic Press, Boston, 1994).
4. S. Haroche, et al., these proceedings.
5. C.A. Blockley, D.F. Walls, and H. Risken, *Europhys. Lett.* 17, 509 (1992).
6. D. J. Wineland, J. J. Bollinger, W. M. Itano, F. L. Moore, and D. J. Heinzen, *Phys. Rev.* A46, R6797 (1992).
7. J.I. Cirac, R. Blatt, A.S. Parkins, and P. Zoller, *Phys. Rev. Lett.* 70, 762 (1993).
8. D.M. Meekhof, C. Monroe, B.E. King, W.M. Itano, and D.J. Wineland, *Phys. Rev. Lett.* 76, 1796 (1996).
9. J.I. Cirac, A.S. Parkins, R. Blatt, P. Zoller, in *Adv. Atomic and Molecular Phys.*, to be published.
10. S. R. Jefferts, C. Monroe, E. W. Bell, and D. J. Wineland, *Phys. Rev.* A51, 3112-3116 (1995).
11. C. Monroe, D. M. Meekhof, B. E. King, S. R. Jefferts, W. M. Itano, D. J. Wineland, and P. Gould, *Phys. Rev. Lett.* 75, 4011 (1995).
12. C. Monroe, D. M. Meekhof, B. E. King, and D. J. Wineland, *Science* 272, 1131

(1996).
13. D. Leibfried, D.M. Meekhof, B.E. King, C. Monroe, W.M. Itano, and D.J. Wineland, submitted.
14. J.H. Eberly, N.B. Narozhny, and J.J. Sanchez-Mondragon, *Phys. Rev. Lett.* 44, 1323 (1980).
15. E. Schrödinger, *Naturwissenschaften* 23, 807 (1935).
16. *Quantum Theory and Measurement*, ed. by J.A. Wheeler, W.H. Zurek (Princeton Univ. Press, Princeton, 1983).
17. W.H. Zurek, *Physics Today*, 44, 36 (1991).
18. J.F. Poyatos, J.I. Cirac, R. Blatt, P. Zoller, *Phys. Rev. A*, to be published.
19. D.E. Pritchard, et al., these proceedings.
20. O. Carnal and J. Mlynek, *Phys. Rev. Lett.* 66, 2689 (1991); D.W. Keith, C.R. Ekstrom, Q.A. Turchette, D.E. Pritchard, *Phys. Rev. Lett.* 66, 2693 (1991); M. Kasevich and S. Chu, *Phys. Rev. Lett.* 67, 181 (1991); J. Lawall, S. Kulin, B. Saubamea, N. Bigelow, M. Leduc, and C. Cohen-Tannoudji, *Phys. Rev. Lett.* 75, 4194 (1995).
21. L.D. Noordam, D.I. Duncan, T.F. Gallagher, *Phys. Rev.* A45, 4734 (1992); R.R. Jones, C.S. Raman, S.W. Schumacher, P.H. Bucksbaum, *Phys. Rev. Lett.* 71, 2575 (1993); M.W. Noel and C.R. Stroud, Jr., *Phys. Rev. Lett.* 75, 1252 (1995).
22. A. Ekert, in *Atomic Physics 14*, ed. by D. J. Wineland, C. E. Wieman, and S. J. Smith, (proc. 14th International Conference on Atomic Physics, Boulder, CO, August, 1994), (AIP Press, NY, 1995), p. 450; A. Ekert and R. Jozsa, *Rev. Mod. Phys.*, July, 1996, to be published.
23. J.I. Cirac and P. Zoller, *Phys. Rev. Lett.* 74, 4091 (1995).
24. P. Zoller, et al., these proceedings.
25. D.P. DiVincenzo, *Phys. Rev.* A51, 1051 (1995).
26. C. Monroe, D. M. Meekhof, B. E. King, W. M. Itano, and D. J. Wineland, *Phys. Rev. Lett.* 75, 4714 (1995).
27. M. Brune, P. Nussenzveig, F. Schmidt-Kaler, F. Bernardot, A. Maali, J.M. Raimond, and S. Haroche, *Phys. Rev. Lett.* 72, 3339 (1994).
28. Q. Turchette, C. Hood, W. Lange, H. Mabushi, H.J. Kimble, *Phys. Rev. Lett.* 75, 4710 (1995).
29. F. Diedrich, J.C. Bergquist, W. M. Itano, and D.J. Wineland, *Phys. Rev. Lett.* 62, 403 (1989).
30. P. Shor, Proc. 35th Ann. Symp. on the Foundations of Comp. Sci. (IEEE Comp. Soc. Press, NY, 1994). p. 124.
31. J. J. Bollinger, D. J. Heinzen, W. M. Itano, S. L. Gilbert, and D. J. Wineland, *IEEE Trans. on Instrum. and Meas.* 40, 126 (1991).
32. P.T.H. Fisk, M.J. Sellars, M.A. Lawn, C. Coles, A.G. Mann, and D.G. Blair, *IEEE Trans. Instrum. Meas.* 44, 113 (1995).
33. M. G. Raizen, J. M. Gilligan, J. C. Bergquist, W. M. Itano, and D. J. Wineland, *Phys. Rev.* A45, 6493 (1992).
34. J. Miller, M. E. Poitzsch, F. C. Cruz, D. J. Berkeland, J. C. Bergquist, W. M. Itano. and D. J. Wineland, Proc., 1995 IEEE Intl. Frequency Control Symp., June

1995, pp. 110-112; M. E. Poitzsch, J. C. Bergquist, W. M. Itano, and D. J. Wineland, *Rev. Sci. Instrum.* <u>67</u>, 129 (1996).

35. W. Nagourney, J. Sandberg, and H.G. Dehmelt, *Phys. Rev. Lett.* <u>56</u>, 2797 (1986); Th. Sauter, R. Blatt, W. Neuhauser, and P.E. Toschek, *Phys. Rev. Lett.* <u>57</u>, 1696 (1989); J.C. Bergquist, R.G. Hulet, W.M. Itano, and D.J. Wineland, *Phys. Rev. Lett.* <u>57</u>, 1699 (1986).

36. D.J. Wineland, J. J. Bollinger, W. M. Itano, and D. J. Heinzen, *Phys. Rev.* A<u>50</u>, 67 (1994).

37. R.J. Hughes, D.F.V. James, E.H. Knill, R. Laflamme, and A.G. Petschek, Los Alamos report LA-UR-96-1266 (1966) (submitted to PRL); Los Alamos eprint archive quant-ph/9604026.

38. D.J. Wineland, et al., in preparation.

39. R. Brewer, R.G. DeVoe, and R. Kallenbach, *Phys. Rev.* A<u>46</u>, R6781 (1992).

40. D. J. Wineland, J. C. Bergquist, J. J. Bollinger, W. M. Itano, D. J. Heinzen, S. L. Gilbert, C. H. Manney, and M. G. Raizen, *IEEE Trans. on Ultrasonics, Ferroelectrics, and Frequency Control* <u>37</u>, 515 (1990).

41. Wayne M. Itano, L.L. Lewis, and D.J. Wineland, *Phys. Rev.* A<u>25</u>, 1233 (1982).

42. M.B. Plenio and P.L. Knight, Phys. Rev. A<u>53</u>, 2986 (1996).

43. We note that the effects of spontaneous emission are significantly reduced if rf transitions between hyperfine levels are induced with inhomogeneous rf fields (Sec. 3).

44. see: *Proc., Fifth Symp. Freq. Standards and Metrology*, ed. by J.C. Bergquist, Woods Hole, MA, Oct. 1995 (World Scientific, Singapore, 1996).

45. R.P. Feynman, *Int. J. Theor. Phys.* <u>21</u>, 467 (1982); *Opt. News* <u>11</u>, 11 (1985); *Found. Phys.* <u>16</u>, 507 (1986).

46. S. Lloyd, *Science,* to be published.

47. J. J. Bollinger, D. J. Wineland, W. M. Itano, and D. J. Heinzen, *Proc., Fifth Symp. Freq. Standards and Metrology*, ed. by J.C. Bergquist, Woods Hole, MA, Oct. 1995 (World Scientific, Singapore, 1996), p. 107.

48. N.F. Ramsey, *Molecular Beams*, (Oxford University Press, London, 1963).

49. W. M. Itano, J. C. Bergquist, J. J. Bollinger, J. M. Gilligan, D. J. Heinzen, F. L. Moore, M. G. Raizen, and D. J. Wineland, *Phys. Rev.* A<u>47</u>, 3554 (1993).

50. J. Sørensen, J Erland, J. Hald, A. Kuzmich, K. Mølmer, and E.S. Polzik, contributed abstract, this meeting.

ENTANGLEMENT AND INDISTINGUISHABILITY: COHERENCE EXPERIMENTS WITH PHOTON PAIRS AND TRIPLETS

A. ZEILINGER

Institut für Experimentalphysik, Universität Innsbruck,
Technikerstraße 25, A-6020 Innsbruck, Austria

Entanglement is one of the most fundamental notions in quantum mechanics. As Erwin Schrödinger put it in 1935, it describes the fact that it is possible for subsystems of a larger quantum system to have properties which are only defined relative to each other while the properties of each subsystem can be quantum mechanically maximally uncertain. In recent years, two-photon states served not only to demonstrate some of the fundamental features of entangled systems, but also led to novel quantum phenomena which might indicate new applications. The present paper presents a simple overview how entanglement arises from indistinguishability of the paths taken by the photons. As a demonstration of the interesting features of entangled states, we discuss the most basic phenomenon of the statistics of two photons at a beam splitter and, as an application, photon dense coding.

1 Introduction

The most important feature of quantum mechanical phenomena is the *superposition principle*[1,3]. In its most basic form, the superposition principle states:

> Whenever a final state might have been arrived at via a number of different ways (paths) such that it is not possible, *not even in principle*, to determine which of the paths the system took, the final state is obtained by adding the probability amplitudes corresponding to all the possible paths.

States which can be superposed in that way usually are called *coherent with respect to each other*[2].

To illustrate the basic notions of superposition, indistinguishability, coherence and entanglement, I will analyze in this paper the behavior of two photons at a single beam splitter. In order to illustrate its quantum mechanical operation we first discuss the case of a single particle incident on a beam splitter.

The beam splitter has two input modes a and b and two output modes c and d (Fig. 1). Quantum mechanically, the action of the beam splitter on the input modes can be written as

$$|a\rangle \to \frac{i}{\sqrt{2}}|c\rangle + \frac{1}{\sqrt{2}}|d\rangle$$
$$|b\rangle \to \frac{1}{\sqrt{2}}|c\rangle + \frac{i}{\sqrt{2}}|d\rangle \quad (1)$$

where, e.g. $|a\rangle$ describes the quantum mechanical state of the particle in input beam a. We should stress here that we do not make any assumption about the nature of the incident particle. In this part of our discussion, it could be a photon, a neutron,

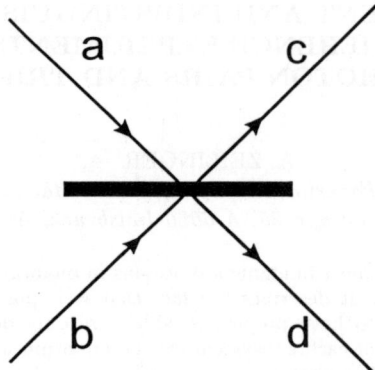

Figure 1: Input beams and output beams at the standard beam splitter.

an electron, an atom or whatever. Eq. 1 describes the fact that the particle can be found with equal probability $p = 0.5 = (\frac{1}{\sqrt{2}})^2$ in either of the output beams c and d, no matter through which input beam it came. The factor i in Eq. 1 is a consequence of unitarity. It describes physically a phase jump upon reflection at the semi-transparent mirror [4].

Let us now assume that our particle is incident in a coherent superposition of beams a and b. This could for example have been achieved by a first beam splitter and passage to the second beam splitter in such a way that it is impossible to decide which path the particle took. The incident photon state

$$|\psi_i\rangle = \frac{1}{\sqrt{2}}|a\rangle + \frac{e^{i\chi}}{\sqrt{2}}|b\rangle \qquad (2)$$

in general contains an additional phase difference between the two input beams. The two outgoing beams c and d then are coherent superpositions of amplitudes incident via a and b. With the beam splitter rules (1) the final state is

$$|\psi_f\rangle = \frac{1}{\sqrt{2}}(i + e^{i\chi})|c\rangle + \frac{1}{\sqrt{2}}(1 - ie^{i\chi})|d\rangle. \qquad (3)$$

This implies that the probabilities p_c and p_d to find the particles in the outgoing beams c and d respectively, are the squares of the corresponding amplitudes.

$$p_c = \frac{1}{2}|i + e^{i\chi}|^2 = \frac{1}{2}(1 - \sin\chi) \qquad (4)$$

$$p_d = \frac{1}{2}(1 + \sin\chi). \qquad (5)$$

Thus, as to be expected from unitarity, the total probability to find the particle in any output is unity. The probability to find it in one of the beams varies sinusoidally with the relative phase of the two input modes. This simple relation is at the basis of all interferometry experiments with single particles, be it photons, electrons, neutrons, or atoms.

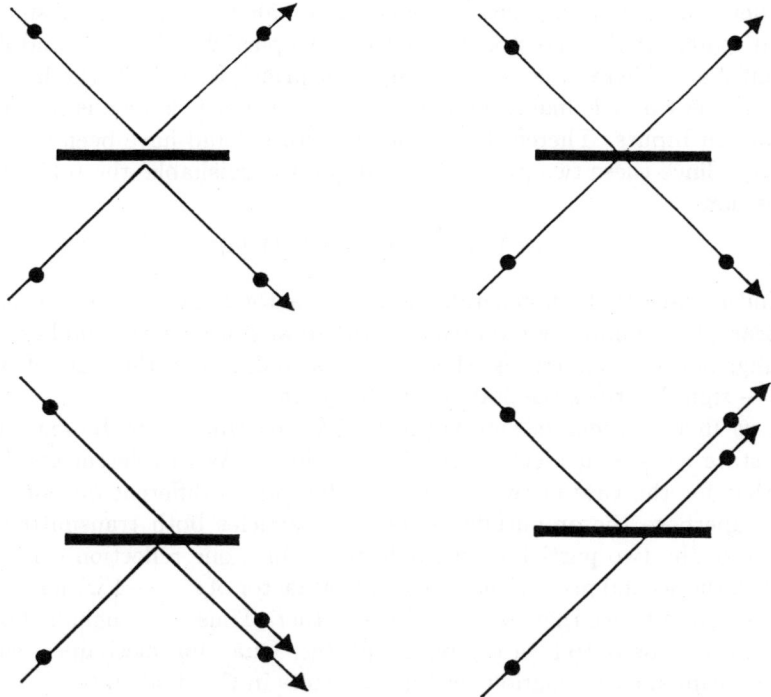

Figure 2: Two particles incident onto a beam splitter, one from each side. Four possibilities exist how the two particles can leave the beam splitter.

2 A beam splitter and two particles

Let us now consider our beam splitter with two incident particles, one in input beam a and one in input beam b. We assume that the two particles have the same momentum and arrive simultaneously. They each have the same probability $p = 0.5$ to transmit the beam splitter or be reflected. Thus, four different possibilities (Fig. 2) arise.

(1) Both particles are reflected, (2) both particles are transmitted, (3) the upper particle is reflected, the lower one is transmitted, and (4) the upper one is transmitted and the lower one is reflected. Each of the four occurs with the same probability, and we have to investigate now whether any interference between these processes is possible. For distinguishable particles, for example for classical ones, no interference arises and we thus arrive at the prediction that in two of the cases, that is, with total probability $p = 0.5$, the two particles end up in different output ports and, with probability $p = 0.25$, both particles end up in the upper output beam and, with the same probability $p = 0.25$, they end up in the lower output beam.

If the particles are quantum mechanically indistinguishable, it is not possible to decide which of the incident particles ended up in a given output port, and we have to consider coherent superposition. In order to calculate the final result, we need the input state. At the detector, we register two particles, 1 and 2, and if they are indistinguishable there is no way, not even in principle, to tell whether particle 1 came from input a or b, likewise for particle 2. All that is known is that they came from different inputs. Therefore, the incident state could have been either $|a\rangle_1|b\rangle_2$ or $|b\rangle_1|a\rangle_2$. Since these two possibilities are indistinguishable, the total state is the superposition

$$|\psi^\pm\rangle = \frac{1}{\sqrt{2}}(|a\rangle_1|b\rangle_2 \pm |b\rangle_1|a\rangle_2. \tag{6}$$

One could remark that, in general, the phase factor between the two terms could be arbitrary, but symmetry requires that the total state is either odd or even upon interchange of the two particles. Hence the + sign describes the state of two bosons and the − sign describes the state of two fermions.

Let us first consider in simple physical terms what is to be expected if the bosonic state $|\psi^+\rangle$ is incident on the beam splitter. With reference to Fig. 2, we realize that for the case of two particles ending up in different output beams, we have to superpose the amplitude of the two particles both transmitted with the amplitude of the two particles both reflected. Since one reflection carries a phase factor of π, the second process has a total phase factor of $i^2 = -1$ with respect to the first process, and hence they extinguish each other. Thus, we conclude, both bosons end up in the same output of the beam splitter. Quantum mechanical calculation, using the beam splitter relations of Eq. 1, results in the final state

$$|\psi^+\rangle \to \frac{1}{\sqrt{2}}(|c\rangle_1|c\rangle_2 + |d\rangle_1|d\rangle_2) \tag{7}$$

which implies that the outgoing bosons are in a coherent superposition of both being in the c beam and both being in the d beam. Each of the two cases happens with the same probability $p = 0.5$. Eq. 7 describes an entangled state such that it is quantum mechanically maximally uncertain in which of the two output beams either of the two bosons will be found, but once one particle is detected with probability $p = 0.5$ in one beam, the other particle will be found with certainty in the same beam.

Considering now the incident state $|\psi^-\rangle$ describing two fermions [5], we realize again that both fermions together pick up the phase shift -1 upon reflection. Yet the state ψ^- prescribes that the state |fermion 1 incident via beam a and fermion 2 incident via beam $b\rangle$ already carries a phase factor -1 relative to the phase state |fermion 2 incident via beam a and fermion 1 incident via beam $b\rangle$. Thus, the total relative phase between the two upper graphs in Fig. 4 is $-1 - 1 = +1$, that is, constructive interference results between the two possibilities. Likewise, we find by similar consideration that the possibilities for the two fermions exiting together via the same output beam extinguish each other. Again, calculation using the beam splitter rules of Eq. 1 results in the output state

$$|\psi^-\rangle \to \frac{1}{\sqrt{2}}(|c\rangle_1|d\rangle_2 - |d\rangle_1|c\rangle_2) \tag{8}$$

Interestingly, as opposed to the single particle case, introduction of a phase shifter into either of the input beams of the beam splitter does not at all change the statistics observed. The reason is that a phase shifter, say, in beam a, applies to both terms in the state of Eq. 5 and thus only results in a total common phase shift.

Another most important comment is that the observed phenomena, i.e. the fact that bosons always leave a beam splitter together via the same output port, and fermions never, is not a consequence of interaction between the two particles, but merely a consequence of interference of various possibilities how the two particles can arrive in the beams behind the beam splitter. It is most important to realize that the behavior of each particle at the beam splitter is only described by the beam splitter rules of Eq. 1, which do not take into account at all whether a second particle is present or not. It is the symmetry of the incident entangled state which then finally determines the output. The resulting distribution of the particles is therefore a direct consequence of quantum statistics and not of any interaction between the two fermions or bosons.

3 Polarization-entangled photon pairs

Interestingly, there are not so many possibilities to create states of entangled particles in the laboratory. The main reason is that we require superposition of states of more than one particle, and such states are more sensitive to external disturbances than single-particle states. Fortunately, the process of spontaneous parametric down-conversion provides mechanisms by which pairs of entangled photons can be produced with reasonable intensity and in good purity. In the down-conversion process, one uses a non-centrosymmetric crystal with nonlinear electric susceptibility. In such a medium, an incoming photon can decay, unfortunately with relatively small probability, into two photons in such a way that energy and momentum inside the crystal are conserved. The details depend on the specific shape of the dispersion surfaces. It is possible to have non-collinear down-conversion, that is, the two photons created are emitted in directions not parallel to the incoming photon.

Recently, in our laboratory in Innsbruck, we[6] have succeeded to produce polarization entangled photons using the process of type-II parametric down-conversion. In that process, the two photons created are emitted with different polarizations (Fig. 3). Calculating the emission directions of the photons[7], one notices that photons of each polarization are emitted into one cone in such a way that the momenta of two photons always add up to the momentum of the pump photon. Thus, the emission direction of each individual photon is completely uncertain within the cone, but once one photon is registered, and thus its emission direction is defined, the other photon is found just exactly opposite from the pump beam on the other cone. The total quantum mechanical state is extremely rich and is a superposition of all such pairs of emission modes.

The interesting point now is that the crystal can be cut and arranged such that the two cones intersect, as shown in Fig. 3. Then, along the lines of intersection, the polarization of neither photon is defined, but what is defined is the fact that the two photons have to have different polarizations. This contains all the interesting and surprising features of entanglement in a nutshell, as pointed out by Schrödinger[8]. Measurement on each of the photons separately is totally random and gives with

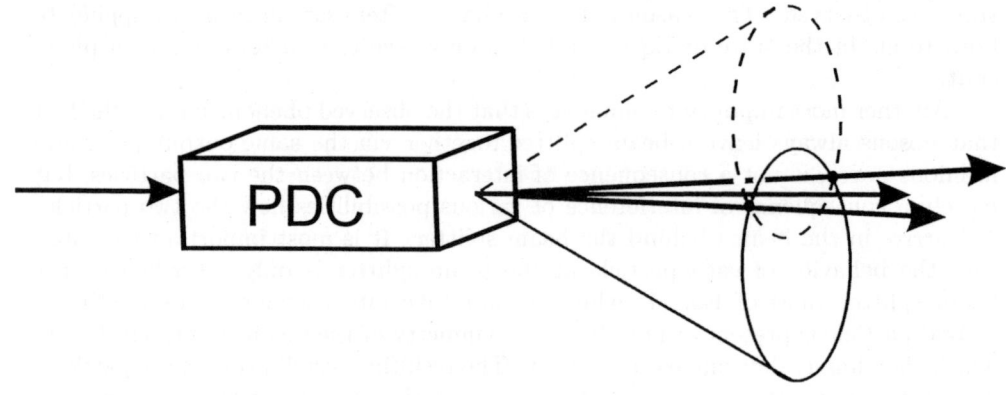

Figure 3: Principle of type-II down-conversion. Two photons arise polarized orthogonally from each other. Each photon is emitted into a cone and along the directions where the two cones intersect, a polarization entangled state results.

50% probability vertical or horizontal polarization. But once one photon is measured, the polarization of the other photon is orthogonal! In the present paper, I do not wish to enter into all the discussions about reality and nonlocality, which arise as a consequence from this simple observation. The reader might simply be referred to the relevant literature about Bell's theorem and its consequences [9].

Choosing an appropriate basis, e.g. $|H\rangle$ and $|V\rangle$, the state emerging through the two beams 1 and 2 thus is a superposition of $|H\rangle|V\rangle$ and $|V\rangle|H\rangle$.

$$\frac{1}{\sqrt{2}}(|H\rangle_1|V\rangle_2 + |V\rangle_1|H\rangle_2) \qquad (9)$$

The huge advantage of the new source is that it provides polarization entangled photon pairs of very high purity, high intensity, and well-defined direction. The quality of the source was demonstrated by us recently in such a way that it was possible to violate Bell's inequality by more than 100 standard deviations in a measurement time of less than 5 minutes [6].

4 Bell states in experiment

Consider two photons which can be either horizontally (H) or vertically (V) polarized. Classically, four possibilities exist:

$$H_1H_2, \quad H_1V_2, \quad V_1H_2, \quad V_1V_2 \qquad (10)$$

that is, either both photons are horizontally polarized, or the first one is polarized horizontally and the second one vertically, or the first one is polarized vertically and the second one horizontally, or, finally, both are vertically polarized. We realize therefore, not unexpectedly, that, using two photons, we can transmit four bits of information, one bit for each photon, depending on whether it is horizontally or vertically polarized.

Consider again two photons, but now contemplate their quantum states. Suppose all we know is that the polarizations of the two photons are equal. Then, the quantum state can be

$$|\phi^+\rangle = \frac{1}{\sqrt{2}}(|H\rangle_1|H\rangle_2 + |V\rangle_1|V\rangle_2) \tag{11}$$

or

$$|\phi^-\rangle = \frac{1}{\sqrt{2}}(|H\rangle_1|H\rangle_2 - |V\rangle_1|V\rangle_2) \tag{12}$$

These two states form the complete basis of the two-dimensional Hilbert space of two photons under the condition that both photons have the same polarization. A general state of two photons with equal polarization can be written is an arbitrary superposition $|\phi^+\rangle$ and $|\phi^-\rangle$.

The states $|\phi^+\rangle$ and $|\phi^-\rangle$ are both maximally entangled, This represents the fact that all we know is that the two photons have the same polarization. The measurement of the polarization of one photon is quantum mechanically maximally uncertain, but once the result for one photon is obtained, the polarization of the other one becomes well defined.

Another possibility is that we know the two polarizations to be different. Then, we can write the two-photon state either as

$$|\psi^+\rangle = \frac{1}{\sqrt{2}}(|H\rangle_1|V\rangle_2 + |V\rangle_1|H\rangle_2) \tag{13}$$

or as

$$|\psi^-\rangle = \frac{1}{\sqrt{2}}(|H\rangle_1|V\rangle_2 - |V\rangle_1|H\rangle_2) \tag{14}$$

Again, any state for which we know that the two photons are differently polarized is a superposition of Eq. 13 and Eq. 14.

Again, Eqs. 11 - 14 only represent the spin part of the state. The full state would have to include its spatial part. This is particularly important with respect to the state of Eq. 14. Apparently, the spin part of that state is antisymmetric, that is, it changes sign upon interchange of the two photons. Since the total state is a product of the spin part with the spatial part, the latter also has to be antisymmetric in order to insure correct bosonic symmetry.

The four states of Eqs. 11 - 14 form a complete orthonormal basis of the spin part of the four-dimensional Hilbert space of the two photons. It is instructive to compare the four quantum states of Eqs. 11 - 14 with the ways how two classical particles can be defined with respect to their polarization (Eq. 10). We note that there, too, we have four different possibilities, and the maximally entangled quantum basis of Eqs. 11 - 14 is just a superposition of the individual states corresponding to the classical possibilities of Eq. 10.

5 Two-photon coding

As mentioned above, the four classical possibilities of Eqs. 11 - 14 imply that we can encode two bits of information into two photons, one bit per photon. This can most easily be seen by identifying, for example, H-polarization with "0" and V-polarization with "1". Analysis of the quantum situation tells us that there, too, we can encode four bits of information into our two photons, because the corresponding Hilbert space is four-dimensional. A most remarkable point now is that one can change from either state of Eqs. 11 - 14 to any other one by manipulating one photon only. This is clearly distinct from the classical situation.

Suppose we start from the state of Eq. 9, as emitted by the source. Then it is clearly possible to switch to state 14 by just introducing a spin-dependent phase shift on one of the two photons. Explicitly, by shifting the phase of the state $|H\rangle_2$ by π with respect to the state $|V\rangle_2$, we generate the state 14. Such a polarization dependent phase shift can easily be obtained by inserting an appropriate birefringent phase shifter into the path of photon 2. If its axis is parallel to either H or V, then the two polarizations of photon 2 experience a different phase shift.

Likewise, the states 11 and 12 can easily be produced out of the states 13 or 14 by just interchanging the polarizations of photon 2. The transformation will turn the state of Eq. 13 into 11 and the state 14 into 12. Thus, by simple transformations, we have obtained all entangled states of the Bell basis.

The observation that it is possible to switch to all states in an entangled basis of two particles defined in a two-dimensional Hilbert space by operations on one particle only has led Bennett and Wiesner [10] to propose what they call quantum dense coding. This means that a sender, commonly called Bob (Fig. 4), can encode four different messages into the photon pair by manipulating only one of the two photons. A receiver, Alice, can read out the corresponding four bits of information if she has (a) access to both photons and (b) is in possession of a Bell state analyzer. The problem now is to identify a proper Bell state analyzer, since we are already in the position of an appropriate source, and of a way to experimentally realize the encoding into the four Bell states.

6 Bell-state analysis

Let us reconsider the behavior of two particles at the beam splitter. We have noticed above that, if two particles are incident onto a beam splitter from the two sides, the further statistical behavior depends on whether the state is bosonic or fermionic in nature. It turned out that the two particles proceed after the beam splitter in different emerging beams if, and only if, the state has fermionic symmetry. If the state has bosonic symmetry, they will always be found in the same outgoing beam. This feature can now be used as a first stepping stone towards a Bell state analyzer. We note that it is only the state $|\psi^-\rangle$ of Eqs. 11 - 14 which has fermionic symmetry in its spin part. This, as already mentioned, implies that its spatial part also has fermionic symmetry and thus this state will be such that the two photons emerge after the beam splitter in different beams. The other three states of Eqs. 11 - 14 have bosonic symmetry, and thus we expect the two photons to emerge in the same

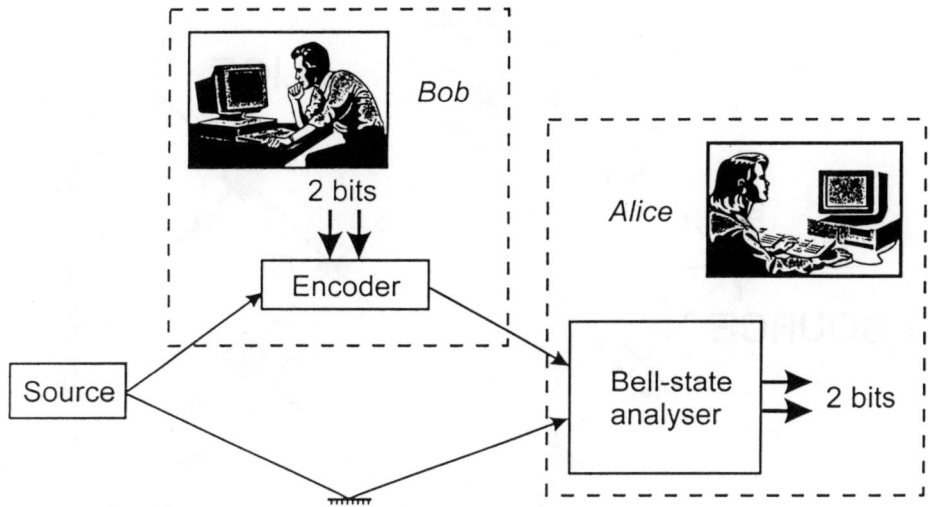

Figure 4: Principle of dense coding. The source emits an entangled pair of particles. Bob then encodes two bits of information onto his particle, and Alice can read out the two bits by joint Bell-state measurement on both particles.

beam. Clearly, these arguments just presented are conceptual ones only, the reader is challenged to write down the full state and demonstrate that it actually behaves in the way just indicated.

Thus, we already arrived at a possibility to identify one of the four Bell states, $|\psi^-\rangle$, uniquely on the basis that it is the only one which gives rise to detection of one photon in each of the outgoing beams of the beam splitter. For a full analysis, we further need a way to distinguish between the other three states, $|\psi^+\rangle$, $|\phi^+\rangle$ and $|\phi^-\rangle$. Careful inspection of these states reveals that it is only in the state $|\psi^+\rangle$ that the two emerging photons have different polarizations. In the two $|\phi\rangle$-states, they always share the same polarization. Thus, a further step in Bell state analysis implies that one inserts a two-channel polarizer into each of the outputs of the beam splitter. Then, only the state $|\psi^+\rangle$ will give a coincidence count between the two outputs of the polarizer on either side of the beam splitter.

Both ϕ-states will give rise to joint detection of the two photons in either detector after the final polarizer. Thus we can distinguish three of the four Bell states [11].

7 Experimental Dense Coding

The various concepts introduced in the chapters above lead to a number of novel experimental possibilities. As the first one, we [11] have realized the experimental

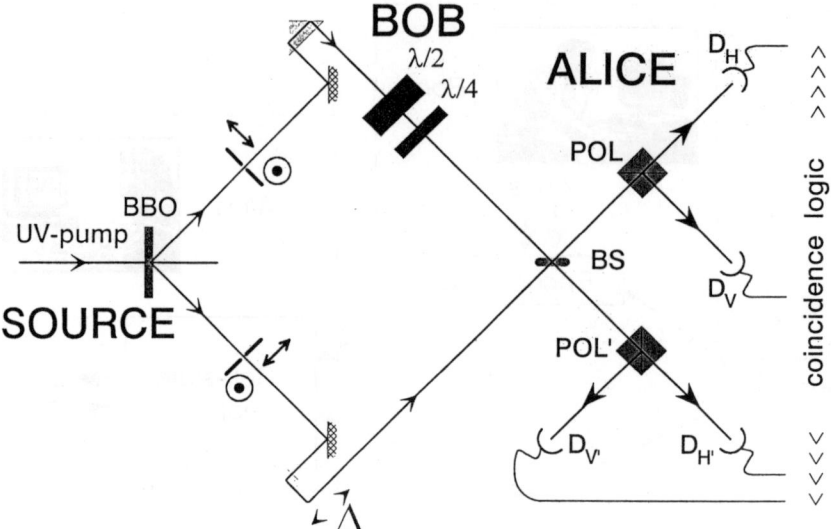

Figure 5: Experimental dense coding [11]. The source emits entangled photon pairs. Bob uses $\lambda/2$ and $\lambda/4$ plates to encode these his two bits, and Alice can read out three of four possible measurements using a beam-splitter based Bell state analyzer.

dense coding scheme (see Fig. 5). A $\lambda/2$ and a $\lambda/4$ plate serve Bob to implement any of the four Bell states. In practice, as pointed out above, we can discriminate with our scheme of beam splitters and polarizers just three of the four Bell states. This means that Bob can encode three messages into his photon, thus certainly surpassing the classical limit of 2 messages, i.e. one bit. In order to demonstrate this possibility the states $|\psi^+\rangle$, $|\psi^-\rangle$ and $|\phi^-\rangle$ were alternatively encoded and identified by proper detection. Fig. 6 shows as a measurement of $|\psi^-\rangle$ the coincidence rate between two detectors placed on either side of the beam splitter. The parameter varied in the experiment was the path length difference, that is, the difference in arrival time of the two photons at the beam splitter. We see that for a large path length difference a certain coincidence rate arises. This just reflects the classical coincidence rate, because the two photons can be distinguishable by their arrival times. Yet, when the two photons arrive at a beam splitter at the same time, an increase of the coincidence rate by a factor of two occurs. This is clearly indicative of the quantum phenomenon, where we now expect that the two photons always end up in different output channels while in the classical situation this only happens in 50% of the cases. Analogous results were obtained for the states $|\psi^+\rangle$ and $|\phi^-\rangle$ and enabled us to encode $log 3 = 1.58$ bits of information into Bob's photon. This 3-valued unit of information might simply be called "trit". For further details, the reader should consult reference [11].

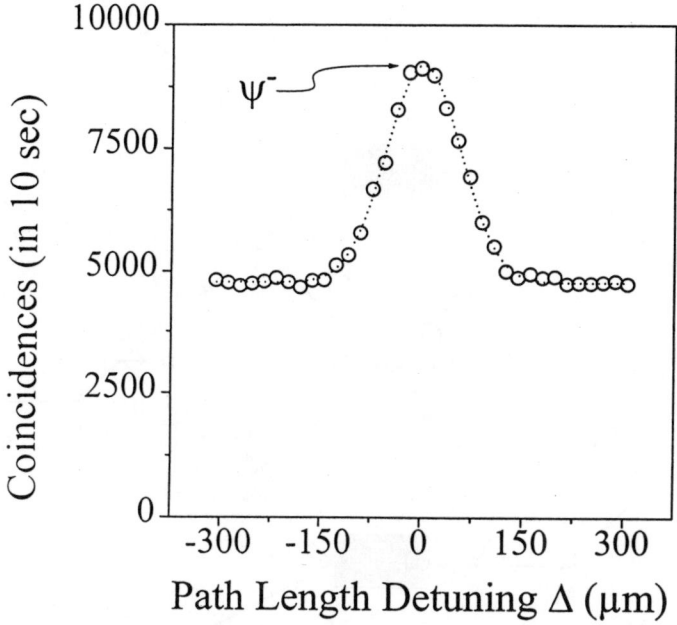

Figure 6: Identification of the ψ^- state after the beam splitter. For large path length differences, the coincidence count rate between the two detectors in the exit beams of the beam splitter is a constant reflecting their distinguishability on the basis of arrival time. For simultaneous arrival (path length detuning = 0), constructive interference doubles the coincidence rate indicating the state ψ^-.

8 Entanglement between more photons

All experiments hitherto have shown entanglement between two particles only. Yet, a number of novel phenomena would be observable once entanglement of more than two particles can experimentally be achieved. The most remarkable of these theoretical expectations is the fact that, with proper states of three or more particles (so-called GHZ states [12]), one can obtain the most striking possible contradiction between quantum mechanics and local realism.

Other interesting possibilities occur when one observes two independent sources. In entanglement swapping [13], it is possible to start with two pairs of photons, each one an entangled pair, but the two pairs independent of each other. Then, by proper measurement of our two photons such that each photon in the pair comes from a different down-conversion, it is possible to entangle the other two photons which never have interacted with each other.

Another interesting case is quantum teleportation [15]. There, one starts with a photon in an unknown quantum state $|\psi\rangle$ and two photons in an entangled state. Then one performs a Bell state measurement on the unknown photon and one of the down-conversion photons. This Bell state measurement gives four different possible results, as pointed out above. Subsequently, one performs one of four different unitary operations, depending on the specific result, on the other down-conversion photon, which then is in exactly the state $|\psi\rangle$ of the first photon. One should note that this does not provide a method for cloning photons, since the initial photon ψ

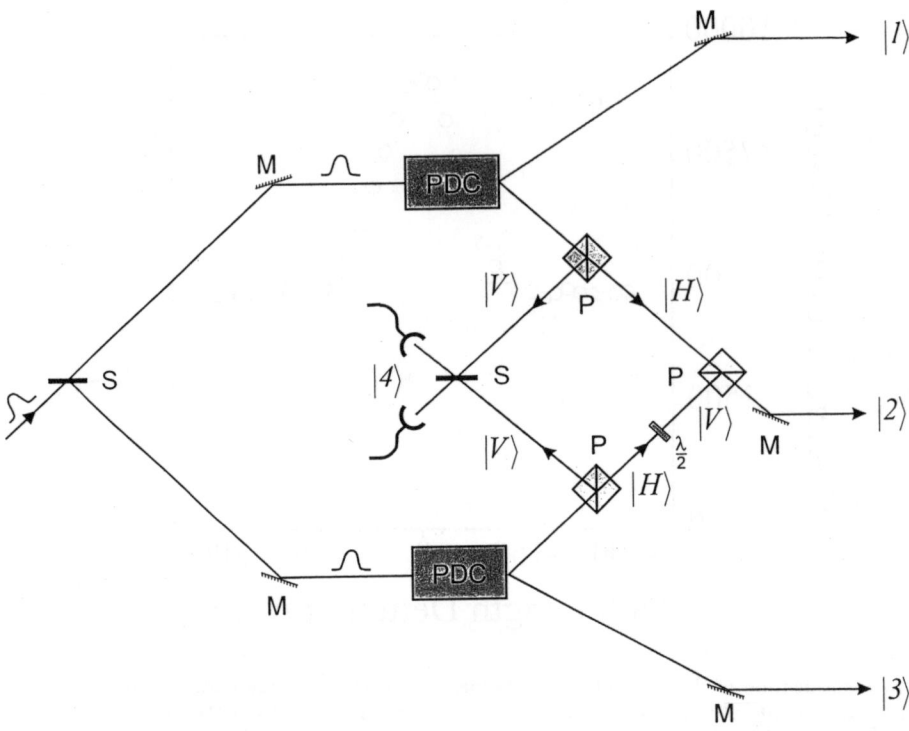

Figure 7: Possible production of a GHZ state. Very short pulses pump two type-II down-conversion crystals. In all of the outgoing beams after the down-conversion crystal, we have very narrow bandwidth filters (not shown). Photon 4 is then split off using two-channel polarizers P and measured behind a beam splitter S in such a way that it is not possible, not even in principle, from which of the two down-conversions it came. $|1\rangle$, $|2\rangle$ and $|3\rangle$ are then in a GHZ state.

disappeared.

It is evident that entanglement swapping and quantum teleportation are closely related to each other. Both necessitate Bell state measurements on two photons which came from independent sources. It is thus an interesting question whether or not it is possible at all to perform the measurement in such a way that one cannot tell from which of the two sources these two photons came. This information is clearly not available if the spectral bandwidth of the two photons is narrow enough such that the coincidence detection in the Bell state analyzer can happen in a time very short compared to the inverse of the bandwidth. This condition has been termed by us [13] to be the condition of ultra-coincidence. It can easily be seen that it cannot be achieved directly in the laboratory today, because existing detectors are not fast enough. Yet another interesting possibility exists [14], and this is actually pursued in our laboratory in Innsbruck. Consider two sources which are simultaneously pulsed, that is, they emit photons within a very short time δt. Subsequently, the photons are passed through narrowband filters, whose inverse bandwidth $1/\delta\omega$ is significantly larger than δt. Then, if the photons are detected behind the filters, it is not possible to determine which source they came from. This procedure is realizable with existing filter bandwidths and using pulsed down-conversion with a pulse duration of the order of 200 to 300 femtoseconds.

Since the ultra-coincidence condition can thus actually be met in the laboratory, it is reasonable to contemplate other new measurement possibilities on photons emerging from two down-conversions. In two down-conversions, we have four photons available altogether, and it is possible to measure these photons in such a way that three-particle GHZ states arise. One such explicit example is shown in Fig. 7. Two down-conversion crystals are pumped by the same pulse of short duration. We assume that the crystals are set such that each one emits the state $|\psi^+\rangle$. The pump intensity is such that with some probability two pairs are emitted. Now, let us assume that we subject the photons emerging in two of the beams to a polarizing beam splitter, a two-channel polarizer. Subsequently, superposing the two V-components at a regular beam splitter implies that detection of the photon behind that beam splitter does not result in determination which of the two crystals it came from. The other components, the H-parts of the amplitude of both photons are then subject to the following operations. Firstly, one of the two photons is subject to a spin rotation to V using a $\lambda/2$ plate and then this V-amplitude is superposed with the H-amplitude emerging from the other polarizer, again on a polarizing beam splitter, in such a way that one beam results.

One possibility is that two photons take their H route after the respective polarizing beam splitters. Thus, no photon emerges in the detectors behind the non-polarizing beam splitter. Secondly, both photons could take their V route, which means that no photon arises at the detector stations 1 and 3, and thirdly, one photon each arises in detector station 1, 2, 3 and 4. Discarding the first two possibilities, it is easy to see, using the standard quantum mechanical rules written down above, that the final state is one of the two states

$$(|H\rangle_1|V\rangle_2|V\rangle_3 - |V\rangle_1|H\rangle_2|H\rangle_3)|V\rangle_4 \qquad (15)$$

$$(|H\rangle_1|V\rangle_2|V\rangle_3 + |V\rangle_1|H\rangle_2|H\rangle_3)|V\rangle_4 \qquad (16)$$

depending on which detector registered photon 4. This result implies that whenever one and only one photon[16] is registered in detector station 4 and three other photons are registered at their respective stations 1, 2 or 3, these latter three photons are in a GHZ state. Existing technology with pulsed lasers and precision down-conversion experiments will make such experiments possible in the near future. It is obvious that an analogous scheme can be invented for beam entanglement in type-I down-conversion[17].

It is to be expected that the experiments discussed in the present paragraph provide further striking evidence for the counter-intuitive features of quantum mechanics[18] and open up perspectives for new insight into its foundations.

Acknowledgments

I would like to acknowledge the fruitful collaboration over the years with D.M. Greenberger, M.A. Horne, P.G. Kwiat, K. Mattle, H. Weinfurter and M. Zukowski on topics presented in or related to this paper. This work was supported by the Austrian Science Foundation FWF, Project No. S6502, and by NSF, grant No. PHY92-13964.

References

1. R.P. Feynman, R.B. Leighton and M.L. Sands in *The Feynman Lectures on Physics* (Addison-Wesley Publishing Co., Inc., Reading, 1989).
2. This should not be confused with the notion of coherent states in quantum optics, which imply specific coherent superpositions of number states.
3. D.M. Greenberger, M.A. Horne and A. Zeilinger, *Phys.Today* (August 1993) 22.
4. This phase factor is required by unitarity. It is not the only possible choice, but the most simple one resulting in a spatially symmetric description of the beam splitter. A. Zeilinger, *Am. J. Phys.* **49**, 882 (1981).
5. A. Zeilinger, H.J. Bernstein, M.A. Horne, *J. Mod. Opt.* **41**, 2375 (1994); R. Loudon in *Coherence and Quantum Optics VI*, eds. J.H. Eberly et al. (Plenum Press, New York, 1990); R. Loudon in *Disorder and Condensed Matter Physics*, eds. J.A. Blackman and J. Taguena (Clarendon Press, Oxford, 1991).
6. P.G. Kwiat, K. Mattle, H. Weinfurter, A. Zeilinger, A.V. Sergienko, Y.H. Shih, *Phys. Rev. Lett.* **75**, 4337 (1995).
7. P.G. Kwiat, *Phys. Rev. A* **52**, 3380 (1995); L. De Caro and A. Garuccio, *Phys. Rev. A* **50**, R2803 (1994).
8. E. Schrödinger, *Naturwissenschaften* **23**, 807, 823, 844 (1935); English translation in *Proceedings of the American Philosophical Society* **124**, 323 (1980).
9. for a review, see e.g D.M. Greenberger, M.A. Horne, A. Shimony and A. Zeilinger, *Am. J. Phys.* **58**, 1131 (1990).
10. C.H. Bennett and S.J. Wiesner, *Phys. Rev. Lett.* **69**, 2881 (1992).
11. K. Mattle, H. Weinfurter, P.G. Kwiat, A. Zeilinger, *Phys. Rev. Lett.* **76**, 4656 (1996).
12. D. Greenberger, M.A. Horne, A. Zeilinger in *Bell's Theorem, Quantum Theory, and Conceptions of the Universe* ed. M. Kafatos (Kluwer, Dordrecht, 1989).
13. M.Zukowski, A. Zeilinger, M.A. Horne and A.K. Ekert, *Phys. Rev. Lett.* **71**, 4287 (1993).
14. M. Zukowski, A. Zeilinger and H. Weinfurter in *Fundamental Problems in Quantum Theory*, eds. D.M. Greenberger, A. Zeilinger (Annals of the New York Academy of Sciences, vol. 755, 1995).
15. C.H. Bennett, G. Brassard, C. Crépeau, R. Josza, A. Peres and W.K. Wootters, *Phys. Rev. Lett.* **70**, 1895 (1993).

16. It is clear that instead of one pair, each produced at each down-conversion crystals, the two pairs could have been created with the same probability in just one crystal. We notice that such events can trivially be discarded, because then no photon emerges in the beam, which directly comes from the other crystal.
17. M.A. Horne, A. Zeilinger in *Symposium on the Foundations of Modern Physics, Joensuu*, eds. P. Lahti and P. Mittelstaedt (World Scientific Publ., Singapore, 1985); M.A. Horne and A. Zeilinger in *New Techniques and Ideas in Quantum Measurement Theory*, ed. D. Greenberger (Annals of the New York Academy of Sciences, 1986); M. Horne, A. Zeilinger in *Microphysical Reality and Quantum Formalism*, eds. A. van der Merwe, F. Selleri, G. Tarozzi (Kluwer, Dordrecht, 1988); M.A. Horne, A. Shimony and A. Zeilinger, *Phys. Rev. Lett.* **62**, 2209 (1989). J.G. Rarity, P.R. Tapster, *Phys. Rev. Lett.* **64**, 2209 (1989).
18. D. Greenberger and A. Zeilinger, *Physics World* (September 1995) 33.

ATOM OPTICS AS A TESTING GROUND FOR QUANTUM CHAOS

C. F. BHARUCHA, J. C. ROBINSON[a], F. L. MOORE[b], K. W. MADISON,
S. R. WILKINSON, BALA SUNDARAM, AND M. G. RAIZEN
*Department of Physics, The University of Texas at Austin,
Austin, TX 78712, USA*

This paper summarizes our recent work on the role of classical dynamics in atom optics with time-dependent dipole potentials. We measure momentum transfer in parameter regimes for which the classical dynamics are chaotic and observe a wide range of phenomena. These include classical mechanisms such as the resonance overlap route to global chaos as well as quantum suppression due to dynamical localization. The high degree of experimental control enables detailed comparison with theory and opens up new avenues for testing ideas in quantum chaos.

1 Introduction

The past few years have seen a resurgence in the use of classical mechanics in the description of strongly perturbed and strongly coupled quantum systems in atomic physics[1,2], where the traditional perturbative treatment of the Schrödinger equation breaks down. In particular, recent advances in classical nonlinear dynamics and chaos have had important applications in the description of the photo-absorption spectrum of Rydberg atoms in strong magnetic fields[3], the microwave ionization of highly excited hydrogen atoms[4], and the excitation of doubly excited states of helium atoms[5]. These examples together with recent work on mesoscopic systems[6] explore classical-quantum correspondence in situations where the classical limit exhibits chaos, an area of study referred to as 'quantum chaos'[7].

Parallel progress in laser cooling and trapping techniques have led in recent years to spectacular advances in the manipulation and control of atomic motion[8]. At the ultra-cold temperatures that are now attainable, the wave nature of the atoms becomes important, leading to the development of the new field of atom optics[9].

Until recently, the primary focus in atom optics has been the development of optical elements such as atomic mirrors, beamsplitters and lenses for atomic de Broglie Waves. Our recent work, reviewed in this paper, has emphasized the novel regime of time-dependent potentials and hence dynamics in atom optics. In particular, we study momentum distributions of ultra-cold atoms exposed to time-dependent one-dimensional dipole forces which are, typically, highly nonlinear. Thus, the classical equations of motion can become chaotic and as dissipation can be made negligibly small in this system, quantum effects can become important. Our work has established that these features together make atom optics a very simple and controlled setting for the experimental study of quantum chaos[10].

As this work deals with momentum transfer from light to atoms, it is important to review some basic concepts. The relevant unit of momentum is one-photon recoil ($\hbar k_L$), the momentum change experienced by an atom when it scatters a photon. For sodium atoms, this velocity change is 3 cm/s. The desired process for atom optics is stimulated scattering, where the atom remains in the ground state, and coherently scatters the photon in the direction of the laser beam. Spontaneous scattering, on the other hand, is a dissipative process and must be minimized. In a single beam (traveling wave) the atom scatters in the forward direction, and there is no net momentum transfer. However in a standing wave of light, created by the superposition of two counter-propagating beams, the atom can also back-scatter. This process leads to a momentum change of two photon recoils, or 6 cm/s velocity change for sodium. The effective dipole potential that the atom experiences scales with intensity I and detuning δ_L from atomic resonance as I/δ_L while spontaneous scattering goes as I/δ_L^2 [11]. Therefore, by detuning farther from resonance, it is possible to make the probability of spontaneous scattering negligible, while still having a substantial dipole potential.

2 Experimental method

The experimental study of momentum transfer in time-dependent interactions consists of three important components: initial conditions, interaction potential, and measurement of atomic momentum. The initial distribution should ideally be narrow in position and momentum, and should be sufficiently dilute so the atom- atom interactions can be neglected. The time-dependent potential should be one-dimensional (for simplicity), with full control over the amplitude and phase. In addition, noise must be minimized to enable the study of quantum effects. Finally the measurement of final momenta after the interaction should have high sensitivity and accuracy. Using techniques of laser cooling and trapping it is possible to realize all these conditions.

A schematic of the experimental set-up is shown in Fig. 1 [12,13]. Our initial conditions are a sample of ultra-cold sodium atoms which are trapped and laser-cooled in a magneto-optic trap (MOT)[8,14]. The atoms are contained in an ultra-high vacuum glass cell at room temperature. The cell is attached to a larger stainless steel chamber which includes a 20 l/s ion pump. The source of atoms is a small sodium ampoule contained in a copper tube that is attached to the chamber. The ampoule was crushed to expose the sodium to the rest of the chamber. Although the partial pressure of sodium at room temperature is below 10^{-10} Torr, there are enough atoms in the low-velocity tail of the velocity distribution that can be trapped. The trap is formed using three pairs of counter-propagating, circularly polarized laser beams (2.0 cm beam diameter) which intersect in the middle of the glass cell, together with a magnetic field gradient which is provided by current-carrying wires arranged in an anti-Helmholz configuration. This configuration is now fairly standard, and is used in many laboratories. These beams originate from a dye laser that is locked 20 MHz to the low frequency (red) side of the $(3S_{1/2}, F = 2) \longrightarrow (3P_{3/2}, F = 3)$ sodium transition at 589 nm. Approximately 10^5 atoms are trapped in a cloud which has an RMS size of 0.12 mm, with an RMS momentum spread of $4.6\hbar k_L$.

The interaction potential is provided by a second dye laser that is tuned typically 5 GHz red of resonance. Different beam configurations were used in the experiments described here utilizing acousto-optic and electro-optic modulators to control the time-dependent amplitude and phase.

The detection of momentum is accomplished by allowing the atoms to drift in

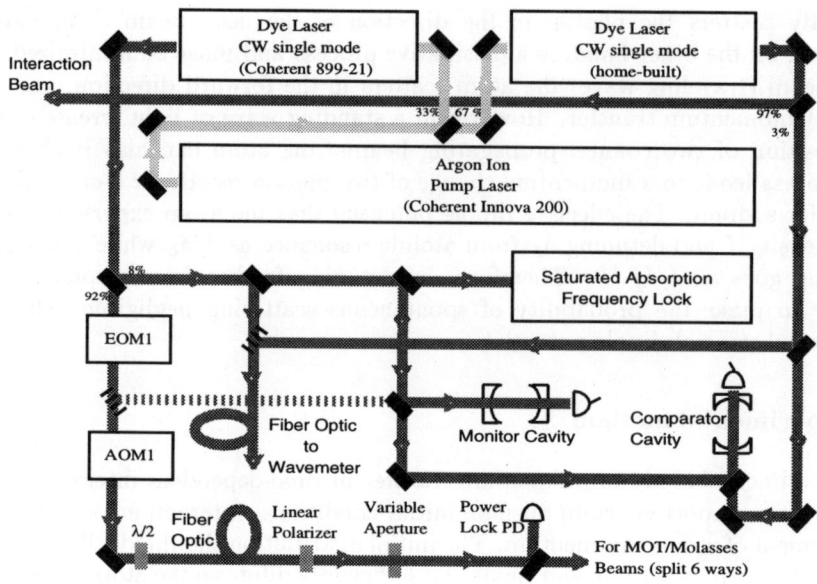

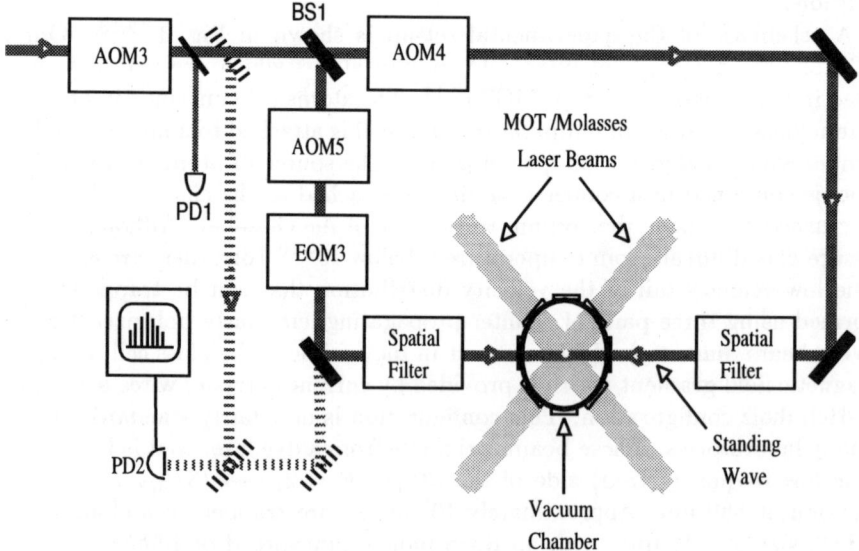

Figure 1: Schematic of the experiment. The upper part of the figure shows the laser set-up while the lower part shows the beam configuration in the modulated standing wave experiment. The acousto-optic modulators (AOM) are used for frequency shifts and as fast shutters while the electro-optic modulators (EOM) are used as phase shifters and frequency modulators.

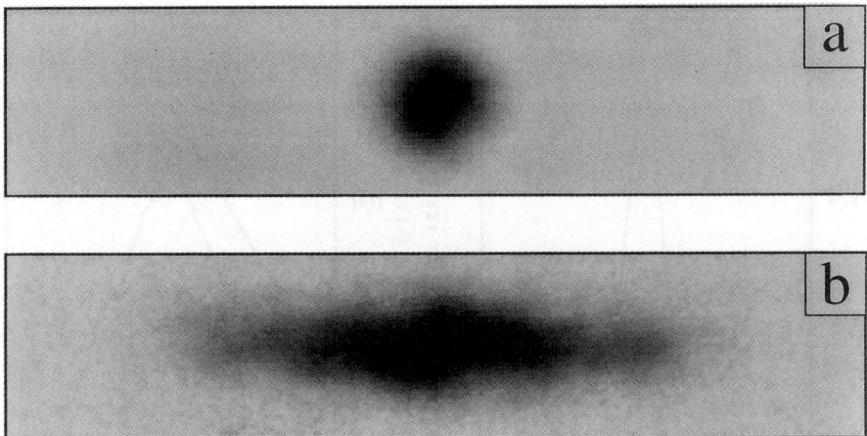

Figure 2: Two-dimensional atomic distributions after free expansion. (a) Initial thermal distribution with no interaction. (b) Localized distribution after interaction with the potential.

the dark for a controlled duration, after the interaction with the standing wave. Their motion is frozen by turning on the optical trapping beams in zero magnetic field to form optical molasses [8]. The motion of the atoms is overdamped, and for short times (tens of ms) their motion is negligible. The position of the atoms is then recorded via their fluorescence signal on a Charge Coupled Device (CCD) and the time of flight is used to convert position into momentum. The entire sequence of the experiment is computer controlled.

In Fig. 2, typical 2-D images of atomic fluorescence are shown. In Fig. 2(a) the initial MOT was released, and the motion was frozen after a 2 ms free-drift time. This enables a measurement of the initial momentum distribution. The distribution of momentum in Fig. 2(a) is Gaussian in both the horizontal and vertical directions. The vertical direction is integrated to give a one dimensional distribution as shown in Fig. 3(a). In Fig. 2(b), the atoms were exposed to a particular time dependent potential. The vertical distribution remains Gaussian, but the horizontal distribution becomes exponentially localized due to the interaction potential, as shown in Fig. 3(b). The significance of the lineshape and other characteristics are analyzed below.

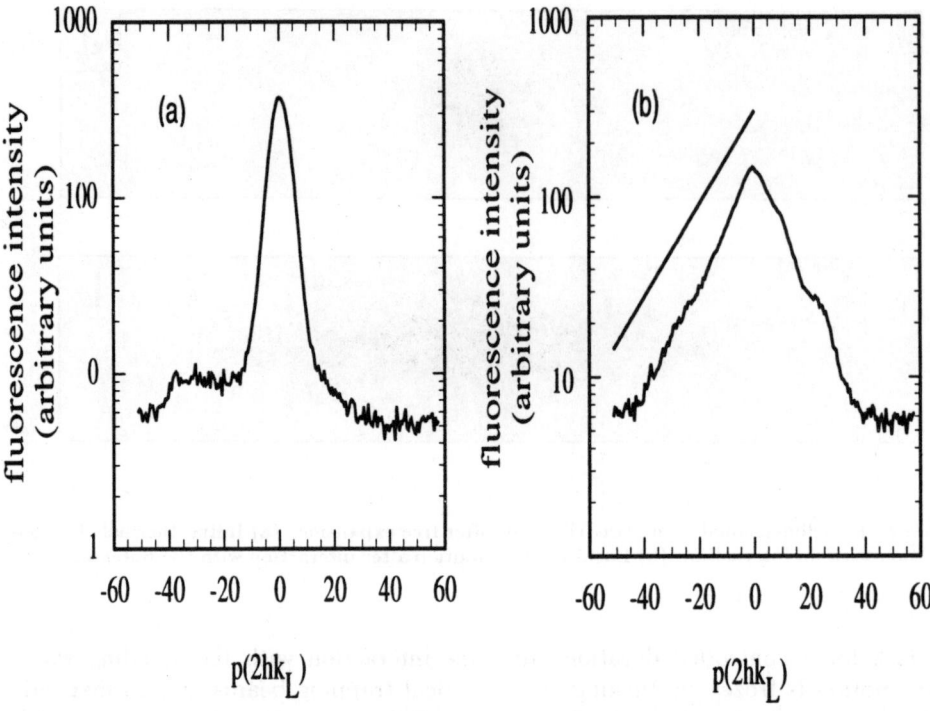

Figure 3: One-dimensional atomic momentum distributions. They were obtained by integrating along the vertical axes of the 2 − D distributions in the previous figure. The horizontal axes are in units of two recoils, and the vertical axes show fluorescence intensity on a logarithmic scale. (a) Initial thermal distribution with no interaction. (b) Localized distribution after interaction with the potential. The significance of the characteristic exponential lineshape is discussed in the text.

3 Single Pulse Interaction

Consider a two level atom of transition frequency ω_0 interacting with a standing-wave of near-resonant light (frequency ω_L). For sufficiently large detuning $\delta_L = \omega_0 - \omega_L$, the excited state amplitude can be neglected and the atom remains in the ground state. The Hamiltonian is then given by

$$H = p^2/2M - (\hbar\Omega_{eff}/8)\cos 2k_L x , \qquad (1)$$

where the effective Rabi frequency is $\Omega_{eff} = \Omega^2/\delta_L$, and k_L is the wavenumber. Ω_{eff} is proportional to the laser intensity I. The effective one-dimensional potential neglects variations of the potential in the two transverse directions. This is justified when the beams are sufficiently large compared to the initial atomic cloud. Quantum mechanically, the atom can exchange energy with the standing wave only in units of $2\hbar k_L$ which results in a ladder of equally spaced momentum states.

The classical analysis of this Hamiltonian is the same as for a pendulum, except that here the conjugate variables are position and momentum instead of angle and angular momentum. A Poincaré surface of section is shown in Fig. 4. The position coordinate is shown for one period of the standing wave which is half a wavelength of light. The stable fixed point corresponds to the bottom of the potential well, and the unstable fixed point corresponds to the top. Our initial conditions are a band in phase space, with small spread in momentum, and uniformly distributed in position on the scale of one period.

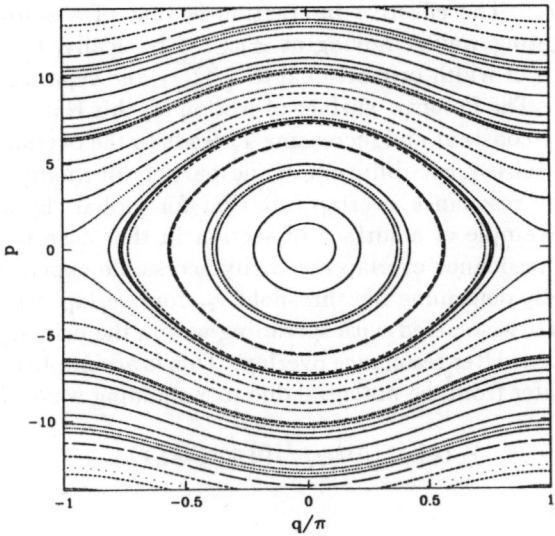

Figure 4: Poincaré surface of section for a single resonance. Momentum (vertical axis) is in units of two recoils, and position is in units of one period of the standing wave potential.

The simplest time-dependent potential is the turning on and off of an interaction, and one expects that for slow turn on/off the evolution will be adiabatic. The conditions for adiabatic behavior are usually very clear for linear potentials (the harmonic oscillator is an example) since there are only a few relevant time scales to consider. The difficulty with nonlinear potentials such as that occuring in the pendulum and the standing wave of light is that there are many time scales, so the conditions for adiabaticity must be examined much more carefully. The opposite extreme of fast passage is generally simpler to understand. We show that for time-scales intermediate to fast passage and adiabatic, mixed phase space dynamics and chaos can be seen even with the mere act of turning an interaction on and off. In the context of atom optics, this type of time-dependent interaction is ubiquitous and occurs, for example, whenever an atomic beam passes through a standing wave of light.

The generic time dependent potential in this case is

$$V(x,t) = (\hbar\Omega_{eff}/8)f(t)\cos 2k_L x \qquad (2)$$

One common case is for $f(t) = \exp -(t/\tau)^2$ corresponding an atomic beam traversing a Gaussian beam waist [15]. We consider here the case $f(t) = \sin^2 \pi t/T_s$, which is turned on for a single period T_s.

This Hamiltonian can be expanded as

$$\begin{aligned} H &= p^2/2M - (\hbar\Omega_{eff}/8) \sin^2 (\pi t/T_s) \cos(2k_L x) , \\ &= p^2/2M - (\hbar\Omega_{eff}/16) [\cos 2k_L x \\ &\quad - (\cos 2k_L(x - v_m t) + \cos 2k_L(x + v_m t))/2] , \end{aligned} \quad (3)$$

where $v_m = \lambda_L/2T_s$. The effective interaction is that of a stationary wave with two counter-propagating waves moving at $\pm v_m$. Classically, there are now three resonance zones each of width proportional to $\sqrt{\Omega_{eff}}$ and separation in momentum proportional to T_s^{-1}. The Poincaré surface of section for this Hamiltonian is shown in Fig. 5. Keeping Ω_{eff} constant and increasing T_s leads to the overlap of these isolated resonances and the subsequent diffusion of the particle in momentum. This is the well known Chirikov resonance overlap criterion for global chaos in Hamiltonian systems [16,17]. An example of a surface of section in that case is shown in Fig. 6. The parameters for resonance overlap are easily accessible experimentally [18].

To experimentally determine the threshold τ_{cr} for overlap, we must distinguish the momentum growth associated with spreading within the primary resonance from diffusion that can occur after resonance overlap. This is accomplished by contrasting the momentum transfer from the potential due to a standing wave of fixed amplitude

$$V'(x) = (\hbar\Omega_{eff}/16) \cos(2k_L x) , \quad (4)$$

for duration T_s with

$$\begin{aligned} V(x,t) &= (\hbar\Omega_{eff}/16) [\cos 2k_L x \\ &\quad - (\cos 2k_L(x - v_m t) + \cos 2k_L(x + v_m t))/2] , \end{aligned} \quad (5)$$

resulting from the sin^2 amplitude modulated standing wave. The key to the interpretation of the experimental results is the realization that *for values of T_s below the threshold for resonance overlap $V'(x)$ and $V(x,t)$ should give the same result.* After overlap of the resonances, $V(x,t)$ will result in significantly larger momentum transfer than $V'(x)$. The experimental results in Fig. 7(b) show the RMS momentum for both cases as a function of pulse duration (rise and fall times of 25 ns are included in the square pulse duration). These agree well with numerical classical simulations shown in Fig. 7(a) as well as with the estimated resonance overlap threshold [18].

How does the predicted quantum behavior compare with experiment and classical simulations? As seen from the dashed curve in Fig. 7(c), we find close agreement between all three for the square pulse potential $V'(x)$. This is an interesting result in its own right, since the coherent oscillations that occur for short times are seen in the experiment with a large ensemble of independent atoms, and in the quantum simulation which uses a single wavepacket approach. For the case of $V(x,t)$ there is also good agreement between the three cases over the entire range of pulse times. However, the quantum widths are slightly lower than the corresponding classical values near the large peak in the RMS width. Although this difference is too small

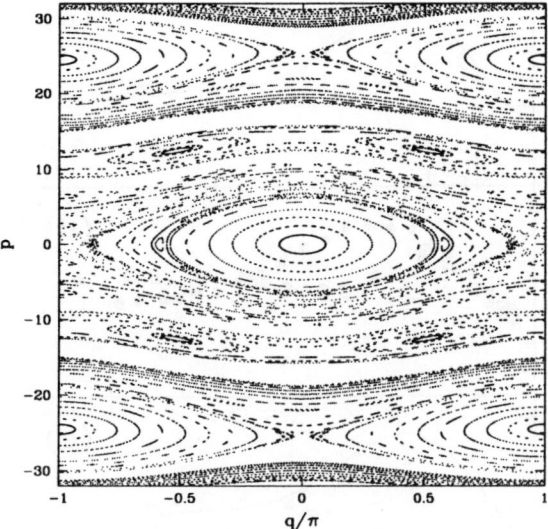

Figure 5: Poincaré surface of section for the sin^2 potential. In this case there are three isolated resonances at 0 and ±25.

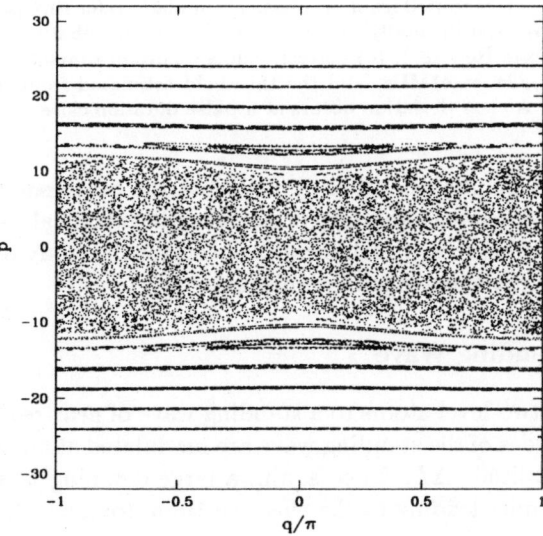

Figure 6: Poincaré surface of section for the sin^2 potential after resonance overlap has occured. There is a bounded region of global chaos.

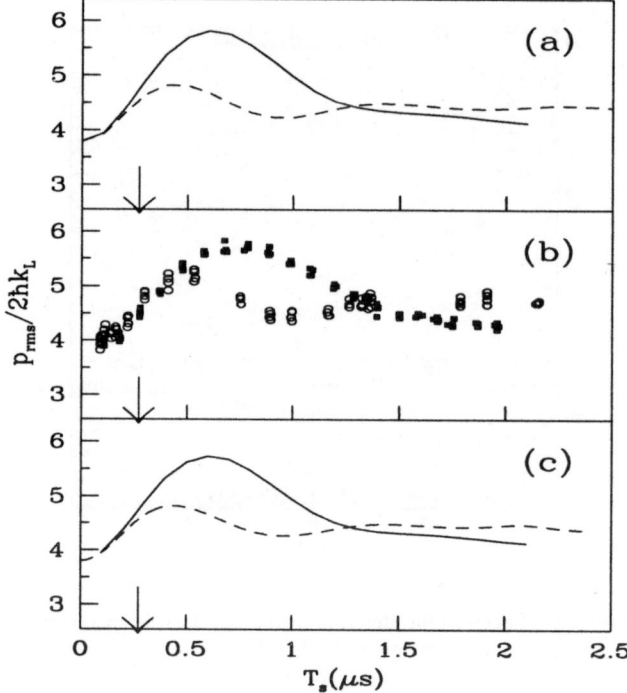

Figure 7: RMS momentum computed from (a) classical simulations for sin^2 (solid line) and square (dashed line) pulses; (b) experimentally measured momentum distributions for sin^2 (solid) and square (open) pulses (from Ref. 18); (c) corresponding quantum simulations, solid and dashed lines respectively. $\Omega_{eff}/2\pi = 41$MHz. and the threshold estimated from resonance overlap is indicated by the arrow. A clear deviation occurs at a pulse duration close to the predicted value.

to be of quantitative significance, it is nevertheless the precursor for differences in quantum and classical behavior that can occur when the classical dynamics are globally chaotic. These differences, which form the basis for the study of quantum chaos, are the focus of the next experiments we discuss.

4 Modulated Standing Wave

We now subject our two-level atoms to a standing wave of near-resonant light where the position of the nodes of the standing wave are modulated at an angular frequency ω_m and with an amplitude ΔL. Once again, a large detuning is used to eliminate the upper level dynamics leading to the effective Hamiltonian[19]

$$H = p^2/2M - (\hbar\Omega_{eff}/8)\cos\left[2k_L(x - \Delta L \sin\omega_m t)\right], \quad (6)$$

Although this Hamiltonian may look somewhat different than the sin^2 case that was discussed earlier, it can also be expanded as a sum of nonlinear resonances using the well known Bessel function expansion. The Hamiltonian then has the form

$$\begin{aligned} V(x,t) &= (\hbar\Omega_{eff}/8)\left[J_0(\lambda)\cos(2k_L x) + J_1(\lambda)\cos 2k_L(x - v_m t)\right.\\ &\quad \left. + J_{-1}(\lambda)\cos 2k_L(x + v_m t) + J_2(\lambda)\cos 2k_L(x - 2v_m t)\cdots\right]\\ &= (\hbar\Omega_{eff}/8)\sum_{m=-\infty}^{\infty} J_m(\lambda)\cos 2k_L(x - mv_m t)\,, \end{aligned} \qquad (7)$$

where J_m are ordinary Bessel functions and $v_m = \omega_m/2k_L$. $\lambda = 2k_L\Delta L$ is the control parameter which takes the dynamics from integrable ($\lambda = 0$) to chaotic. Unlike the single pulse case that was considered earlier, here we take the interaction time to be independent of the modulation period.

Now the classical dynamics involves several resonances equally spaced in momentum with widths proportional to $\sqrt{|J_m(\lambda)|}$. For a given value of λ there are substantial resonances only for $m \leq \lambda$ therefore the interaction turns off for $v > \lambda v_m$, leading to a bounded region in momentum spanned by the resonances. The oscillatory behavior of the Bessel functions leads to the recurrence of stable classical structures within a bounded region of chaos, as λ is varied.

The classical dynamics can also be understood in terms of resonant-kicks that occur twice during each modulation period: consider an atom subjected to the modulated standing wave. When the standing wave is moving with respect to the atom, the time-averaged force is zero, since the sign of the force is changing as the atom goes over 'hill and dale' of the periodic potential. However, twice during each modulation period, the standing wave is stationary in the rest frame of the atom, and the atom gets a resonant-kick which changes its momentum. The magnitude and direction of the resonant-kick depends on where the atom is located with respect to the standing wave at that time. In this system, an atom experiences two resonant-kicks every modulation period, although they are not equally spaced in time. The boundary in momentum can be understood from this picture, since for each value of λ there is a maximum velocity of the standing wave. Once an atom is moving faster than this maximum valocity, the resonant-kicks cannot occur, and the atom is essentially free.

The variation of the classical RMS momentum width as a function of λ is shown in Fig. 8 (dot-dashed line)[20]. The parameters are $\omega_m/2\pi = 1.3$ MHz. and $\Omega_{eff}/2\pi \approx 25$ MHz.. The interaction time was chosen to be 20 μs which is sufficiently long for the experimentally observed momentum spread to saturate. At small values of λ, the distribution of the classical simulation saturates near the resonant-kick boundary. As λ is increased, oscillations occur with the dips corresponding to zeros of the Bessel functions. Notice that the overall amplitude of the oscillations decreases as λ is increased due to the reduction of the size of each resonant-kick. This can be understood from the impulse approximation, since the maximum classical force is fixed, but the time that the standing wave potential is stationary in the rest frame of the atom is inversely proportional to λ. We have run the classical simulation for longer times and find that the peaks grow until the resonant-kick boundary, while the dips grow much more slowly. This difference in rates is explained by the phase portraits shown in Fig. 9 (top panel). In this figure, the peaks in RMS momentum are at values of λ for which the dynamics are primarily chaotic. In contrast, at the dips, a

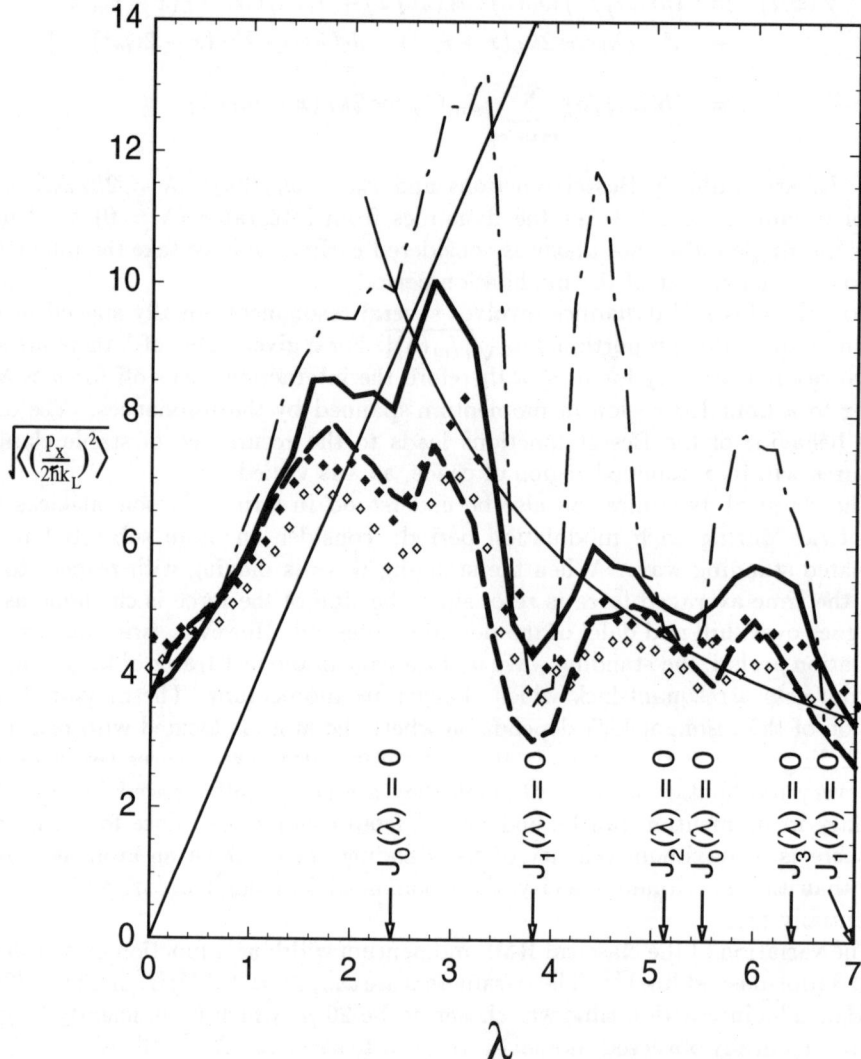

Figure 8: The RMS momentum width as a function of the modulation amplitude λ. Experimental data is denoted by diamonds and have a 10% uncertainty associated with them. The empty diamonds are for an interaction time of 10 μs and the solid diamonds are for 20 μs; classical simulation for 20 μs (dash-dot line); quantum Schrödinger for 20 μs (heavy dashed line); quantum Floquet in the long-time limit (heavy solid line). The light solid lines denote the resonant-kick boundary and the curve proportional to λ^{-1} predicted in Ref. 19.

primary island determines the momentum transfer. The classical lineshapes in Fig. 9 (middle panel) clearly show these features as well as the effect of the resonant-kick boundary. Initial conditions contained within an island remain trapped, while those in the chaotic domain diffuse up to the boundary, leading to 'boxlike' distributions. A clear example of the stability at the dips is at $\lambda = 3.8$ where J_1 has its first zero. The final momentum spread in this case is governed by the surviving island due to J_0 and the system is nearly integrable. Note that the oscillations of the Bessel functions are reflected in the exchange of the location of hyperbolic and elliptic fixed points, which is clearly visible on contrasting the phase portraits for $\lambda = 0$ and $\lambda = 3.0$, beyond the first zero of J_0 at $\lambda = 2.41$.

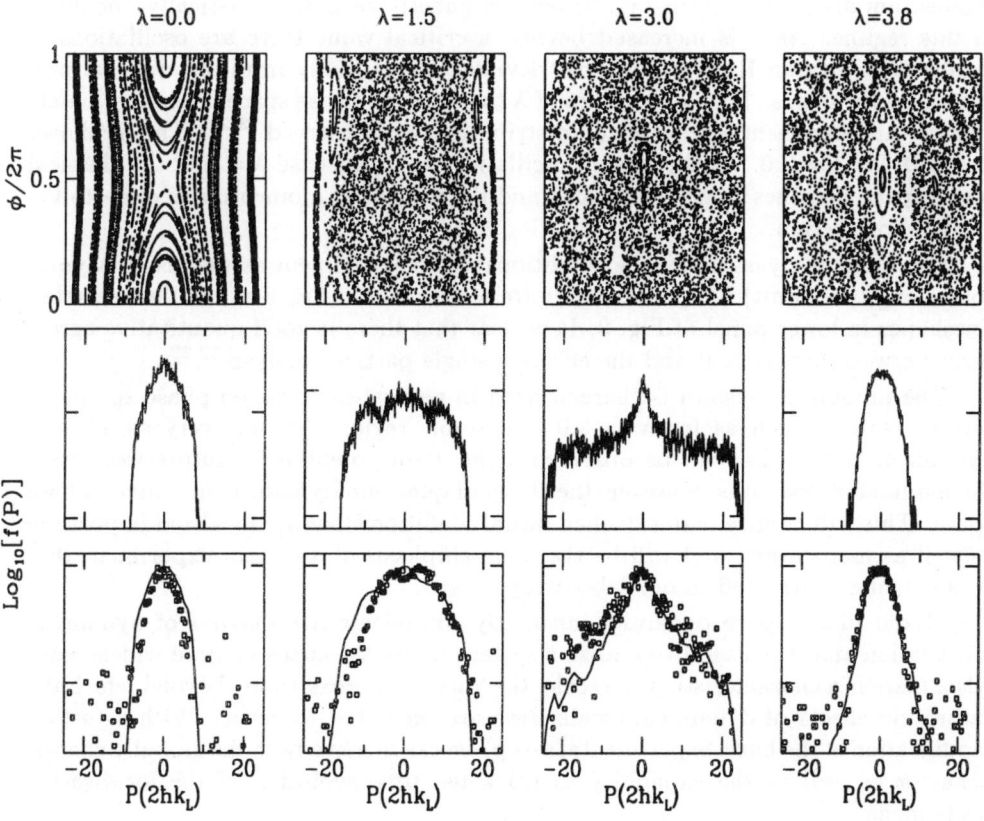

Figure 9: Poincaré surfaces of section (upper panel), classical momentum distributions (middle panel), and experimentally measured momentum distributions with Floquet theory (bottom panel, theory marked by lines) for runs with parameters similar to those in Fig. 8. Note that the vertical scales for the distributions are logarithmic and are marked in decades.

It is well known that classically diffusive behavior can be suppressed quantum mechanically by a mechanism analogous to Anderson localization [21]. Referred to as dynamical localization, it predicts saturation in the energy transfer (momentum, in our case) and a resulting exponential lineshape with a characteristic localization length ξ (in momentum). In the experiments, we have to ensure that the location of the resonant-kick boundary is much further than ξ to observe this lineshape. As this boundary scales linearly with λ, we expect to see the appearance of dynamical localization only beyond some value of λ.

The measured RMS momenta vs. λ are shown in Fig. 8 (diamonds). The empty and solid diamonds are for two different interaction times showing that these results are close to saturation for the range of λ shown. Note that for small values of λ there is good agreement with the classical prediction. At $\lambda = 0$ the system is integrable and momentum is trivially localized. As λ is increased the phase space becomes chaotic, but growth is limited by the resonant-kick boundary. Our measured momentum distributions (in Fig. 9, bottom panel) are characteristically "boxlike" in this regime. As λ is increased beyond a critical value there are oscillations in localization with an RMS spread that deviates substantially from the classical prediction at the peaks. For those values of λ the classical phase space is predominately chaotic, and exponentially localized distributions are observed [13,20]. This is shown in Fig. 9 for $\lambda = 3.0$. At the dips in oscillation, as in the case $\lambda = 3.8$, the classical phase space becomes nearly integrable and the measured momentum is close to the classical prediction.

Quantum analyses under the conditions of the experiment as well as an asymptotic (long-time limit) Floquet analysis are also shown in Fig. 8 as are the predicted lineshapes in lower panel of Fig. 9. It is clear that there is good quantitative agreement between experiment and the effective single particle analysis [20,22].

The modulated system is characterized in general by a mixed phase space. In certain regimes such as for $\lambda = 3.0$, the stable regions become very small, and dynamical localization can be observed. The main potential for future work with the modulated system is, however, the study of quantum dynamics in a mixed phase space. This will require better defined initial conditions that are localized in position as well as momentum, and will be the main emphasis of a cesium experiment that is now being contructed in our laboratory.

Mixed phase space dynamics inherently complicate the analysis of dynamical localization and it is useful to realize a system where the chaos is more widespread. Also, there is a characteristic time scale, the 'quantum break time', beyond which the saturation effects of dynamical localization are predicted to occur. With a further modification of the basic experimental setup, we can achieve both the globally chaotic behavior as well as the capability to track the time evolution of the localization phenomena.

5 Kicked Rotor

The classical kicked rotor or the equivalent standard mapping is a textbook paradigm for Hamiltonian chaos. The Hamiltonian for the problem is given by

$$H = p^2/2 + K \cos q \sum_n \delta(t - nT) \tag{8}$$

The evolution consists of resonant-kicks that are equally spaced in time, with free motion in betwen. K is called the stochasticity parameter, and is the standard control parameter. As K is increased, the size of each resonant- kick grows. Beyond a threshold value of $K = 4$ it was shown that phase space is globally chaotic [21]. The quantum version of this problem has played an equally important role for the field of quantum chaos, and a wide range of effects have been predicted [21]. In our realization, we have the cosine potential of the standing wave multiplied by $f(t)$, a train of N pulses with unit peak heights and period T. This system was previously analyzed in the context of molecular rotation excitation [23]. The nonzero pulse widths lead to a finite number of resonances in the classical dynamics, which limits the diffusion resulting from overlapping resonances to a band in momentum. However, by decreasing the pulse duration with constant area, the width of this band can be made arbitrarily large, approaching the δ- function pulse limit. The boundary in momentum can also be understood using the concept of an impulse. If the atomic motion is negligible while the pulse is on, the momentum transfer is an impulse, similar to the resonant-kick in the modulated system. For a sufficiently large velocity, the atom has time to move over the periodic potential while the pulse is on, averaging the impulse to zero. The result is a momentum boundary which can be pushed out by making each pulse shorter. The practical constraints of available laser power limits the pulse duration, since in these experiments the pulse train is created by turning a continuous-wave laser on and off with an acousto-optic modulator.

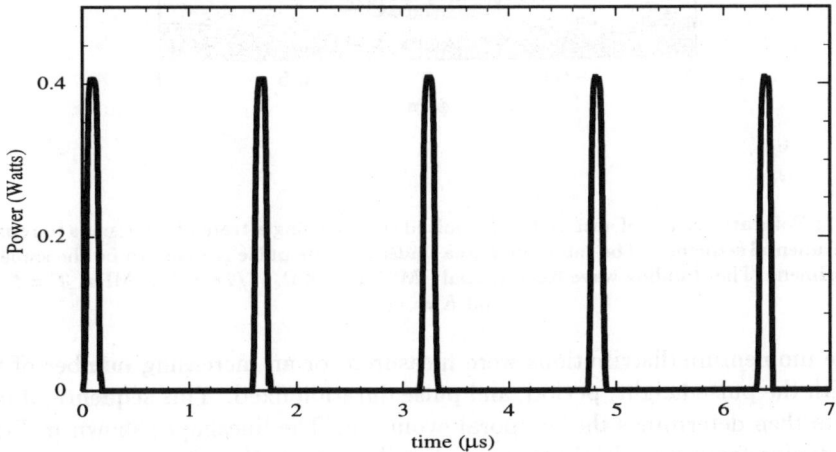

Figure 10: Digitized temporal profile of the pulse train measured on a fast photo-diode. The vertical axis represents the total power in both beams of the standing wave. $f(t)$ and Ω_{eff} are derived from this scan.

A typical experimental pulse train is shown in Fig. 10. Each pulse is typically non-Gaussian and the integrated area is used in the comparison with theory. The pulse period, duration, and number of pulses in a burst are variable parameters in the experiment. The bounded region of chaos arising from the finite pulse duration is illustrated by the classical phase portrait, for typical experimental parameters, shown in Fig. 11. The central region of momentum in this phase portrait is in very close agreement with the delta-kicked rotor model with a stochasticity parameter of $K = 11.6$, which is well beyond the threshold for global chaos.

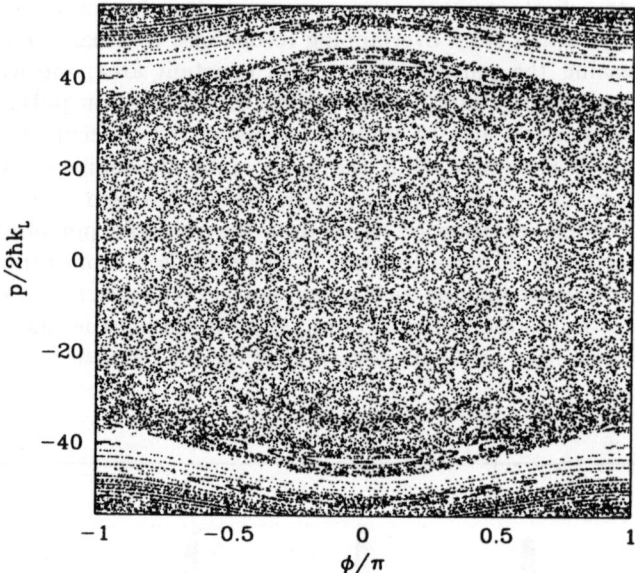

Figure 11: Poincaré surface of section for the pulsed system using a train of Gaussians to represent the experimental sequence. The integrated area under a single pulse is taken to be the same as in the experiment. The standing wave has a spatial RMS value of $\Omega_{eff}/2\pi = 75.6$ MHz. $T = 1.58$ μs, and $K = 11.6$.

The momentum distributions were measured for an increasing number of kicks (N), with the pulse height, period, and pulse duration fixed. This sequence of measurements then determines the temporal evolution. The lineshapes shown in Fig. 12 clearly evolve from an initial Gaussian distribution at $N = 0$ to an exponentially localized distribution after approximately $N = 8$. We have measured distributions out until $N = 50$ and find no further significant change. The growth of the mean kinetic energy of the atoms as a function of the number of kicks was calculated from the data and is displayed in Fig. 13. It shows diffusive growth initially until the quantum break time, after which dynamical localization is observed[24]. Though not shown here, classical and quantum calculations both agree with the data over the diffusive regime. Beyond the quantum break time, the classical energy continues to increase diffusively while the measured lineshapes stop growing, in agreement with

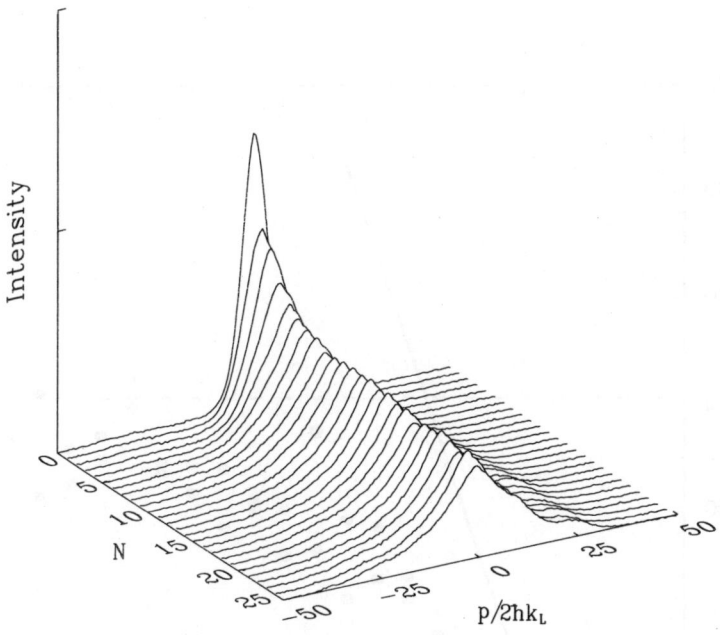

Figure 12: Experimental time evolution of the lineshape from the initial Gaussian until the exponentially localized lineshape (from Ref. 24). The break time is approximately 8 kicks. Fringes in the freezing molasses lead to small asymmetries in some of the measured momentum lineshapes as seen here and in the inset of Fig. 13. The vertical scale is measured in arbitrary units and is linear.

the quantum prediction. The observed lineshape is shown (Fig. 13 inset), and is clearly exponential. These results are the first experimental observation of the onset of dynamical localization in time, and the quantum break time [24].

Between kicks the atoms undergo free evolution for a fixed duration. The quantum phase accumulated during the free evolution is $e^{-ip^2T/2M\hbar}$. An initial plane wave at $p = 0$ couples to a ladder of states separated by $2\hbar k_L$. For particular pulse periods, the quantum phase for each state in the ladder is a multiple of 2π, a condition known as a *quantum resonance* [21]. More generally, a quantum resonance is predicted when the accumulated phase between kicks is a rational multiple of 2π. We have scanned T from 3.3 µs to 50 µs and find quantum resonances when the quantum phase is an integer multiple of π. For even multiples, the free evolution factor between kicks is unity, and for odd multiples, there is a flipping of sign between each kick. Quantum resonances have been studied theoretically, and it was shown that instead of localization, one expects energy to grow quadratically with time [25]. This picture, however, is only true for an initial plane wave. A general analysis of the quantum resonances shows that for an initial Gaussian wavepacket, or for narrow distributions not centered at $p = 0$, the momentum distribution is actually smaller than the exponentially localized one, and settles in after a few kicks [26]. Our experimental results are shown in Fig. 14. Ten quantum resonances are found for T ranging between 5 µs (corresponding to a phase shift of π) and 50 µs (10π) in steps of 5 µs. The saturated momentum lineshapes as a function of T are shown in Fig. 14(a). The narrower, non-exponential profiles are the resonances between

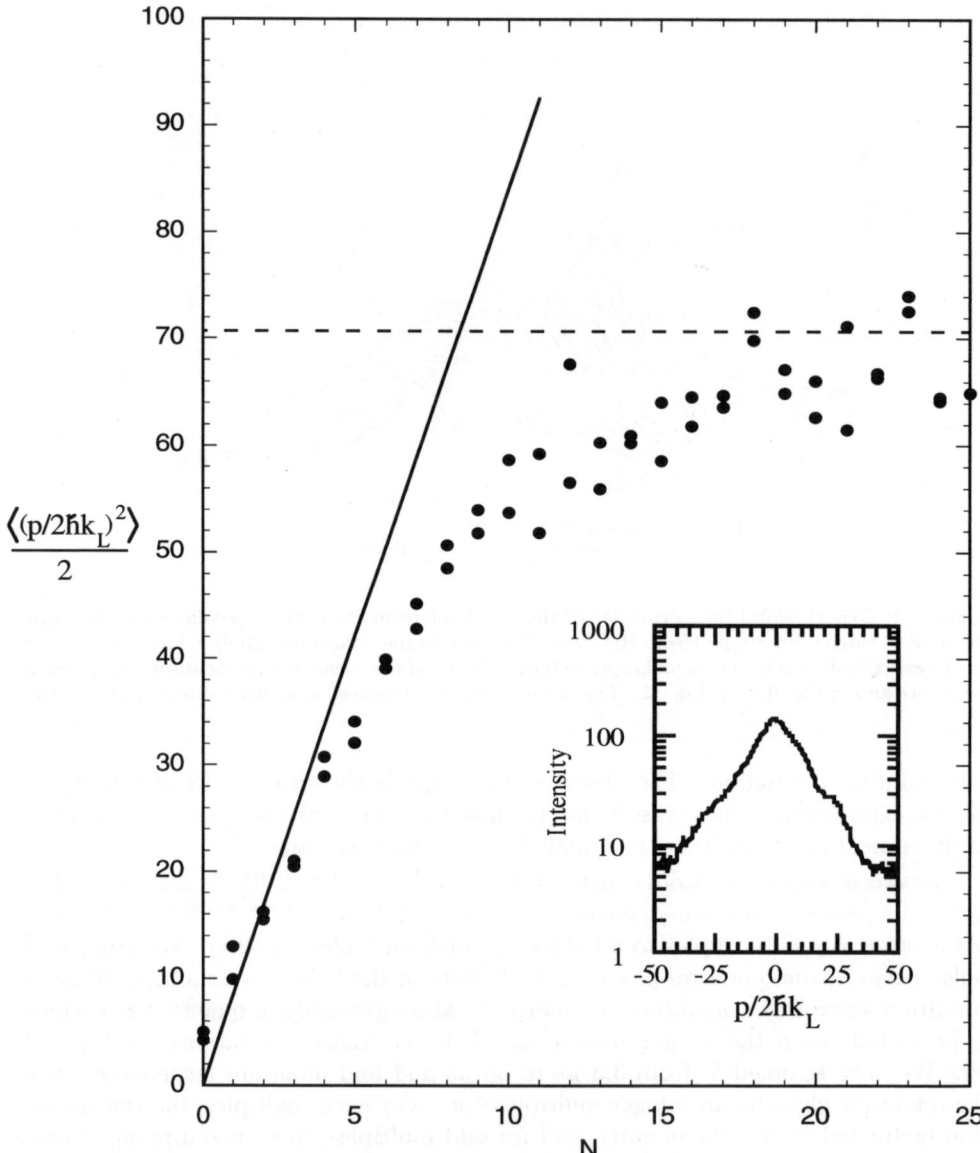

Figure 13: Energy $< (p/2\hbar k_L)^2 > /2$ as a function of time (from Ref. 24). The solid dots are the experimental results. The solid line shows the calculated linear growth from the classical dynamics. The dashed line is the saturation value computed from the theoretical localization length ξ. The inset shows an experimentally measured exponential lineshape on a logarithmic scale which is consistent with the theoretical prediction.

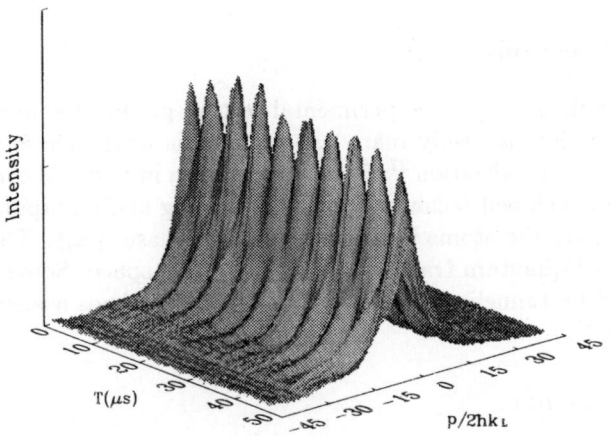

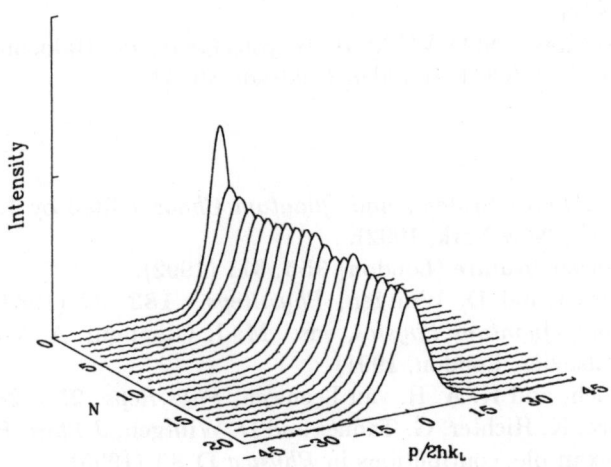

Figure 14: Experimental observation of quantum resonances: (a) Occurrence as a function of the period of the pulses. The surface plot is constructed from 150 lineshapes measured, for each T, after 25 kicks. This value of N ensures that the lineshapes are saturated for the entire range of T shown. At resonance, the profiles are non-exponential and narrower than the localized shapes which appear off-resonance. Note that the vertical scale is linear. (b) Time evolution of a particular resonance ($T = 10~\mu$s).

which the exponentially localized profiles are recovered. The time evolution of the lineshape at a particular resonance is shown in Fig. 14(b) from which it is clear that the distribution saturates after very few kicks.

6 Future Directions

This work establishes a new experimental testing ground for quantum chaos, and it should be possible to study many aspects of this field. These include the study of noise-induced delocalization [27,28] and localization in two and three dimensions [29]. Using recently developed techniques of atom cooling and manipulation it should be possible to prepare the atoms in a small region of phase space. This would enable a detailed study of quantum transport in mixed phase space. Some interesting topics to study would be tunneling from islands of stability, chaos assisted tunneling, and quantum scars [30].

Acknowledgements

This work was supported by the Office of Naval Research, the R.A. Welch Foundation, and the National Science Foundation.

[a] present address: Motorola, Semiconductor Technologies Laboratory, Austin, Texas 78721.
[b] present address: NOAA/CMDL, Nitrous Oxide and Halocompounds Division, Mail Stop R/E/CG1, Boulder, Colorado 80303.

References

1. *Irregular Atomic Systems and Quantum Chaos*, edited by J. C. Gay (Gordon and Breach, New York, 1992).
2. R. V. Jensen, *Nature* (London) **355**, 311 (1992).
3. H. Friedrich and D. Wintgen, *Phys.Reps.* **183**, 37 (1989); D. Delande in *Chaos and Quantum Physics*, eds. M.-J. Giannoni, A. Voros and J. Zinn-Justin (Elsevier, London, 1991).
4. P. M. Koch, and K. A. H. van Leeuwen, *Phys.Reps.* **255**, 289 (1995).
5. G. S. Ezra, K. Richter, G. Tanner and D. Wintgen, *J.Phys.* **B24**, L413 (1991).
6. see, for example, contibutions in *Physica* **D 83** (1995).
7. see, for example, *Quantum chaos : between order and disorder : a selection of papers*, edited by G. Casati and B. V. Chirirov (Cambridge Univ Press, New York, 1995).
8. Laser cooling and trapping is reviewed by Steven Chu in *Science* **253**, 861 (1991).
9. C. S. Adams, M. Sigel, and J. Mlynek, *Phys. Reps.* **240**, 145 (1994).
10. This work was reviewed in "Search and Discovery", *Physics Today* June 1995, pg. 18.
11. C. Cohen-Tannoudji, in *Fundamental Systems in Quantum Optics*, Les Houches, 1990; J. Dalibard, J. M. Raimond and J. Zinn-Justin, eds., (Elsevier, 1992).

12. J. C. Robinson, Ph.D. thesis, University of Texas at Austin, 1995 (University Microfilms, Ann Arbor, Michigan).
13. F. L. Moore, J. C. Robinson, C. Bharucha, P. E. Williams, and M. G. Raizen, *Phys. Rev. Lett.* **73**, 2974 (1994).
14. E. Raab, M. Prentiss, A. Cable, S. Chu, and D. Pritchard, *Phys. Rev. Lett.* **59**, 2631 (1987).
15. see, for example, P. J. Martin, P. L. Gould, B. G. Oldaker, A. H. Miklich, and D. E. Pritchard, *Phys. Rev.* **A36**, 2495 (1987).
16. B. V. Chirikov, *Phys. Reps.* **52**, 265 (1979).
17. G. H. Walker, and J. Ford, *Phys. Rev.* **188**, 416 (1969).
18. J. C. Robinson, C. F. Bharucha, K. W. Madison, F. L. Moore, Bala Sundaram, S. R. Wilkinson, and M. G. Raizen, *Phys. Rev. Lett.* **76**, 3304 (1996).
19. R. Graham, M. Schlautmann, and P. Zoller, *Phys. Rev.* **A45**, R19 (1992).
20. J. C. Robinson, C. Bharucha, F. L. Moore, R. Jahnke, G. A. Georgakis, Q. Niu, M. G. Raizen, and Bala Sundaram, *Phys. Rev. Lett.* **74**, 3963 (1995).
21. L. E. Reichl, *The Transition to Chaos in Conservative Classical Systems: Quantum Manifestations* Springer-Verlag (1992).
22. P. J. Bardroff, I. Bialynicki-Birula, D. S. Krähmer, G. Kurizki, E. Mayr, P. Stifter, and W. P. Schleich, *Phys. Rev. Lett.* **74**,3959 (1995).
23. R. Blümel, S. Fishman, and U. Smilansky, *J. Chem. Phys.* **84**, 2604 (1986).
24. F. L. Moore, J. C. Robinson, C. F. Bharucha, Bala Sundaram, and M. G. Raizen, *Phys. Rev. Lett.* **75**, 4598 (1995).
25. F. M. Izrailev, and D. L. Shepelyansky, *Sov. Phys. Dokl.* **24**, 996 (1979); *Theor. Math. Phys.* **43**, 553 (1980).
26. Q. Niu and Bala Sundaram, to be submitted for publication.
27. T. Dittrich and R. Graham, *Europhys. Lett.* **4**, 263 (1987).
28. S. Fishman, and D. L. Shepelyansky, *Europhys. Lett.* **16**, 643 (1991).
29. Giulio Casati, Italo Guarneri, and D. L. Shepelynansky, *Phys. Rev. Lett.* **62**, 345 (1989).
30. E. J. Heller, and S. Tomsovic, *Physics Today* **46**, 38 (1993).

COHERENT ULTRA-BRIGHT XUV LASERS AND HARMONICS

J.S. Wark, K. Burnett, D. Chambers, M.H. Key, S.G. Preston, A. Sanpera, C.G. Smith,
J. B. Watson, M. Zepf, J. Zhang
*Department of Physics, Clarendon Laboratory, Parks Road,
University of Oxford, Oxford OX1 3PU, United Kingdom*

C.L.S. Lewis, A.G. McPhee
*Department of Pure and Applied Physics, Queens University of Belfast,
Northern Ireland, United Kingdom*

A.E. Dangor, P. Lee, A. Dyson
*Blackett Laboratory, Imperial College of Science, Technology and
Medicine, London, United Kingdom*

G.J. Pert, P.B. Holden, J.A. Ploues
Department of Computational Physics, University of York, York, United Kingdom

G.J. Tallents, L. Dwivedi, M. Holden
Department of Physics, University of Essex, Colchester, Essex, United Kingdom

P.A. Norreys, D. Neely
*Central Laser Facility, Rutherford Appleton Laboratory, Chilton, Didcot, Oxon, United
Kingdom*

P. Fews
H.H. Wills Physics Laboratory, University of Bristol, Bristol, United Kingdom

S. Moustazis, M. Bakarezos, P. Loukakos
IESL/FORTH, University of Crete, Crete, Greece

Over the past decade considerable advances have been made in the generation of coherent XUV radiation. Progress has been made on three main fronts: the development of both collisional and recombination XUV lasers, the generation of high odd-order harmonics from the interaction of intense sub-picosecond lasers with noble gases, and the generation of both odd and even harmonics from ultra-intense sub-picosecond laser interactions with ponderomotively-steepened high-density plasmas.

1 Introduction

The extension of laser action into the x-ray region has been an area of increasingly successful experimental investigation for over a decade, and the subject of theoretical inquiry for more than twice that length of time. The frequency scaling of spontaneous and stimulated emission rates dictates that direct XUV laser action can only be achieved at extremely high power densities: thus the focused output of high power optical lasers has been the main method used to pump x-ray lasers.

In addition to the direct manufacture of x-ray laser media, significant advances have been made over the past few years in the production of coherent XUV radiation by the generation of ultra-high harmonics of high intensity, sub-picosecond laser pulses. Two methods of such harmonic generation have been explored. Firstly, the interaction of intense laser pulses with gaseous targets (primarily noble gases). In the intense laser field, comparable to the Coulomb field in the atom, electrons can tunnel through the Coulomb barrier, giving rise to a large non-linear susceptibility. Symmetry of the atomic potential dictates that only odd-order harmonics are generated. Such harmonics are typically generated with sub-picosecond laser pulses at irradiances of between 10^{14} and 10^{16} Wcm^{-2}. Note that atomic unit of field (i.e. the field experienced by the electron in the first Bohr orbit of hydrogen) is 5.142 x 10^{11}Vm^{-1}, corresponding to a laser intensity of 3.52 x 10^{16} Wcm^{-2}. Secondly, if an extremely intense pulse (10^{17} - 10^{19} Wcm^{-2}) is incident onto a solid target, both odd and even harmonics can be produced. These harmonics are associated with the electron current being dragged back and forth across the density step. At the lower end of the irradiance range indicated above, and for ultrashort (say, 100-fsec) high contrast-ratio laser pulses, the relevant density step is that of the solid-vacuum interface itself, whereas for higher irradiances and longer pulselengths (of order a picosecond or longer), the density step is produced by ponderomotive steepening of the pre-plasma, formed before or during the leading edge of the pulse.

The three methods of XUV generation outlined above produce sources with widely different divergences and coherence properties. In this paper comparison of the present status of these source characteristics will be made, and a preliminary assessment of their applicability for other areas of science outlined.

2 X-ray Lasers

The first XUV laser scheme to achieve high gains, produced using the Nova laser facility at Lawrence Livermore National Laboratory, relied on collisional excitation of the Ne-like Se ion.[1] Production of the high density (10^{21}cm^{-3}) high temperature laser medium was achieved by irradiation of a thin film target by nanosecond pulses of optical (0.53-μm) light at intensities of order 5 x 10^{13} Wcm^{-2}. Soon after, laser schemes that relied on achieving population inversion by preferential recombination of electrons into upper states of ions in adiabatically cooling plasmas were demonstrated.[2,3] Since this ground breaking work the field has advanced on several fronts: the reduction of laser wavelength towards the so-called water-window at 44-Å;[4] the development of oscillator-amplifier modes of operation and the improvement in laser efficiency by the reduction of the effects of refraction due to electron density gradients within the gain region;[5-7] and the demonstration of "table-top" schemes utilising picosecond lasers[8] or capillary discharges.[9]

The majority of work in this area has concentrated on the collisionally pumped schemes, with Neon-like ions being used for wavelengths in excess of about 100-Å, and Nickel-like ions below this wavelength. For many years the output of these Neon-like systems was poorly understood: in particular, detailed modeling predicted higher gain on the $J = 0 - 1$ transition compared to that on the $J = 2 - 1$ doublet. This anomaly has recently been resolved. It has been found that the reduction of the $J = 0 - 1$ output was due to severe refraction of x-rays out of the lasing medium due to the steep electron-density gradient close to the surface of the target. These refraction effects are more severe for the $J = 0 - 1$ transition, as its gain region is at high densities, and correspondingly higher refractive index gradients. Reduction of

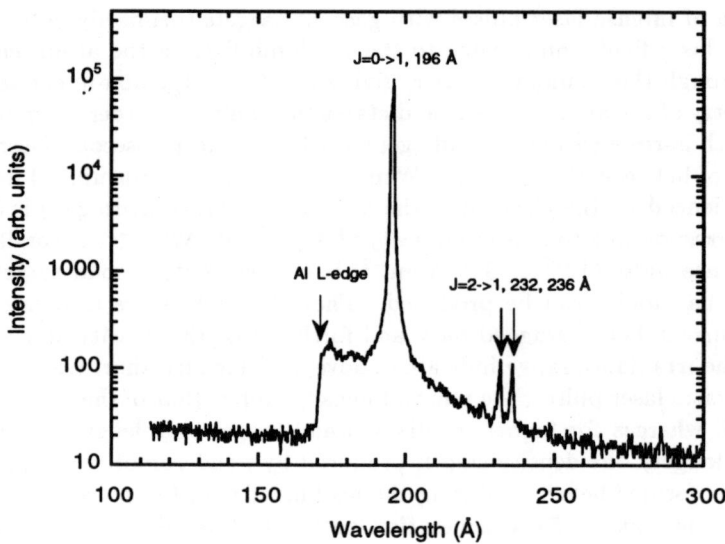

Figure 1: Spectrum of a Neon-like Germanium laser operated with a prepulse, illustrating the domination of the 19.6-nm line.

the density gradients, and consequent alleviation of refraction problems, has been achieved by operating the optical drive laser with a small prepulse (10^{-4}-10^{-1} in power) a few nanoseconds before the main pulse. The prepulse forms a relatively long scalelength plasma, which is then heated by the main pulse. This reduction in density gradient has been shown to increase the power of the $J = 0 - 1$ to over two orders of magnitude that of the $J = 2 - 1$ line, with a significant reduction in pulse length.[5-7] Fig. 1 shows the output spectrum of a Neon-like Germanium laser operated with such a prepulse.[10]

Use of this prepulse technique has greatly reduced the energy of the optical laser needed to produce saturated laser output. To increase the brightness of the source further it will be necessary to reduce the divergence of the beam towards the diffraction limit. At present, typical beam divergences from the 50-μm diameter laser are of order 10 to 25-mrad, and are not significantly better than calculated divergences based simply on the aspect ratio of the lasing medium - i.e. the beam is far from coherent, although improvements in target geometries are improving this situation.

3   Harmonics from gaseous targets

In a nanosecond high-power laser pulse, as used for the most of the XUV laser production described above, the oscillatory motion of the electrons in the laser-field is randomised by collisions with ions, thus heating the electrons to high temperatures (typically up to a keV). The impact of the electrons with the ions is the mechanism by which further ionization proceeds, and thus, though by no means in thermal equilibrium, the ionization stage reached is characteristic of the electron

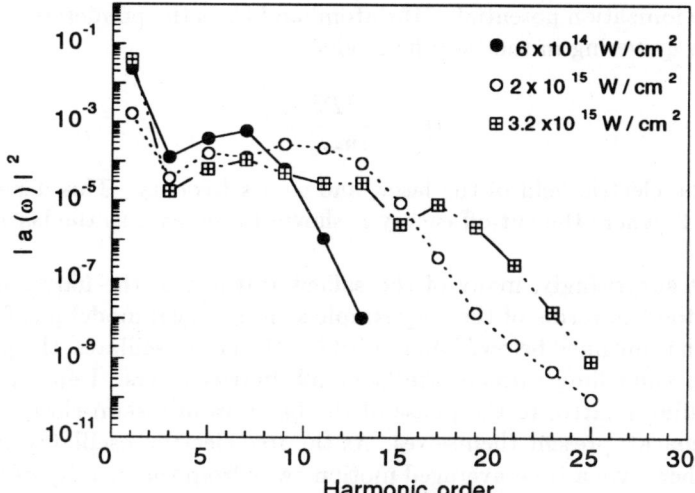

Figure 2: Simulations of the harmonic spectra of Helium for a short 248-nm pulse with a linear turn on, taken from the work of Sanpera et al (Ref. 11)

temperature. In contrast, if a very intense short pulse interacts with a target of sufficiently low density (e.g. a gas), then there is insufficient time for electron-ion collisions to occur. Under these circumstances the only mechanism available for ionization is direct ionization due to the laser field itself - i.e. at high fields the electron quantum-mechanically tunnels out of the Coulomb potential and continues to oscillate freely in the laser field: this is known as optical ionization. If the laser pulselength is sufficiently short, electrons in neutral atoms and low ionization stages can experience very high laser fields before significant optical ionization to the next ion stage occurs. This allows us to study the highly non-linear response of such neutrals and singly-ionized ions to extreme laser fields comparable to the Coulomb field of the atom, and far beyond the point where perturbation theory is valid.

The probability of the electron tunneling through the Coulomb barrier is a highly non-linear function of the laser intensity. Furthermore, after tunneling, the amplitude of the electron oscillation in the laser field is large - typically of order 10 - 100 bohr radii. Thus the susceptibility of the atom is extremely large and non-linear, and harmonics are produced. A typical harmonic spectrum, in this case simulated by a time-dependent solution to the Schrödinger equation in the single active electron approximation, is shown in Fig. 2, taken from the work of Sanpera et al.[11] Only odd orders are produced, due to the symmetric nature of the atomic potential. The generic features of harmonic spectra are a rapid decrease in production efficiency over the first few (say 5-th to 7-th) harmonics, followed by a plateau in the response, with a relatively sharp cut-off. The cut-off has been shown to occur at a photon energy $\hbar\omega$ corresponding to approximately

$$\hbar\omega = I_p + 3.17 U_p \qquad (1)$$

where I_p is the ionization potential of the atom, and U_p is the ponderomotive energy of the electron quivering in the laser field, given by

$$U_p = \frac{e^2 E_0^2}{4m\omega_0^2} \qquad (2)$$

where E_0 is the electric field of the laser, and ω_0 its freqency. This dependence is shown in Fig. 2, where the cut-off energy is shown to increase as the laser intensity increases.

Somewhat surprisingly, many of the salient features of the harmonic spectra can be understood in terms of the very simple semi-classical model put forward by Corkum, which is outlined below.[12] As the field of the laser oscillates, the probability of the electron tunnelling through the Coulomb barrier alters. Depending on the time of tunnelling relative to the phase of the laser, from a semi-classical point of view three scenarios present themselves. As the free electron oscillates in the laser field it can either have a time-averaged motion away from the vicinity of the parent ion, such that it never recrosses the ion, or, secondly, it can follow a trajectory which causes it to recross the ion core in the first (and perhaps several subsequent) laser cycles. In this simple semi-classical view, it is during such recollisions that the electron has a probability of returning to the ground state of the atom, emitting a harmonic photon with an energy of the ionization potential of the atom plus the kinetic energy of the electron in the oscillating field at the time of recollision. Finally, if the electron tunnels through the barrier exactly at the peak of the laser field, it will oscillate with zero time-averaged velocity, with one of its extrema being the ion core (if we neglect that fact that the free electron is not actually 'born' at the centre of the ion - i.e. we assume the amplitude of electron motion in the laser field is significantly greater than atomic dimensions - a good approximation at these laser intensities).

Corkum demonstrated that the cut off in photon energy described in equation (1) can be interpreted using the classical equations of motion of an electron. If we assume that the tunnel-ionized electron is produced at the ion core at some time in the laser cycle, and then solve the equations of motion to determine the kinetic energy of the electron upon its first recollision with the core, we find that the maximum kinetic energy an electron can achieve is $3.17U_p$, for electrons that are produced at a phase angle of 17° relative to the laser field.

From the above description it can be seen that the processes of harmonic generation and optical ionization are inextricably linked. After the electron tunnels through the Coulomb barrier it has some finite probability of recombining with the core. This probability is greatest for the first recollision: in Corkum's semi-classical model the electron wavefunction diffuses in a direction transverse to the oscillation axis, and thus the probability of recombination rapidly decreases. If the electron does not recombine with the core, it drifts away from the parent ion, and we can consider the atom ionized. Thus it can be seen that for a given atom the ponderomotive energy of the freed electron cannot increase without limit with laser intensity due to depletion of the parent atoms by ionization. Thus the magnitude of the second term in equation (1) is dictated by the maximum laser intensity that an atom can experience before significant optical ionization occurs. In quantum

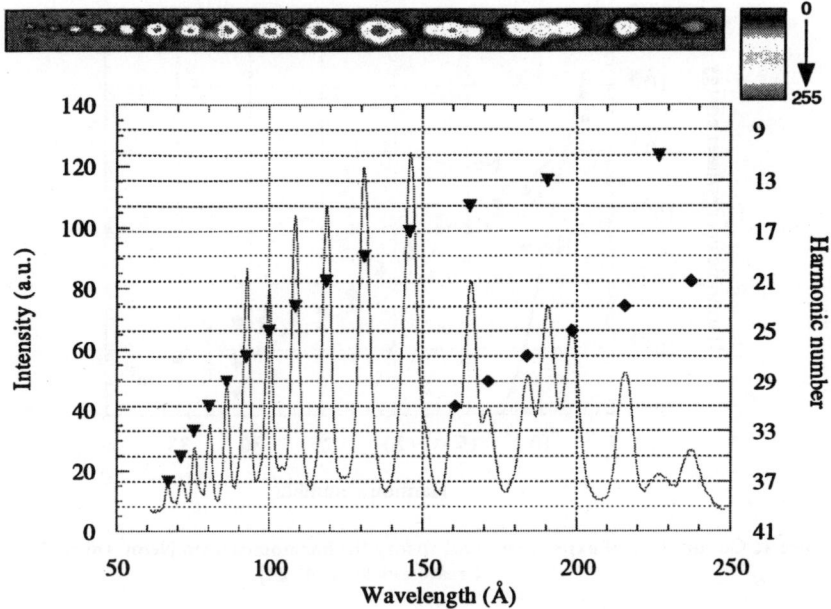

Figure 3: Harmonics generated from the interaction of a KrF (248-nm) laser with a Helium target, taken from the work of Preston et al (Ref. 24)

mechanical terms, the probability of the electron wavefunction remaining in a state corresponding to a bound level in the atom decreases with increasing laser intensity. However, those small number of neutral atoms that do survive to high intensities will generate high energy harmonics efficiently once they do eventually ionize. Furthermore, it can be seen why most work has been performed using noble gases: they can produce the highest harmonic photon energies and efficiencies as they have the highest neutral atom ionization potentials. This effects both terms on the right hand side of equation (1), in that the higher the ionization potential, the higher the laser intensity (and thus pondermotive energy) required for optical ionization.

As well as considering the response of single atoms, to find the overall conversion of laser light to XUV harmonics we must also consider phase matching effects. The fundamental laser and harmonic photon will dephase in the gaseous medium. Such effects will be far more severe in the presence of free electrons due to the far higher (negative) susceptibility at the fundamental wavelength for free electrons than the (positive) susceptibility of the neutral gas. Indeed, for neutral gases at low density the phase matching is often dominated by the geometrical phase factor of a Gaussian beam. This immediately leads to the question of the best means of maximising harmonic yield: is it better to use relatively low laser intensities, such that both the harmonic response and thus ionization of atoms is low, but the phase matching relatively good (the fundamental and harmonic propagate in a mainly neutral gas), or use high laser intensities, such that those atoms that do survive to high intensities have an extremely non-linear response, but the harmonics that are produced by each atom remain in phase with the fundamental over a significantly

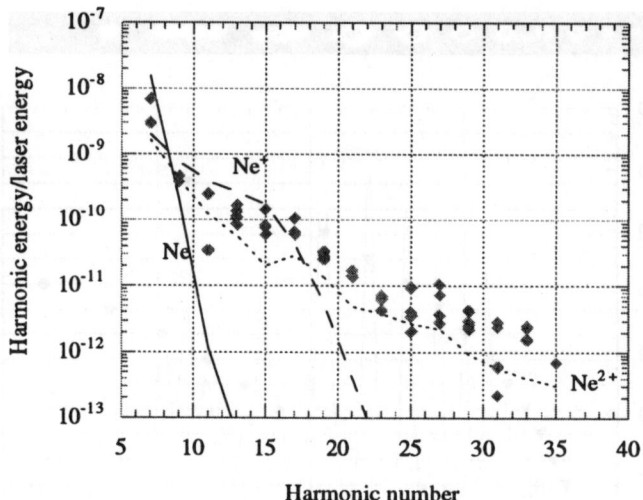

Figure 4: Comparison of experiment and theory for harmonics from Neon, taken from the work of Preston et al (Ref. 24)

shorter length, being dephased by the free electrons?

In most of the early pioneering work in the field the former conditions were used,[13–18] and detailed studies of the effects of the geometrical phase factor were performed. More recently, the scaling of conversion efficiency with atomic density has been investigated.[19] However, the work of Ditmire and co-workers established that the highest conversion efficiencies would indeed be produced at saturation intensities and above,[20] thus in the light of the above discussion, for efficient harmonic production we necessarily are operating in the presence of free electrons. Furthermore, in this regime Ditmire also showed that shorter wavelength drivers were more effective at producing harmonics, although the cutoff in photon energy given by equation (1) was necessarily reduced to the the quadratic scaling of the ponderomotive term (equation (2)) with fundamental wavelength. The increased efficiency of shorter wavelength drivers is attributable to two effects. Firstly, for a high harmonic of a given photon energy the dephasing length between fundamental and harmonic scales as ω_0^2, and thus the scaling of harmonic yield due to phase effects scales as ω_0^4. The second effect can be understood once again in terms of Corkum's semi-classical model: the electron that has tunneled through the Coulomb barrier returns to the core more rapidly (in approximately one half of a laser cycle) for a short wavelength drive, thus its wavefunction has had less time to diffuse, and the cross section for harmonic production increases. This increase in conversion efficiency for shorter primary wavelengths has been verified by detailed modeling.[21]

Thus far we have been considering the response of neutral atoms. Once the neutral has been optically ionized, we would expect that harmonics could be produced from the action of the laser field on the subsequent ionization stage. As the ion is more tightly bound, we would expect a reduction in the harmonic conversion efficiency compared to the neutral due to the reduced cross section of the electron with the ion core. However, as the inherent efficiency of harmonic production with a short wavelength driver is higher, harmonics from ions may still be observed: the higher ionization potentials of the ions will increase the cut-off energy. Thus it is interesting to enquire whether higher energy harmonics can be produced using short wavelength lasers interacting with ions than with long wavelength lasers interacting with neutrals. The majority of groups working in this area have used relatively long wavelength lasers (e.g. Ti-Sapphire or Neodymium glass). The highest harmonics produced with such long-wavelength drivers are the 109-th of Ti-Sapphire (800-nm) at 74-Å[22] and the 141-st of Nd:glass (1053-nm) at 75-Å.[23]

Thus far, harmonic generation from ions has not been identified in the experiments with 1.05-μm drivers. Although there is some indication of ion response with 0.53-μm light, it is with 0.248-μm radiation that the effect of ions has been definitively recorded. In Fig. 3 we show results by Preston et al, where they observed upto the 37-th harmonic of a KrF (248-nm) laser - a harmonic wavelength of 67-Å- in interactions with a Helium target, and upto 35-th harmonic in interactions with a Neon target.[24] Comparison with simulations has shown that the highest energy harmonic radiation for Neon was produced by the doubly-ionized atom (see Fig. 4). Simular conclusions of evidence of ion response have been reached by Krause et al,[25] in their analysis of the work of Sarakura.[26]

4 Harmonics from solid surfaces and ponderomotively steepened plasmas

In addition to harmonic generation from gaseous targets, there has recently been a renewal in interest in generating high order harmonic radiation from high-power laser interactions with solid targets.[27-31] Such high harmonics were first observed in nanosecond experiments using CO_2 lasers at irradiances of order 10^{15} Wcm^{-2} (where upto the 46th was observed), where the long laser wavelength (10.6-μm) ensured significant ponderomotive steepening of the plasma density profile.[32-36] Both odd and even order harmonics are generated via the relativistic current associated with the electrons being dragged back-and-forth across this asymmetric density step. Due to the λ_0^2 scaling of the ponderomotive force, we would expect to observe similar phenomena using 1.05-μm lasers at irradiances in excess of 10^{17} Wcm^{-2}.

However, it should be stressed that for these ultra-short pulses we can conceive of two ways in which the laser can be incident upon a steep density profile. If there is no significant plasma expansion during the laser pulse, then the laser effectively interacts with the target-vacuum boundary. If, however, plasma expansion does take place, then if the ponderomotive force is sufficiently great it may be possible for significant ponderomotive steepening of the density profile to take place during the laser pulse. These two situations can be classed as true interaction with solids, and interaction with a ponderomotively steepened plasma.

Recently von der Linde et al reported the observation of the 15th harmonic from a 130-fs laser-solid interaction using a Ti:Sapphire laser at 800-nm with intensities

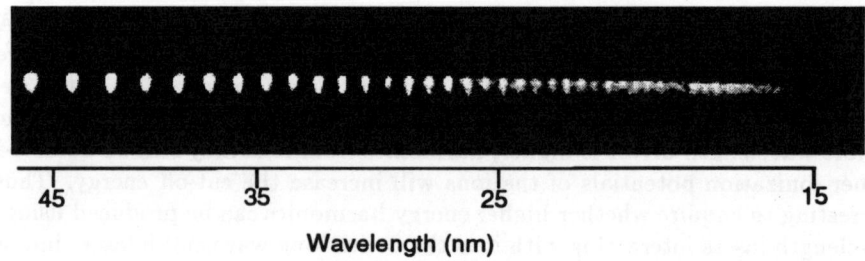

Figure 5: A spectrum of harmonics from ponderomotively steepened plasmas taken from the work of Norreys et al (Ref. 38)

upto 10^{17}Wcm^{-2}.[31] They interpreted their results as an interaction with a vacuum-solid step, conclude that the harmonics were produced in a specularly reflected narrow beam, and reported conversion efficiencies of order 10^{-8} to 10^{-9}.

For the situation of pondermotively steepened plasmas, Gibbon has recently performed PIC code simulations of harmonic generation for sub-picosecond pulses.[37] He concludes that for $I\lambda^2 > 10^{19}$ $\text{W}\mu\text{m}^2\text{cm}^{-2}$, and modest shelf densities of order $N_e/N_{critical} = 10$, upto 60 harmonics can be generated with power conversion efficiencies of 10^{-6}. Importantly, Gibbon's simulations predict that the harmonic order is simply determined by $I\lambda^2$, thus short wavelength lasers should produce shorter absolute wavelengths for a given value of $I\lambda^2$. Thus short wavelength, intense lasers may eventually provide a route to shorter wavelength, higher conversion efficiency harmonics than have hitherto been generated.

The most spectacular work to date in this area has been performed by Norreys and co-workers.[38] They observed upto the 68th harmonic of 1.05-μm light in first order diffraction, with indications of 75th in second order with laser intensities on target upto 10^{19}Wcm^{-2}, and with energy conversion efficiencies estimated at ranging from 10^{-4} to 10^{-6}. The experiment was performed using the Chirped Pulse Amplification beam line on the VULCAN laser at the Central Laser Facility of the Rutherford Appleton Laboratory.[39] The laser produced pulses of 2.5 picoseconds duration and energies of around 20 J on target. The contrast ratio was measured to be better than 10^{-6} using a third order auto-correlator. A single shot autocorrelator allowed individual pulse lengths to be measured. The laser beam was focused onto the target by an f/4.2, 44 cm focal length off-axis parabolic mirror.

Fig. 5 shows a spectrum taken when 20.7 J of p-polarised laser energy in 2.6-psec was incident on a target consisting of 2-μm CH coating onto a metal sandwich target (25-μm Mo on 50-μm Pd). The maximum entropy deconvolved x-ray penumbral images established that the spot diameter was $\sim 9 - \mu$m full width half maximum (FWHM), yielding an intensity on target of $9 \times 10^{18} \text{Wcm}^{-2}$.

The harmonics were found to be emitted into a wide angular range, to be independent of additional prepulse, and to be insensitive to the polarisation of the incident beam. With the level of prepulse inherent in this laser, we would expect significant pre-plasma to be formed. These effects - no observable difference in harmonic generation between s and p polarisations, the very large angular distribution and the relative insensitivity to prepulse levels - suggest that the critical density surface is rippled during the interaction, as this blurs the distinction between s and p polarisation. The development of a Rayleigh–Taylor like instability at the critical surface has been observed in 2.5 dimensional PIC simulations when a high intensity, picosecond laser pulse interacts with a pre-formed plasma.[40,41]

5 Comparison of XUV sources

In comparing these three different XUV sources, we must first decide upon an appropriate figure of merit: this will generally be the spectral brightness - i.e. the power per unit area, per unit solid angle, per unit frequency interval. Accurate comparisons of the spectral brightness of the various sources is difficult, as in many cases the coherence of the XUV radiation has not been measured. In the case of XUV lasers, some spatial coherence measurements have been made,[42] but these were generally without the prepulses that have recently been shown to vastly improve the gain length per unit energy input by negating some of the deleterious effects of refraction (see section (2)) - although some work has started in this area.[43] By the definitions within the Van Cittert Zernicke theorem the best spatial coherence was found to be approximately equivalent to a 15-μm diameter incoherent source. This spatial coherence is not particularly good, when one considers that it is only a few times smaller than the laser aperture, and for the Neon-like Ge laser represents a system which is a few hundred times the diffraction limit! Despite this far from optimal coherence, the spectral brightness of such lasers is still remarkably impressive. For these collisional x-ray lasers, the fractional linewidth is of order 10^{-4}, which is mainly determined by thermal Doppler broadening. A comparison of the various sources, compiled by M.H. Key,[44] can be seen in Fig. 6, where we compare the lower bounds of the spectral brightnesses.

Again, for the harmonics generated from gaseous targets, information on the spatial coherence is sparse. Some coherence measurements have been made for the situation where little ionization takes place.[45] However, to our knowledge, no such equivalent measurements have been published for those situations where the laser intensity was close to, or exceeded, the saturation intensity, such as in the work of Ditmire and Preston cited above.[20,24] We would expect the coherence to be worse in these latter cases, as the electron-density, and hence refractive index, in the gas will depend on the laser intensity, thus degrading the beam quality. Thus, despite the fact that these measurements have yielded the greatest conversion efficiencies, it is not known for certain how the spectral brightnesses compare. In Fig.6 we have used the measured cone angle of the incident laser to define the brightness. Thus the figures given should be treated as a lower bound on the true spectral brightness. In general, due to pulselength considerations, the fractional linewidth of the harmonics from gaseous targets is of order 10^{-3}. If the gas ionizes during the pulse, some further spectral broadening and blue shifting can occur due to the time-dependent ionization (and hence time-dependent refractive index).

For the harmonics produced from pondermotively-steepened plasmas the situation is different again. In this case, the divergence of the source is large, as is the

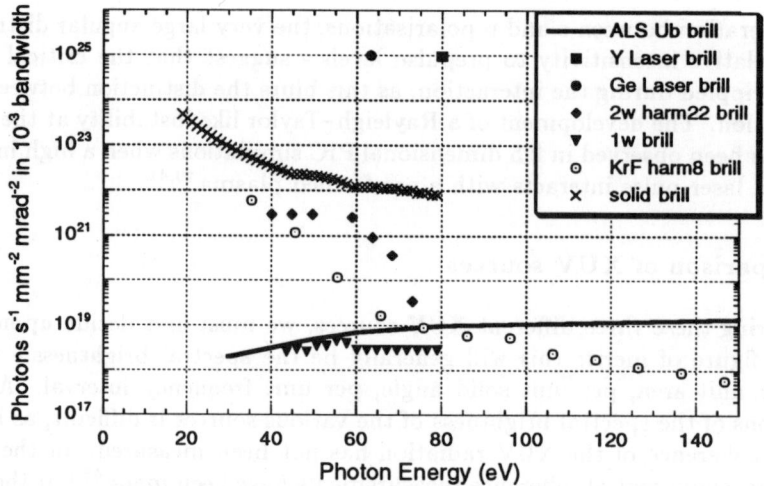

Figure 6: Comparison of the spectral brightness of various sources from the work of Key (Ref. 44). The KrF data is taken from Ref. 24, and the 1ω and 2ω data refers to the 1.05-μm and 0.53-μm data from Ref. 20.

fractional linewidth - which is of order 10^{-2}. This large linewidth is thought to be due to self phase modulation of the primary laser pulse as it traverses the plasma before reaching the critical density surface. There will also be some degree of spectral broadening due to the Doppler effect, as the critical surface is accelerated towards the target surface, with a peak velocity approaching 0.02 of the speed of light at an irradiance of 10^{19}Wcm^{-2}.[46] It should be noted that the spectral brightness is for most of these sources, at present, a few orders of magnitude greater than those available from synchrotron sources.

One particularly interesting feature to note from Fig.(6) is the superior instantaneous spectral brightness of harmonics from ponderomotively-steepened solids compared to harmonics from interactions with gaseous targets: the high conversion efficiency and small source size (the spot size of the high harmonics has been measured to be of order 2-μm) more than compensates for the high divergence and bandwidth of the source. With the large spectral coverage that these harmonics afford compared to x-ray lasers, they may prove to be a highly useful source for applications, such as non-linear optics in the XUV.

6 Applications

The two main areas in which x-ray lasers have currently been applied are interferometry of high-density (i.e. in excess of 10^{20}cm^{-3}) laser-plasmas[47] and radiography of laser accelerated foils.[48] The brightness of the Yttrium x-ray laser beam is equivalent to that from a several GeV blackbody - making it ideal for the probing of hot (keV) laser-produced plasmas. Furthermore, in such plasmas absorption and refraction render conventional optical techniques unsuitable.

Recent advances in multilayer mirror technology now allow XUV mirrors to be manufactured with reflectivities as high as 0.65, with a high degree of uniformity.[49] Furthermore, beamsplitters have been also been developed, with transmission and reflection coefficients of 0.15 and 0.2 at the Yttrium x-ray laser wavelength (155Å).[47] Using such optics, da Silva and co-workers have constructed an XUV Mach-Zehnder interferometer, and used it to diagnose the density profiles of laser-produced plasmas of relevance to laser fusion.[47]

The high brightness of the x-ray laser beam also allows it to be used to detect small thickness modulations in high opacity foils. If a beam is passed through a rippled foil, the intensity modulation due to the small variations in thickness is proportional to the thickness modulation and the product of the absorption coefficient and the thickness. Using such a technique Key and co-workers have measured the imprint pattern of a laser on a laser-accelerated thin silicon foil.[48,50] These measurements are crucial for direct drive laser-fusion research, as they help determine the degree of uniformity of illumination necessary for maintainance of the integrity of implosion of the fusion target.

High order harmonic radiation has also been used recently to diagnose high density plasmas.[51] The plasmas of interest, with densities in excess of 10^{23}cm^{-3} were themselves generated with a sub-picosecond laser pulse and thus, due to their highly transient nature, could not be probed with a conventional x-ray laser, which has a typical pulse length greater than 50-psec. Harmonics have also been used to measure the radiative lifetime of the 1s2p ^{1}P state of Helium.[52] It is also extremely encouraging to note that high harmonic radiation is now being used as a tool in condensed matter physics, with studies of antibonding states on the Ge(111):As surface being reported,[53] and more recently the application to high resolution atomic core level spectroscopy, which can be used to study chemistry at surfaces.[54]

At this early stage, it is not clear what additional specific applications such sources may have. However, for harmonics in particular, the cost of the necessary optical laser systems required to generated such high brightness XUV radiation has fallen dramatically over the past few years, so that such sources can truly be described as table-top systems of moderate cost. This wider availability will aid in the realisation of the potential of these high brightness XUV sources.

7 Conclusions

In summary, we have discussed recent improvements in the efficiency of x-ray lasers by use of the prepulse technique. The conversion of laser light into high energy harmonics has been presented, with specific results of harmonic generation from ions using a KrF laser given. Finally, harmonic generation from the interaction of a short intense pulse with a ponderomotively-steepened plasma has been demonstrated.

The spectral brightness of these three distinct types of source has been compared. At present, x-ray lasers are the brightest source in the XUV, followed by harmonics from ponderomotively-steepened plasmas. The latter source having the advantage of some degree of tunability. Harmonics from gaseous targets are also of import, as they can be generated using table-top equipment, and such sources are starting to find application in other areas of research, such as condensed matter physics. We believe that these short pulse, bright XUV sources will become a useful, complementary source to more conventional sources such as synchrotrons.

Acknowledgments

Much of the work described in this article was jointly funded by the United Kingdom Engineering and Physical Sciences Research Council and the European Communities Large Facilities Access Programme. The authors gratefully acknowledge the support of laser operations, target preparation and engineering support groups of the Central Laser Facility of the Rutherford Appleton Laboratory.

References

1. M.D. Rosen, P.L. Hagelstein, D.L. Matthews, E.M. Campbell, A.U. Hazi, B.L. Whitten, B. MacGowan, R.E. Turner, R.W. Lee, G. Charatis, Gar E. Busch, C.L. Shepard, and P.D. Rockett, Phys. Rev. Lett. **54**, 106 (1985).
2. S. Suckewer, C.H. Skinner, H. Milchberg, C. Keane, and D. Voorhees, Phys. Rev. Lett. **55**, 1753 (1985).
3. C. Chenais-Popovics, R. Corbett, C.J. Hooker, M.H. Key, G.P. Kiehn, C.L.S. Lewis, G.J. Pert, C. Regan, S.J. Rose, S. Saadat, R. Smith, T. Tomie, and O. Willi, Phys. Rev. Lett. **59**, 2161 (1987).
4. B.J. MacGowan, S. Maxon, L.B. DaSilva, D.J. Fields, C.J. Keane, D.L. Matthews, A.L. Osterheld, J.H. Scofield, G. Shimkaveg, and G.F. Stone, Phys. Rev. Lett. **65**, 420 (1990).
5. Nilsen and J.C. Moreno, Opt. Lett. **20**,1386 (1995).
6. H. Daido, Y. Kato, K. Murai, S. Ninimiya, R. Kodama, G. Yuan, Y. Oshikane, M. Takagi, and H. Takabe, Phys. Rev. Lett. **75**, 1074 (1995).
7. M. Nantel et al, Opt. Lett. **20** 2333 (1995).
8. B.E. Lemoff, G.Y. Yin, C.L. Gordon III, C.P.J. Barty, and S.E. Harris, Phys. Rev. Lett. **74**, 1574 (1994).
9. J.J. Rocca, V. Shyaptsev, F.G. Tomasel, O.D. Cortazar, D. Hartshorn, and J.L.A. Chilla, Phys. Rev. Lett. **73**, 2192 (1994).
10. J. Zhang, Private Communication (1996).
11. A. Sanpera, P. Jonsson, J.B. Watson, and K. Burnett, Phys. Rev. A **51** 3148 (1995).
12. P.B. Corkum, Phys. Rev. Lett. **71**, 1994 (1993).
13. M. Ferray, A. L'Huillier, X.F. Li, L.A. Lompre, G. Mainfray, and C. Manus, J. Phys. B: At. Mol. Opt. Phys. **21** L31 (1988).
14. X.F.Li et al., Phys. Rev. A **39**, 5751 (1989).
15. A. L'Huillier, K. J. Schafer, and K. C. Kulander, Phys. Rev. Lett. **66**, 2200 (1991).
16. Anne L'Huillier, Philippe Balcou, and L.A. Lompre, Phys. Rev. Lett., **68**, 166 (1992).

17. Ph. Balcou and Anne L'Huillier, Phys. Rev. A **47**, 1447 (1993).
18. Ph. Balcou et al, J. Phys. B: At. Mol. Opt. Phys. **25**, 4467 (1992).
19. C. Altucci, T. Starczewski, E. Mevel, C.G. Wahlstrom, B. Carre, and A. L'Huillier, J. Opt. Soc. Am. B **13** 148 (1996).
20. T. Ditmire, J. K. Crane, H. Nguyen, L. B. DaSilva and M. D. Perry, Phys. Rev. A **51**, R902 (1995).
21. J.B. Watson, A. Sanpera, and K. Burnett, Phys. Rev. A **51**, 1458 (1995).
22. J.J. Macklin, J.D. Kmetec, and C.L. Gordon III, Phys. Rev. Lett. **70**, 766 (1993).
23. M. D. Perry and G. Mourou, Science **264**, 917 (1994).
24. S.G. Preston, A. Sanpera, M. Zepf, W.J. Blyth, C.G. Smith, J.S. Wark, M.H. Key, K. Burnett, M. Nakai, D. Neely, and A.A. Offenberger, Phys. Rev. A Rap. Comm. **53**, R31 (1996).
25. J.L. Krause, K.J. Schafter, and K.C. Kulander, Phys. Rev. Lett. **68**, 3535 (1992).
26. N. Sarukura et al., Phys. Rev. A **43**, 1669 (1991).
27. S.C.Wilks, W.L.Kruer and W.B.Mori. IEEE Trans. Plasma Sci. **21**, 120 (1993).
28. S.V.Bulanov, N.M.Naumova and F.Pegoraro. Phys. Plasmas **1**, 745 (1994).
29. S. Kohlweyer et al., Optics Commun. **117** 431 (1995).
30. R.Lichters and J.Meyer-ter-Vehn. Gesellschaft für schwerionenforschung, Darmstadt Report GSI-95-06 p64 (1995). ISSN 0171-4546.
31. D. von der Linde et al., Phys. Rev. A., **52**, R25, (1995).
32. R.L.Carman, D.W.Forslund and J.M.Kindel. Phys. Rev. Lett. **46**, 29 (1981).
33. B.Bezzerides, R.D.Jones and D.W.Forslund. Phys. Rev. Lett. **49**, 202 (1982).
34. R.L.Carman, C.K.Rhodes and R.F.Benjamin. Phys Rev A **24**, 2649 (1981).
35. N.H. Burnett et al. Appl. Phys. Lett. **31**, 172 (1977).
36. C Grebogi et al., Phys. Fluids **26**, 1904 (1983).
37. P.Gibbon, Phys. Rev. Lett., **76**, 50 (1996).
38. P.A. Norreys, M. Zepf, S. Moustaizis, A.P. Fews, J. Zhang, P. Lee, M. Bakarezos, C.N. Danson, A. Dyson, P. Gibbon, P. Loukakos, D. Neely, F.N. Walsh, J.S. Wark, and A.E. Dangor, Phys. Rev. Lett. **76**, 1832 (1996).
39. C.N.Danson et. al. Optics Commun. **103**, 392 (1993).
40. S.C.Wilks et al., Phys.Rev. Lett. **69**, 1383 (1992).
41. A.Pukhov and J.Meyer-ter-Vehn. Gesellschaft für schwerionenforschung, Darmstadt Report GSI-95-06, p42- 43 (1995). ISSN 0171-4546.
42. J.E. Trebes, K.A. Nugent, S. Mrowka, R. London, T.W. Barbee, M.R. Carter, J.A. Koch, B.J. MacGowan, D.L. Matthews, L.B. DaSilva, G.F. Stone, and M.D. Feit, Phys. Rev. Lett. **68** 588 (1992).
43. R.E. Burge, G.E. Slark, X. Cheng, M.T. Browne, D. Neely, C.L.S. Lewis, A. MacPhee, SPIE **2520** 256 (1995).
44. M.H. Key, Private Communication, (1996).
45. Pascal Salieres, Anne L'Huillier, and Maciej Lewenstein, Phys. Rev. Lett. **74** 3776 (1995).
46. M. Zepf, M. Castro-Colin, D. Chambers, S.G. Preston, J.S. Wark, J. Zhang, C.N. Danson, D. Neely, P.A. Norreys, A.E. Dangor, A. Dyson, P. Lee, A.P.

Fews, P. Gibbon, S. Moustazis, and M.H. Key, Phys. Plasmas (to be published).
47. L.B. Da Silva, T.W. Barbee, R. Cauble, P. Celliers, D. Ciarlo, S. Libby, R.A. London, D. Matthews, S. Mrowka, J.C. Moreno, D. Ress, J.E. Trebes, A.S. Wan, and F. Weber, Phys. Rev. Lett. **74** 3991 (1995).
48. D.H. Kalantar, M.H. Key, L.B. Da Silva, S.G. Glendinning, J.P. Knauer, B.A. Remmington, F. Weber, and S.V. Weber, Phys. Rev. Lett. **76** 3574 (1996).
49. T.W. Barbee Jr., J.C. Rife, W.R. Hunter, M.P. Kowalski, R.G. Cruddace, and J.F. Seely, Appl. Opt. **32** 4852 (1993).
50. M.H. Key, T.W. Barbee Jr., L.B. Da Silva, S.G. Glendinning, D.H. Kalantar, S.J. Rose, and S.V. Weber, J. Quant. Spectrosc. Radiat. Transfer **54** 221 (1995).
51. W. Theobald, R. Häßner, C. Wülker, and R. Saubrey, Phys. Rev. Lett. **77** 298 (1996).
52. J. Larsson, E. Mevel, R. Zerne, A. L'Huillier, C-G. Wahlstrom, and S. Svanberg, J. Phys. B: At. Mol. Opt. Phys. **28** L53 (1995).
53. R. Haight and D.R. Peale, Phys. Rev. Lett. **70** 3979 (1993).
54. R. Haight and P.F. Seidler, Appl. Phys. Lett. **65** 517 (1994).

HOLLOW ATOMS

R. MORGENSTERN
KVI Atomic Physics, Rijksuniversiteit Groningen,
Zernikelaan 25, 9747 AA Groningen, The Netherlands

The formation and decay of hollow atoms resulting from the interaction of multiply charged ions with metal and insulator surfaces is discussed. While evidence for the formation of hollow atoms comes from measurements of the image charge acceleration, various steps of their decay can be observed by measuring yields of low energy electrons and energy spectra of Auger electrons and X-rays. It is shown that formation and decay are not well separated sequential processes, but that projectile and solid state surface form a highly dynamic system, in which the high potential energy of the projectile is converted into the energy of a large number of emitted electrons and photons. It is like huge fireworks, taking place within a few femtoseconds!

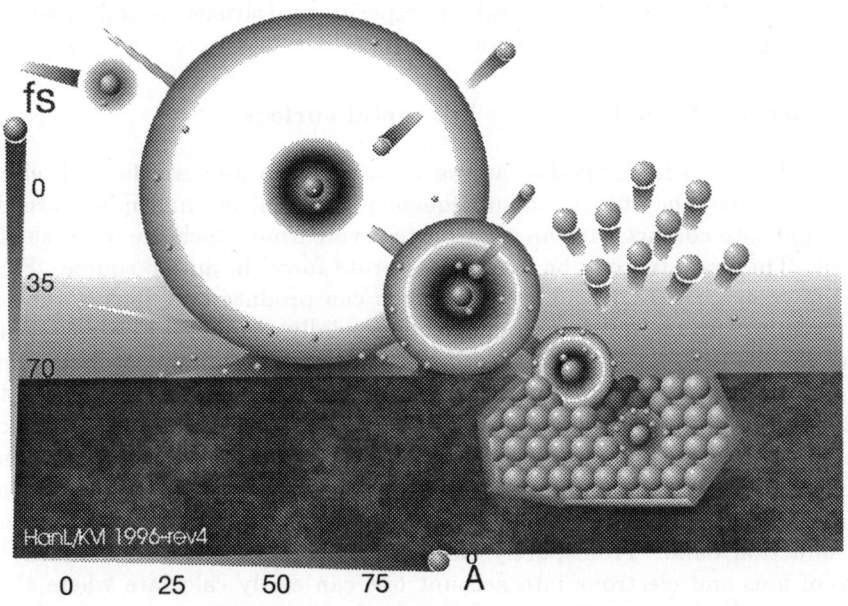

Figure 1: An artist's view of hollow atom formation and decay when a multiply charged ion approaches a metal surface

1 Hollow atoms: an invention or a discovery?

Can an atom be hollow? Like a nut with only the outer hard shell and without the delicious nucleus? Of course not – every atom needs the nuclear charge to keep the electrons together! However, the electrons from inner shells can in principle all be removed and placed in higher shells. The result of that operation – an atom with empty or scarcely filled inner shells and all electrons in outer shells – is what is called a hollow atom. It might have intriguing properties. Imagine an Ar atom with all its 18 electrons in the orbital with principal quantum number $n = 18$. All electrons could, e.g., have their spins parallel, resulting in an Ar atom with huge spin S. Also, since these $n = 18$ orbitals are nearly degenerate, a strong configuration interaction would result and this could give rise to a nearly classical collective electron motion. Would it be possible to observe plasmons in hollow atoms like those excited in the conduction band of metals? Last but not least such an atom carries a lot of excitation energy, which could possibly be released during the interaction with radiation or with matter. There is in fact a lot of speculation that hollow atoms might be an appropriate medium for X-ray lasers, for atomic lithography and for other fancy applications [1,2]. Although "hollow atom" is certainly a very descriptive term, and several general articles on this topic have been published in recent years [1,3,4,5,6,7], one might wonder whether this term indicates the discovery of a real object or rather an invention which only exists as an idea in our fantasy. In the following it will be shown that there is strong experimental evidence that hollow atoms are really formed. Moreover it will be discussed in which configurations they are formed in realistic experimental situations and experimental evidence of their decay will be shown.

2 Formation of "hollow atoms" at metal surfaces

The easiest way to form "hollow atoms" is a two-step process: first all electrons are removed from the atoms and subsequently the resulting multiply charged ions are brought into contact with an electron reservoir from which the outer shells can be filled. The first step can be done with 'brute force' in an ion source. Electron cyclotron resonance (ECR) ion sources which can produce bare ions of the second row elements – up to Ne^{10+} – are now commercially available. Electron beam ion sources (EBIS) or traps (EBIT) [8] – although delivering only very low beam intensities – can routinely deliver ions up to U^{82+} ! When such ions come into contact with metal surfaces they are rapidly neutralised in a quasi resonant way, i.e., electrons end up in orbitals with binding energies that approximately equal the workfunction of the metal. Fig.1 shows an artist's view how in the first place an electron cloud is formed around the ion, which subsequently shrinks and eventually merges into the conduction band. From purely classical considerations which take the image charge of ions and electrons into account one can easily calculate where the first electron transfer occurs during the approach of an ion to the surface. However it is rather difficult to properly predict the exact electronic configuration of the resulting atom. The distance at which the potential barrier between the metal surface and the approaching ion has sufficiently dropped to allow a transfer of electrons from the conduction band (with workfunction W) to the ions is approximately given by [9]

$$R_c = \frac{\sqrt{2q}}{W} \tag{1}$$

whereby the electron is transferred into an orbital with principal quantum number:

$$n = \frac{q}{\sqrt{2W(1 + \frac{q-1/2}{\sqrt{8q}})}} \tag{2}$$

The further evolution of the neutralisation can be viewed in two slightly different ways:

[i] One can consider the binding energy of the electron in a given orbital n around a given charge q, using the well known formula (hydrogenic energy plus image shift)

$$E_b = -\frac{1}{2}\left(\frac{q}{n}\right)^2 \frac{q}{2d} \qquad (3)$$

This implies that upon closer approach to the surface the electron energy is shifted upwards to smaller binding energies (see fig.2a), that the potential barrier drops and that subsequent electrons are transferred into orbitals with lower n-values.

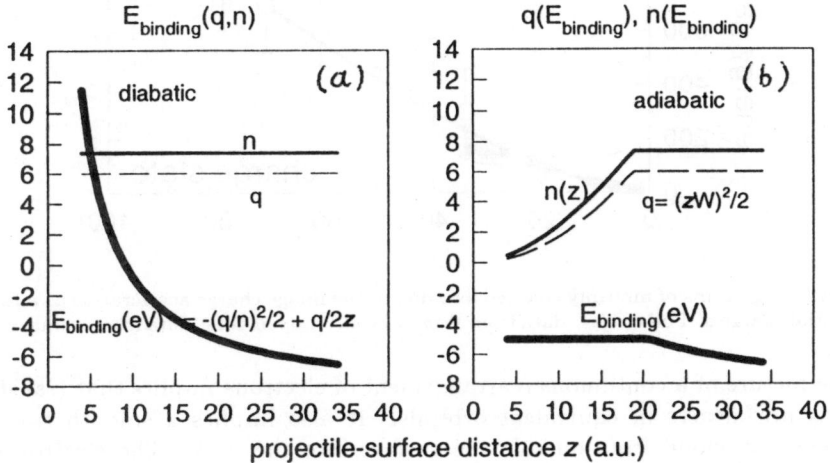

Figure 2: Binding energy, effective charge state q_{eff} and principal quantum number n as function of projectile-surface distance z.

[ii] One presumes that the effective charge of the projectile and the principal quantum number n of all the electrons – earlier and newly transferred ones – are continuously adjusted such that the binding energy of the electrons is kept constant, and that the potential barrier remains at a level close to the top of the conduction band. The latter situation is depicted in fig.2b: E_b is fixed at the workfunction (here $W = 5$ eV), whereas q and n decrease continuously until the projectile is completely neutralised.

The dependence of the effective charge q_{eff} on the projectile-surface distance can be obtained by inverting eq.1, yielding

$$q_{eff} = \frac{1}{2}(zW)^2 \qquad (4)$$

Direct quantitative evidence for such a decrease of q_{eff} comes from measurements of the energy gain of ions in front of metal surfaces due to the image charge acceleration [10]. This gain is given by

$$\Delta E = \int_\infty^0 \frac{(q_{eff})^2}{(2z)^2}dz = \frac{\sqrt{2}}{6}Wq^{\frac{3}{2}} \qquad (5)$$

Fig 3 shows measured energy gains for ion charges up to $q = 80$ in comparison with eq.5. Obviously eq.5 gives a rather accurate description of the observations, which implies that the neutralisation process can be properly described by the classical overbarrier model. It should be noted that Burgdörfer and Meyer [11] have originally given a more detailed description in which a quantized decrease of the charge was assumed (staircase model), thereby eventually arriving at the result of eq.5 as well.

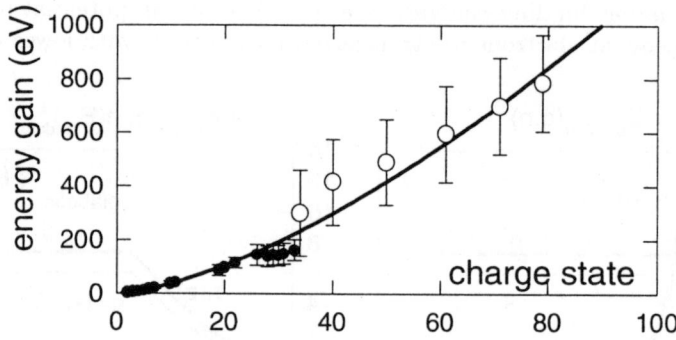

Figure 3: Energy gains of multiply charged ions due to the image charge acceleration as a function of the initial charge q. Full circles: data from Winter et al.[10]; open dots: data from Aumayr et al.[5]

The picture of a continuous rearrangement of electrons implies that transferred electrons are always in equivalent orbitals. It also implies a smooth transition of the electron cloud from outside the surface into the bulk. The electron cloud of the hollow atom merges into the conduction band. This view is supported by calculations of Arnau et al. [12], who for a stationary situation have studied the rearrangement of conduction band electrons when an ion is brought into the electron gas. They could show that the electrons are arranged such that they "mimic" atomic states with a binding energy close to that of the conduction band electrons. For N^{6+} ions brought into the bulk of an Al crystal this implies that the conduction band electrons mimic the M-shell, leaving the L-shell vacant, and therefore hollow nitrogen atoms can still exist in the bulk of aluminum.

3 Decay of hollow atoms

Unfortunately hollow atoms are by no means formed in stationary states. First of all the approach to the surface requires a continuous adjustment of the number of electrons and of the principal quantum number n of the orbitals in which the electrons reside. Moreover, as soon as the first electrons transit to the projectile, they "see" all the vacant inner orbitals and therefore a cascade of Auger and radiative transitions starts. Typical rates for Auger transitions are $10^{13} - 10^{16}\,\mathrm{s}^{-1}$, much higher than those for radiative transitions ($10^9 - 10^{12}\,\mathrm{s}^{-1}$, largely dependent on the charge). The main three experimental methods to observe the decay of hollow atoms are

(i) measurement of yields for low energy electron emission

(ii) spectroscopy of (high energy) Auger electrons, and

(iii) spectroscopy of (high energy) photons.

In principle the low energy electrons carry the most direct information about the hollow atoms because they are emitted at the beginning of the decay phase. High energy electrons and photons are emitted in a later stage of decay and therefore provide only indirect information on the initially formed hollow atom.

4 Low energy electron yields

The group of HP. Winter and F. Aumayr in Vienna has specialised in the measurement of low energy electron yields. They have constructed a detector in which the number of emitted electrons per incident ion is converted into a pulseheight. The pulseheight spectrum then allows a rather direct determination of the average number of emitted electrons. Moreover the statistics of electron emission can be measured, yielding extra information on the initial decay process.

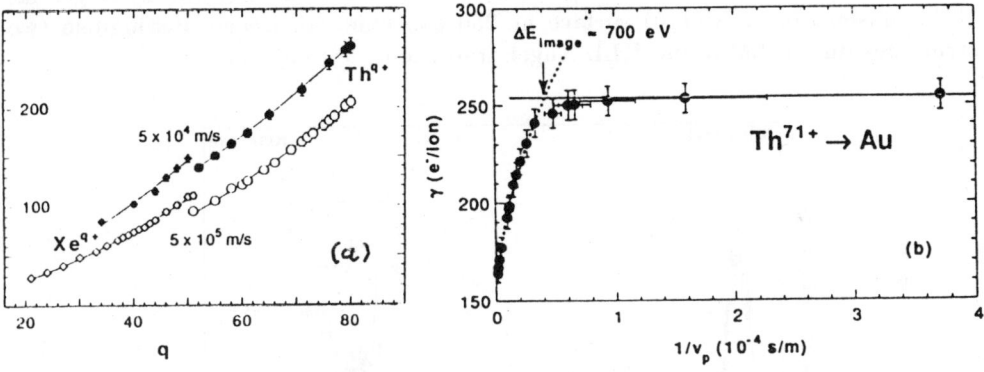

Figure 4: Yields of low energy electrons from a Au surface as a function of (a) the initial projectile charge q for two different projectile velocities, and (b) as a function of the inverse projectile velocity for charge $q = 71$. (From Aumayr et al.[13,14])

Fig.4 shows measured yields for a Au surface as a function of the initial projectile charge for two different projectile velocities. Yields of more than 200 electrons per incident ion are observed for projectile charges $q > 70$. Apparently the time available in front of the surface plays an important role. Lower yields are observed for higher velocities. In Fig.4b the yield for Th^{71+} is shown as a function of $1/v_p$ (with v_p the velocity perpendicular to the surface), i.e., as a function of the time available in front of the surface. A drastic increase of yield with increasing time is observed, until saturation takes place at the point where the lower limit of the velocity imposed by the image charge acceleration is reached. From these numbers one can estimate an average electron emission rate and finds values as high as 20 electrons per fs! Vaeck and Hansen [15] have calculated Auger rates for various configurations and indeed find values in this range. One has of course to keep in mind that in front of the surface not really stationary states are formed, since the ion-surface distance is changing so quickly that an orbiting wave packet of electrons "sees" a significantly modified effective nuclear charge after each revolution. Moreover there is a strong Stark mixture of various states due to the dipolar field of the charge - image charge constellation. Due to this Stark mixture the fast decaying states will also act as a loss channel for the more stable ones, and this implies that the Auger decay cascade is governed by the highest transition rates.

There is no reason to believe that slow electron emission could not proceed further if more time was available. In fact 200 electrons with 10 to 20 eV each do not exhaust the potential energy of more than 100 keV available in a $q = 70$ ion. It is just the limited time! But what is the time limiting process? We will come to this point in the next section.

5 Spectra of high energy Auger electrons from hollow atoms

Whereas in the initial phase of hollow atom decay low energy electrons are emitted, in a later phase also the inner shells are involved in Auger processes, yielding higher energy electrons. Fig.5. shows a spectrum of electrons resulting from hydrogenlike N^{6+} collisions on a Ni(110) surface at 250 eV. One can clearly distinguish two structures due to LMM and KLL Auger transitions.

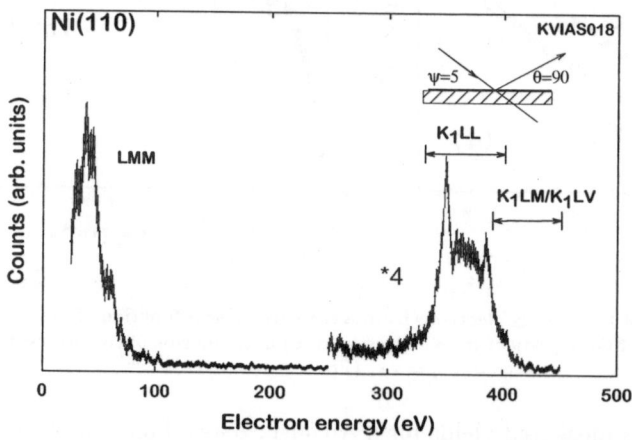

Figure 5: Energy spectrum of Auger electrons resulting from N^{6+} colliding on a Ni(110) surface at 250 eV. One can distinguish electrons from LMM and KLL Auger processes (from [16]).

In order to identify the L-shell configurations responsible for the various peaks in the KLL Auger structure we have performed Hartree-Fock type atomic structure calculations using the Cowan code [17]. Especially the configurations responsible for the two sharp peaks at the low- and the high-energy side of the KLL structure could be identified: the peak on the low energy side (at 350 eV) corresponds to a doubly filled L-shell ($1s2s^23l^4$) configuration, the one on the high energy side corresponds to a $(1s2s^22p^4)$ configuration (in all calculations a completely neutralised projectile was assumed by placing additional electrons in the M-shell).

One can suspect that electron emission from the sparsely filled L-shell at 350 eV occurs in an earlier stage of the collision than emission of those from the completely filled L-shell at 385 eV. This is supported by measurements of the Doppler shift which the electrons suffer because they are emitted from the moving projectile. The position of the 350 eV peak varies as a function of the detection angle, and this variation agrees exactly with the Doppler shift of electrons ejected from the projectile moving on the inbound trajectory with its original velocity. Moreover, the contribution of this peak to the whole KLL structure seems to be limited by the time available in front of the surface in a similar way as the yield of low energy electrons. This is shown in fig.6, where the relative fraction of this peak is shown as a function of the inverse perpendicular velocity $1/v_p$.

What is it, that upon reaching the surface makes an end to the emission of low energy electrons and to those KLL Auger transitions that originate from a doubly filled L-shell? As discussed above, the outer shells of captured electrons make a smooth transition into the conduction band and so the surface should not have a drastic influence. Also absorption effects in the solid are not such that the target surface could introduce a steplike modification of the electron emission. Nevertheless, a rapid change of the decay characteristics seems to occur when the close collision region is reached, and there has been a lot of discussion in the past to what extent the decay takes place either above or below the surface [18,3,19,20,21,22].

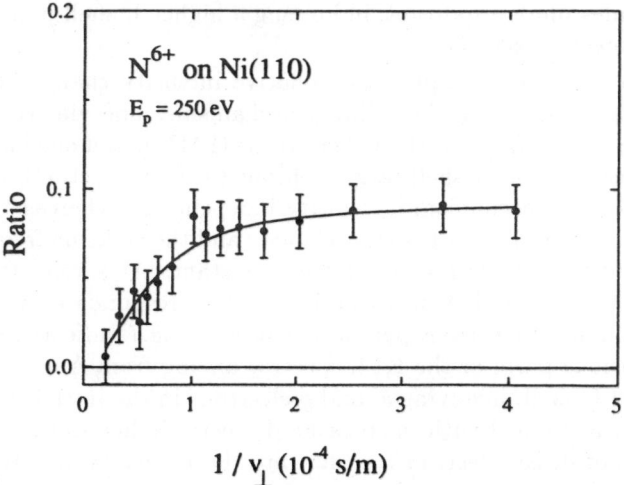

Figure 6: Relative fraction of the low energy peak in the KLL Auger structure of the electron spectrum as function of the inverse vertical projectile velocity, i.e., the time available in front of the surface. The saturation is caused by the image charge acceleration which poses a lower limit to the vertical velocity (from Das and Morgenstern[21]).

The property which seems to rapidly change upon a close approach is the population of the initially vacant L-shell orbitals. As soon as the close collision region between projectile and target atoms is reached, i.e., a distance of less than about

2 a.u., L-shell vacancies can be filled in binary collisions by direct electron transfer from target to projectile orbitals. Folkerts and Morgenstern [23] have invoked such a "side-feeding" process to explain the relatively low LMM Auger intensity as compared with the KLL intensity. This side-feeding abruptly terminates the Auger cascades from higher levels. Folkerts et al.[24] have found that during N^{6+} +Au collisions at 3.75 keV/amu a practically complete relaxation of the inner-shell vacancies takes place within less than 30 fs.

The initially captured electrons of the projectile in outer shells thus "see" a sudden change of the effective nuclear charge. The well-known "shake-off" effect, which is often invoked to explain the emission of a second electron after inner-shell photo-ionization, probably plays an important role here. The fast filling of inner-shell vacancies initiates a shake-off of more loosely bound electrons, thus giving an extra contribution to the low energy electron yield. As opposed to the normal shake-off, here the shake process starts from a multiply excited atom with several electrons at binding energies around the workfunction. Moreover, the effective inner charge is not only changed by one, but by several units within a very short time interval. So far no atomic physics calculations exist which quantitatively describe such a super shake process.

The fast filling of inner shell vacancies thus has two consequences: (i) a sudden emission of shake-off electrons which contribute to the low energy electron yield, and (ii) a shift of the KLL-Auger lines to higher energies. Fig.7 shows some measured spectra resulting form N^{6+} + Al collisions at various impact energies. One can see that with increasing collision energy the high energy part of the KLL Auger structure becomes more important, indicating a higher L-shell population for most of the KLL Auger transitions.

We have attempted to explain the relative intensity changes within a simple model[25] by assuming two L-shell filling mechanisms: one via Auger cascades involving valence band (LVV) or M-shell electrons (LMM), and one via quasi resonant electron transfer from inner shell target orbitals to the projectile L-orbitals. We assumed that the first mechanism is velocity independent, whereas the latter one is proportional to the number of L-shell vacancies and the collision frequency, i.e., proportional to v/d with d the relevant lattice constant. At a velocity corresponding to a collision energy of 60 keV nitrogen ions this corresponds to an average L-shell filling rate of roughly 3 electrons per fs. The bars in Fig.7 indicate relative contributions to the various parts of the KLL-spectra, arising from Auger transitions with differently filled L-shell, involving s- and p-electrons in the initial state respectively. The qualitative agreement with the measured spectra indicates that indeed the side-feeding process of direct electron transfer into the L-shell is mainly responsible for the L-shell filling in the close collision region. A more quantitative description of L-shell filling in close collisions, in which the relevant cross sections are calculated within a molecular orbital treatment, was recently given by Stolterfoht et al.[26].

A closer look at the electron energy spectra reveals that especially in the initial phase of L-shell filling the population of various states is by no means a purely statistical one. Our calculations show that the two low energy peaks of the KLL-Auger structure arise from a $(2s^2)$ and a $(2s2p)$ L-shell configuration respectively. A purely statistical population would yield a peak ratio of 1:9 for the two lowest

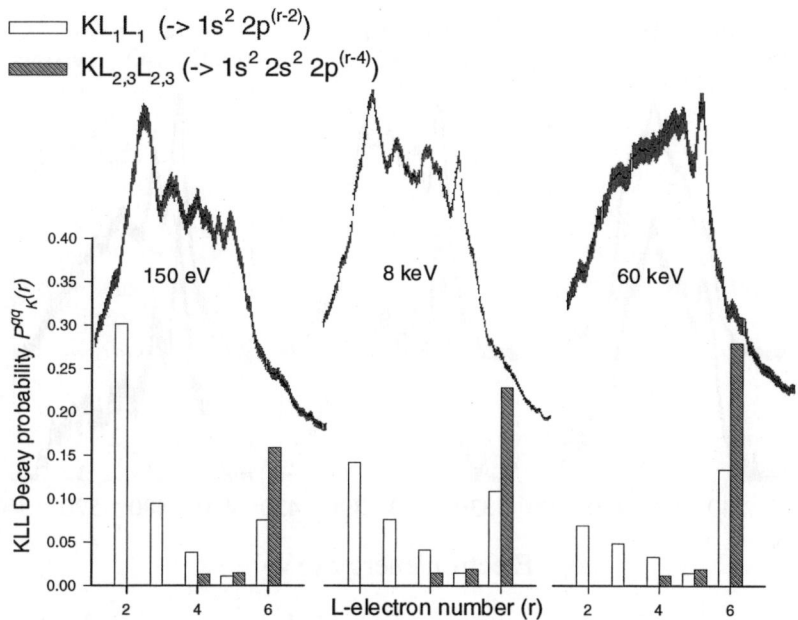

Figure 7: Velocity dependence of KLL Auger spectrum. The measured spectra are compared with simulated relative intensities obtained from a model description of L-shell filling and decay (data from Limburg et al.[25]).

peaks. The actually observed ratio is close to 3:1, which implies that in many more cases than statistically expected, decay from the $(2s^2)$ configuration takes place. This can not be explained by the higher decay rates of $(2s^2)$ as compared to $(2s2p)$ and is most likely due to a relaxation of the $(2p^2)$ and $(2s2p)$ configuration into $(2s^2)$ via Coster Kronig processes, during which L-shell electrons are rearranged and again slow electrons from the M-shell or conduction band are ejected. This has been discussed in more detail by Limburg et al.[27]. In order to investigate the role of such Coster Kronig processes they have compared electron energy spectra arising from H-like $(1s)$ and metastable He-like $(1s2s)$ ions of C, N and O as projectiles. For the He-like projectiles – having a 2s-electron from the beginning – decay from the $2s^2$ configuration might be much more likely than for the H-like projectiles. As can be seen from the comparison of the experimental spectra in fig.8 this is in fact the case for C ions as projectiles. For N and O ions however the spectra from hydrogenlike $(1s)$ and metastable He-like $(1s2s)$ projectiles are very similar to each other. This can be ascribed to Coster Kronig processes, which do occur in N and O, and eventually result in a preferential $2s^2$ population, but which do not occur in C ions, since in that case the rearrangement energy is not sufficient to eject electrons with a binding energy of about 5 eV.

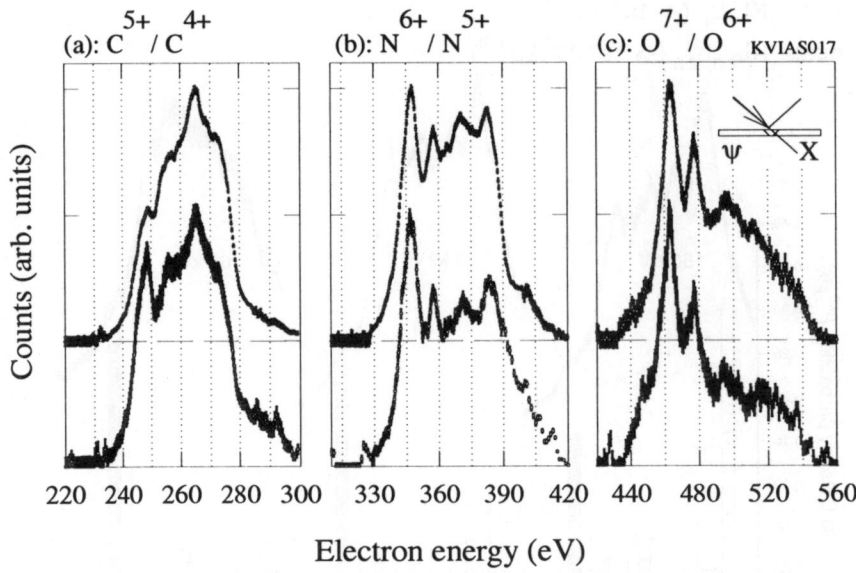

Figure 8: Comparison of KLL Auger spectra obtained with H-like (upper spectra) and metastable He-like (lower spectra) projectiles of C, N and O ions colliding on a Si(100) surface. For C projectiles a pronounced emission from the $2s^2$ configuration (low energy peak) is only observed for the metastable $C^{4+}(1s2s)$ ions. For N and O projectiles the spectra from both the He- and H-like projectiles show the preference for $2s^2$ emission, indicating that Coster Kronig relaxation into the $2s^2$ configuration takes place before the KLL Auger transition (from Limburg et al.[27]).

6 X-ray spectra from hollow atoms

An alternative method to observe the decay of hollow atoms - or at least the final steps of this decay - is the detection of X-rays. For low-Z atoms fluorescence yields are rather low, but since radiative transition rates are proportional to Z^4, an appreciable photon yield of several percent can, e.g., be expected for radiative transitions into the K-shell of hollow Ar atoms. For this reason hydrogenlike Ar^{17+} or bare Ar^{18+} ions were in the past often used as projectiles to study formation and decay of hollow atoms by photon analysis. As compared to the emission of KLL Auger electrons, the emission of K_α radiation starts earlier since only one electron is needed in the L-shell. K X-rays are therefore especially suitable to study the early stages of L-shell filling. Fig.9 shows two K X-ray spectra resulting from Ar^{17+} collisions on a Au surface at two different angles of incidence. The K_α line consists of several contributions corresponding to different numbers of L-shell electrons. At small impact angles the approach towards the surface and thus also the L-shell filling is significantly slower. Therefore K_α X-ray emission on the average takes place in the presence of fewer L-electrons, which is reflected in the shift of the K_α line to higher energies.

An even more detailed picture can be obtained from high resolution spectra, obtained with a crystal spectrometer. Such spectra are shown in fig.10. Each peak represents a K_α transition in the presence of a well defined number of L-electrons

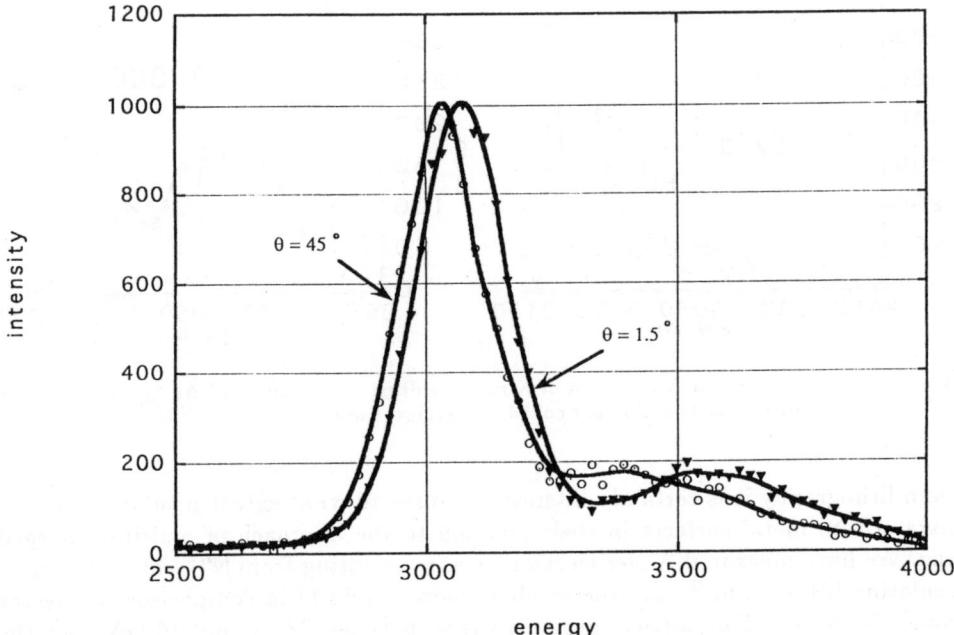

Figure 9: Spectrum of K_α and K_β X-rays resulting from impact of Ar^{17+} ions on a SiO_2 target for two different angles of incidence (from Briand et al.[28]). The shift to higher photon energies at small angles of incidence indicates decay from a projectile with incompletely filled L-shell.

and the degree of L-shell filling can rather directly be deduced from such spectra. Such a one-by-one relation between measured energy and L-shell filling is not the case for Auger electrons, because the binding energy of two electrons is involved. Therefore X-ray spectra are easier to interpret and can be regarded to have a "clock-property" on the fs scale [29]: they easily allow to compare the (well known) X-ray transition rate with the (unknown) L-shell filling rate. As long as the decay rate is fast as compared to the filling rate, most of the transitions will start from an L-shell with only one electron, whereas in the other extreme situation – slow decay and fast filling – transitions from a completely filled L-shell will dominate. In fig.10 obviously these two situations are rather well approximated with the low- and high energy collisions.

Recently Winecki et al.[30] have exploited this "clock property" of X-ray spectra to determine the time- and distance dependent population of K-, L- and M-shell populations during collisions of 51 keV Ar^{17+} ions on a graphite surface. In this way valuable details on the ion-surface interaction could be obtained.

7 Hollow atom formation at insulating surfaces?

As we have seen from the examples above there are clear indications that hollow atoms are formed outside metal surfaces. In view of possible application for ion

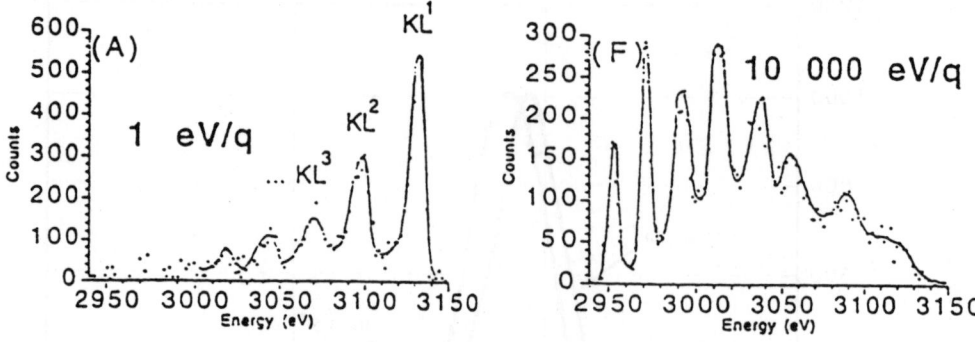

Figure 10: High resolution K_α spectra of X-rays resulting from impact of Ar^{17+} ions on a Au target for two different collision energies (from Briand et al.[31]).

beam lithography it is certainly interesting to see to what extent insulator surfaces deviate from metal surfaces in their reaction to the approach of multiply charged ions. We have measured Auger electron spectra resulting from N^{6+} collisions on an insulating LiF crystal[32], and the result is shown in fig.11 in comparison to spectra from a Si target. The impact energy is varied between 78 eV and 16 keV, but the angle of incidence is simultaneously adjusted such that the approach to the surface always takes place with the same low perpendicular velocity for each spectrum. The most remarkable feature of the LiF spectra is the absence of the low energy peak - which for metal targets was found to be a signature for above-surface processes. This suggests that there is no slow L-shell filling process via Auger cascades in front of the LiF surface, and that L-shell population proceeds only via direct transfer of target electrons. It should be pointed out that Briand et al. - investigating Ar^{17+} collisions on insulating SiO$_2$ and SiH - did find evidence for a slow L-shell filling in these systems. However, it should be kept in mind that in these cases direct electron transfer in binary projectile target collisions preferably populates the Ar M- or even higher shells. In this way indeed a hollow Ar-projectile might be formed, the L-shell of which is only slowly filled via subsequent cascades. A similar effect was observed by Limburg et al.[33] using Ne^{9+} projectiles on LiF targets.

8 Can hollow atoms survive their creation?

So far we have seen that formation and decay of hollow atoms near surfaces are processes which can hardly be separated from each other. Decay via Coster Kronig and Auger processes is extremely fast, so that many cycles of neutralisation, ionisation, reneutralisation etc. take place as long as the surface can supply electrons at a sufficiently high rate. Since the projectiles are accelerated toward the surface by their image charge, they normally enter the close collision region where their inner shells are filled, and this terminates their "hollow atom" phase. Yamazaki et al.[34] attempted to observe "hollow atoms" which survived the interaction with a surface. To realise multiple electron transfer and at the same time avoid violent collisions

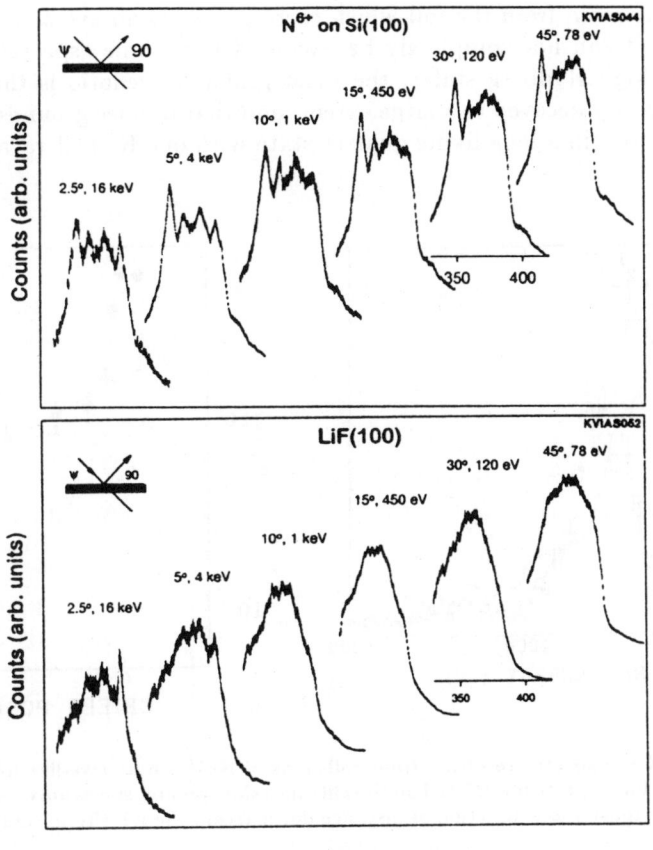

Figure 11: Comparison of KLL Auger spectra obtained from collisions of N^{6+} ions on a metallic Al and an insulating LiF target respectively. Most remarkably the peak at the low energy side – which for metal and Si targets is ascribed to above-surface decay – is missing for the LiF target (from Limburg et al.[32]).

they prepared a thin (10 mm) foil of Al_2O_3 with straight microcapillaries of about 100 nm diameter in a honeycomb pattern. This foil was bombarded with Ne^{9+} ions and the resulting hollow atoms or ions were observed via their X-ray decay on the entrance- and exit-side of the capillaries. In view of the limited length of these capillaries it was hoped to observe hollow atoms behind the foil that survived the interaction with the capilllary surfaces.

Fig.12 shows two X-ray spectra, measured on the entrance and the exit side of the foil respectively. The fact that the latter one is significantly shifted to higher photon energies already indicates that the radiative transitions observed here take place in a highly ionised projectile. In fact Yamazaki et al. could identify $Ne^{7+}(1s2s2p)^4P$ and $Ne^{8+}(1s2p)^3P$ ions as the most probable candidates for the observed emission. This identification is supported by the observed intensity decrease downstream from the foil which corresponds to an average lifetime of 0.8 ns. Although it can not completely be excluded that some "spectator electrons" still reside in high Rydberg states, the most probable scenario is that the hollow atoms have already decayed to a large extent, and that besides groundstate particles mostly Li-like ions in a long living quartet state with one K-shell vacancy leave the capillaries.

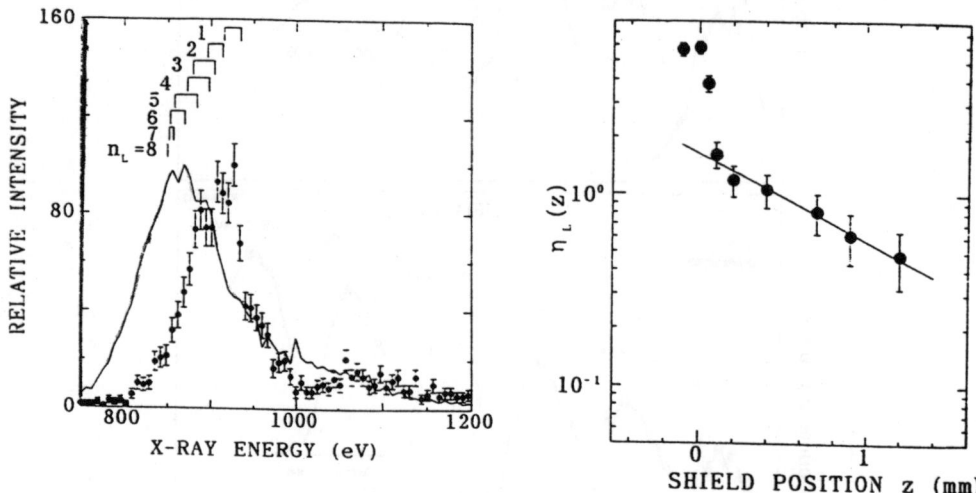

Figure 12: (a) X-ray spectra resulting from collisions of Ne^{9+} on a sieve-like foil with 100 nm capillaries. Solid line: spectrum emitted on the entrance side; points: spectrum emitted on the exit side. (b) X-ray intensity as a function of distance downstream the foil (from Yamazaki et al.[34]).

9 Outlook

Are there more promising ways to produce "free hollow atoms"? The two directions in which this could be attempted are: (i) decrease the formation time and/or (ii) increase the decay time. To realise the first option one can try to use metallic clusters or C_{60} bucky balls as targets instead of solid state surfaces. Since electron transfer occurs at large impact parameters hollow atoms might be formed 'en passent' without the complications arising from direct filling of innershell vacancies. First experiments in this direction have been performed by Briand et al.[35]. To realise the second option one could try to use highly magnetised targets such that high spin states are formed with a higher probability which might have significantly longer lifetimes.

Regarding applications of "hollow atoms" one has to realise that so far most experiments have been done with the aim to observe and understand their formation and decay. It is still a long way to handle them in a controlled way.

Acknowledgements

I would like to thank R. Hoekstra, J. Limburg and S. Schippers for their essential ideas and contributions to the "hollow atom" work in Groningen, and for the lively discussions we have had. The work performed in Groningen is part of the research program of the 'Stichting voor Fondamenteel Onderzoek der Materie' (FOM) which is financially supported by the 'Nederlandse Organisatie voor Wetenschappelijk Onderzoek' (NWO).

References

1. I.Hughes, Physics World 8 (1995) 43.
2. R.W. Schmieder, R.J. Bastasz, in AIP Conf. Proceedings 274: VIth Int.Conf. on the Phys. of Highly Charged Ions, eds. P. Richard, M. Stöckli, C.L. Cocke, C.D. Lin, (1993) 675.
3. H.J. Andrä, A. Simionovici, T. Lamy, A. Brenac, G. Lamboley, J.J. Bonnet, A. Fleury, M. Bonnefoy, M. Chassevent, S. Andriamonje, A. Pesnelle, Z.Phys. D - Atoms, Molecules and Clusters 21 (1991) S135.
4. F. Aumayr, HP. Winter, Comments At. Mol. Phys. 29 (1994) 275.
5. F. Aumayr, in "The Physics of Atomic and Electronic Collisions" (L.J. Dube, J.B.A. Mitchell, J.W. McConkey, C. Brion eds. AIP Conf. Proceedings 360) (1995), 631.
6. J. Das, R. Morgenstern, Comments on Atomic and Molecular Physics 29 (1993) 205-227.
7. R. Morgenstern, J. Das, Europhysics News 25 (1994) 3-6.
8. D.H.G. Schneider, M.A. Briere, Physica Scripta 53 (1996) 228-242.
9. J. Burgdörfer, P. Lerner, F.W. Meyer, Phys. Rev. A44 (1991) 5674.
10. H. Winter, C. Auth, R. Schuch, E. Beebe, Phys. Rev. Lett. 71 (1993) 1939.
11. J. Burgdörfer, F.W. Meyer, Phys. Rev. A47 (1993) R20-R22.
12. A. Arnau, P.A. Zeijlmans von Emmichoven, J.I. Juaristi, E. Zaremba, Nucl. Instr. Meth. Phys. Res. B 100 (1995) 279.
13. F. Aumayr, H. Kurz. D. Schneider, M.A. Briere, J.W. McDonald, C.E. Cunningham, HP. Winter, Phys. Rev. Lett.72 (1993) 1943.
14. H. Kurz. F. Aumayr, D. Schneider, M.A. Briere, J.W. McDonald, C.E. Cunningham, HP. Winter, Phys. Rev. A 49, (1994) 4693.
15. N. Vaeck, J. Hansen, J. Phys. B: At.Mol.Opt.Physics 28 (1995) 3523.
16. J. Limburg, J. Das, S. Schippers, R. Hoekstra, R. Morgenstern, Surf. Sci. 313 (1994) 355-364.
17. S. Schippers, J. Limburg, J. Das, R. Hoekstra, R. Morgenstern, Phys. Rev. A 50 (1994) 540-552 and 4429-4430.
18. F.W. Meyer, S.H. Overbury, C.C. Havener, P.A. Zeijlmans van Emmichoven, D.M. Zehner, Phys. Rev. Lett. 67 (1991) 723.
19. J. Das, L. Folkerts, S. Bergsma, R. Morgenstern, Phys. Rev. A 45 (1992) 4669-4674.
20. R. Morgenstern, J. Das, Physica Scripta T 46 (1993) 231-235.
21. J. Das, R. Morgenstern, Phys. Rev. A 47 (1993) R755-758.

22. H.J. Andrä, A. Simionovici, T. Lamy, A. Brenac, A. Pesnelle, Europhys. Lett. 23 (1993) 361.
23. L. Folkerts and R. Morgenstern, Europhys. Lett. 13 (1990) 377.
24. L. Folkerts, S. Schippers, D.M. Zehner, F.W. Meyer, Phys. Rev. Lett 74, (1995) 2204.
25. J. Limburg, S. Schippers, I. Hughes, R. Hoekstra, R. Morgenstern, S. Hustedt, N. Hatke, W. Heiland, Phys. Rev. A51, (1995) 3873-3882.
26. N. Stolterfoht, A. Arnau, M. Grether, R. Köhrbrück, A. Spieler, R. Page, A. Saal, J. Thomaschewski, J. Bleck-Neuhaus, Phys. Rev. A 52 (1995) 445-456.
27. J. Limburg, J. Das, S. Schippers, R. Hoekstra, R. Morgenstern, Phys. Rev. Lett.73 (1994) 786-789.
28. J.-P. Briand, S. Thuriez, G. Giardino, G. Borsoni, M. Froment, M. Eddrief, C.Sébenne, Phys. Rev. Lett. 77 (1996) 1452-1455.
29. J.P. Briand, L. de Billy, P. Charles, S. Essabaa, P. Briand, R. Geller, J.P. Desclaux, S. Bliman, and C. Ristori, Phys. Rev. A 43 (1991), 565.
30. S. Winecki, C.L. Cocke, D. Fry, M.P. Stöckli, Phys. Rev. A 53 (1996) 4228-4237.
31. J.-P. Briand, G. Giardino, G. Borsoni, M. Froment, M. Eddrief, C. Sébenne, S. Bardin, D. Schneider, J. Jin, H. Khemliche, Z. Xie, M. Prior, Phys. Rev. A 54 (1996) in press.
32. J. Limburg, S. Schippers, R. Hoekstra, R. Morgenstern, H. Kurz, F. Aumayr, HP. Winter, Phys. Rev. Lett. 75 (1995) 217 -220.
33. J. Limburg, S. Schippers, R. Hoekstra, R. Morgenstern, H. Kurz, M. Vana, F.Aumayr, HP. Winter Nucl. Instr. and Meth. in Phys. Research B 115 (1996) 237.
34. Y. Yamazaki, S. Ninomiya, F. Koike, H. Masuda, T. Azuma, K. Komaki, K. Kuroki, M. Sekiguchi, J. Phys. Soc. Japan 65 (1996) 1199.
35. J.-P. Briand, L. de Billy, J. Jin, H. Khemliche, M. Prior, Z. Xie, M. Nectoux, D.H. Schneider, Phys. Rev. A 53 (1996) R2925.

INTERDISCIPLINARY EXPERIMENTS WITH POLARIZED NOBLE GASES

E.W. Otten

Institut für Physik, Johannes Gutenberg-Universität Mainz, FRG

The paper discusses recent developments in polarizing large quantities of noble gases by optical pumping for use in different fields of physics and applied science. Among these are (i) scattering experiments of polarized beams from polarized He-3 targets which serve as substitute for polarized neutron targets, (ii) He-3 as a neutron spin filter to polarize neutron beams at research reactors, and (iii) polarized noble gases inhaled into the lungs and dissolved in the blood in order to perform magnetic resonance imaging of the lungs, the brain, etc.

1 INTRODUCTION

Polarizing noble gases is an art which stems from the golden age of optical pumping in the sixties. This is witnessed for instance by many papers on optical pumping which were read at the 1st Zeemann Centennial Conference held here in Amsterdam in 1965 on occasion of Zeeman's 100th birthday. Within 31 years Zeeman proceeded from birth to the discovery of his effect, but we, celebrating this discovery today by a 2nd Zeeman Centennial, are still talking on optical pumping. Where is the progress?

Indeed, most of the atomic physics on which our present technique relies has been cleared up already in the sixties and even some of the interdisciplinary questions which I will be dealing with today have been asked already then, but the final answer had not yet been given. Anyway, you will be amazed by the great variety of fields which have opened up in recent years after we have learned to pump up large quantities of noble gases to a high degree of polarization. It is so large, in fact, that I cannot give a comprehensive overview on all the technical and scientific aspects connecting to these new lines of research. I have to confine myself to showing just the general lines.

2 HOW TO POLARIZE AND HANDLE LARGE QUANTITIES OF NOBLE GASES

Optical pumping (OP) has always attracted the interest of experimentalists dealing with polarized particles by its virtue of reaching a polarization degree P of the sample far beyond the Boltzmann equilibrium

$$P = \frac{N^+ - N^-}{N^+ + N^-} = e^{-\mu B / kT} \qquad (1)$$

determined by the ratio of the magnetic dipole energy μB over the thermal energy kT. We will restrict ourselves to spin $I = 1/2$ systems and count atoms in the upper (lower) spin states as N^+ (N^-) However, this advantage of OP is necessarily accompanied by the danger of the system relaxing back to the Boltzmann equilibrium (1) much faster than it might be pumped up. Consequently the dynamic equilibrium

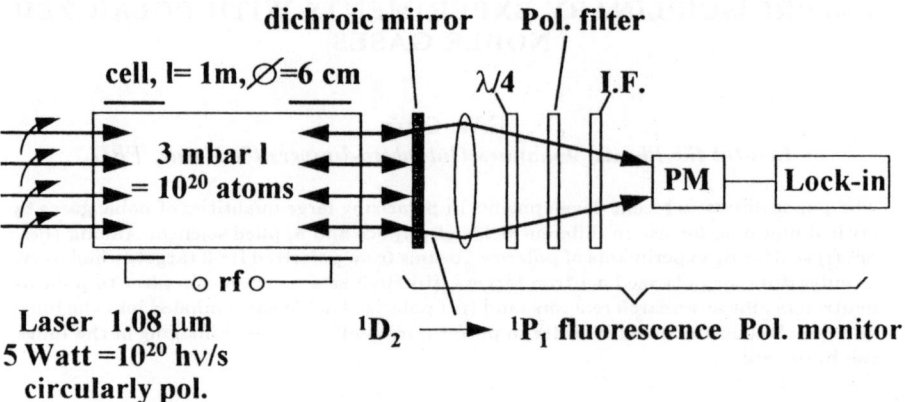

Figure 1: Scheme of ^{3}He pumping via metastability exchange

between polarzing and relaxing action would stop at a polarization far below the aspired value of order unity. In this respect, noble gases are particularly privileged, since their closed electron shell protects the spins to a high degree from relaxing through hyperfine coupling to external spins during collisions with other gases or the walls. On the other hand, their high lying levels preclude OP by resonance absorption from the atomic ground state. This difficulty has been circumvented by the follwoing classical roundabouts:

1. Exciting the noble gas in a discharge to a metastable state which is then pumped by a convenient resonance line and transfers its polarization to the ground state by metastability exchange collisions. This method is particularly suited for ^{3}He [1].

2. Mixing an alkali vapour with the noble gas, pumping the alkali and waiting for the polarization of the alkali spin to be transferred to the nuclear spin of the noble gas by spin exchange collisions [2].

2.1 Metastability exchange method

I will first describe the recent developments in the former scheme which our group in Mainz is persuing in collaboration with the group of Michèle Leduc at ENS, Paris.

Fig. 1 shows the principle set-up. Since the metastables in the discharge which runs at a pressure $\simeq 1$ mb typically have a relative concentration of 10^{-6} only and hence present little optical opacity, one chooses very long ($\simeq 1$ m) pumping vessels in order to increase the efficiency of transfering polarization from photons to atoms. The standard laser in use now is the so-called LNA laser developed at ENS [3]. It is a relative of the YAG laser and can be tuned to the $1.08\,\mu$ resonance line of helium by 2 etalons housed in a commercial cw-YAG-laser cavity. The output power of 5 - 10 watts is sufficient to almost saturate the transition over a cross section of $\simeq 10\,\text{cm}^2$ and over the Doppler broadened absorption width of about 2 GHz. Hence, the polarization in the hyperfine states of the metastable $^3S_1$ state builds up within microseconds. On the same time scale, on the other hand, the metastability is transferred to the next atom by an exchange collision.

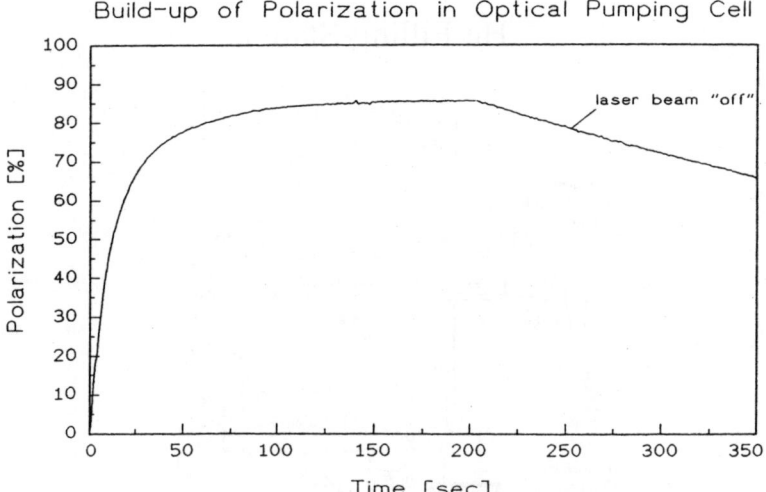

Figure 2: Build up of a ^{3}He polarization by optical pumping the metastable state in a plasma.

It is the electrostatic exchange interaction between "gerade" ($^3\Sigma_g$) and "ungerade" ($^3\Sigma_u$) molecular states of the intermediate excimer which is responsible for the large reaction constant of this exchange process; it is of order

$$c = \langle \sigma_{\text{m.ex}} v_r \rangle \simeq 10^{-11}\,\text{cm}^3/\text{s} \qquad (2)$$

which is the average of the product of the cross section and the relative velocity of the partners. Exchange collisions conserve the total spin of the partners and, therefore, establish a kind of spin temperature equilibrium between ground and metastable state, such that the occupation numbers of Zeeman levels in both states are distributed as

$$N(m_F) \propto e^{-\beta m_F} \qquad (3)$$

Fig. 2 shows the buildup of nuclear polarization in our pumping cell as monitored by the polarization degree of the flourescence light. Within 20 s it rises up to 65% corresponding to a production rate of $\vec{R}_s \simeq 3 \times 10^{18}$ polarized spins/s ($\simeq 0.5$ bar liter/h). The gas is then compressed by a 2-step titanium compressor into a target cell up to a pressure of 3 to 10 bar (see Fig. 3)[4]. The target may then be closed off by a valve and removed for some remote operation. We now reach up to $P_{\text{He}} = 55\%$ in the target. Serious polarization losses during compression which have been reported earlier[4] have been mostly overcome in the meantime.

Long relaxation time in the target is essential for any remote type of operation. Apparently relaxation by paramagnetic impurities of the glass walls dominates. It could be largely suppressed by a Cs coating[5]. This may be understood by the extremely small absorption energy of helium on alkaline and particularly Cs surfaces[6] which minimizes the interaction time with the surface. Thus, relaxation times ≥ 100 h have been observed.

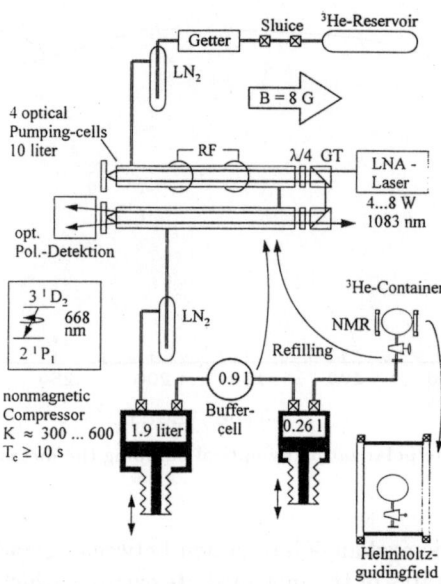

Figure 3: Schematic design of ^{3}He polarizer and compressor.

2.2 Rb spin exchange method

In the Rb-pumping scheme, on the other hand, the polarization is transferred from the Rb valence electron to the nuclear spin of He-3 through their mutual hyperfine coupling during the time of collision. This is much weaker than the electrostatic interaction mentioned earlier and yields a reaction constant of only $\langle \sigma_{s.ex} v_r \rangle = 1.6 \times 10^{-19}$ cm^3/s for the case of He-3. Nevertheless, this bottleneck can be partly compensated by pumping optically very thick Rb vapours of density [Rb] in excess of 10^{14} atoms/cm^3 with the help of powerful Ti-Al$_2$O$_3$ lasers (TRIUMF group [7], Princeton/Ann Arbor/SLAC collaboration [8]) or, more recently, of diode laser arrays [9,10] delivering now $\geq$ 100 watt DC output power. Moreover, the Rb-exchange method allows to polarize the noble gas directly at the desired pressure, e.g., at 10 bar in the case of He-3. These two gain factors in density now fully outweigh the much smaller reaction constant as compared to metastability exchange, such that the production rate of polarized spins per volume (V) reaches the order of

$$\frac{\vec{R}_s}{V} = P_{Rb}[Rb][^3He]\langle \sigma_{s.ex} v_r \rangle = \left(10^{16}/\text{s} \cdot \text{cm}^3\right) P_{Rb} \tag{4}$$

which is even somewhat higher than in the former case. Another advantage of the Rb scheme lies in the fact that it shortcuts the cumbersome mechanical compression after pumping. On the other hand, this scheme cannot be scaled up to large pumping volumes because of the tremendous opacitiy of the Rb vapour. If maximum production rate is really asked for, one probably has to choose the metastability

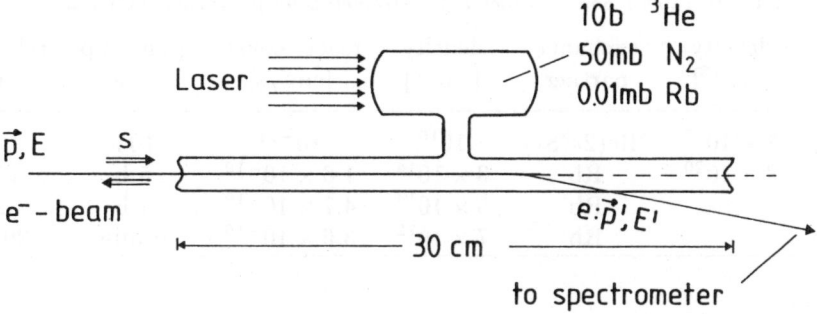

Figure 4: High pressure polarized ^{3}He target used at SLAC.

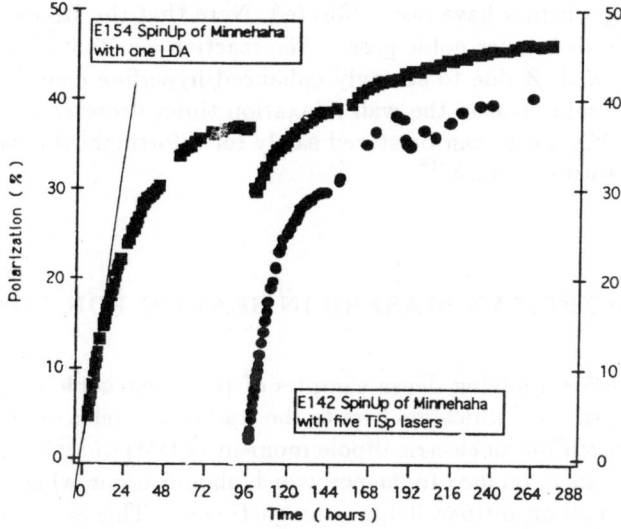

Figure 5: Build up of ^{3}He polarization in the SLAC target by Rb spin exchange.

exchange method eventually.

Fig. 4 shows the scheme of the polarized He-3 target which the Princeton group [11] has built for measuring the polarized structure function $g_{1n}(x)$ of the neutron in the SLAC experiments E142[8] and E154[11]. The upper part is the pumping cell, the lower one the 30 cm long target which is crossed by the electron beam. Besides 10 bar of He-3 and a few droplets of Rb it contains 50 mbar of nitrogen which is necessary to quench Rb fluorescence and hence to avoid depolarizing radiation trapping. In Fig. 5 we see the buildup of the polarization in the SLAC target; it takes almost 100 h until a saturation polarization of 40 - 45% is reached. The initial slope corresponds to a production rate of $\vec{R}_s \simeq 2 \times 10^{17}$/s. Note that one laser diode array (LDA) used in the later E154 experiment does a better job now than the titanic effort of 5 Ti-Al$_2$O$_3$ lasers did in E142 before. In the compressed phase the polarization is usually measured by comparing the NMR signal of the noble gas (in nondestructive, rapid adiabatic passage) to the one of protons in an equally shaped water filled cell at the same frequency. More convenient and probably safer

Table 1: Table 1 Some characteristic parameters for polarizing noble gases.

Gas	density [cm^{-3}]	exchange partner	density [cm^{-3}]	react. const. [cm^3/s]	pump up time	relaxation time
^{3}He	3×10^{16}	^{3}He($2s^3$S$_1$)	10^{10}	10^{-11}	10 s	100 h
^{3}He	3×10^{20}	Rb	3×10^{14}	1.6×10^{-19}	5 h	100 h
^{21}Ne		Rb	5×10^{14}	4.7×10^{-19}	1 h	2 h
^{129}Xe		Rb	7×10^{12}	3.6×10^{-16}	6 min	20 min

is a direct measurement of the static field of the order of mGauss which is produced by the polarized spins at the cell surface [12]. In Table 1 some characteristic data of both pumping schemes have been collected. Note that the Rb-exchange method is also applicable to heavier noble gases. The reaction constants for spin exchange increase rapidly with Z due to strongly enhanced hyperfine coupling during collision; but for the same reason the wall relaxation times decrease as well. Still, the polarization of ^{129}Xe, e.g., can be stored safely for a fortnight if it is solidified and kept at a temperature of 4.2 K [13].

3 ^{3}He-/^{129}Xe-ZEEMAN MASERS IN SEARCH FOR EDM

As first example for applying dense samples of polarized noble gases I choose an experiment in progress [14] which fits best to the traditional fields of interest of ICAP. It aims at the search for an electric dipole moment (EDM) of ^{129}Xe to be measured by the difference in resonance frequency which should occur when one applies an electric field parallel or antiparallel to a magnetic one. This is the usual approach well known from EDM research on the neutron or ^{199}Hg (see, e.g., Ref. 15). The Ann Arbor group has chosen a set-up where polarized ^{3}He and ^{129}Xe simultaneously fill the same resonance cell. Since the EDM-effect in the heavy ^{129}Xe atom is supposed to be enhanced by several orders of magnitude as compared to ^{3}He, the ratio of their resonance frequencies will mainly show the EDM of ^{129}Xe. A very precise way of measuring these frequencies is by running a maser on each of them. For this purpose the maser cell has to be supplied continously with freshly polarized gas. This is achieved by connecting it to a pumping cell in which it is polarized by Rb spin exchange. At present the group has managed to run both masers separately and to determine their inherent frequency precision by comparing to a precise and noise free reference in an Allen diagram (Fig. 6).

The frequency error slopes down with integration time τ, as $\tau^{-3/2}$ which is expected when white phase noise dominates frequency noise. The level of 10^{-8} Hz achieved after $\tau = 10^4$ s is an encouraging number which, if improved by another factor of 10 and realized in the full EDM experiment finally would set a competitive limit below 10^{-27} e·cm.

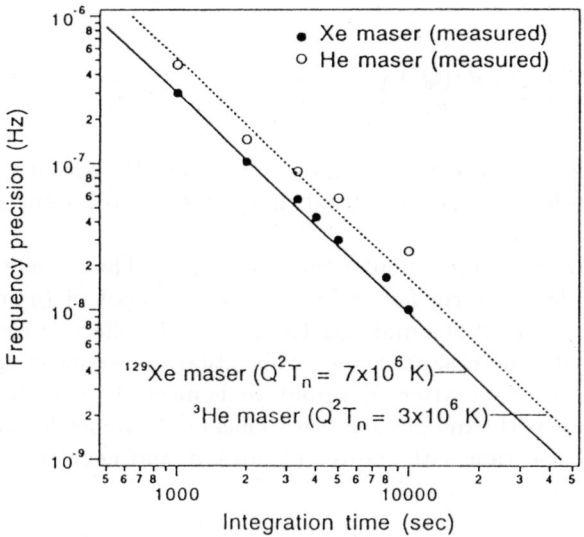

Figure 6: Allen-diagram of ^{3}He and ^{129}Xe masers

4 SCATTERING POLARIZED BEAMS FROM POLARIZED ^{3}He

Polarized targets of simple systems like ^{1}H, D, or ^{3}He have always attracted the interest of nuclear and particle physicists for testing symmetry properties of the fundamental interactions and, more recently, for studying the structure of the nucleon. In this game polarized ^{3}He plays the role of a substitute for a polarized neutron target since the spin of this nucleus is dominantly made up by the one of the single neutron, whereas the proton spins are paired off. Within this context I will describe three recent experiments, one devoted to the determination of the electric form factor of the neutron, the second clarifying the nuclear structure of ^{3}He, and the third measuring the already mentioned deep inelastic, polarized structure function of the neutron $g_{1n}(x)$.

4.1 Determination of the electric neutron form factor in $^3\vec{\mathrm{He}}(\vec{e}, e'n)$ scattering

The internal structure of hadrons has always been a central issue in particle physics since R. Hofstadter discovered in the fifties that the charge and magnetism of the proton are spread over a certain volume. Nowadays the form factors of the nucleons are cornerstones for checking effective quark models and QCD calculations. In this respect the electric form factor of the neutron is particularly interesting, since it should vanish identically in any first order quark model where the constituent quarks occupy the same spatial wave function. Charge distribution and magnetization of a neutron enter the elastic cross section of electron scattering through their Fourier-transforms, the elastic form factors $G_E^2(Q^2)$ and $G_M^2(Q^2)$, which are functions of the square of the four-momentum transfer, Q^2. They modify the Mott cross section of a point-like charge by

$$\frac{d\sigma(Q^2)}{d\Omega} = \left(\frac{d\sigma(Q^2)}{d\Omega}\right)_{Mott} \cdot [aG_E^2(Q^2) + bG_M^2(Q^2)] \quad (5)$$

where a and b are kinematical factors known from the Rosenbluth formula. At $Q^2 = 0$ the two form factors yield the total charge (i.e., zero for the neutron) and magnetic moment, respectively.

$G_{En}(Q^2)$ is a hundred times smaller than $G_{Mn}(Q^2)$. Thus it is extremely difficult to extract its electric form factor from a measurement of (5) reliably. The problem can only be solved if one manages to enhance the effect of electric scattering by interference with the magnetic one. This situation occurs in spin polarized scattering, since the electric scattering amplitude is insensitive to the relative orientation of spins whereas the magnetic one is. Thus one searches for an asymmetry of the scattering cross section with respect to positive and negative helicity of the electron beam given by

$$A = \frac{\sigma^+ - \sigma^-}{\sigma^+ + \sigma^-} = P_e \frac{P_n^\perp c G_{En} G_{Mn} + P_n^\| d G_{Mn}^2}{a G_{En}^2 + b G_{Mn}^2} = A^\perp + A^\| \quad (6)$$

where P_e is the polarization of the electron beam and $P_n^\perp$ and $P_n^\|$ are the components of the neutron polarization parallel and perpendicular to the three-momentum transfer. The coefficients c and d are two more known kinematical factors. We see that the asymmetry measured for perpendicular neutron polarization $A^\perp$ is indeed proportional to the interference term between electric and magnetic scattering whereas $A^\|$ is of purely magnetic origin. Measuring the ratio of these two asymmetries gives directly the electric form factor in terms of the known magnetic one.

$$G_{En}(Q^2) = \left(c A^\perp / d A^\|\right) G_{Mn}(Q^2). \quad (7)$$

How to get a target of free polarized neutrons of sufficient density? We can only dream of it. But a polarized ^{3}He target is a realistic substitute since its spin and magnetic moment is almost entirely carried by the unpaired neutron (see Fig. 7).

To make sure that it is the neutron and not the proton from which the electrons are scattered quasi-elastically one asks for coincidence between the scattered electron and the neutron. We have performed this experiment with the 840 MeV beam of the Mainz Microtron which was fed by $10\,\mu A$ of polarized electrons with $P_e \approx 50\%$ [16] (in the meantime 80%!). Regarding the ^{3}He-target we did not yet dispose of the detachable high pressure cells described above, but used a 1 bar target cell with $P_{He} = 50\%$, where the gas was polarized and compressed on-line by a Toepler pump in a closed circuit [17]. Since the cross section as well as the asymmetries are small we use large arrays of lead glass Cherenkov detectors for the electrons and plastic scintillators for the neutrons, in order to collect sufficient statistics. For the case of parallel neutron polarization $P^\|$ Fig. 8 shows the electron spectrum for positive and for negative electron helicity. In the quasi-elastic peak on the righthand one recognizes the asymmetry of this scattering.

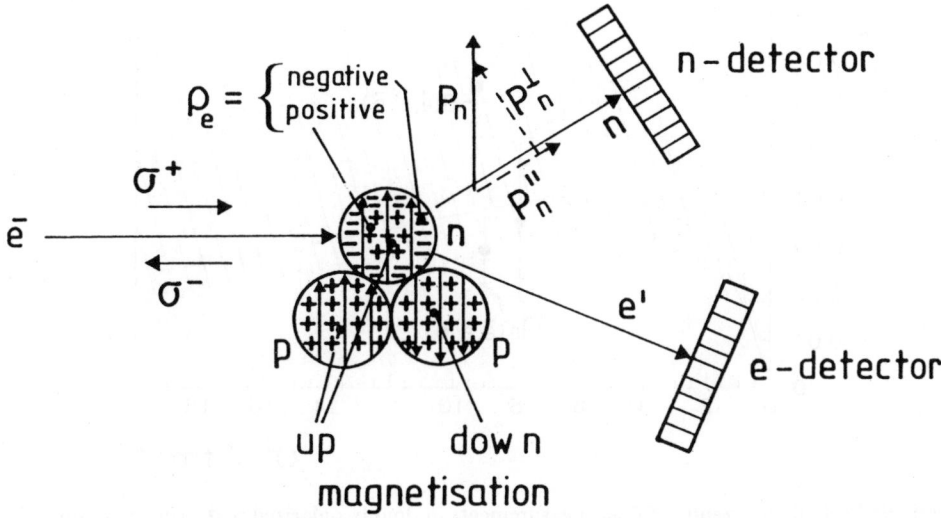

Figure 7: Principal of determining neutron form factors by exclusive, doubly polarized, quasielastic $^3\vec{\mathrm{He}}(\vec{e}, e'n)$ scattering.

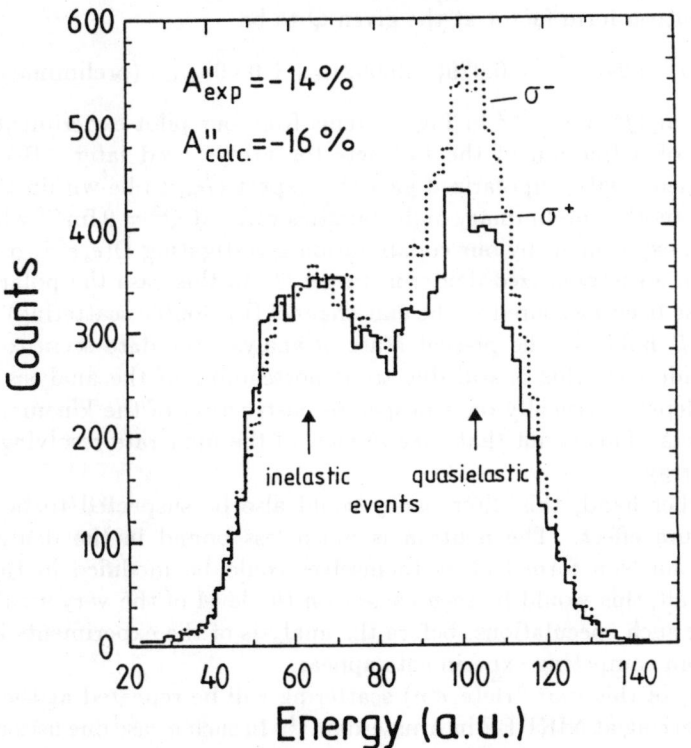

Figure 8: Asymmetry in the quasielastic peak of scattered electrons in the $^3\vec{\mathrm{He}}(\vec{e}, e'n)$ reaction.

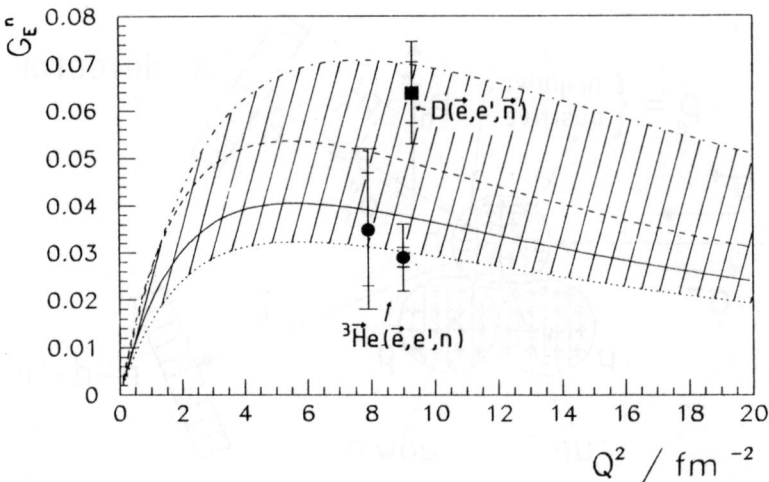

Figure 9: Preliminary results of G_{En} measurements in doubly polarized scattering experiments.

The preliminary analysis of the full data set yields for the decisive ratio of asymmetries in (7) the value [18] $A_{exp}^{\perp}/A_{exp}^{\parallel} = -0.146(13)$, from which we determine the electric neutron form factor at the given Q to be

$$G_{En}(Q^2 = 9\,\text{fm}^{-2}) = 0.029(\pm)0.0021_{stat} \pm 0.005_{syst} \quad \text{(preliminary)} \qquad (8)$$

The point at $Q^2 = 8\,\text{fm}^{-2}$ in Fig. 9 stems from our pilot experiment [19] which disposed only of a fraction of the full detector arrays used later. Results from earlier experiments with unpolarized particles expect G_{En} to lie within the dashed area. There is another preliminary high statistics value at $Q^2 = 9\,\text{fm}^{-2}$ which stems from a parallel experiment by our collaboration investigating $D(\vec{e}, e'\vec{n})$ quasielastic scattering from an unpolarized deuteron target [20]. In this case the polarization of the neutron has been measured in the exit channel (by double scattering) for which also (6) and (7) hold. In the present state of analysis the data seem to disagree. We dont exclude that this is still due to a shortcoming of the analysis since the asymmetries depend critically on a proper reconstruction of the kinematics of the scattering events. This is not that easy in view of the moderate resolving power of our detector arrays.

On the other hand, this discrepancy could also be suspected to be due to a nuclear structure effect: The neutron is much less bound in the deuteron than in He-3. The nucleon form factors themselves could be modified in the nuclear medium; if at all, this would be seen easiest on the level of the very small G_{En}. It is too early for such speculations, before the analysis of the experiments is finished and results from competing experiments appear.

By the end of this year $^3\vec{\text{He}}(\vec{e}, e'n)$ scattering will be repeated at the 600 MeV electron storage ring at NIKHEF in Amsterdam [21]. In such a case one uses a windowless internal gas target at very low pressure in order not to disturb the circulating beam too much. The loss in target thickness is to some extent compensated by the increase of the average beam current. In case of ^{3}He it is convenient to use the metastability exchange method since no compression is needed. Such targets have

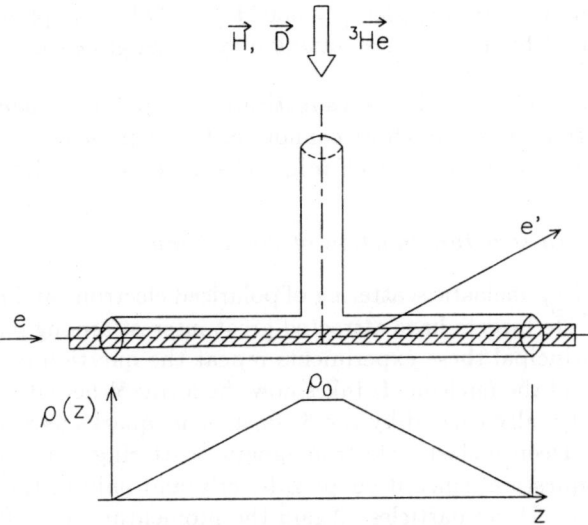

Figure 10: Principal of the open, internal, polarized gas target in use at the DESY storage ring. Below, the density profile.

been developed first by a Caltech/MIT collaboration and have been used before at the proton storage ring at Indiana [22] and the Hera-Ring at DESY; Fig. 10 shows the example of the DESY target [23].

4.2 Checking the nuclear wave function of $^3\vec{He}$

Before drawing definite conclusions from G_{En} measurements from a polarized ^{3}He target one has to assess the structure of this nucleus as well as the quasielastic reaction mechanism. The problem may be focussed to the following questions:

1. To what extent is the nuclear spin of ^{3}He indeed carried by the neutron?

2. To what extent do final static interactions of the primary hit nucleon with the two others interfere?

These questions have been investigated in polarized proton scattering experiments from $^3\vec{He}$ at TRIUMF in Vancouver [7] and at the above-mentioned proton ring at Indiana [22]. The idea was to study the knockout of a neutron or a proton in quasielastic kinematics, where momentum and energy transfer are measured accurately in coincidence between the scattered proton and the knocked out nucleon. The asymmetry of the cross section with respect to spin reversal has been measured as a function of scattering angle, momentum, and energy transfer. The data were analysed with the help of the known polarization dependent scattering amplitudes in free nucleon-nucleon scattering. The quasielastic reaction gives in particular access to the momentum wave function of the bound nucleons. The results corroborate extensive Fadejev calculations of recent years. In short they state:

1. At the peak of its momentum wave function at $\Psi_n(p=0)$ the neutron carries all of the nuclear spin ($P_n/P_{^3He} = 0.98(8)$) and in the average over all momenta 84%. The missing fraction is shared between proton spins and an orbital d-wave.

2. At high momentum transfer $q > 500\,\text{MeV/c}$ (which applies to the scattering experiments) final state interactions play a negligible role.

Thus our approach of using $^3\vec{\text{He}}$ as a substitute for a polarized neutron target seems to be on solid grounds, although we cannot yet tell at present, whether and to what extent the neutron form factor itself might change by the nuclear binding.

4.3 The polarized structure function of the neutron

In recent years deep inelastic scattering of polarized electrons and muons from polarized $\vec{\text{H}}$, $\vec{\text{D}}$, and $^3\vec{\text{He}}$ targets have attracted great interest among particle and nuclear physicists. In principal these experiments repeat the question of the foregoing section on the level of the nucleon. It takes now the form: Where does the nucleon spin come from? Is it really carried by the 3 constituent quarks as a naive quark model would suggest? Deep inelastic electromagnetic scattering is particularly suited for answering this question, since it couples directly and only to the quarks as if they were free, charged Dirac particles. Again the momentum wave function of the hit particle, here a quark, is determined in quasi-free kinematics by measuring the four momentum transfer Q and energy transfer ν. In this highly relativistic regime the quark momentum is usually expressed by the Björken variable

$$x = \frac{Q^2}{2M\nu} \qquad (9)$$

(M = nucleon mass). It is the fraction of the total momentum of the nucleon (measured, e.g., in the cg system) that is carried by the hit quark. Hence x ranges from 0 to 1. These momentum distributions are usually called structure functions of the nucleon. In case of an asymmetry measurement from longitudinally polarized particles it is called the polarized structure function $g_1(x)$.

Fig. 11 shows this structure function for the case of the neutron as measured from a polarized ^{3}He target in the reaction $^3\vec{\text{He}}(\vec{e}, e')$. The dots are the yet unpublished results from the before-mentioned SLAC experiment E154[11], the circles are preliminary data from the competing Hermes experiment at DESY[23]. The latter is again a storage ring experiment with an internal $^3\vec{\text{He}}$ target (Fig. 7). Judged from statistical grounds the storage ring experiment seems to be handicapped as compared to the external beam experiment; but it has the chance to go for more detailed information by coincidence experiments.

At present the detailed x-dependence of $g_1(x)$ is barely understood in terms of quantitative nucleon models. But one can check certain sum rules. One of them, the so called Ellis-Yaffe-sum rule, concerns $g_1(x)$ and predicts explicitly for the neutron

$$\int_0^1 g_{1n}(x) dx = -0.021(18) \qquad (10)$$

in agreement with the published result from E142 at SLAC of $-0.021(11)$[8].

However, a significant failure of the Ellis-Jaffe sum rule was reported for the proton a few years earlier which gave rise to the so-called spin crisis of the proton

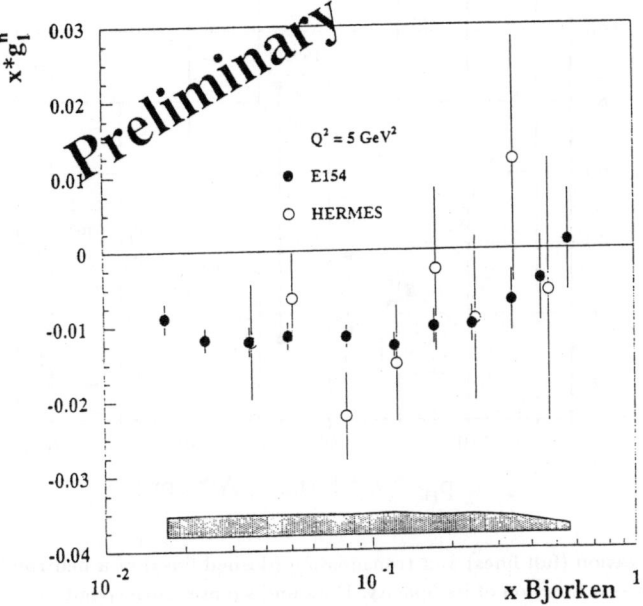

Figure 11: Preliminary results for the polarized neutron structure function measured at SLAC (dots) at DESY (circles).

which means the following: The Ellis-Jaffe sum rule was derived under the assumption that the spin of the nucleons is made up entirely from the spins of its u and d quarks as well as their antiparticles $\bar{u}$, $\bar{d}$ which are known to occur as so-called sea quarks in the nucleon. Hence the failure of the Ellis-Jaffe sum rule for the proton had to be interpreted in the sense that these quarks apparently carry only a minor fraction of the nucleon spin. Where can it go elsewhere? Into orbital angular momentum, into strange quark pairs $(s, \bar{s})$ and into gluon spins. At present this question remains open.

5 POLARIZED ^{3}He AS A NEUTRON SPIN FILTER

A very promising application of samples of dense polarized ^{3}He opens up for thermal and epithermal neutron beams. Neutrons with spin opposite to that of ^{3}He are absorbed with a cross section of 6000 barn; this enormous number applies for neutrons with a wavelength of 1 Å and decreases in proportion to this. The parallel component, on the other hand, is hardly attenuated by elastic scattering which has a cross section of a few barn only. Thus, polarized ^{3}He can serve as a neutron spin filter for a broad band of energies and for any direction of the neutron momentum. This is a great advantage over traditional neutron polarizers or analyzers which accept only a very limited phase space, like Bragg reflexes or total reflexion from magnetized materials. First successful attempts to polarize a neutron beam by a ^{3}He spin filter have been undertaken with a sample polarized by the Rb spin exchange method [24].

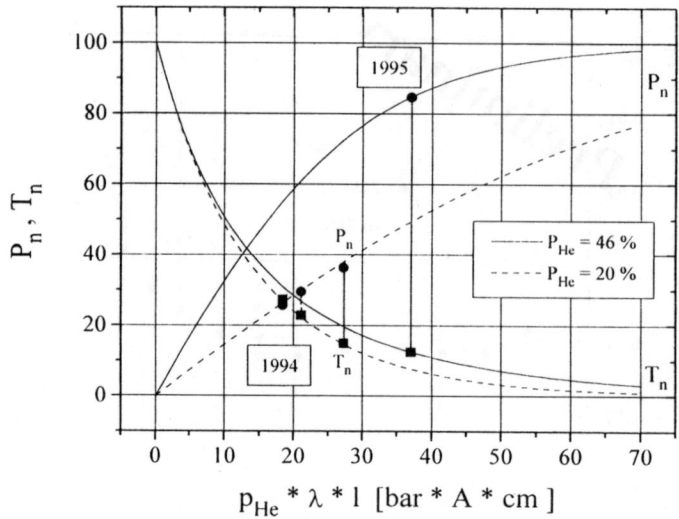

Figure 12: Polarization (full lines) and transmission (dashed lines) of a neutron beam through a $^3\vec{\text{He}}$ spin filter as a function of its opacity. Dots and squares correspond to measurements.

Fig 12 shows polarization and transmission of a neutron spin filter as a function of its opacity which is proportional to the product of its pressure, its length and the wave length of the neutrons; it is given in practical units bar · Å · cm. The contrast of the filter rises of course with ^{3}He polarization. Two curves are plotted for P_{He} equals 20% and 46%, respectively.

The lower value, 20%, was achieved in our first experiment at the Mainz TRIGA reactor in 1994 [25]. The second test with 46% polarization by the end of last year yielded a neutron polarization of 84% at a total transmission of 14% [26]. This is not too far anymore from the ideal values $P_n = 100\%$ at 50% transmission.

For this experiment the target was polarized at the piston compressor in our institute and then transferred to the broad band neutron beam from a TRIGA reactor on campus. The polarization was analyzed by a Bragg-reflex from a magnetized cobalt iron single crystal. In collaboration with the Institut Laue Langevin, Grenoble, we have built another two-piston ^{3}He polarizer and compressor. It will serve solid state as well as nuclear physics experiments at the ILL. ^{3}He spin filters would be particularly useful as polarizers and analyzers for pulsed neutron beams from spallation sources. Time of flight analysis and a large detector array would allow simultaneous measurement of doubly differential cross sections regarding scattering energy and angle. A ^{3}He analyzer close to the scatterer could analyze the polarzation of the scattered neutrons simultaneously at full solid angle of scattering. A single ^{3}He cell would thus replace the extremely expensive arrangement of an array of super mirrors used otherwise for polarization analysis of scattered neutrons.

6 MAGNETIC RESONACE TOMOGRAPHY WITH HYPERPOLARIZED NOBLE GASES

Two years ago W. Happers group at Princeton demonstrated in co-operation with the magnetic resonance imaging group at Duke University in a seminal paper the possiblity of imaging lung tissue filled with hyperpolarized noble gas [27]. The term "hyperpolarized" indicates that the noble gas has been polarized (by means of optical pumping) to a degree far beyond the ordinary Boltzmann equilibrium (1) which is of order 10^{-6} to 10^{-5} under usual NMR conditions. The first pilot experiments were performed with small samples of ^{129}Xe and ^{3}He polarized by rubidium spin exchange. The gas was inserted into the lungs of small test animals and NMR-pictures taken. The enormous gain in polarization degree by 5 orders of magnitude as compared to conventional NMR outweighs not only the loss in spin density between gas and tissue. It also offers enough signal capacity in order to feed the many hundreds consecutive NMR pulses which each pixel suffers in the process of image taking. For that purpose the magnetization is tilted only by a few degrees out of the field axis by each rf-pulse, such that the longitudinal magnetization is affected only marginally per pulse. The induction signal is still large enough! This procedure is enforced by the circumstance that the hyperpolarization can be used up only once during image taking. Once destroyed by resonance it will recover only to the tiny value of the Boltzmann equilibrium which is useless at these small densities.

On the other hand this circumstance allows to use a quite fast sequence of pulses separated only by the short transverse relaxation time T_2 which is shortened by diffusion through the field gradient to the order of milliseconds. In ordinary NMR the sequence would instead be governed by the much longer longitudinal relaxation time T_1 necessary for recovery of the magnetization after applying a strong resonant pulse. Thus it takes about half a minute for taking a highly resolved 3D picture of the human lung. This time matches well to the observed T_1 of ^{3}He and ^{129}Xe under these conditions. It also corresponds to the time a patient is capable of holding his breath.

NMR imaging works through encoding each pixel by a sequence of many different values of the field gradient in all three dimensions whereas the induction loop picks up the integral signal from the whole sample. The wanted spatial distribution of spins is then obtained by applying a Fourier transform to the whole sequence of induction signals which has been recorded as a function of the field gradients applied. Its not the place here to discuss all the details and tricks of this procedure. Instead one may memorize the following general remark: The primary signal observed in MRI is nothing but the form factor of the object in the space of gradient dependent Larmor frequencies – another application of the Zeeman effect !

In the meantime the new method has already been applied to humans by a Princeton Duke collaboration [28] and a Mainz/Heidelberg collaboration [29]. Fig. 13 shows a slice of the lungs of a volunteer who has inhaled about half a liter hyperpolarized ^{3}He [28]. Fig. 11 compares the lungs of a healthy volunteer on the left to the heavily damaged ones of a smoker on the right. The latter images were taken by a collaboration of our group with the radiology department at our university [30]. The healthy lungs show an almost constant signal intensity, i.e., homogenous ventilation

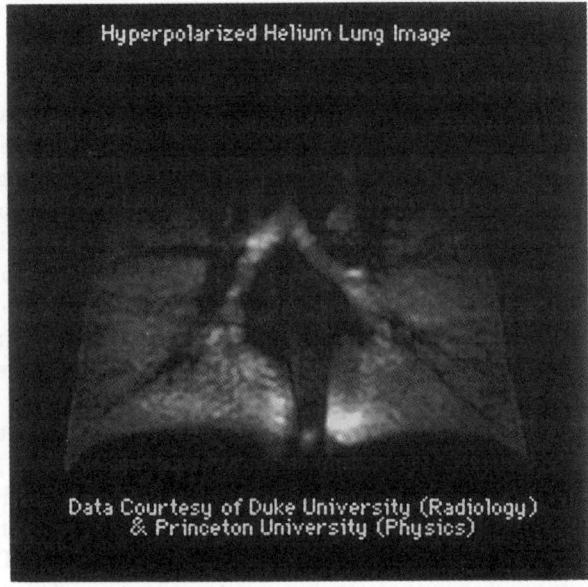

Figure 13: Magnetic resonance image of the human lung inflated with ^{3}He.

except for the blood vessels which show no signal (black). The smoker's lungs, on the other hand, show a large fraction of destroyed or obstructed regions, a diagnosis of high clinical relevance. Also tumors and tuberculosies have been identified in a first survey of patients.

The large interest raised in the medical community by this method stems from the fact that porous tissue like the lungs is difficult to image by conventional NMR. Also X-raying or scintigraphy from inhaled radioactive gas do not give truly satisfactory results for that task. X-raying suffers from poor contrast and scintigraphy from marginal resolution.

In contrast to Helium, Xenon is resorbed easily by the blood and transported to the whole body where it attaches in particular to lipids, for instance in the brain. Fortunately the relaxation time in the arterial blood is long enough ($T_1 \simeq 10\,\text{s}$) [31] for the Xenon to reach its destination still in a polarized state. Spectroscopic signals in the human brain have been obtained by several groups which include the observation of large chemical shifts of diagnostic relevance [31,32,33]. The Ann Arbor collaboration has managed to image the enflading of the brain of a rat by hyperpolarized ^{129}Xe [33]. Although ^{129}Xe - MRI is handicapped against ^{3}He - MRI by a much smaller signal power (a factor of 7.5 by the magnetic moments, a factor of 5 by the nuclear polarization reached at present and another factor of 4 by the isotopic abundancy), modern tomographs are still capable of picking up a reasonable signal. So it seems that MRI with hyperpolarized noble gases will soon conquer the clinics as well as biomedical research in a number of fields.

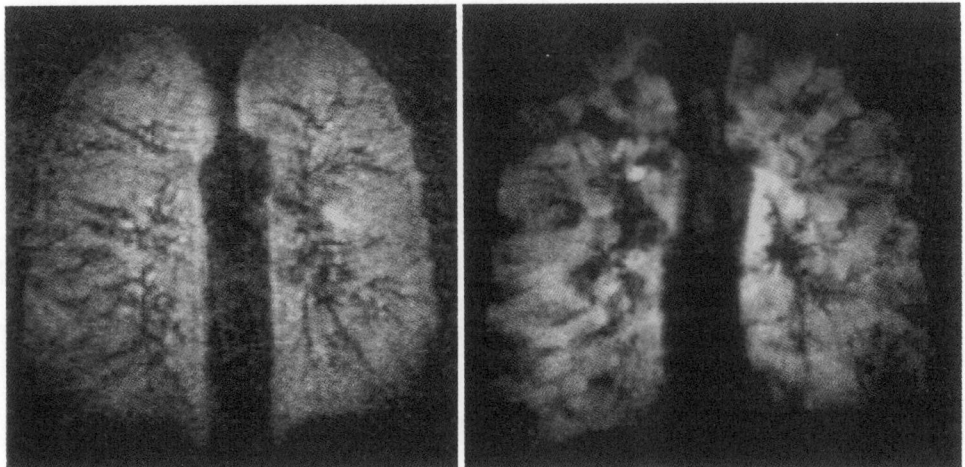

Figure 14: ^{3}He-MRI slices of human lungs, to the left of a healthy volunteer, to the right of a heavy smoker.

7 CONCLUSION

Optical pumping of noble gases has been subject of studies in atomic physics for very many years. With the advent of powerful pumping lasers it became possible and worthwhile to start a research and development program towards production of large quantities of spin-polarized noble gases. Both methods in the field, metastability as well as Rb spin exchange, have met the goal, partly in a competing, partly in a complementary way. Three major interdisciplinary fields have been started by this development:

1. The investigation of neutron structure by scattering electrons from polarized ^{3}He targets.

2. Establishing a ^{3}He spin filter for polarizing and analysing thermal and epithermal neutrons in a large phase space, serving all kinds of neutron scattering experiments.

3. An important spin-off arose from this fundamental research with the advent of magnetic resonance imaging of the lungs and other organs by hyperpolarized noble gases.

References

1. F.D. Colegrove, L.D. Schearer and G.K. Waltes, Phys. Rev. **132** (1963) 2561.
2. M. Bouchiat, T.-R. Carver and C.M. Varnum, Phys. Rev. Lett. **5** (1960) 463.
3. C.G. Aminoff, C. Larat, M. Leduc, B. Viana and D. Vivien, J. Luminescence **50** (1991) 21.

4. J. Becker W. Heil, B. Krug, M. Leduc, M. Meyerhoff, P.J. Nacher, E. W. Otten, Th. Prokscha, L.D. Schearer and R. Surkau, Nucl. Instrum. Meth. A **346** (1994) 45.
5. W. Heil H. Humblot, E.W. Otten, M. Schfer, R. Surkau and M. Leduc, Phys. Lett. A **201** (1995) 337.
6. P.J. Nacher and J. Dupont-Roc, Phys. Rev. Lett. **67** (1991) 2966.
7. E. J. Brash et al., Phys. Rev. C **47** (1993) 2064.
8. P.L. Anthony et al., Phys. Rev. Lett. **71** (1993) 959.
9. T.E. Chupp and M.E. Wagshul, Phys. Rev. A **40** (1989) 4447.
10. W.J. Cummings, O. Husser, W. Lorenzon, D.R. Swenson, and B. Larson, Phys. Rev. A **51** (1995) 4842.
11. G.D. Cates, Princeton, priv. comm.; M.V. Romalis et al., contributed to this conference.
12. E. Wilms, M. Ebert, W. Heil, and R. Surkau, subm. to Phys. Rev. A.
13. M. Gatzke, G.D. Cates, B. Driehuys, D. Fox, W. Happer, and B. Saam, Phys. Rev. Lett. **70** (1993) 690.
14. T.E. Chupp, R.J. Hoare, R.L. Walsworth, and Bo Wu, Phys. Rev. Lett. **72** (1994) 2363.
15. J.P. Jacobs, W.M. Klipstein, S.M. Lamoreaux, B.R. Heckel, and E.N. Fortson, Phys. Rev. A **52** (1995) 3521.
16. E. Reichert in *Proc. Int. Workshop on Polarized Beams and Polarized Gas Targets*, Köln, June 1995, ed. H. Paetz, gen. Schieck and L. Sydow, in press at World Scientic C., Singapore.
17. G. Eckert, W. Heil, M. Meyerhoff, E.W. Otten, R. Surkau, M. Werner, M. Leduc, P.J. Nacher and L.D. Schearer, Nucl. Instrum. Meth. A **320** (1992) 53.
18. J. Becker et al., Mainz, priv. comm.
19. M. Meyerhoff et al., Phys. Lett. B **327** (1994) 201.
20. M. Ostrick et al., Mainz, priv. comm.
21. C. de Jager, NIKHEF, Amsterdam, priv. comm.
22. L.A. Miller et al., Phys. Rev. Lett. **74** (1995) 502.
23. R.G. Milner, MIT and DESY, priv. comm.
24. K.P. Coulter, A.B. Mc Donald, W. Happer, T.E. Chupp and M. Wagshul, Nucl. Instr. and Meth. in Phys. Res. A **288** (1990) 463.
25. W. Heil and R. Surkau, *Int. Workshop on New Tools for Neutron Instrumentation*, Les Houches, France, June 95, in print at Journal of Neutron Research.
26. R. Surkau, J. Becker, M. Ebert, T. Großmann, W. Heil, D. Hofmann, H. Humblot, M. Leduc, E.W. Otten, D. Rohe, K. Siemensmeyer, M. Steiner, F. Tasset and N. Trautmann, submitted to NIM.
27. M.S. Albert, C.D. Cates, B. Driehuys, W. Happer, B. Saam, C.S. Springer Jr. and A. Wishnia, Nature, **370** (1994) 199.
28. G.D. Cates, Princeton, priv. comm.
29. M. Ebert, T. Großmann, W. Heil, E.W. Otten, M. Leduc, P. Bachert, M.V. Knopp, L.R. Schad, M. Thelen, The Lancet **347** (1996) 1297.
30. H.U. Kauczor, M.Ebert, K.F. Kreitner, H. Nilgens, R. Surkau, W. Heil, D. Hofmann, E.W. Otten, and M. Thelen, in print at Radiology.

31. M. Albert, V. Schepkin, and T. Budinger, J. Comput. Assist. Tomogr. **19** (1995) 975.
32. G.D. Cates, Princeton, priv. comm.
33. S.D. Swanson, M.S. Rosen, B.W. Agranoff, K.P. Coulter, R.C. Welsch, T.E. Chupp, preprint Ann Arbor.

THE CREATION AND STUDY OF BOSE-EINSTEIN CONDENSATION IN A COLD ALKALI VAPOR

C. E. WIEMAN, E. A. CORNELL, D. JIN, J. ENSHER,
M. MATTHEWS, C. MYATT, E. BURT, R. GHRIST
JILA, National Institute of Standards and Technology and University of Colorado,
and Department of Physics, University of Colorado, CB 440, Boulder, CO 80309

Bose-Einstein condensation in a gas resulted from the combination of research in laser cooling and trapping, the study of spin polarized hydrogen, and the investigation of atomic collisions at very low temperatures. After discussing how these advances led to BEC, we will cover recent measurements on BEC and experimental progress.

1 Bose-Einstein condensation in a gas (1924-1995)

In June of 1995, our group at JILA first observed Bose-Einstein condensation (BEC) in a gas of very cold rubidium atoms.[1] In the first part of this paper we will discuss how this was accomplished, and the previous work by many other groups that made it possible. In the second portion of the paper we will discuss work on BEC which has been carried out at JILA in the past year.

Although all the constituents of atoms (neutrons, protons, and electrons) are Fermions, if they are assembled such that the total spin of the atom is an integer, and the atoms remain far apart compared to the size of their electron clouds, they will behave as weakly interacting Bosons. The first discussion of the energy distribution of such Bosonic atoms when placed in a container was given by Einstein in 1924.[2] For a macroscopic container, the spacing between quantized energy levels is extremely small and thus at normal temperatures the atoms are distributed over a large number of different levels. At finite but very low temperatures however, the Bose-Einstein distribution formula predicts a large fraction of the atoms will go into the lowest energy level of the container (see Fig. 1). This is known as Bose-

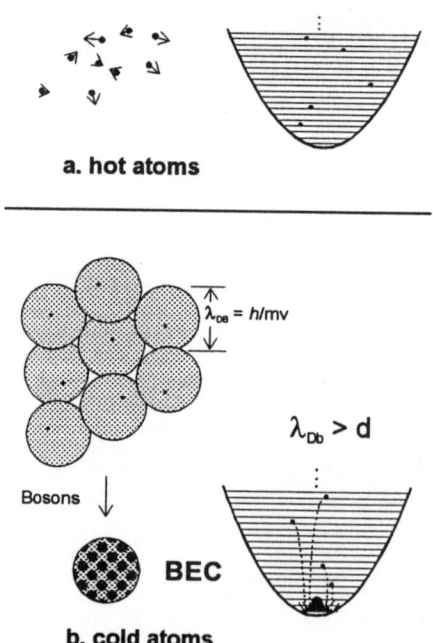

Figure 1) a. Distribution of hot atoms over the quantized energy levels in a macroscopic container. b. When Bosons are cooled sufficiently that the Debroglie wavelength, l_{Db}, is larger than the spacing between atoms, d, the atoms fall into the lowest energy state in the potential.

Einstein condensation. The condition for this to happen is that the atomic phase space density must be so large that the Debroglie wavelengths of the cold atoms are greater than the interparticle spacing.

London was the first to appreciate the real physical significance of BEC, and in particular that it would result in a large number of atoms in a single quantum state. He suggested that this macroscopic quantum behavior could explain the remarkable properties of superfluid helium and superconductivity. The primary motivation for our work was to explore this macroscopic quantum behavior of BEC in a gas, and to compare and contrast it with that of the other macroscopic quantum states we know,[3] particularly superfluid helium. Being a liquid, helium is quite different from the ideal gas discussed by Einstein. Because the atoms are very close together in the liquid, it is a strongly interacting system. These strong interactions are responsible for much of the interesting behavior we associate with superfluid helium, but the interactions also make it much more complicated to understand the macroscopic properties of superfluid helium in terms of the microscopic interaction between two helium atoms.

BEC in an alkali vapor is nearly the perfect tool for exploring how the microscopic interactions between atoms lead to the macroscopic properties of the many atom quantum state. Because the atoms in the condensate are far apart compared to their atomic size, the interactions are weak and well understood. Furthermore, we can readily adjust these interactions in experiments by a variety of means, the simplest of which is just to change the density of the gas. Finally, we have very good optical diagnostics for looking at the condensate and measuring its properties. This combination of factors makes this an excellent system for studying in detail how one goes from the microscopic to the macroscopic. Much of the design of our experiment was motivated by the desire to produce BEC in a simple manner which would facilitate carrying out experiments on it to explore this area of physics.

A gaseous BEC is also interesting in that it is the atomic analogue to laser light and shares the primary feature which makes laser light useful, namely very high phase space number density. This suggests the possibility of many laser-like interference measurements and applications.

The principle difficulty in producing BEC in a gas is that, at the desired low densities, the BEC transition temperature will be very low, on the order of 100 nK. Obviously this is a major technical challenge, but there is also the fundamental difficulty that at this temperature all atoms want to be a solid, not a gas. The solution to this problem is to "cheat" thermodynamics by creating a system with two very different time scales. The first scale is the time for the gas to come to thermal equilibration as a gas, and this should be very short. The second time scale is the time it takes for the vapor to go to its true equilibrium ground state (a solid), and this must be very long. Thus the gas will remain in its metastable super-saturated vapor state for a long time, during which it can Bose condense.

These considerations led us to the idea that these conditions could be satisfied by cooling alkali atoms by the combination of two different technologies: laser cooling and trapping, and magnetic trapping and evaporative cooling. Very briefly, we first produced BEC as follows. The heart of the apparatus is shown in Fig. 2. A magneto-optic trap (MOT) is created inside a glass vapor cell in the usual manner using light from diode lasers. The laser cooled and trapped atoms are then loaded into a magnetic trap; the laser light is turned off, and the atoms are then cooled through the BEC transition by evaporation.[4] To observe the cooled sample, we turn off the magnetic fields, allowing the atoms to fly apart. We then take a "shadow snapshot" of the expanded cloud. This

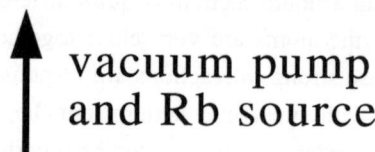

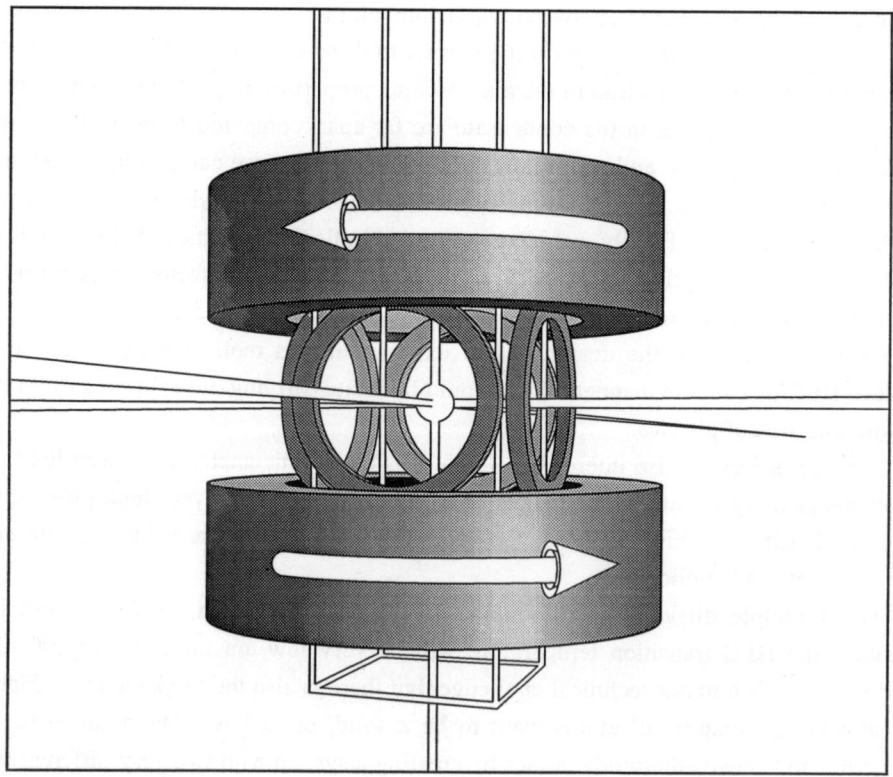

Figure 2) BEC trapping cell. A rectangular glass cell (2.5 cm square by about 10 cm high) is attached to a vacuum pump and rubidium reservoir (not shown). Laser beams coming from all six directions go through the cell. The magnetic fields are produced by the two large coils and the four smaller coils.

image is obtained by illuminating the expanded cloud with a very short pulse of laser light which is tuned to the resonant frequency of the atoms. The atoms absorb the light, thereby casting a shadow in the illuminating laser beam, and this shadow is imaged onto a CCD array (TV camera). This shadow image is the two-dimensional projection of the velocity distribution of the original cloud of atoms in the magnetic trap. From the velocity distribution we can extract the temperature and various other properties of the sample.

A set of three such pictures are shown in Fig. 3.[1] These correspond to three repetitions of the experiment, where the only difference is the amount of evaporative cooling. In the left-most picture, we have cooled the atoms only down to 200 nK, and what we see is a round hill, which looks like the familiar Maxwell-Boltzmann velocity distribution. At higher temperatures (not shown here), the cloud has the same shape with a larger width. The middle picture shows a cloud (~10,000 atoms) where the sample was cooled further, down to about 100 nK. On top of the rounded hill, a narrow spire has emerged which is centered at zero velocity. If we cool even further (right), we can produce a sample (~2000 atoms) in which the hill is completely gone, and only the narrow spire remains.

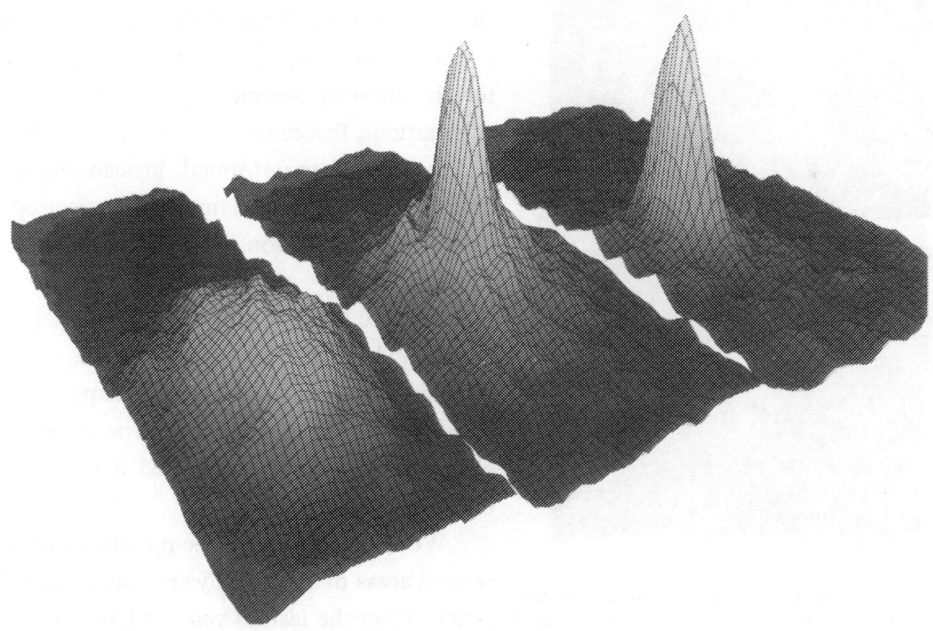

Figure 3) Two dimensional velocity distributions of the trapped cloud for three experimental runs, corresponding to temperatures of (left to right) 200 nK, 100 nK, and ~0 nK. The axes are the x and z velocities, and the number density of atoms per unit volume. This density is extracted from the measured optical thickness of the shadow. The original color versions can be seen on the JILA WWW home page at http://jilav1.colorado.edu/www/bose-ein.html, and on the 1996 APS calendar.

You can see how this behavior is exactly what one expects for BEC if you go back to the original concept illustrated in Fig. 1. The normal atoms that are distributed over many energy levels form the Maxwell-Boltzmann–like hill. The atoms in the lowest energy state of the potential are the most localized in both position and velocity space. Thus as atoms condense into that state, they form a very narrow peak in the velocity distribution, which sits on top of the broader hill of noncondensed atoms.

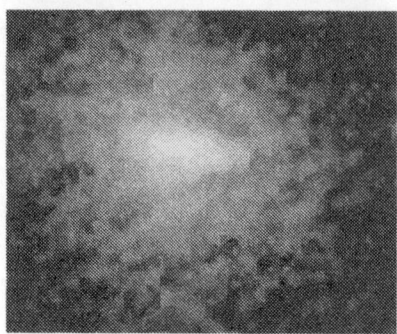

Other features of these velocity distributions indicate we are seeing BEC. One is the peak density of the trapped cloud as a function of temperature. This density is nearly constant as the temperature is lowered, until the transition temperature is reached. It then changes dramatically, increasing by a factor of 100 within 75 nK. This provides a strong indication of a phase transition. Another interesting aspect of the condensate is revealed by looking down on the peaks of Fig. 3 from above, as shown in Fig. 4. Figure 4a shows the contour lines of the rounded hill as circular, indicating an isotropic distribution for the thermal sample, as required by the equipartition theorem. Figs. 4b and 4c show that the spires are not round; instead they are quite elliptical, indicating an anisotropic velocity distribution. This elliptical distribution is an actual image of a macroscopic quantum wavefunction. It is elliptical because the shape of the wave function reflects the anisotropic shape of our harmonic trapping potential. These various observations clearly established that we had observed BEC.

Figure 4) Two D plot of x and z velocity distributions of the samples shown in Fig. 3.[1] Images shown are negatives of actual data, so brighter corresponds to more atoms (less transmitted lights).

We will now discuss the previous work in several areas of atomic physics that led to this result. Over the last 20 years, a large number of groups have been actively developing remarkable new capabilities for using laser light to both cool and confine atoms.[5] This includes major programs at ATT, NIST, JILA, ENS, Stanford, Institute of Spectroscopy (Moscow), MIT, and others. While many of these ideas and techniques are incorporated into our experiment, these advances have been discussed in many previous ICAP talks, and so we will not review them here.

Over roughly the same 20 year period, there was an active effort to produce BEC in a gas of spin polarized hydrogen. This work was inspired by the theoretical prediction of Stwalley and Nosanow that a gas of spin polarized hydrogen should remain a gas down to zero temperature, and therefore would be a possible candidate for producing gaseous BEC. This led to a number of groups attempting to cool and compress hydrogen by various means to get it to Bose condense.[6] There has been a large amount of experimental and theoretical work on this subject, involving groups at MIT, Amsterdam, Harvard, UBC, Cornell, Kurchatov, Eindhoven and others, with the most extensive efforts being at the first two. Many problems were encountered. These involved various processes by which the hydrogen atoms would lose their polarization and recombine. The groups developed increasingly clever methods for overcoming these processes, ultimately with the best performance being achieved through the use of magnetic trapping of the hydrogen atoms, and the invention of a new type of cooling called evaporative cooling. Using this technique, the MIT group achieved very impressive results, cooling nearly to the condensation temperature.[6]

The magnetic trap uses the interaction between the magnetic moments of the atoms and an appropriately configured inhomogeneous magnetic field to confine the atoms. In evaporative cooling, the most energetic atoms are allowed to escape out of the potential well, and in doing so they carry off more than their share of the energy. This leaves the remaining atoms colder. This is exactly how a cup of coffee cools – the energetic coffee molecules leap out of the cup into the room. In the hydrogen experiments, as in ours, the edge of the potential well over which the atoms escape is slowly lowered as the sample is cooled, thereby continually cooling the sample to lower and lower temperatures. The final temperature in the evaporative cooling is set by the final level of the potential.

Our interest in BEC arose from our investigations as to what was limiting the densities and temperatures which could be reached in laser traps, specifically MOTs. We learned that there were several relevant processes, all of which involved the scattering of photons between the atoms.[7] This made us realize we could get colder trapped atom samples by first laser cooling and trapping the atoms, and then removing the light and confining the atoms in a purely magnetic trap.[8] Although the magnetic confining force is weak, even modest fields can easily confine atoms which have been laser cooled to well below a milliKelvin.

In fact, it was the ease of creating such cold magnetic trapped samples that encouraged us to think about what more could be done with them. This brought to our attention the extensive work that had been done on spin polarized hydrogen in an effort to achievement Bose-Einstein condensation. So in about 1989 we started considering if it would be possible to apply these techniques to our cold alkali systems. When we looked carefully at the extensive work which had been done on hydrogen, we realized that hydrogen was not a special case. Actually all the processes that had hindered BEC in hydrogen were just examples of the spin polarized gas quenching to a lower energy state, and in this sense it was equivalent to any other atoms. This led to the realization that the key requirement for achieving BEC was to get the largest possible difference in equilibration time scales mentioned above, and this should determine the choice of atomic species. From looking at the hydrogen studies it was also possible to know what atomic processes were responsible for the two time scales. The equilibration of the atoms in the vapor came from the elastic scattering of atoms, and so this rate should be as large as possible. The quenching of the metastable state arose from inelastic collisions, which for the case of magnetically trapped, spin polarized atoms was dominated by dipole spin flip collisions.

With this guide, we realized that hydrogen was not only not unique, it was probably not even the best choice. Our hunch was that heavy alkali atoms, being large and fluffy, would have much larger elastic scattering rates, while having comparable spin flip rates. However, these ideas could only be hunches, because the relevant low temperature atoms physics was completely unknown in 1989. Over the subsequent several years there was a great deal of work done on understanding the very low energy collisions of alkali atoms. Notable contributions to this subject have come from many groups including those at Texas, Eindhoven, JILA, NIST, Connecticut, Wisconsin, MIT, and Maryland. This work made it clear that our original hunch was correct, and it also gave a much better idea as to what atomic densities and magnetic trap lifetimes were required to evaporatively cool to BEC. With these in mind we then developed a few optical and magnetic trapping and squeezing techniques that produced the necessary conditions.[9]

2 Bose-Einstein condensation results (1996)

The remainder of this paper will discuss recent results from our group. We have just finished construction of a new apparatus which has some advantages over the original system. This apparatus uses the same basic principle but has two separate MOTs.[10] The first one is in an upper chamber which has a relatively high pressure of rubidium vapor. The second is in a differentially pumped lower chamber which has very low pressure. The two chambers are connected by a long narrow tube, and after the atoms are trapped in the upper chamber they are then given a small push which sends them down the tube to be caught in the lower MOT. The atoms will stay in the lower MOT for many hundreds of seconds and so we can load many such bunches of atoms into it. This allows us to start with many more atoms, and have a longer trap lifetime for evaporative cooling. We then magnetically trap these atoms and evaporatively cool them. For the sake of variety we have used our old "baseball coil" magnetic trap in this apparatus rather than the TOP trap used in our first BEC machine. This is not a fundamental difference in that the two types of traps appear to be fairly similar in their ability to produce condensates. They each have technical advantages and disadvantages for any particular experiment. In the new BEC apparatus, we have now produced condensates containing about 2×10^6 atoms and further improvements are likely. Probably the most important aspects of this apparatus, relative to our first BEC apparatus, are that it still has the advantages of using simple and low cost diode laser and vapor cell MOT technology, and it also provides a much greater (some orders of magnitude) margin of error in the operating conditions under which condensates can be produced. As well as making experiments much easier, this has allowed us to easily produce condensates in both the F=2, m=2 and F=1, m=-1 spin states of ^{87}Rb. We are currently using these condensates to study heating processes and dipole spin flips, both of which involve interesting and somewhat subtle low temperature atomic physics.

We have also used the original BEC apparatus to carry out a series of measurements on condensates over the past year. These include studies of the collective excitations[11] of the condensate, measurement of the fraction of atoms in the condensate as a function of temperature,[12] and a determination of the specific heat of the sample as a function of temperature.[12] To observe the phonon-like collective excitations, we use a technique which is very similar to the free induction decay method of NMR. We first apply a periodic perturbation to the condensate and then watch the condensate oscillate freely after the perturbation is removed. The periodic perturbation is produced by applying time dependent magnetic fields which gently squeeze the condensate. We have excited modes with two different symmetries. The monopole or m=0 mode is excited by uniformly squeezing and stretching the condensate in the isotropic x-y plane. The quadrupole or m=2 mode is excited by applying sinusoidal squeezes along the x and y directions which are 90 degrees out of phase. This excites a mode which corresponds to an elliptical distortion of the condensate that rotates in the plane. By waiting varying amounts of times after the end of the perturbation before we release the cloud and look at it, we can determine the time dependence of the spatial distortion of the condensate. Figure 5 shows the distortion as

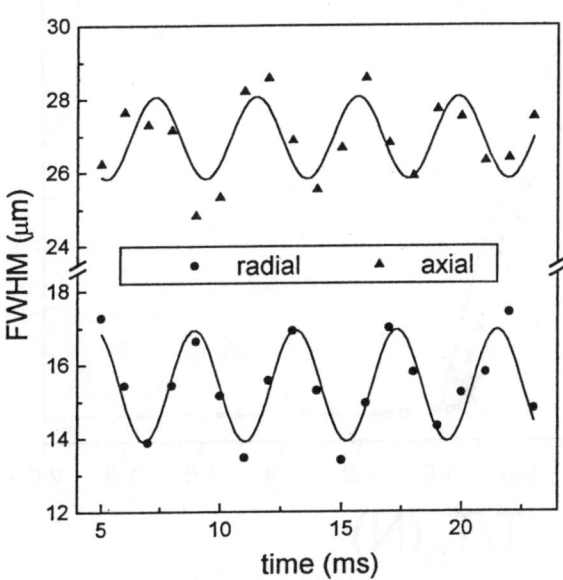

Figure 5) Time dependence of the radial and axial width of the condensate during the free oscillation period, after the m=0 mode has been excited.

a function of time. It can be seen that it is a very clean damped sinusoidal oscillation. By fitting this curve to a damped sine wave we determine the oscillation frequency and the damping time of the excitation.

We have measured the oscillation frequencies of the m=0 and m=2 modes as a function of the interactions in the condensate. The excitation frequencies are dramatically shifted from the value corresponding to a noninteracting gas, which is also the excitation frequency of the noncondensed trapped atoms. The observed condensate excitation frequencies agree well with theory over the full range of interaction strengths.[11] We have measured the damping of the m=0 mode and compared it with the damping we observe in a corresponding excitation of a cloud just above the transition temperature. We find the condensate excitation persists for a factor of 4 longer than a thermal cloud of the same density. Currently there is no theoretical prediction for the damping (or even a clear understanding of the relevant mechanism), although it is a subject of considerable current theoretical activity. It will be interesting to study these excitations further to examine such features as how the damping depends on temperature,

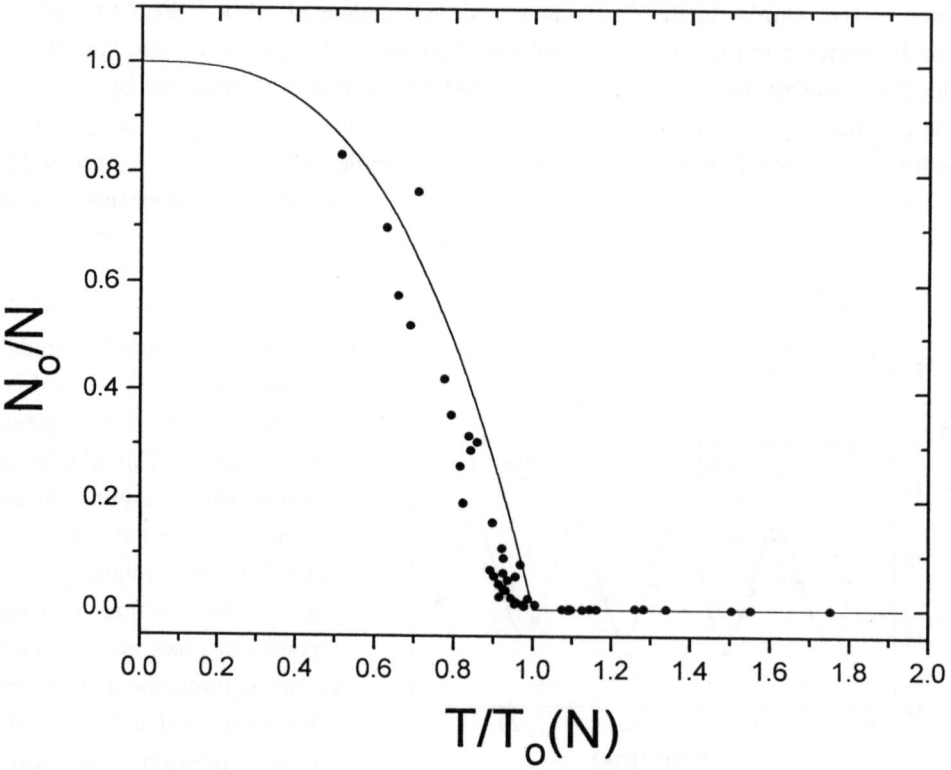

Figure 6) The fraction of atoms in the condensate as a function of temperature. The temperature is in units of the critical temperature of an ideal gas in a harmonic potential in the thermodynamic limit. The solid line on the graph shows fraction for this ideal gas case, $1-(T/T_c)^3$.

the behavior of higher frequency modes, and the excitation of vortices. Ultimately it should be possible to completely understand the quantum fluid dynamics of this system.

In a second set of measurements, we have studied the thermodynamics of the phase transition by looking at both the population of the ground state as a function of temperature and the total energy in the trapped cloud, again as a function of temperature. Surprisingly, the major difficulties and time required in this work are in data analysis, rather than data acquisition. These measurements require accurate absolute determinations of detailed shape of the velocity distribution. This is a major problem in imaging and requires careful handling of a host of subtle calibration and distortion issues such as focusing, magnification, saturation, absorption, optical aberrations, light polarization, atomic polarization, resolution, detuning, backgrounds, atomic motion, and time dependent magnetic field effects. We have carefully studied each of these effects and corrected the images as required. From these processed absorption images we then obtain the density distribution, the total number of atoms, and the temperature of the sample. We determine the temperature by fitting the wings of the cloud to a Bose-Einstein distribution (which for the wings looks nearly identical to a Gaussian) with the temperature as a fitting parameter. We determine the fraction of atoms in the condensate by integrating over the full image to get the total number of atoms in the sample and comparing that with the number of atoms in the central narrow peak. The results are shown in Fig. 6. The solid line is the value predicted for an ideal gas in a harmonic potential in the thermodynamic limit. It should be emphasized that this is not a fit; there are no adjustable parameters. It can be seen that there is good agreement with the general shape, but the measured condensation temperature is 0.94(5) of this ideal gas case. It has been proposed that interactions and the finite size of the sample will affect this curve, but the detailed predictions have varied considerably. At this point, we can only conclude that any such effects are small, although there may be a slight lowering of the transition temperature. In the future we hope to reduce our uncertainties to the 1% level to further explore these issues.

In Fig. 7 we show the total energy of the cloud as a function of temperature. For this measurement we find the second moment of the cloud to get the kinetic energy after it was released, which gives the total energy in the cloud before release. To remove the trivial dependence of the energy on number of atoms (which varies), we have plotted the energy per atom scaled by the ideal gas critical temperature. The specific heat is the derivative of this curve, although here it is the specific heat at constant trapping potential, as opposed to the usual constant volume or pressure. Although the data is noisier than one would like, it is clear that there is the discontinuous change in the specific heat at the transition temperature predicted for the BEC transition, but unlike the case of liquid helium it does not diverge. For comparison we have calculated the corresponding total energies for samples of Maxwell-Boltzmann particles and noninteracting Bosons. As shown, they both agree well with the data at high temperatures, but differ considerably

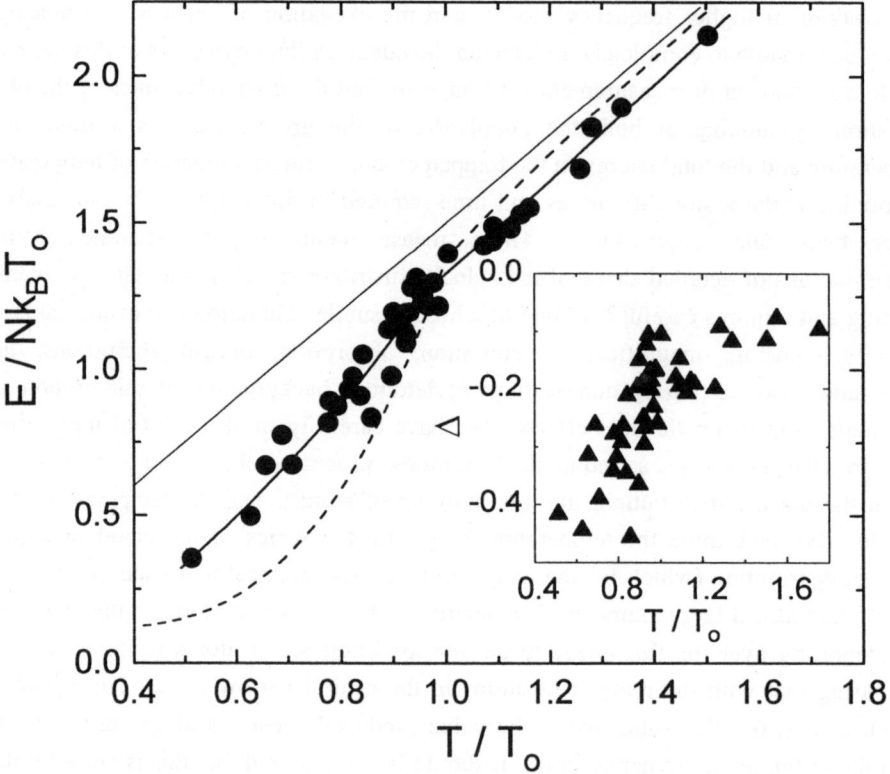

Figure 7) Scaled energy per atom in the cloud versus temperature. The straight solid line is the calculated curve for Maxwell-Boltzmann particles, the dashed line is the calculated curve for an ideal Bose gas. The inset shows the difference between the data and the M.-B. curve. The curved solid line is an arbitrary polynomial fit to the data.

from the observations at low temperature. It will be interesting to calculate the energy for an interacting Bose gas and compare with this data.

Many different contributions to atomic physics have made it possible to produce BEC in atomic vapors. In the future it should be possible to routinely create condensates and study their many remarkable properties. The work reported here is a small step in that direction.

Acknowledgments
This work has been supported by NSF, ONR, and NIST.

References

1) M. H. Anderson, J. R. Ensher, M. R. Matthews, C. E. Wieman and E. A. Cornell, *Science* **269**, 198-201 (1995).

2) A. Einstein, Sitzber. Kg. Preuss. Akad. Wiss. (1924), p. 261; (1925), p. 3.

3) In addition to superfluid helium, two other examples are superconductivity and BEC in excitons. The latter work is discussed in J.-L. Lin and J. P. Wolfe, *Phys. Rev. Lett.* **71**, 1222-1229 (1993). For an extensive discussion of BEC in many different systems, see "Bose-Einstein Condensation," Eds., A. Griffin, D. Snoke and S. Stringari (Cambridge University Press, Cambridge, 1995).

4) Since this conference is dedicated to the discovery of the Zeeman effect, we might note that this effect plays a major role in this experiment. The MOT is based on using the Zeeman effect to control the radiation pressure force and give it a spatial dependence, while the magnetic trap is based entirely on the shift of the energy levels of the atoms due to a magnetic field.

5) See the following volumes for numerous articles and references on the subjects of laser cooling and trapping. N. R. Newbury and C. E. Wieman, *Am. J. Phys.* **64**, 18-20 (1996) (this article also provides extensive references on magnetic trapping); C. Wieman and S. Chu, Eds., Special Issue on Laser Trapping and Cooling, *J. Opt. Soc. Am. B* Vol. **6**, No. 11 (1989); Proceedings, Enrico Fermi International Summer School on Laser Manipulation of Atoms and Ions, Varenna, Italy, Eds., E. Arimondo, W. Phillips, and F. Strumia (North Holland, Amsterdam, 1992).

6) For a review of the hydrogen work, see T. Greytak, pg. 131 in "Bose-Einstein Condensation," and references therein, Eds. A. Griffin, D. Snoke and S. Stringari (Cambridge University Press, Cambridge, 1995).

7) D. Sesko, T. Walker, C. Monroe, A. Gallagher and C. Wieman, Phys. Rev. Lett. **63**, 961-964 (1989); T. Walker, D. Sesko and C. Wieman, Phys. Rev. Lett. **64**, 408-411 (1990); D. Sesko, T. Walker and C. Wieman, J. Opt. Soc. Am. B **8**, 946-958 (1991).

8) C. Monroe, W. Swann, H. Robinson and C. Wieman, Phys. Rev. Lett. **65**, 1571-1574 (1990).

9) M. H. Anderson, W. Petrich, J. R. Ensher, and E. A. Cornell, Phy. Rev. A. **50**, R3597-3600 (1994); N. Petrich, M. H. Anderson, J. R. Ensher and E. A. Cornell, J. Opt. Soc. Am. B **11**, 1332-1335 (1994); W. Petrich, M. H. Anderson, J. R. Ensher and E. A. Cornell, Phys. Rev. Lett. **74**, 3352-3355 (1995).

10) C. J. Myatt, N. R. Newbury, R. W. Ghrist, S. Loutzenhiser and C. E. Wieman, Optics Letters **21**, 290 (1996).

11) D. S. Jin, J. R. Ensher, M. R. Matthews, C. E. Wieman and E. A. Cornell, *Phys.. Rev. Lett.* **77**, 416 (1996).

12) J. R. Ensher, D. S. Jin, M. R. Matthews, C. E. Wieman and E. A. Cornell, Univ. of Colorado. To be published.

Macroscopic quantum phenomena in trapped Bose-condensed gases

Yu. Kagan[1], E.L. Surkov[1], and G.V. Shlyapnikov[1,2]
(1) *Russian Research Center Kurchatov Institute,*
Kurchatov Square, 123182 Moscow, Russia
(2) *Van der Waals - Zeeman Institute, University of Amsterdam,*
Valckenierstraat 65-67, 1018 XE Amsterdam, The Netherlands

We discuss the possibilities of investigating macroscopic quantum phenomena in low-temperature Bose-condensed gases, which is one of the main reasons to study these systems. Attention is focused on dynamic and kinetic properties of trapped gases, such as those manifesting themselves in the response of the system to time-dependent variations of the confining field. We put an emphasis on the difference in the behavior of Bose-condensed gases from that of ordinary thermal samples. Correlation properties of the evolving Bose-condensed gas in time-dependent traps and the phenomenon of stochastization in the condensate evolution are analyzed.

1 Introduction

The recent successful experiments on Bose-Einstein condensation (BEC) in trapped ultra-cold alkali atom gases [1-3] have generated a lot of interest in the physics of dilute Bose-condensed systems. The second set of experimental studies in Rb [4] and Na [5-7] vapors already provides us with more detailed information on the properties of Bose-condensed trapped gases, and we are now well positioned to understand why and how they behave differently compared to ordinary thermal gases. Macroscopic quantum phenomena in trapped Bose-condensed gases can be discussed in various aspects. For example, one can consider thermodynamic and optical properties of a static trapped gas or the phenomenon of superfluidity. But we should emphasize here that the presence of the trapping field makes the situation unique, since the field variations can strongly influence trapped condensates and, hence, generate new phenomena emphasizing the macroscopic quantum nature of a Bose-condensed state. Just this circumstance allowed one to reveal the presence of a condensate in Rb [1] and Na [3] experiments through studying the gas expansion after switching off the trap and is now actively used in ongoing investigations. Therefore, we will focus our attention on macroscopic quantum phenomena manifesting themselves in the response of a Bose-condensed gas to time-dependent variations of the confining field.

At present related theoretical studies include the analysis of ground-state properties [8-13] and elementary excitations [14-17] of a static trapped Bose-condensed gas and numerical solutions of mean field equations for the condensate and elementary excitations [9,18-22]. The mean field theory describing the condensate evolution under arbitrary frequency variations in isotropic harmonic traps has been developed in [23]. Generalizing this theory to the case of anisotropic time-dependent traps we will show that the evolution of a condensate is completely different from the evolution of collisionless thermal gases. On the contrary, in the hydrodynamic regime the evolution of a thermal gas is in many aspects similar to that of the condensate. In this case the most distinct feature of BEC remains the characteristic velocity of the evolution: As well as in the collisionless regime, at temperatures much larger than the mean field interaction between particles the condensate evolves much slower than the thermal cloud. We will discuss the phenomenon of stochastization in the condensate evolution and address the problem of relaxation of the evolving condensate. It will be shown that clear signatures of BEC and the effects of the loss of coherence can be found in local correlation properties of the evolving gas through the measurement of the rates of intrinsic or light-induced inelastic collisional processes.

2 Static trapped condensate

We first give a brief outline of the properties of a static trapped condensate. The character of BEC is influenced by the presence of discrete trap levels and the interaction between particles. Densities $n \sim 10^{12} - 10^{14}$ cm^{-3} achieved in current BEC experiments are well in the regime of a weakly interacting gas, where the mean interparticle separation is much larger than the characteristic radius of interatomic interactions (~ 100 Å). Hence, the energy of interparticle interaction in the system will be a sum of pair interactions, and the latter can be found from the solution of a pair scattering problem. This leads to the interaction energy per particle in the condensate, $n_0(\mathbf{r})\tilde{U}$, where $n_0(\mathbf{r})$ is the condensate density, and $\tilde{U} = 4\pi\hbar^2 a/m$, with m being the atom mass and a the scattering length (scattering amplitude in the zero energy limit, which is roughly of the order of above mentioned radius of interaction). We will further assume $|a| \ll l_0$, where l_0 is the amplitude of zero point oscillations in the trap. This makes the condition of a weakly interacting gas, $n_0|a|^3 \ll 1$, compatible with the assumption of a macroscopically large number of particles in the condensate,

$$N_0 \gg 1. \tag{1}$$

We now clearly see that the character of BEC depends on the dimensionless parameter (see, e.g., [11])

$$\eta = \frac{n_0|\tilde{U}|}{\varepsilon_0}, \tag{2}$$

which is the ratio of the mean interaction energy per particle to the characteristic level spacing $\varepsilon_0 \approx \hbar^2/ml_0^2$, n_0 being the maximum condensate density. The presence of a macroscopic number of particles in the condensate allows us to describe the condensate in terms of macroscopic wavefunction $\Psi_0(\mathbf{r})$ which is a c-number and in some sense represents the mean value of the field operator of atoms (see, e.g., [24]). The condensate density is given by $n_0(\mathbf{r}) = |\Psi_0(\mathbf{r})|^2$. For $\eta \ll 1$ the interaction is not important, and BEC can be regarded as macroscopic occupation of the ground state of the trapping potential. In a spherically symmetric harmonic trap this condition requires $1 \ll N_0 \ll l_0/|a|$.

If $\eta > 1$, the elastic interaction between particles comes into play. A question of principal importance concerns the stability of the condensate with respect to this interaction. Repulsive interaction ($a > 0$) makes the condensate stable, as in this case the transfer of a particle from the condensate to any other state should lead to increasing energy of the system. The shape of the condensate wavefunction strongly changes with increasing N_0. For $\eta \gg 1$, which in a spherically symmetric harmonic trap requires $N_0 \gg l_0/|a|$, the structure of trap levels becomes unimportant as they will be smeared out by the interaction between particles. In this case the correlation length

$$l_c = \frac{\hbar}{(2mn_0\tilde{U})^{1/2}} \approx \frac{l_0}{(2\eta)^{1/2}} \ll l_0, \qquad (3)$$

and the kinetic energy of the condensate can be neglected. Accordingly, the condensate density profile is determined by the condition [25,26]

$$n_0(\mathbf{r})\tilde{U} + V(\mathbf{r}) = \mu, \qquad (4)$$

where V is the trapping potential, and the chemical potential $\mu = n_0\tilde{U}$.

For attractive interaction between particles ($a < 0$) the picture drastically changes. A Bose condensate with $\eta > 1$, for which the discrete structure of trap levels is not important, can not be formed at all, since in this case the accumulation of particles in one quantum state would be associated with an increase of energy. On the other hand, the case $\eta \ll 1$ is characterized by the presence of an energy gap ε_0 for one-particle excitations. In this case it is possible to form a metastable Bose-condensed state [9,11]. This state is separated by a large energy barrier from lower states, which ensures a long characteristic lifetime of the metastable condensate [11].

We will mainly discuss the case of repulsive ($a > 0$) and large ($\eta \gg 1$) interaction between particles. In this case the boundary of the condensate spatial region is determined by the condition $\mu = V(\mathbf{r}_0)$, which in a spherically symmetric harmonic trap with frequency ω_0 gives a sphere with radius $\bar{r}_0 = (2\mu/m\omega_0^2)^{1/2}$. At temperatures $T \gg n_0\tilde{U}$ it is much smaller than the characteristic thermal size of the gas $\bar{r}_T$ which can be found from the relation $T \approx V(\bar{r}_T)$. For example, in a harmonic trap we have $\bar{r}_T/\bar{r}_0 \approx (T/n_0\tilde{U})^{1/2}$.

3 Evolution of a condensate and scaling dynamics

We first consider the evolution of a condensate with fixed number of particles in an anisotropic harmonic potential $V(\mathbf{r}) = m\sum_i \omega_i^2 r_i^2/2$ with time-dependent frequencies $\omega_i(t)$. The equation for the condensate wavefunction, with above-condensate particles neglected, reads

$$i\hbar\frac{\partial\Psi_0}{\partial t} = -\frac{\hbar^2}{2m}\Delta\Psi_0 + \frac{m}{2}\sum_i \omega_i^2(t)r_i^2\Psi_0 + \tilde{U}|\Psi_0|^2\Psi_0. \qquad (5)$$

To solve Eq.(5) we generalize the method developed in [23] for isotropic traps and introduce d scaling parameters $b_i(t)$ (instead of one), where d is the dimension of the system. A similar procedure was recently used in [27]. Turning to new coordinates $R_i = r_i/b_i(t)$ we search for the solution in the form

$$\Psi_0(\mathbf{r},t) = \frac{1}{\sqrt{\mathcal{V}(t)}}\chi_0(\mathbf{R},\tau(t))\exp{(i\Phi(\mathbf{r},t))}, \qquad (6)$$

where the dimensionless volume $\mathcal{V}(t) = \prod_i b_i(t)$ and $\tau(t) = \int^t dt'/\mathcal{V}(t')$. Substituting Eq.(6) into Eq.(5) we require the cancellation of the $\nabla_R\chi_0$ terms, which gives the phase

$$\Phi(\mathbf{r},t) = \frac{m}{2\hbar} \sum_i r_i^2 \left[\frac{\dot{b}_i(t)}{b_i(t)}\right]. \tag{7}$$

Along the lines of [23], introducing the equations for the scaling parameters:

$$\ddot{b}_i + \omega_i^2(t) b_i = \frac{\omega_{0i}^2}{b_i \mathcal{V}(t)}, \tag{8}$$

with initial conditions $b_i(0) = 1$, $\dot{b}_i(0) = 0$ and initial frequencies ω_{0i}, we arrive at the equation of motion

$$i\hbar \frac{\partial \chi_0}{\partial \tau} = -\frac{\hbar^2}{2m} \sum_i \frac{\mathcal{V}(t)}{b_i^2(t)} \frac{\partial^2 \chi_0}{\partial R_i^2} + \frac{m}{2} \sum_i \omega_{0i}^2 R_i^2 \chi_0 + \tilde{U} |\chi_0|^2 \chi_0. \tag{9}$$

For strong interaction between particles, where the initial chemical potential $\mu = n_0 \tilde{U} \gg \hbar \omega_{0i}$, the ratio of the kinetic energy term to the non-linear interaction term in Eq.(9) is initially very small and scales as $\varepsilon(t) = \sum_i (\hbar \omega_{0i}/\mu b_i(t))^2 \mathcal{V}(t)$. Hence, if the condition $\varepsilon(t) \ll 1$ is satisfied with increasing t, the kinetic energy term can be omitted and in the variables R_i, τ the problem is reduced to an interacting Bose gas in a harmonic well with constant (initial) frequencies. With $\chi_0(\mathbf{R}, \tau(t)) = \overline{\chi}_0(\mathbf{R}) \exp(-i\mu\tau(t))$, Eq.(9) goes over into Eq.(4) in which one should substitute $|\overline{\chi}_0(\mathbf{R})|^2$ for the density. Accordingly, we have

$$\overline{\chi}_0(\mathbf{R}) = \frac{1}{\tilde{U}^{1/2}} \left(\mu - \frac{m}{2} \sum_i \omega_{0i}^2 R_i^2\right)^{1/2}; \quad R_i = \frac{r_i}{b_i(t)} \tag{10}$$

in the spatial region where the argument of the square root is positive and zero otherwise. Thus, Eqs. (6) and (10) give a universal scaling solution for $\Psi_0(\mathbf{r},t)$ under arbitrary variations of the frequencies and anisotropy of the external harmonic potential.

In contrast to thermal gases, in the case of a condensate we are dealing with the evolution of a macroscopic wavefunction which conserves the phase coherence. On the other hand, the evolution dynamics itself is governed by *classical* equations (8) which are in fact equations of motion following from the *classical* Hamiltonian of "scaling dynamics"

$$\mathcal{H}_{sd} = \frac{1}{2} \sum_i (p_i^2 + \omega_i^2(t) q_i^2) + \frac{\overline{\omega}_0^2}{\prod_i q_i}. \tag{11}$$

Here $\overline{\omega}_0 = (\prod_i \omega_{0i})^{1/d}$ is the geometrical mean of initial frequencies and $q_i = (\overline{\omega}_0/\omega_{0i}) b_i$, $p_i = \dot{q}_i$. Notice that $\prod_i q_i = \mathcal{V}$. The Hamiltonian $\mathcal{H}_{sd}$ describes d classical harmonic oscillators coupled to each other through the non-linear term of volume scaling $\overline{\omega}_0^2/\mathcal{V}$.

Remarkably, in order to understand macroscopic quantum behavior of a Bose-condensed gas in the course of evolution one should solve a non-linear classical problem.

4 BEC signatures in the evolution picture. Collisionless regime

The evolution of a condensate is very different from that of collisionless thermal gases where the mean free path of a particle is much larger than the sample size. Thus, if the thermal component of a Bose-condensed gas is in the collisionless regime the evolution picture shows a number of signatures of BEC. We will demonstrate this by using two simple examples.

The first one deals with resonance frequencies of small density (shape) oscillations, such as those observed in the recent JILA [4] and MIT [7] experiments. In fact, they are the eigenfrequencies of small oscillations, corresponding to the Hamiltonian (11) with $\omega_i(t) = \omega_{0i}$. This Hamiltonian is minimized at $q_i = \overline{\omega}_0/\omega_{0i}$, $p_i = 0$, and in the vicinity of this point we arrive at the quadratic form which gives the third-order secular equation for the frequencies

$$Det\left[1 + \left(2 - \frac{\Omega^2}{\omega_{0i}^2}\right)\delta_{ij}\right] = 0. \quad (12)$$

In the case of cylindrical symmetry we obtain one uncoupled oscillation with the projection of orbital angular momentum on the symmetry axis, $M = 2$, and two coupled oscillations with $M = 0$. The corresponding frequencies are given by

$$\Omega_0 = \sqrt{2}\omega_{0r}; \quad \Omega_\pm = \frac{\omega_{0r}}{\sqrt{2}}\left[\left(4 + 3\beta^2 \pm \sqrt{9\beta^4 - 16\beta^2 + 16}\right)\right]^{1/2}, \quad (13)$$

where $\beta = \omega_{0z}/\omega_{0r}$ is the ratio of the axial to radial trap frequency. This result was also found in [15] from the analysis of elementary excitations in the hydrodynamic approach and in [27] by considering the condensate evolution under weak modulation of the trap frequencies. Resonance frequencies Ω_0 and Ω_- were observed for Rb condensate in the JILA experiment [4] for the ratio $\beta = \sqrt{8}$ ($\Omega_- \approx 1.8\omega_{0r}$) and calculated numerically in [21] for this trapping geometry. The frequencies $\Omega_\pm$ were observed in the Na experiment at MIT [7] for $\beta = 0.08$ ($\Omega_+ \approx 2\omega_{0r}$, $\Omega_- \approx 1.58\omega_{0z}$).

In the collisionless thermal cloud axial oscillations occur at frequency $2\omega_{0z}$, and radial oscillations with both $M = 0$ and $M = 2$ at frequency $2\omega_{0r}$. Significant difference between these frequencies and the condensate frequencies (13) is promising for experimental identification of the presence of the condensate and measuring its oscillation spectrum.

The frequencies (12),(13) are independent of the interaction between particles, although we are considering the case of strong interaction. This is a consequence of harmonicity of the external potential. In any other trapping field the dependence on the interaction will be pronounced.

Another example is the expansion of the condensate after abrupt switching off the trap ($\omega_i(t > 0) = 0$). As follows from the analysis of Eqs. (8), on a time scale $t \gg \omega_{max}^{-1}$, where ω_{max} is the highest initial frequency, the expansion becomes free in all directions and for each scaling parameter we have $\dot{b}_i = \gamma_i \omega_{0i}$. In the spherically symmetric 3-d case the numerical coefficient $\gamma = \sqrt{2/3}$ [23], and the instantaneous size of the condensate $r_0(t) = \overline{r}_0 b(t) = (4\mu/3m)^{1/2} t$. Since $\mu = n_0 \tilde{U}$, the velocity of expansion of the condensate boundary, $v_0 = \dot{r}_0(t) = (4n_0\tilde{U}/3m)^{1/2}$, is determined by the interaction between particles and governed by the characteristic sound velocity in the condensate $c_s = (n_0 \tilde{U}/m)^{1/2}$. The same statement holds for anisotropic traps. But in this case due to anisotropy in the initial density gradients the velocity depends on the direction of expansion: $v_{0i} = \gamma_i \sqrt{2} c_s$. In cylindrically symmetric traps the asymmetry of free expansion is characterized by the ratio of the axial to radial size, γ_z/γ_r. Our results for this quantity as a function of β are presented in Fig.1. In the limiting case $\beta \ll 1$, investigated in the recent MIT experiment [5], we obtain $\gamma_z/\gamma_r = \pi\beta/2$ (see also [27]). The expansion predominantly occurs in the radial direction, and the initially cigar-shaped condensate becomes pancake-shaped.

A collisionless thermal cloud, for any initial anisotropy of the trap, undergoes symmetric free expansion with thermal velocities $v_T \approx \sqrt{2T/m}$. For this reason, the observation of asymmetry in the expansion of the (low density) gas in the JILA experiment [1] was the key evidence for BEC. Theoretically, asymmetric expansion of the condensate was first shown in numerical solution of Eq.(5) [20].

At temperatures $T \gg n_0 \tilde{U}$ the thermal velocity $v_T \gg c_s$, and, hence, the condensate expands much slower than the thermal cloud. This was another signature of BEC, observed in the experiment [1].

5   Hydrodynamic regime

At sufficiently high densities the thermal cloud will be in the hydrodynamic regime where the mean free path of a particle is much smaller than the sample size. Already, current Na experiments at MIT [5,6,7] ($n \sim 10^{14}$ cm^{-3}) are close to this limit. As it turns out, the evolution of hydrodynamic thermal gases is in many aspects similar to the evolution of the condensate. In particular, the frequencies of small shape oscillations and the asymmetry of free expansion are almost the same. Neglecting dissipation, this can be found from the Euler equation of motion for the gas in the anisotropic parabolic potential:

$$\frac{\partial v_i}{\partial t} + \sum_j v_j \frac{\partial v_i}{\partial r_j} + \omega_i^2(t) r_i + \frac{1}{mn(\mathbf{r},t)} \frac{\partial P(\mathbf{r},t)}{\partial r_i} = 0, \qquad (14)$$

where $n(\mathbf{r},t)$ and $P(\mathbf{r},t)$ are the density and pressure profiles, and $v_i(\mathbf{r},t)$ the velocity field. We will again search for the scaling solution, now directly for the density profile:

$$n(\mathbf{r},t) = \frac{\tilde{n}(\mathbf{R})}{\mathcal{V}(t)}, \qquad (15)$$

with $\tilde{n}(\mathbf{r})$ being the initial density distribution and $R_i = r_i/b_i(t)$ (cf. Eq.(6)). Then the local velocity satisfying the continuity equation can be written as $v_i = r_i[\dot{b}_i(t)/b_i(t)]$. Assuming that the evolution occurs adiabatically, the "continuity equation" for the entropy in the case of a Boltzmann gas leads to the profile of pressure $P(\mathbf{r},t) = \tilde{P}(\mathbf{R})/\mathcal{V}^{5/3}(t)$. In the initial static gas the profiles of density and pressure are related to each other by

$$m\tilde{n}(\mathbf{R})\omega_{0i}^2 R_i + \frac{\partial \tilde{P}(\mathbf{R})}{\partial R_i} = 0.$$

This reduces the Euler equation (14) to the equations for the scaling parameters:

$$\ddot{b}_i + \omega_i^2(t) b_i = \frac{\omega_{0i}^2}{b_i \mathcal{V}^{2/3}(t)}, \qquad (16)$$

somewhat different from Eqs.(8) for the evolution of the condensate.

Eqs.(16) can be derived from the classical Hamiltonian of scaling dynamics

$$\tilde{\mathcal{H}}_{sd} = \frac{1}{2} \sum_i (p_i^2 + \omega_i^2(t) q_i^2) + \frac{3\overline{\omega}_0^2}{2 \left(\prod_i q_i \right)^{2/3}}, \qquad (17)$$

which immediately allows us to find the secular equation for the resonance frequencies of small shape oscillations:

$$\mathrm{Det} \left[\frac{2}{3} + \left(2 - \frac{\tilde{\Omega}^2}{\omega_{0i}^2} \right) \delta_{ij} \right] = 0. \qquad (18)$$

In the case of cylindrical symmetry we again have one uncoupled ($M=2$) and two coupled ($M=0$) oscillations, with frequencies

$$\tilde{\Omega}_0 = \sqrt{2} \omega_{0r}; \quad \tilde{\Omega}_\pm = \frac{\omega_{0r}}{\sqrt{3}} \left[(5 + 4\beta^2 \pm \sqrt{16\beta^4 - 32\beta^2 + 25}) \right]^{1/2}. \qquad (19)$$

The former is the same as in the condensate, and the difference between $\tilde{\Omega}_\pm$ and $\Omega_\pm$ is less than 10% for any β. Moreover, numerical analysis of Eq.(18) shows that also in the absence of cylindrical symmetry the maximum difference between $\tilde{\Omega}$ and the corresponding Ω does not exceed 10% for any combination of the trap frequencies. Thus, if the thermal cloud is in the hydrodynamic regime, experimental determination of the oscillation spectrum of the condensate is rather difficult. In axially long traps ($\beta \ll 1$) Eq.(19) gives $\tilde{\Omega}_+ = 1.83\omega_{0r}$, $\tilde{\Omega}_- = 1.55\omega_{0z}$. In the MIT experiment for $\beta \approx 0.08$ [7], corresponding to the hydrodynamic regime in the axial direction and intermediate regime in the radial direction, one would expect $\tilde{\Omega}_+$ in the range $(2 - 1.8)\omega_{0r}$, and $\tilde{\Omega}_-$ in the range $(2 - 1.55)\omega_{0z}$. This is in qualitative agreement with the values $2\omega_{0r}$ and $1.8\omega_{0z}$ found experimentally for the thermal cloud at $T/T_c \approx 2$.

The asymmetry of free expansion of a hydrodynamic thermal cloud follows directly from Eqs. (16) with $\omega_i(t) = 0$. At times $t \gg \omega_{\max}^{-1}$ we obtain $\dot{b}_i = \tilde{\gamma}_i \omega_{0i}$. As the initial thermal size of the gas in the i-th direction $\bar{r}_i \approx (2T/m\omega_{0i}^2)^{1/2}$ the velocity of expansion $v_i = \bar{r}_i \dot{b}_i(t) \approx v_T \tilde{\gamma}_i$. The asymmetry of expansion is close to that for the condensate. This is demonstrated in Fig.1 where in the case of cylindrical symmetry we compare the quantities $\tilde{\gamma}_z/\tilde{\gamma}_r$ and γ_z/γ_r. It is important that due to inertia and significant decrease of temperature the expansion retains its asymmetry after the gas leaves the hydrodynamic regime and becomes collisionless. Thus, at initially high density of the thermal cloud the asymmetry of free expansion is not a signature of BEC.

However, as well as in the collisionless regime, at $T \gg n_0 \tilde{U}$ the thermal cloud expands much faster than the condensate ($v_T \gg c_s$). In the expanding hydrodynamic

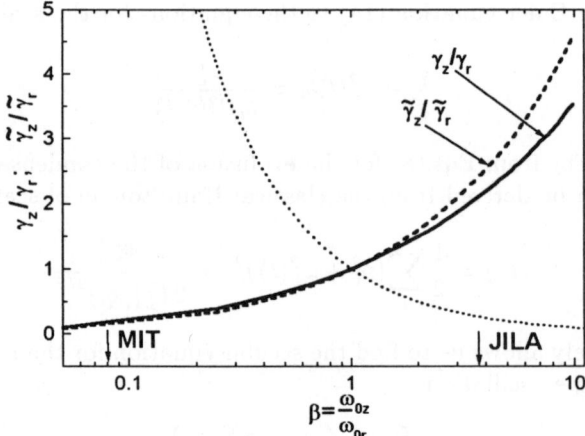

Figure 1: The quantities γ_z/γ_r (solid curve) and $\tilde{\gamma}_z/\tilde{\gamma}_r$ (dashed curve) versus β. The dotted curve is the initial ratio of the axial to radial size, β^{-1}. Vertical arrows indicate the values of β in the JILA [4] and MIT [5-7] experiments.

gas this is the most distinct feature of BEC. It is worth emphasizing that the much slower expansion of the condensate is a universal property characteristic of any trapping geometry, irrespective of the ratio between the initial size of the condensate and the thermal gas.

6 Stochastization in the condensate evolution.

Classical equations (8) can be used for describing the evolution of the condensate wavefunction for arbitrary magnitude and character of the frequency variations. If the frequencies change from ω_{0i} to ω_{1i} on a large time scale $\tau_0 \gg \omega_{0i,1i}^{-1}$, the initial condensate is adiabatically transformed to ground state of the system in the final trapping field. For fast and significant change of the frequencies (on a time scale $\tau_0 < \omega_{0i,1i}^{-1}$) there will be large undamped oscillations of the scaling parameters $b_i(t)$ and, hence, of the condensate density and phase. In the isotropic case at times where the frequency is already constant the parameter $b(t)$ oscillates with a constant amplitude. Accordingly, the expansion of the condensate is followed by a compression to the original shape [23].

In anisotropic traps we have a new ingredient, i.e., the coupling between different degrees of freedom through the non-linear term of volume scaling in the Hamiltonian $\mathcal{H}_{sd}$ (11). This can lead to stochastization of motion of the scaling parameters b_i and, hence, to a new phenomenon - stochastic evolution of the condensate.

We will give an example of the calculation where without changing the initial cylindrical symmetry the frequencies were abruptly changed from ω_{0i} to $\omega_{1i} = \{1.7, 1.7, 1\}$. In Fig.2 we present the Poincare map for 3 phase-space trajectories corresponding to 3 different sets of initial frequencies. The mapping points for the initial set $\omega_{0i} = \{5.4, 5.4, 6\}$ describe almost regular quasiperiodic motion in the

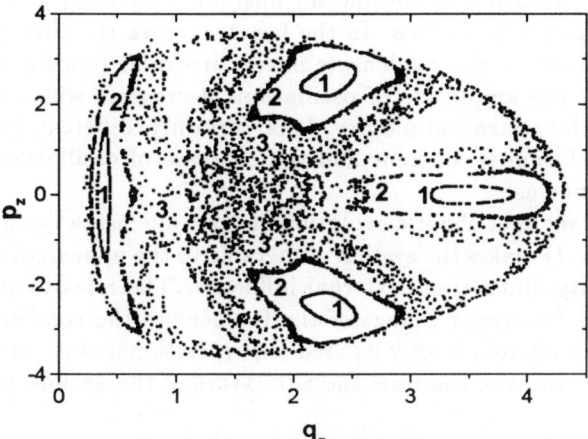

Figure 2: Poincare map for the case of abrupt change of the frequencies $\omega_{0i} \to \omega_{1i} = \{1.7, 1.7, 1\}$. The phase-space trajectories for the initial sets $\omega_{0i} = \{5.4, 5.4, 6\}$, $\omega_{0i} = \{6.5, 6.5, 5\}$ and $\omega_{0i} = \{6.6, 6.6, 4.2\}$ are labeled as 1, 2 and 3, respectively.

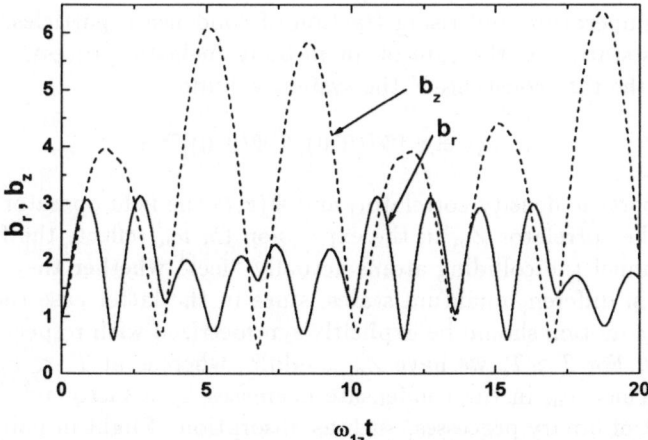

Figure 3: Time evolution of the scaling parameters b_r and b_z for the case of abrupt change of the frequencies from $\{6.5, 6.5, 5\}$ to $\{1.7, 1.7, 1\}$. The solid curve corresponds to b_r, and the dashed curve to b_z.

vicinity of a second-order non-linear resonance. Points for $\omega_{0i} = \{6.6, 6.6, 4.2\}$ show that a large part of the phase space is occupied by stochastic motion. For the initial set $\omega_{0i} = \{6.5, 6.5, 5\}$ we find an unstable trajectory intermediate between quasiperiodic motion and chaos. In the last two cases the time dependence of the axial and radial size of the condensate is very irregular (see Fig.3). During certain periods of time the amplitude of oscillations decreases, which can be treated as "imitation" of the relaxation picture. It is, certainly, different from real relaxation because the evolution is coherent and the decrease of oscillations after some time changes to an increase.

It is important that chaotic evolution of the condensate density and, especially, of the phase $\Phi(\mathbf{r}, t)$ makes the system vulnerable to the appearence of real relaxation and irreversibility under small external influence. The relaxation can be promoted by the fact that for strong change of the frequencies the condensate wavefunction $\Psi_0(\mathbf{r}, t)$ corresponds to a highly excited superpositional state of the final trapping potential, where most of the time the admixture of the ground state of the system is very small.

7 Correlation properties

As one can see from the results obtained above, it is crucial to find phenomena which demonstrate the difference of the evolution of the condensate from that of the thermal gas in the hydrodynamic regime and allow us to study the loss of coherence and relaxation in the course of evolution. We believe that in this respect the most promising is the measurement of local correlation properties, such as those manifesting themselves in the rates of inelastic collisional processes. As has been found in [28], below the BEC transition point T_c the inelastic rate continuously decreases with decreasing temperature and rising fraction of condensate particles. In a spatially homogeneous static gas the rate of an m-body inelastic process, $\nu_m = \alpha_m Z_m V$, where α_m is the rate constant, V the system volume,

$$Z_m = <[\hat{\Psi}^\dagger(0,0)]^m [\hat{\Psi}(0,0)]^m>, \qquad (20)$$

the local m-particle density correlator, and $\hat{\Psi}(\mathbf{r}, t)$ the field operator of atoms. The presence of the correlator Z_m in the expression for ν_m reflects the fact that in the incoming channel the colliding atoms actually "feel" whether they are in one and the same or in different quantum states, since in the latter case the wavefunction of the relative motion should be explicitly symmetrized with respect to interchange of the atoms. For $T > T_c$ we have $Z_m = m! n^m$, whereas at $T \ll T_c$ the correlator $Z_m = n^m$. Thus, ν_m in the condensate decreases by a factor $m!$ [28]. Already the measurement of binary processes, such as absorption of light in pair collisions [29] or spin relaxation, allows us to determine the condensate density. In this case

$$Z_2 = 2n^2 - n_0^2(T), \qquad (21)$$

and to a certain extent we have an analog of the well-known Hanbury Brown-Twiss effect for photons, but now inherent in the system of interacting particles and depending on temperature.

The process of three-body recombination is more sensitive to the presence of the condensate. The rate is determined by the correlator

$$Z_3 = 6n^3 - 9nn_0^2 + 4n_0^3, \qquad (22)$$

which leads to the "1/6 law" at $T \ll T_c$.[28]

There is the question of what happens with these correlation properties in the spatially inhomogeneous evolving Bose-condensed gas. Obvious generalization of the above expression for the rate reads

$$\nu_m(t) = \alpha_m \int d\mathbf{r} Z_m(\mathbf{r}, t). \qquad (23)$$

If almost all atoms are in the condensate, using the scaling solution (6) for $\Psi_0(\mathbf{r}, t)$ in Eq.(20), we find

$$\nu_m(t) = \nu_m(0) \left[\frac{V(0)}{V(t)}\right]^{m-1}, \qquad (24)$$

where $\nu_m(0)$ is the inelastic rate in the initial static condensate, and the appearance of the quantity $[V(0)/V(t)]^{m-1}$ is a trivial consequence of the changing system volume: $V(t) = V(0)\mathcal{V}(t)$.

These results show that coherent evolution retains the effect of reduction of inelastic processes, characteristic of the static condensate. This is the case for any coherent evolution of the condensate, however irregular and stochastic it is. The loss of coherence will increase local density correlators and, hence, lead to an increasing inelastic rate. Thus, one can identify the presence of the condensate and study its relaxation in the course of evolution through the measurement of the rates of intrinsic or light-induced inelastic collisional processes.

8 Concluding remarks

In spite of the existence of many similar features, the behavior of trapped Bose-condensed gases is truly different from that of ordinary thermal samples. This should attract a great interest to finding new macroscopic quantum phenomena in future investigations. Of particular interest is the dynamics and kinetics of gases with attractive interaction between particles ($a < 0$), especially in view of the recently revealed possibility to switch the sign of a by using nearly resonant light [30]. This opens prospects to study various aspects of the fundamental problem of "collapse" of the condensate. In addition, there can be a possibility of stabilizing BEC in gases with initially negative a.

Acknowledgments

We acknowledge discussions with J.T.M. Walraven, T.W. Hijmans, M.W. Reynolds, and P. Zoller. This work was supported by the Dutch Foundation FOM, by NWO (project NWO-047-003.036), by INTAS and by the Russian Foundation for Basic Studies.

References

1. M.H. Anderson, J.R. Ensher, M.R. Matthews, C.E. Wieman, and E.A. Cornell, Science, **269**, 198 (1995).
2. C.C. Bradley, C.A. Sackett, J.J. Tolett, and R.G. Hulet, Phys. Rev. Lett., **75**, 1687 (1995).
3. K.B. Davis, M.-O. Mewes, M.R. Andrews, N.J. van Druten, D.S. Durfee, D.M. Kurn, and W. Ketterle, Phys. Rev. Lett., **75**, 3969 (1995).
4. D.S. Jin, J.R. Ensher, M.R. Matthews, C.E. Wieman, and E.A. Cornell, Phys. Rev. Lett., **77**, 420 (1996).
5. M.-O. Mewes, M.R. Andrews, N.J. van Drutten, D.M. Kurn, D.S. Durfee, and W. Ketterle, Phys. Rev. Lett., **77**, 416 (1996).
6. M.R. Andrews, M.-O. Mewes, N.J. van Drutten, D.M. Kurn, D.S. Durfee, C.G. Townsend, and W. Ketterle, Science, **273**, 84 (1996).
7. M.-O. Mewes, M.R. Anderson, N.J. van Drutten, D.M. Kurn, D.S. Durfee, C.G. Townsend, and W. Ketterle, Phys. Rev. Lett., **77**, 988 (1996).
8. T.W. Hijmans, Yu. Kagan, G.V. Shlyapnikov, and J.T.M. Walraven, Phys. Rev. B, **48**, 12886 (1993).
9. P.A. Ruprecht, M.J. Holland, K. Burnett, and M. Edwards, Phys. Rev. A, **51**, 4704 (1995).
10. G. Baym and C.J. Pethick, Phys. Rev. Lett., **76**, 6 (1996).
11. Yu. Kagan, G.V. Shlyapnikov, and J.T.M. Walraven, Phys. Rev. Lett., **76**, 2670 (1996).
12. F. Dalfovo and S. Stringari, Phys. Rev. A, **53**, 2477 (1996).
13. R.J. Dodd, M. Edwards, C.J. Williams, C.W. Clark, M.J. Holland, P.A. Ruprecht, and K. Burnett, Phys. Rev. A, **54**, 661 (1996).
14. A.L. Fetter, Phys. Rev. A, **53**, 4245 (1996).
15. S. Stringari, Phys. Rev. Lett., **77**, 2360 (1996).
16. K.G. Singh and D.S. Rokhsar, Phys. Rev. Lett., **77**, 1667 (1996).
17. L. You, W. Hoston, and M. Lewenstein, Preprint, 1996.
18. M. Edwards and K. Burnett, Phys. Rev. A, **51**, 1382 (1995).
19. M. Edwards, R.J. Dodd, C.W. Clark, P.A. Ruprecht, and K. Burnett, Rhys. Rev. A, **53**, R1950 (1996).
20. M. Holland and J. Cooper, Phys. Rev. A, **53**, 1954 (1996).
21. M. Edwards, P.A. Ruprecht, K. Burnett, R.J. Dodd, and C.W. Clark, Phys. Rev. Lett., **77**, 1671 (1996).
22. P.A. Ruprecht, M. Edwards, K. Burnett, and C.W. Clark, Preprint, 1996.
23. Yu. Kagan, E.L. Surkov, and G.V. Shlyapnikov, Phys. Rev. A, **54**, R1753 (1996).
24. E.M. Lifshitz and L.P. Pitaevskii, *Statistical Physics, Part 2* (Pergamon Press, Oxford, 1980).
25. V.V. Goldman, I.F. Silvera, and A.J. Leggett, Phys. Rev. B, **24**, 2870 (1981).
26. D.A. Huse and E.D. Siggia, J. Low Temp. Phys., **46**, 137 (1982).
27. Y. Castin and R. Dum, Preprint, 1996.
28. Yu.Kagan, B.V. Svistunov, and G.V. Shlyapnikov, Pis'ma Zh. Eksp. Teor. Fiz., **42**, 169 (1985) [JETP Lett., **42**, 209 (1985)].

29. Yu. Kagan, B.V. Svistunov, and G.V. Shlyapnikov, Pis'ma Zh. Eksp. Teor. Fiz., **48**, 54 (1988) [JETP Lett., **48**, 56 (1988)].
30. P.O. Fedichev, Yu. Kagan, G.V. Shlyapnikov, and J.T.M. Walraven, Phys. Rev. Lett., **77**, 2913 (1996).

Doppler-Free Spectroscopy of Trapped Atomic Hydrogen

Thomas C. Killian, Dale G. Fried, Claudio L. Cesar *, Adam D. Polcyn, Thomas J. Greytak, and Daniel Kleppner

Department of Physics and Center for Materials Science and Engineering, Massachusetts Institute of Technology, Cambridge, Massachusetts 02139

Abstract

Trapped atomic hydrogen in the sub-millikelvin temperature regime provides an opportunity for significantly increasing the resolution of the 1S–2S transition. Initial experiments displayed a line width of 3 kHz, apparently limited by instability in the excitation laser. We summarize these results, discuss the major expected sources of line broadening and frequency shifts, and outline the analysis of two-photon resonance of trapped atoms in a regime where the periodic motion of atoms in the trap must be considered. The results suggest that a line width for the 1S–2S transition approaching its natural line width, 1.3 Hz, may be attainable. The experiment operates close to the regime for Bose-Einstein condensation, and the expected effect of BEC on the spectrum is described. Condensate atoms can be ejected from the trap by photon recoil, forming an intense coherent atomic beam.

1 Introduction

The steadily increasing accuracy of hydrogen spectroscopy has provided a continuing stimulus for extending the precision of QED calculations over the years [1,2,3,4]. It has also played a central role in remarkable advances in laser spectroscopy. Two-photon Doppler-free spectroscopy of the 1S–2S transition is prominent in both of these streams of development. The 1S–2S transition has now replaced the 2S–2P transition in yielding the most accurate value for the Lamb shift. Through comparisons of the frequencies of the 1S–2S and 2S–4S/4P/4D transitions [2,5] and of the 1S–3S and 2S–6S/6D [1], which are almost exactly in the ratio of 4:1, the 1S Lamb shift can be calculated. A direct measurement of the 1S–2S transition frequency then determines the Rydberg constant [6].

The 1S–2S transition is unique because its intrinsic resolution is higher than other UV or optical transitions in hydrogen by a factor of $\sim 10^6$. Because the 2S state is metastable—it decays by two-photon emission with a lifetime of 0.12 sec.—the natural linewidth for the transition is 1.3 Hz, in contrast to optical transitions which typically have linewidths of many MHz.

As one expects for an electric dipole forbidden transition, the two-photon excitation rate of the 2S state is extremely small. The low rate is to some extent compensated experimentally by the possibility of Doppler-free excitation, which eliminates first order Doppler broadening and allows all the atoms to interact with the laser simultaneously [7]. (The two-photon 1S–2S transition is excited by a standing wave at half the transition frequency. Atoms that absorb one photon from each of the two counter-propagating beams experience no net first order Doppler shift.) This method was first demonstrated by Hänsch and his co-workers in 1974[8]. In the initial experiments the linewidth was hundreds of MHz, due chiefly to the pulsed laser source then in use. The development of CW ultraviolet sources greatly decreased the laser linewidth, and by employing atomic beam methods the atom temperature was reduced, the interaction time was extended, and other line broadening mechanisms were suppressed. Recently, Hänsch and his colleagues were able to exploit the low velocity tail of the energy distribution in a 10 K atomic hydrogen beam to yield a linewidth of only 1 kHz[9]. However, to achieve the natural linewidth, the resolution must be further reduced by a factor 1000.

We describe here the first results of a technique that greatly extends the interaction time of the atom with the radiation field and should yield a linewidth close to the natural limit. The atoms are confined in a magnetic trap at temperatures in the sub-millikelvin regime. In our first efforts [10] the linewidth appeared to be limited by fluctuations in our laser source, which currently results in about a 3 kHz spectral width, but it is believed that much narrower lines are possible.

2 Cold Trapped Hydrogen

The key to extending the interaction time is to confine the atoms without significantly perturbing their electronic structure. This is accomplished by evaporatively cooling hydrogen in a magnetic trap, as first proposed by Hess [11]. The trapping method has been described elsewhere [12,13,14] and we here summarize only its chief features.

Atoms are confined in the "low-field seeking" $F = 1, m_F = 1$ hyperfine state for which the energy increases with magnetic field. The trap, in the Ioffe-Pritchard configuration [15,11], has a field minimum at the center. Four elongated coils produce a quadrupole field in the plane perpendicular to the z-axis, and short "pinch" solenoids create a magnetic barrier along the z-axis at either end of the trap. In order to inhibit non-adiabatic spin flips, a separate solenoid provides a uniform bias field along the z-axis.

The trap is loaded from a pulsed RF discharge at 0.7 K in a magnetic field of 4 T. The atoms flow into the cell (250 mK while loading), cool on the superfluid helium coated cell walls, and collide with other hydrogen atoms in the trap region. Some of these collisions result in atoms with low enough energy that they become trapped in the 600 mK deep magnetic potential. The sample is then evaporatively cooled [16] by slowly lowering a potential barrier. This is done on a time scale that is long compared to the thermalization time of the gas, allowing hot atoms to escape and forcing the remaining atoms to cool. The resulting samples consist of $10^{13} - 10^{10}$ atoms at temperatures of 25 mK – 100 μK, with atomic densities up to 8×10^{13} cm^{-3}.

The magnetic field strength midway between the pinch coils at a distance ρ from the axis of the trap is $B = \sqrt{B_\rho^2 + B_{z,o}^2}$. The quadrupole field strength is given by $B_\rho = \rho B_w/a$, where the cell radius in our trap is $a = 22$ mm and the field at the cell wall is B_w (maximum 0.91 tesla). The axial bias field at the center of the trap is $B_{z,o} \approx 2 \times 10^{-4}$ tesla. The magnetic potential energy of an

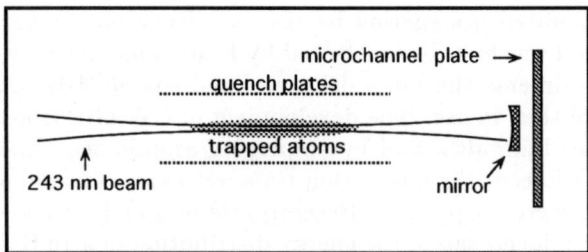

Figure 1: Schematic diagram of optical excitation and detection apparatus. The UV beam is aligned along the axis of the magnetic trap. The beam traverses the cloud of trapped atoms and is retro-reflected. After a UV excitation pulse, the (metastable) $2S$ atoms are detected by applying an electric field across the quench plates which mixes the $2S$ state with the $2P$ state (1.6 ns lifetime). The resulting L_α photons are detected with the microchannel plate.

atom is $V(\vec{r}) \approx \mu_B |\vec{B}(\vec{r})|$. The density varies as $\exp(-V(\vec{r})/k_B T)$, resulting in a characteristic thermal radius $\rho_{th} = 2ak_B T/\mu_B B_w$ for the cylindrically shaped sample. Depending on the temperature and conditions of confinement, ρ_{th} varies between 40 and 1100 μm. The sample length is roughly 15 cm.

3 Optical Excitation and Detection

The $1S$–$2S$ transition is induced by laser light at 243 nm which passes along the axis of the trap and is retro-reflected to produce the standing wave required for Doppler-free absorption. The $2S$ atoms are detected by applying a quenching electric field of about 8 V/cm which Stark mixes the $2S$ and $2P$ states, causing rapid radiative decay with the emission of Lyman-α fluorescence (122 nm). The radiation is detected by a microchannel plate at the end of the cell. Figure 1 depicts the primary features of the trap and optical detection system.

The 243 nm radiation is generated by doubling the output of a 486 nm CW dye laser[17]. The dye laser is frequency stabilized to a reference cavity and its frequency is initially calibrated using the Te_2 spectrum[18]. With ~500 mW of blue, up to 40 mW of 243 nm radiation can be produced, but due to constraints associated with performing spectroscopy in a dilution refrigerator, the power delivered to the atoms is typically 3 mW. The beam is focused in the fundamental Gaussian mode with a waist radius of $w_o = 40$ μm and divergence length $z_o = 2$ cm. The 243 nm light is controlled by an acousto-optical modulator which provides pulses of adjustable duration.

Because the cell was not initially designed for optical excitation, the collection solid angle is only 4×10^{-4} ster rad. The detection efficiency is further decreased by the 10% transmission of a UV filter, 60% transmission of a cell window, and 25% detector efficiency, so that the over-all efficiency is estimated to be only 6×10^{-6}. Nevertheless, signal rates as large as 1000 counts/sec have been observed.

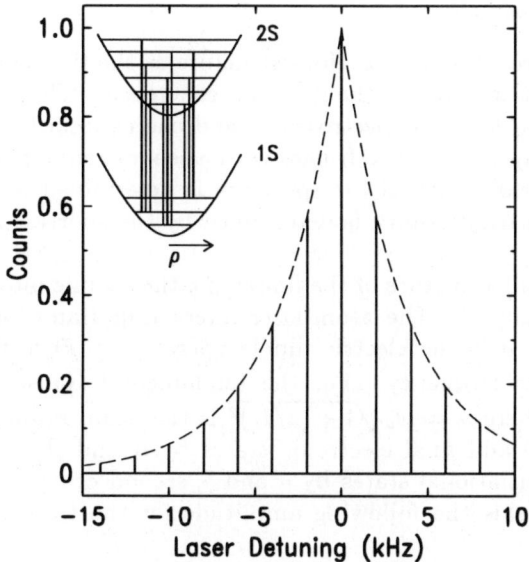

Figure 2: $1S$–$2S$ excitation spectrum displaying a time of flight profile. The UV detuning (at 243 nm) is δ. The density is 3×10^{12} cm^{-3}, the temperature is 1.7 mK, and the UV power is ≈ 1.5 mW. The total UV exposure time at each point is 2.7 s. Here the dominant source of broadening is the finite interaction time of an atom moving across the UV beam, which leads to an exponential spectrum: $\exp(-|\delta|/\delta_0)$. The solid line corresponds to $\delta_0 = 11$ kHz, which yields a full width at half maximum of 15 kHz. (From reference 9)

4 Two-Photon Spectroscopy for Trapped Atoms

4.1 Overall Structure

At temperatures above a few mK, the excitation spectrum is similar to that of a classical gas. We observe the characteristic exponential shape for time-of-flight broadened two-photon absorption in a Gaussian beam [19], $F(\delta) \sim \exp(-|\delta|/\delta_o)$. Here, $\delta = \nu - \nu_o$ is the frequency detuning from resonance of the 243 nm radiation. The linewidth parameter is $\delta_o = u/4\pi w$, where $u = \sqrt{2k_B T/M}$ is the most probable speed for an atom of mass M in a gas at temperature T ($u = 130\sqrt{T}$ m/s K$^{1/2}$ for H), and w is the radius of the light beam. This behavior can be seen in Fig. 2. The temperature inferred from the linewidth can be independently verified by observing the energy distribution of the trapped gas as the trap is dumped [13]. We have studied spectra of samples in the temperature range between 10 mK and 150 µK and have measured linewidths from 30 kHz to 3 kHz. Our estimated error in inferring the temperature is about 25%, arising chiefly from uncertainty in the size of the beam waist in the excitation region.

4.2 Underlying Structure

At lower temperatures, the line develops structure as the trapped atoms repeatedly pass through the laser beam without losing coherence. This is similar to Ramsey separated oscillatory field spectroscopy, and one expects the 1S–2S excitation probability to display interference fringes with a separation of twice the vibrational frequency. The overall width of the spectrum is determined by the time of flight linewidth, while the fringe width is determined by the shortest relevant coherence time.

To obtain a detailed picture of the lineshape, the atomic motion can be treated quantum mechanically [20]. The atom-laser interaction Hamiltonian is $H = -\vec{\mu} \cdot \vec{E}(\vec{r})e^{-i\omega t}/2$, where $\vec{\mu}$ is the electric dipole operator, $\vec{E}(\vec{r})$ is the laser field amplitude and ω is its frequency. For the fundamental Gaussian mode, $\vec{E}(\vec{r}) = \vec{E}_o \exp(-\rho^2/w^2)$, where $w = w_o\sqrt{1+(z/z_o)^2}$ is the beam radius and $z_0 = \pi w_0^2/\lambda$. Denoting the initial and final electronic states by α and β respectively, and the initial and final translational states by a and b, second order time-dependent perturbation theory yields the following amplitude for the transition from $|\alpha, a\rangle$ to $|\beta, b\rangle$:

$$C_{\alpha,a}^{\beta,b}(t) = \left[\sum_\gamma \frac{\mu_{\beta,\gamma}\mu_{\gamma,\alpha}}{(i\hbar)^2} \frac{E_0^2}{2i(\omega_{\gamma,\alpha}-\omega)}\right] \frac{e^{i\delta(\omega)t}}{i\delta(\omega)} \langle b| e^{-2\rho^2/w^2} |a\rangle$$

Here, $\delta(\omega) = \omega_{\beta,\alpha} + \Omega_{b,a} - 2\omega$, where $\hbar\omega_{\beta,\alpha}$ ($\hbar\Omega_{b,a}$) is the transition energy between the electronic (translational) states.

To understand the physical content of this expression, consider transitions between free momentum states. The translational matrix element $I_a^b = \langle b| e^{-2x^2/w^2} |a\rangle$ takes the form

$$I_a^b \sim \int e^{-i(p_b-p_a)x/\hbar} e^{-2x^2/w^2} dx$$

I_a^b is the spatial Fourier transform of the beam intensity profile. When this Gaussian amplitude for a single atom is averaged over a thermal sample the result is the familiar exponential time-of-flight lineshape.

In the case of confined atoms with quantized translational states, the matrix elements depend on the overlap of the wave functions with the laser profile. If the atoms are confined to the small region $\rho \ll w$ inside the beam, then

$$I_a^b = \langle b|a\rangle = \delta_{a,b},$$

and no change in the momentum state is possible. The situation is reminiscent of Dicke narrowing in which motional broadening is suppressed when an atom which is confined to a distance less than $\lambda/2\pi$ experiences neither amplitude nor phase variations in the field.

In our situation, the atoms are confined near the axis where the potential is harmonic. We neglect longitudinal motion so that $|a\rangle$ and $|b\rangle$ are two-dimensional simple harmonic oscillator states with frequency Ω_o, and use a Cartesian basis with occupation numbers $|a\rangle = |a_x, a_y\rangle$. Transitions occur when the laser frequency satisfies $2\omega_l = \omega_{\beta,\alpha} + \Omega_{b,a}$, where $\Omega_{b,a} = \Omega_o[(b_x+b_y)-(a_x+a_y)]$. Parity considerations require that the total change in occupation numbers be even. The spectrum

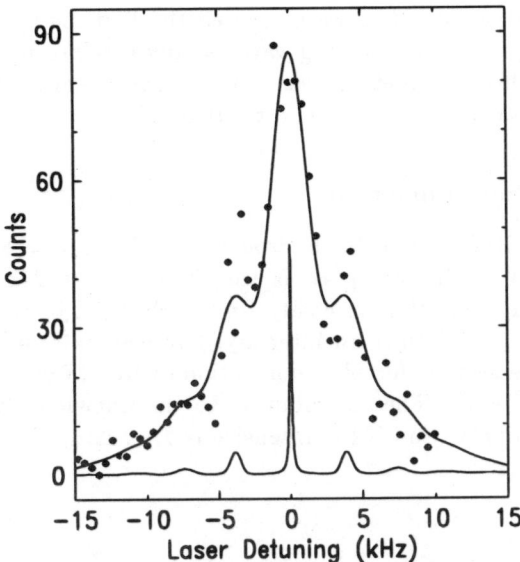

Figure 3: Calculated excitation spectrum for a 200 μK sample in the absence of broadening mechanisms. The solid line is the picket fence lineshape described in the text, with sideband spacing given by the trap oscillation frequency. The dashed line is the exponential envelope, due to time-of-flight broadening. Inset: schematic diagram of trap vibrational states on the 1S and 2S electronic manifolds, indicating allowed transitions from the lowest three 1S trap states.

takes a "picket fence" appearance, centered at $\omega_{\beta,\alpha}/2$, with sidebands separated by Ω_o. When an average is made over a thermal set of oscillator occupation numbers, the result is that the picket fence lies under an exponential envelope function like that for untrapped atoms, discussed above. Figure 3 shows the characteristic sideband structure. The sideband spacing is equal to the radial trap oscillation frequency, 2 kHz in this case, which can be varied by changing the trapping fields.

4.3 Line Broadening Mechanisms

Natural Line Width

Because no damping has been included, the "picket fence" spectrum consists of delta functions. In the absence of other broadening mechanism, the underlying lineshape is governed by two-photon radiative decay, and consists of a Lorentzian of width $\Delta\nu_o = 1/4\pi\tau_{2S} = 0.66$ Hz. (Following convention, we express line widths and frequency shifts in terms of the laser frequency, so that all frequencies refer to 243 nm radiation.) $\Delta\nu_o$ provides the natural scale for considering broadening and frequency shift mechanisms.

A primary concern is that the 2S atoms actually last for their natural lifetime. To our knowledge, this lifetime has never been directly observed. The possibility that the 2S lifetime in the trap could be significantly shortened by Stark quenching or some other process was of some concern, for it could frustrate our fluorescence detection scheme. To study this, we measured the lifetime by exciting the atoms with a short laser pulse and delaying various times before applying the quenching pulse [10]. The observed lifetime was close to the natural lifetime, though the measurement was too crude to provide a reliable value.

Saturation and photoionization

The two photon excitation rate for hydrogen is [21] $R_{1S-2S} = 84\ I^2/\gamma$ s^{-1}, where I is expressed in W/cm^2, and $\gamma/4\pi$ is the linewidth at 243 nm. It is convenient to introduce a saturation intensity for which $R_{1S-2S} = \tau_{2S}^{-1}$. The result is $I_{sat} = 0.9$ W cm^{-2}. At such an intensity, however, photoionization cannot be neglected. The cross section for photoionization of the 2S state by 243 nm radiation is[22] 7.9×10^{-18} cm^2. The contribution to the linewidth from this process is $\Delta\nu_{p.i.} = 1.5\ I$ Hz cm^2/W, and if the intensity is I_{sat}, $\Delta\nu_{p.i.}$ is close to the natural linewidth.

Zeeman Shift

The 1S–2S transition frequency for $F = 1, m_F = 1$ hyperfine states is independent of magnetic field except for a small relativistic effect on the electron g-factors[23]. The resulting frequency shift of the 1S–2S transition is given by $\delta\nu_Z = \alpha^2\ \mu_B B/8h = 9.3 \times 10^4\ B$ Hz/T where B is the magnetic field and α is the fine structure constant. In the trap, the atoms experience a characteristic magnetic field $B = 2k_B T/\mu_B$, so that at a temperature of 100 μK the shift is about 27 Hz. Because of the distribution of energies, there is an apparent broadening comparable to the shift. The Zeeman shift can be reduced by a number of strategies, including operating at lower temperature and restricting attention to the central part of the line where the lowest energy atoms make the dominant contribution.

Density Effects

One expects a frequency shift due to the difference between atom–atom interaction energies in the 1S–1S and 2S–1S systems—the low temperature equivalent of a pressure shift. A recent calculation [24] predicts a shift of $\delta\nu_{dens} = -1.5 \times 10^{-10}\ n(\vec{r})$ Hz cm^3. We have not yet observed this effect in our spectra although it should be barely detectable in our highest density samples. This shift should not present a fundamental limit to spectroscopic accuracy since at a density of 10^{10} cm^{-3} the predicted shift is comparable to the natural linewidth. Conversely, above this value the frequency shift can be used to measure density.

A broadening accompanies the shift. The contribution due to elastic collisions has been calculated to be $\Delta\nu_{dens} = 2.4 \times 10^{-9}\ n\ \sqrt{T}$ Hz cm^3/K$^{1/2}$. For accessible trapping conditions, this effect is much smaller than the shift. This broadening can be explained as the result of decoherence due to collisions. The "hard sphere" collision rate is $\gamma_{col} = 4\pi a^2 \bar{v} n(\vec{r})$ where $a = -2.3$ nm is the 2S–1S scattering length [24], and $\bar{v} = \sqrt{8k_B T/\pi M}$ is the mean speed for an atom of mass M. Then $\Delta\nu_{dens} = \gamma_{col}/8\pi$, indicating that atoms lose coherence in about two collisions.

AC and DC Stark Effects

Electric fields in the trapping region can shift and broaden the transition and quench the 2S state. The DC Stark effect causes a shift $\delta\nu_{DC} = 1800\ E^2$ Hz cm^2/V^2. A 50 mV/cm stray field results in a 4.5 Hz shift.

In an electric field the 2S state is quenched by Stark-mixing with the 2P state. Indeed, to obtain our spectra, a large electric field is applied after the laser excitation pulse to induce fluorescence from the 2P state. The broadening due to a residual electric field is $\Delta\nu_{DC} = 220\ E^2$ Hz cm^2/V^2. A 50 mV/cm field would cause broadening comparable to the natural linewidth. However, direct measurements of the 2S lifetime in our trap indicate that Stark quenching is small.

The AC Stark effect introduces a frequency shift while the atoms are in the laser beam given by [25] $\delta\nu_{AC} = 1.67\ I$ Hz cm^2/W. For intensity I_{sat}, $\delta\nu_{AC} \approx 1.5$ Hz. Because of the spread in atom energy, the AC Stark effect can introduce a broadening comparable to the shift.

4.4 Detailed Lineshape

To calculate the lineshape for the 1S–2S transition, the equations of motion for the the electronic and translational degrees of freedom must be integrated and thermally averaged. Such a treatment accurately predicts motional effects and includes line broadening and frequency shifts from photoionization and collisions as discussed above.

We use an effective two-level Hamiltonian [25] for Doppler-free, two-photon excitation, with the rotating wave approximation.

$$H = \frac{\hbar}{2} \begin{bmatrix} \omega_0 + \Delta(\vec{r}, n, ...) & \Omega_R(\vec{r}) e^{2i\omega_l t} \\ \Omega_R(\vec{r}) e^{-2i\omega_l t} & -\omega_0 - \Delta(\vec{r}, n, ...) \end{bmatrix}$$

where $\Omega_R(\vec{r}) = 9.26\ I$ cm^2/W s is the two-photon Rabi frequency for this transition [26,21]. Here, $\hbar\omega_0$ is the 1S–2S energy splitting, and ω_l is the laser frequency. The energy $\hbar\Delta$ includes all energy shifts of the 1S–2S energy levels, which may depend on position or density, for example.

We have neglected the effects of axial atomic motion and axial spatial variations of the laser beam, which are small, and have assumed linearly polarized light. Results of such a treatment are shown in Fig. 4, accompanied by an observed spectrum. For the particular trap conditions of figure 4, in the calculation for a monochromatic laser, the width of the central peak is due chiefly to photoionization. The sidebands are broadened by trap anharmonicity since more energetic atoms sample more of the linear potential where the energy level spacing decreases with increasing energy.

5 Prospects for ultra-precise spectroscopy

5.1 Lamb Shift

As described in the introduction, 1S–2S two-photon spectroscopy has played a major role in recent advances in determining the Lamb shift. Current measurements

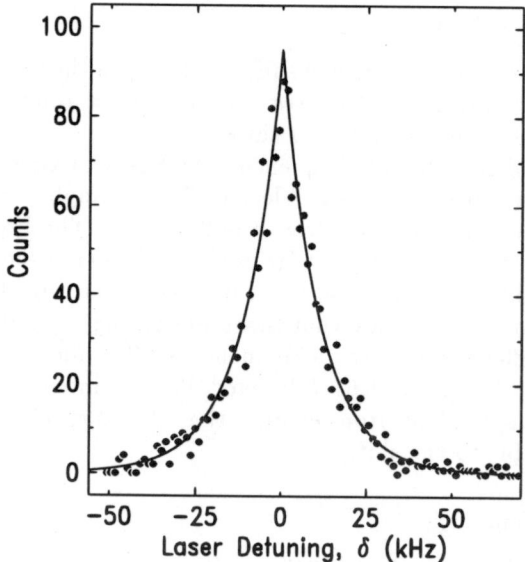

Figure 4: 1S–2S excitation spectrum with sideband structure. Each data point represents an exposure of 350 ms. The upper solid line is the calculated spectrum including the effects of trap anharmonicity and photoionization, convoluted with a 3 kHz laser linewidth. The sample temperature is found to be 150 μK. The lower curve is the line shape calculated for a monochromatic light source. (From reference 9)

of the 1S Lamb shift are limited by effects such as the AC Stark shift and possibly first and second order Doppler shifts, as well as counting statistics. Spectroscopy with magnetically trapped hydrogen could alleviate these problems.

One way to improve upon the current experimental state of the art [27] would be to measure the difference between the frequency of the 2S–4S transition and 1/4 the frequency of the 1S–2S transition to better than 10 kHz. The difficulty lies in the 2S–4S transition. Doppler shifts would be negligible with our 100 μK sample, and residual electric fields and their accompanying DC Stark shift have been shown to be very small. (In any case could be accurately determined by observing the 2S lifetime.) Because we can interact with the trapped 2S atoms for more than 100 ms, we can drive the 2S–4S transition with relatively low power. 50 W/cm^2 would quench 10% of the metastables in 50 ms, while only AC Stark shifting the 2S–4S transition by 3 kHz. Presumably the uncertainty due to this would be below 1 kHz. The signal rate for such a measurement appears to be favorable. The density shift for the 2S–4S transition is not known and is a potential source of uncertainty.

5.2 Optical Frequency Standard

It may be possible to use the 1S–2S frequency as the basis of an optical frequency standard. In comparison with trapped ion devices, cold hydrogen could provide a much higher signal rate with a comparable resonance linewidth, though at the price of greater experimental complexity. The potential of cold hydrogen as a laboratory optical frequency standard deserves consideration.

6 Studies of a Bose Condensate

The two-photon nature of the 1S–2S transition can be exploited as a sensitive and unambiguous probe for detecting a Bose condensate [26]. Normally one excites the 1S–2S transition in a Doppler-free mode by employing counter-propagating beams so that all atoms are resonant simultaneously, but Doppler-sensitive excitation by a single beam can also occur. Due to photon recoil, this line is blue-shifted from the Doppler-free peak by $\delta_{recoil} = h/M\lambda^2 = 6.697$ MHz. For a given laser frequency, only atoms within within a narrow velocity range are excited, and so the signal is usually very weak. When a Bose condensate is present, however, the center of the Doppler sensitive spectrum will display a narrow and intense feature due to the large number condensate atoms, for which $\langle v \rangle \sim 0$. In the absence of other broadening mechanisms, the width of this feature is given by the momentum uncertainty of the condensate, $\Delta p = h/\Delta z$, where Δz is the extent of the condensate parallel to the laser beam. Typically, this will produce a linewidth of ~ 1 kHz.

The atoms excited in this fashion recoil with a speed of 3 m/s, which can eject them from the trap. The beam divergence, expected to be a few milliradians, arises from both the momentum spread of the condensate and the divergence of the laser. (These divergences are comparable under typical experimental conditions.) Consequently, the ejected atoms form a coherent atomic beam—effectively behaving like the output of an atom laser.

If we cross the Bose-Einstein transition at 30μK, and attain a 2.5% condensate fraction [28], we can expect to excite 10^9 condensate atoms per second. They can be directed towards to Lyman-α detector, greatly increasing their detection solid angle, resulting in much higher detection efficiency than the 6×10^{-6} obtained presently with the normal fraction.

This work was supported by the National Science Foundation, the Air Force Office of Scientific Research, and the Office of Naval Research.

[*] Current Address: Escola Tecnica Federal do Ceara (ETFCE); 60040-531 Fortaleza, CE; Brazil.
1. S. Bourzeix, B. de Beauvoir, F. Nez, M. D. Plimmer, F. de Tomasi, L. Julien, and F. Biraben, Phys. Rev. Lett. **76**, 384, (1996).
2. D. J. Berkeland, E. A. Hinds, and M. G. Boshier, Phys. Rev. Lett. **75**, 2470 (1995).
3. M. Weitz, A. Huber, F. Schmidt-Kaler, D. Leibfried, W. Vassen, C. Zimmermann, K. Pachuki, T. W. Hänsch, L. Julien, and F. Biraben, Phys. Rev. A **52**, 2664 (1995).

4. M. Weitz, D. Leibfried, A. Huber, H. Geiger, W. König, M. Prevedelli, T. Udem, T. Heupel, K. Pachucki, and T. W. Hänsch, *Proceedings of the Fifth Symposium on Frequency Standards and Metrology, Woods Hole, Massachusetts, 1995*, edited by James C. Bergquist.
5. F. Schmidt-Kaler, D. Leibfried, S. Seel, C. Zimmermann, W. König, M. Weitz, and T. W. Hänsch, Phys. Rev. A **51**, 2789 (1995).
6. T. Andreae, W. König, D. Wynands, D. Leibfried, F. Schmidt-Kaler, C. Zimmermann, D. Meschede, and T. W. Hänsch, Phys. Rev. Lett. **69**, 1923 (1992).
7. L. S. Vasilenko, V. P. Chebotaev, and A. V. Shishaev, Pis'ma Zh. Eksp. Teor. Fiz. **12**, 161 (1970) [JETP Lett. **12**, 113 (1970)].
8. T. W. Hänsch, S.A. Lee, R. Wallenstein and C. Wieman, Phys. Rev. Lett. **34**, 307 (1975).
9. T. W. Hänsch, private communication.
10. C. L. Cesar, Dale G. Fried, Thomas C. Killian, Adam D. Polcyn, Jon C. Sandberg, Ite A. Yu, Thomas J. Greytak, Daniel Kleppner, and J. M. Doyle, Phys. Rev. Lett. **77**, 255 (1996).
11. H. Hess, Phys. Rev. B **34**, 3476 (1986).
12. J. M. Doyle, J. C. Sandberg, I. A. Yu, C. L. Cesar, D. Kleppner, and T. J. Greytak, Phys. Rev. Lett. **67**, 603 (1991).
13. J. M. Doyle, J. C. Sandberg, N. Masuhara, I. A. Yu, D. Kleppner, and T. J. Greytak, J. Opt. Soc. Am. B **6**, 2244 (1989).
14. J. M. Doyle, Ph.D. thesis, Massachusetts Institute of Technology, 1991.
15. D. Pritchard, Phys. Rev. Lett. **51**, 1336 (1983).
16. N. Masuhara, J. M. Doyle, J. C. Sandberg, D. Kleppner, T. J. Greytak, H. F. Hess, and G. P. Kochanski, Phys. Rev. Lett. **61**, 935 (1988).
17. T. W. Hänsch, in *Atomic Physics 14*, edited by D. J. Wineland, C. E. Wieman, and S. J. Smith (American Institute of Physics, New York, 1995), p. 63; T. W. Hänsch, in *Frontiers of Laser Spectroscopy*, edited by T. W. Hänsch and M. Inguscio, (North Holland, 1994), p. 287.
18. D. H. McIntyre, Ph.D. thesis, Stanford University, 1987.
19. C. Bordé, C. R. Hebd. Séan. Acad. Sci. B **282**, 341 (1976); F. Biraben, M. Bassini, and B. Cagnac, J. Phys. (Paris) **40**, 445 (1979).
20. Claudio L. Cesar, Ph.D. thesis, Massachusetts Institute of Technology, 1995.
21. F. Bassani, J. J. Forney, and A. Quattropni, Phys. Rev. Lett. **39**, 1070, (1977).
22. A. Burgess, Mem. Royal Astron. Soc. **69**, 1 (1965).
23. H. A. Bethe and E. E. Salpeter, *Quantum Mechanics of One- and Two-Electron Atoms* (Plenum, New York, 1977).
24. M. J. Jamieson, A. Dalgarno, and J. M. Doyle, Mol. Phys. **87**, 817 (1996).
25. R. G. Beausoleil and T. W. Hänsch, Phys. Rev. A **33**, 1661, (1986).
26. Jon C. Sandberg, Ph.D. thesis, Massachusetts Institute of Technology, 1993.
27. K. Pachucki, D. Leibfried, M. Weitz, A. Huber, W. Konig and T. W. Hänsch, J. Phys. B: At. Mol. Opt. Phys. **29**, 177, (1996).
28. T. J. Greytak, in *Bose-Einstein Condensation*, edited by A. Griffin, W. W. Snoke, and S. Stringari (Cambridge University Press, Cambridge, 1995), p. 131.

QED AND THE GROUND STATE OF HELIUM

W. HOGERVORST, K.S.E. EIKEMA, W. VASSEN and W. UBACHS
Laser Centre Vrije Universiteit, Department of Physics and Astronomy,
De Boelelaan 1081, 1081 HV Amsterdam, The Netherlands

With a phase-modulated extreme ultraviolet pulsed laser source the frequency of the 1^1S - 2^1P transition of helium at 58 nm has been measured with high accuracy. The phase modulation scheme enabled measurement and reduction of frequency chirp, usually limiting the precision in pulsed spectroscopy. From the measured transition frequency of 5130495083(45) MHz of ^{4}He a value of the ground state Lamb shift of 41224(45) MHz is deduced, in good agreement with a theoretical value of 41233(35) MHz based on QED calculations up to order $\alpha^5 Z^6$. Also an accurate value for the transition isotope shift ^{4}He-^{3}He of 263410(7) MHz has been determined.

I Introduction

For many years one-electron systems like hydrogen [1] and positronium [2] were the only atomic systems in which low-energy quantum electrodynamic theory (QED) could be accurately tested. Progress in theoretical calculations also make multi-electron atoms and ions of interest for QED studies. Energy calculations without QED and higher-order relativistic effects in e.g. helium and helium-like ions are nowadays so accurate that experimental transition frequencies can be used as a test for Lamb shift calculations [3]. The most important contributions to the Lamb shift are the self-energy and vacuum polarization. Helium is particularly interesting because of two-electron contributions to the Lamb shift. They originate from mutual shielding of the nucleus by the electrons, decreasing the one-electron QED shift, and from a proximity effect of both electrons. Accurate measurements of the Lamb shift in helium are now available for the metastable $2^{1,3}$S states [4-7]. Two-electron Lamb shift contributions are best studied in ^{1}S states where the proximity effect (self-energy) is ten times larger than in ^{3}S states, contributing up to 10% to the total Lamb shift [3]. Experimental values for the 2^1S Lamb shift were deduced from high resolution cw laser spectroscopy on transitions to 1snp [4-6] and 1snd states [7] by Drake et al. [8], in reasonable agreement with their theoretical calculations. Recently Shiner et al. [9] measured the isotope shift ^{4}He - ^{3}He in the transition $2^3S_1 - 2^3P_0$ with high precision, from which an accurate value for the nuclear charge radius of ^{3}He could be extracted.

The Lamb shift in the $1\,^1S$ ground state as well as its two-electron contribution are more than one order of magnitude larger than in the $2\,^1S$ state. Also isotope shifts in ground state transitions are largest. However, the ground state is not easily accessible due to the large energy difference with excited states. In 1993 we showed in a preliminary experiment on the $1\,^1S - 2\,^1P$ transition at 58.4 nm that this energy difference can be bridged using high power pulsed laser systems and harmonic up-conversion [10]. This encouraged us to start a dedicated experiment using a CW laser at 584 nm and pulse amplification techniques. However, precision spectroscopy with such a pulsed laser system is hampered by unwanted phase modulation effects in pulse-dye-amplifiers (PDA). Phase modulation results in time-dependent frequency excursions (chirp) which lead to a calibration error for the pulsed output relative to the frequency standards based on cw laser saturation spectroscopy in the visible. This effect turned out to be the limiting uncertainty in this experiment. We recently obtained a $1\,^1S$ Lamb shift of 41260(175) MHz and measured the isotope shift $^4He - ^3He$ to be 263410(7) MHz [11]. This latter value is more than two orders of magnitude better than the classical value of Herzberg [12] and is not sensitive to chirp effects. The early experiments on e.g. hydrogen encountered similar problems, before cw excitation was eventually achieved [see Ref. 13]. Many experiments, however, can at present only be performed by pulsed excitation. As a result the study of frequency chirp has become an active field of research. Heterodyning techniques have been developed to measure chirp phenomena [14,15] and methods to decrease chirp, such as anti-chirping of excimer-laser driven systems (pulse length 20-30 ns), were investigated [16].

Here we present a precision measurement of the $1\,^1S - 2\,^1P$ transition of helium at 58 nm using phase-modulated narrow band extreme ultraviolet radiation (XUV). A fast electro-optic modulator (EOM) system was employed to control the phase (chirp) of a PDA at 584 nm which provides the power for non-linear upconversion to 58.4 nm. Together with an accurate chirp measurement technique for nanosecond pulses and precise modelling of the excitation process, we demonstrate that frequency chirp can be reduced significantly.

2 Experiment

A schematic of the setup is shown in Fig.1. The principle of the measurement is described elsewhere [11]. A PDA, pumped by an injection-seeded Nd:YAG laser (10 Hz, 740 mJ at 532 nm in a 6.5 ns pulse), amplifies 150 mW of light at 584 nm from a

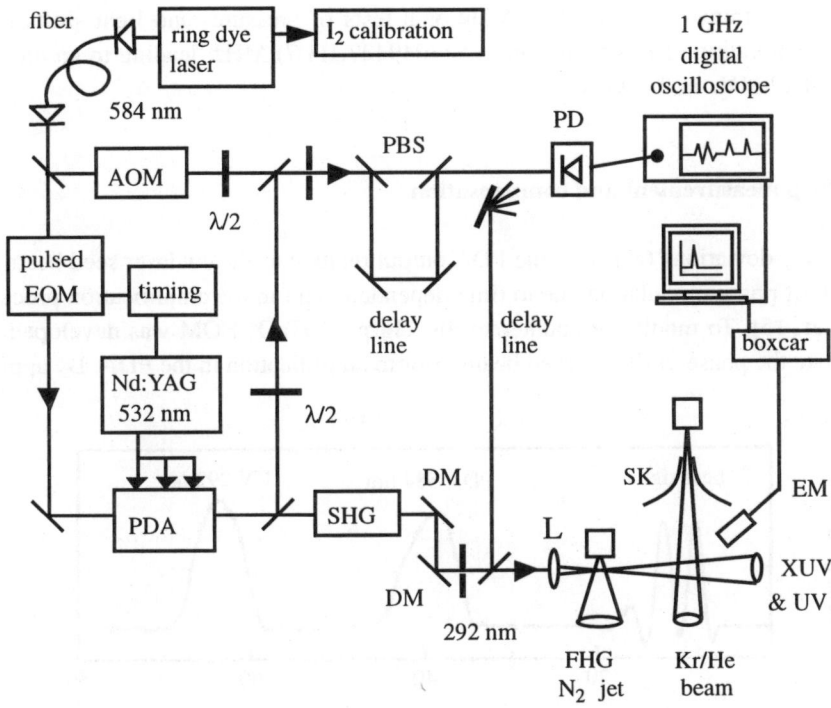

Fig. 1 : Schematic of the experimental setup (AOM=acousto-optic modulator, EOM=electro-optic modulator, PBS=polarizing beamsplitters, PDA=pulse dye amplifier, DM=dichroic mirror, FHG=fifth harmonic generation, PD=photodiode, SK=skimmer, EM=electron multiplier)

ring dye laser to 220 mJ in 6.5 ns pulses. This output is frequency doubled in a KD*P crystal to ~100 mJ at 292 nm in a 6 ns pulse, and subsequently focused (f=24.3 cm) in a pulsed jet of N_2 for fifth-harmonic generation. In this process 10^5-10^6 photons at 58.4 nm are generated in a beam overlapping with the UV. To reduce Doppler effects the 1 ^{1}S - 2 ^{1}P transition is induced in a skimmed and pulsed beam (~18 mrad divergence) of 10% helium seeded in 90% krypton. The seeding slows down helium atoms to 480(100) m/s compared to 1200(300) ms for pure helium. A few percent of the atoms excited to 2 ^{1}P state is ionized by the UV light. A pulsed extraction field is used to collect and detect the ions with an electron multiplier and boxcar integrator. The primary calibration is performed at 584 nm by saturation spectroscopy on the

P88(15-1)-o transition in I_2 [11]. Additional tests of pressure- and light shifts improved calibration of this transition to 513049427.1(1.7) MHz, leading to an uncertainty of 17 MHz in the XUV.

3 Chirp measurement and compensation

Frequency deviations (chirp) in the PDA output relative to the cw laser seed beam is a result of phase modulation due to time-dependent gain in the amplification process [see Ref. 15]. To modify or counteract this chirp a $LiTaO_3$ EOM was developed to modulate the phase of the cw seed beam prior to amplification in the PDA. By apply-

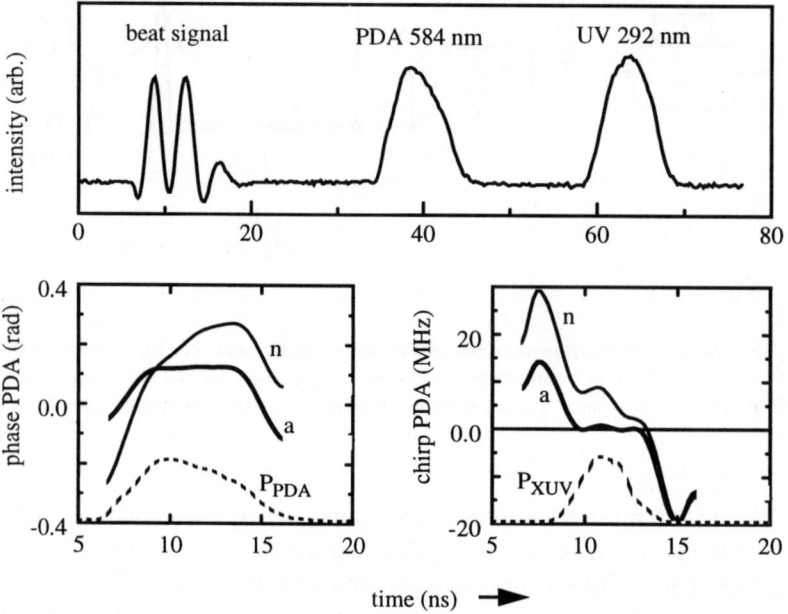

Fig. 2: Chirp measurement of normal (n) and anti-chirped (a) PDA pulses (average over 10 pulses). Upper part: typical oscilloscope readout with beat signal between PDA output and 250 MHz frequency shifted cw beam, the PDA pulse and UV pulse. Lower part: phase changes and resulting frequency chirp in the PDA. Dotted lines represent the PDA and XUV pulse shape ($P_{xuv} \sim P_{uv}^{4.5}$). Under conditions of anti-chirp the reconstructed phase is constant and the frequency excursion close to zero during the time window of XUV-generation.

ing a pulsed driving voltage to the EOM the average frequency chirp can be either decreased or increased by varying the width and slope of the rising and falling edge. The total phase modulation is monitored with a method similar to that of Fee et al.[14] by heterodyning part of the PDA output with the cw seed beam, 250 MHz acousto-opticially shifted (see Fig. 1). We extended this method by measuring also the (time-delayed) PDA and UV pulse on the same photodiode connected to a 1 GHz bandwidth, 5 Gs/s digital oscilloscope. A perpendicular polarization component is used to measure the PDA pulse intensity at exactly the same position in the beam where the heterodyne signal is observed. With this information the heterodyne signal can be corrected for the effects of the pulse envelope, resulting in accurate phase reconstruction even for short pulses of 6.5 nanosecond duration. The phase Φ and chirp $\Delta v = (1/2\pi) \, d\Phi/dt$ of the PDA pulses (see Fig. 2) is monitored on line together with an estimate of the expected frequency shift effect in the XUV. In this manner the chirp can be adjusted interactively with the EOM. Most of the chirp measurements are based on an average of 50–100 laser pulses to reduce the influence of pulse to pulse chirp fluctuations of about 1–10 MHz.

From the measured dependence of the ion signal on UV power (ion signal ~ $P_{uv}^{5.5}$) it follows that chirp compensation with the EOM is only necessary in the central part of the PDA pulse (Fig. 2). However, processes such as ionization of the nonlinear medium may shift XUV generation to the leading edge of the UV pulse. To get a better understanding of the effect of chirp on the resonance transition we also performed measurements with induced chirp. For this purpose a 2 ns wide chirp pulse of ~80 MHz is applied with different time delays relative to the PDA output pulse. In the XUV this results in rapid frequency sweeps of ~800 MHz. In each case the spectrum of the $1\,^1S - 2\,^1P$ transition is recorded, together with a chirp measurement in the visible (averaging over 100 laser pulses). From the chirp measurements, the simultaneously recorded UV pulses, and the measured dependence of the ion signal on UV power a predicted spectrum based on numerical integration of the optical Bloch equations is calculated. Comparison with the experimental lineshapes in Fig. 3 shows that the effect of chirp on the $1\,^1S - 2\,^1P$ is well understood. Due to the intense short additional chirp pulses the transition lineshape (asymmetry and linewidth) is highly sensitive to the effective XUV production timing. As a result the XUV production time window could be set accurately at +0.2(4) ns relative to the UV pulse centre, with a XUV pulse width of ~ 3(0.5) ns. We conclude that XUV is indeed produced only at the peak UV intensity, so chirp reduction is necessary only in the centre of the PDA pulse. This is accomplished by applying a counteracting phase modulation to the PDA seed beam ('anti-chirp'), resulting in a flat phase and considerably reduced

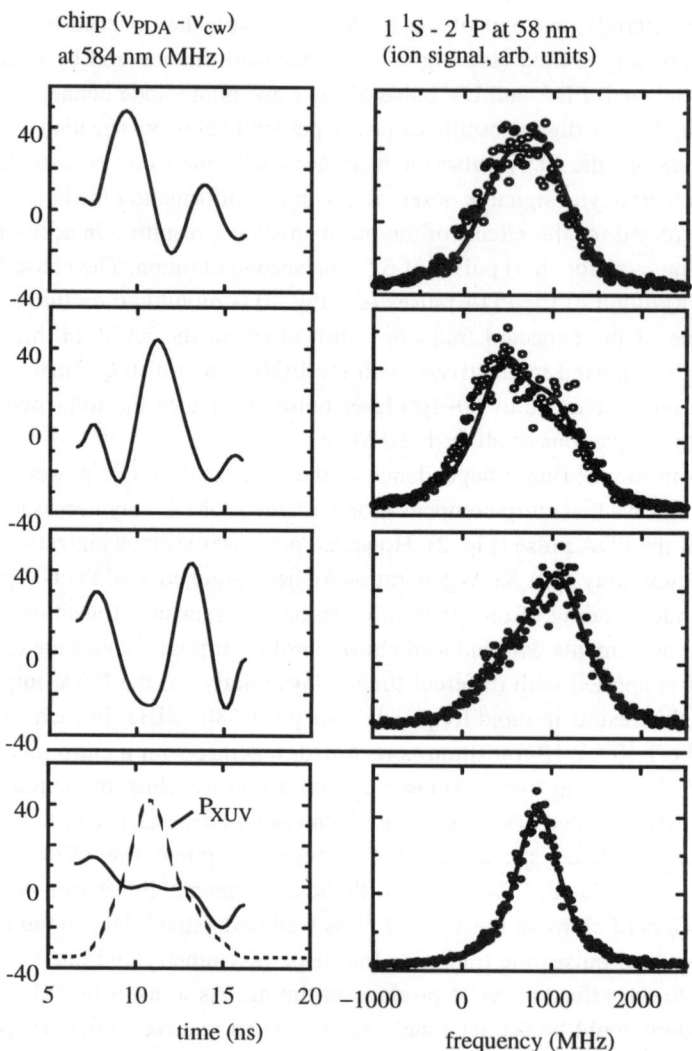

Fig. 3: Dependence of experimental and calculated 1 ^{1}S - 2 ^{1}P transition lineshape (right side, average over 4 scans) on chirp (left side, average over 100 laser pulses). The upper three traces show the distortion due to extra chirp, while the lower trace shows a typical 'anti-chirped' lineshape. The origin is the position of the I_2 calibration. The dotted line in the 'anti-chirped' measurement is the estimated XUV power dependence on UV power.

Table 1: Experimental and theoretical values for the 1 ^{1}S - 2 ^{1}P transition frequency with their 1σ errors. All numbers in MHz.

	value	1σ
Measured, (PDA chirp corrected)	5130495110	5
Corrections:		
Chirp : measurement analysis	*	14
PDA beam inhomogenities	-	20
5th harmonic, measured	10	13
Dynamic Stark shift	−44	15
Doppler shift	7	20
drift (*systematic*)	-	3
Line shape	-	3
I$_2$ calibration	-	17
Corrected value; experimental	5130495083	45
Theory	5130495074	35
* : already included in 'measured'		

chirp in the time window of interest (Fig. 2). Applying such anti-chirp the resonance frequency shifts 65 - 90 MHz upward, but no change in resonance line shape is observed due to the relatively small 'normal' chirp excursions. Calculated residual chirp shift is typically < 10 MHz, for which the measurements were corrected. The difference between calculated and measured anti-chirp shift is always within 10 MHz, determined for a large part by the statistical uncertainty in the XUV transition measurements. The uncertainty in the exact timing and width of the XUV pulse and in the chirp measurement procedure itself generates an uncertainty in the calculated anti-chirp of 14 MHz. Variation of the chirp over the spatial profile of the PDA beam was measured as well, resulting in an additional uncertainty of 20 MHz in the XUV (see Table 1).

At the UV power density used for fifth-harmonic generation ($< 10^{13}$ W/cm^2) frequency chirp may also arise from ionization/excitation and time-dependent UV field-induced refractive index changes in the gaseous nonlinear medium. The ionization/excitation effect has been measured by changing the N$_2$ density in two UV focal power density situations, resulting in a small correction of +10(13) MHz. The UV field-induced chirp is a single-atom effect and can not be measured by changing the

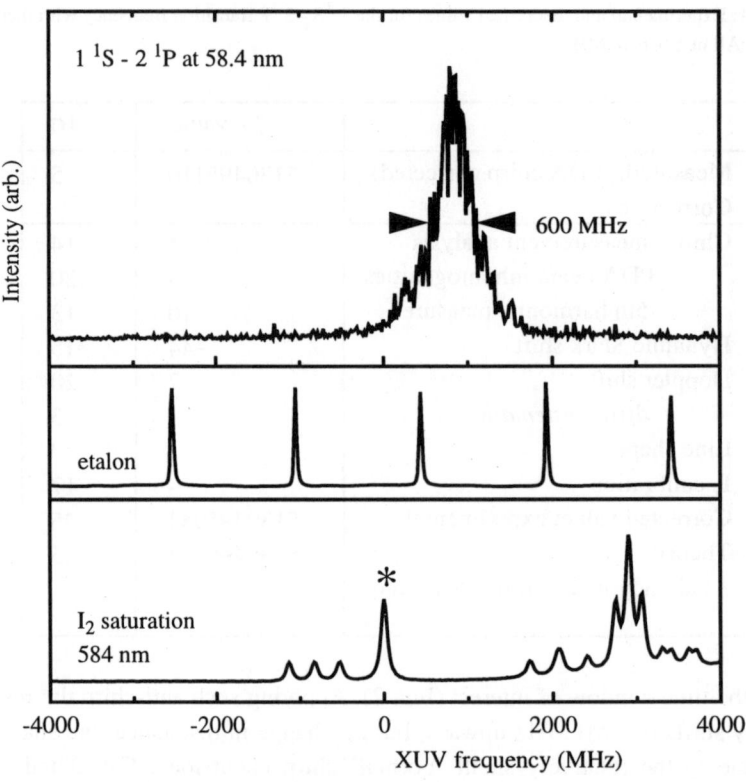

Fig. 4: The 1 ^{1}S - 2 ^{1}P resonance transition for anti-chirped PDA pulses with etalon and I$_2$ saturated absorption spectrum. The * indicates the 'o'-component of the P88(15-1) transition used for absolute calibration.

density of the N$_2$. Due to the increased interest in high-harmonic generation also estimates for this effect are now available [17], which show that the average effect is negligible in our case. The same can be concluded for frequency-doubling-induced chirp in KD*P. This follows from a comparison of a recent realistic calculation [18] with our visible power density of ~72 MW/cm^2 and an estimated maximum phase mismatch of $\Delta k=0.07$ mm^{-1}. The additional chirp from self-phase modulation in KD*P [19] is even smaller. The uncertainty of these small effects are correlated (due to XUV production timing) and are included in the chirp correction uncertainty of 14 MHz.

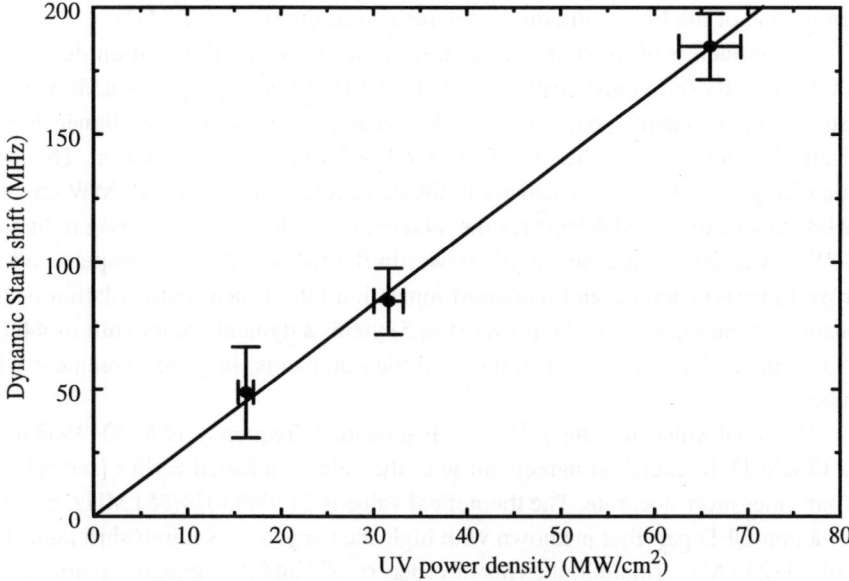

Fig. 5: Dynamic Stark shift measurement in ^{4}He as a function of UV power density in the interaction region. The horizontal error bars indicate the uncertainty in the relative UV power densities. The absolute UV power has an accuracy of 40-50 %.

4 Results

For the final measurement first a Doppler shift of −7(20) MHz was determined from a number of recordings with the He/Kr mixture and with pure helium, all under 'normal' chirp conditions. This was followed by scans with anti-chirped XUV. An example of a scan under anti-chirp conditions is shown in Fig. 4. Included is a saturated absorption spectrum of of the relevant iodine line as well as an etalon spectrum. The 1 ^{1}S- 2 ^{1}P average transition frequency of 5130495110(5) was determined from a weighted average of all normal-chirp and anti-chirped measurements. A small frequency drift with time (attributed to beam alignment changes) introduced a systematic uncertainty of 3 MHz. In our recent experiment [11] backscatter from the skimmer disturbed the atomic beam resulting in a Doppler-broadened asymmetric transition line shape of 950 MHz. With a new skimmer a symmetric transition lineshape of ~600 MHz (natural linewidth is 300 MHz) is now obtained. The uncertainty in the determination of the line centre due to the resonance lineshape is 3 MHz.

The influence of dynamic Stark shift on the $1\,^1S - 2\,^1P$ transition due to the high UV intensity (previously calculated to be 3.1 Hz /Wcm^{-2} [11]) was addressed by measuring the resonance position in quick succession with three pre-aligned lenses (24.3 cm, 33.9 and 49.0 cm) used to focus the UV for harmonic generation. This procedure changes the UV power density in the detection region from ~16 MW/cm^2 for the 24.3 cm lens to ~66 MW/cm^2 for the 49.0 cm lens. The expected strong reduction in XUV power due to decreasing UV power in the focus is largely compensated by improved phase-matching and increased ionization rate. Linear extrapolation of the resonance frequency to zero UV power (Fig.5) yields a dynamic Stark shift of 44(15) MHz for the 24.3 cm lens used in the final measurements, in good agreement with calculations.

Our final value for the $1\,^1S - 2\,^1P$ transition frequency is 5130495083(45) MHz (Table 1), in excellent agreement with the value published earlier [see ref. 11] but four times more accurate. The theoretical value is 5130495074(35) MHz, consisting of a non-QED part that is known with high accuracy, a $1\,^1S$ Lamb shift contribution of –41233 MHz (including terms of order $\alpha^5 Z^6$ and 2-loop corrections) and a $2\,^1P$ Lamb shift contribution of -37.5 (1.8) MHz [3,8]. The error in the theoretical $1\,^1S$ Lamb shift is entirely due to terms of order $\alpha^5 Z^6$. In a 1/Z expansion only the leading term, –68.8 MHz, has been calculated. Higher-order terms are therefore estimated to be smaller by a factor of 2, resulting in a theoretical uncertainty estimate of 35 MHz. The theoretical uncertainty in the $2\,^1P$ level is only 1.8 MHz and therefore not important. As a result an improved ionization potential of He of 198310.6672(15) cm^{-1} can be deduced. The good agreement between the theoretical Lamb shift of 41233(35) MHz and the experimental value of 41224(45) MHz may be interpreted as a verification of the approximation used to calculate a two-electron relativistic contribution of 771.11 MHz of order $\alpha^4 Z^5$ for which no full theory exists. A similar agreement was found for the $2\,^1S$ level in helium [8] and the $2\,^3S$ level in Li$^+$ [20]. With this experiment we achieved an unprecedented accuracy of 9 parts in 10^9 in XUV spectroscopy. It provides an experimental test of QED-phenomena in the ground state of helium at the accuracy level of present day calculations.

Acknowledgements

We wish to thank G.W.F. Drake for making available unpublished calculations and for useful comments, and acknowledge financial support of the Netherlands Foundation for Fundamental Research on Matter (FOM).

References

1. M. Weitz, A. Huber, F. Schmidt-Kaler, D. Leibfried, W. Vassen, C. Zimmermann, K. Pachucki, T.W. Hänsch, L. Julien, F. Biraben, *Phys. Rev.* A **52**, 2664 (1995)
2. K.P. Jungmann in *Atomic Physics* 14, p102; Eds D.J. Wineland, C.E. Wieman and S..J. Smith (American Institute of Physics, 1995).
3. G.W.F. Drake, *Adv. At. Mol. Opt. Phys.* **31**, 1 - 62 (1993), and private communication with G.W.F. Drake. See also: G.W.F. Drake in *Atomic Physics* 13, p3; Eds. H. Walther, T.W. Hänsch and B. Neizert (American Institute of Physics, 1993).
4. D. Shiner, R. Dixson, P. Zhao, *Phys. Rev. Lett.* **72**, 1802 (1994).
5. F.S. Pavone, F. Marin, P. DeNatale, M. Inguscio, F. Biraben, *Phys. Rev. Lett.* **73**, 42 (1994). See also M. Inguscio *et al.* in *Atomic Physics* 14, p81; Eds D.J. Wineland, C.E. Wieman and S..J. Smith (American Institute of Physics, 1995).
6. C. J. Sansonetti and J.D. Gillaspy, *Phys. Rev.* A **45**, R1 (1992).
7. W. Lichten, D. Shiner, Z. X. Zhou, *Phys. Rev.* A **43**, 1663 (1991).
8. G.W.F. Drake, I.B. Khriplovich, A.I. Milstein, A.S. Yelkhovsky, *Phys. Rev.* A **48**, R15 (1993).
9. D. Shiner, R. Dixon and V. Vedantham, *Phys. Rev. Lett.* 74, 3553 (1995).
10. K.S.E. Eikema, W. Ubachs, W. Vassen, and W. Hogervorst, *Phys. Rev. Lett.* **71**, 1690 (1993).
11. K.S.E. Eikema, W. Ubachs, W. Vassen, and W. Hogervorst, *Phys. Rev. Lett.* **76**, 1216 (1996).
12. G. Herzberg, *Proc. Roy. Soc. London* A 248, 309 (1958).
13. C. Wieman and T.W. Hänsch, *Phys. Rev.* A **22**, 192 (1980).
14. M.S. Fee, K. Danzmann, S. Chu, *Phys. Rev.* A **45**, 4911 (1992).
15. N. Melikechi, S. Gangopadhyay, and E.E. Eyler, *J. Opt. Soc. Am.* B **11**, 2402 (1994).
16. I. Reinhard, M. Gabrysch, B. Fischer van Weikersthal, K. Jungmann and G. zu Putlitz, *Appl. Phys.* B. accepted for publication.
17. M. Lewenstein, P. Salières, A. L'Huillier, *Phys. Rev.* A **52**, 4747 (1995).
18. A.V. Smith, M.S. Bowers, *J. Opt. Soc. Am.* B **12**, 49 (1995).
19. R. Adair, L.L. Chase, S. A. Payne, *Phys. Rev.* B **39**, 3337 (1989).
20. E. Riis, A.G. Sinclair, O. Poulsen, G.W.F. Drake, W.R.C. Rowley, and A.P. Levick, *Phys. Rev.* A **49**, 207 (1994).

Towards coherent atomic samples using laser cooling

Jean Dalibard
Laboratoire Kastler Brossel[a], *24 rue Lhomond, 75005 Paris, France*

Abstract

We present in this paper the basic principles of a laser cooling scheme which should lead to a quantum degenerate gas, and we review the various experimental efforts which are presently pursued along those lines. We then explore the coherence properties of such laser cooled gases by studying a *gedanken* experiment in which one would measure the beatnote between two independent degenerate atomic samples.

1 Introduction

Evaporative cooling recently proved to be an impressive technique to obtain strongly degenerate atomic gases and to reach Bose-Einstein condensation (BEC)[1,2,3]. Pursuing with the concurrent cooling scheme, namely laser cooling, may therefore seem at first sight a bit questionable. Our purpose is to show that there are still important aspects of laser cooling which are worth being studied.

First laser cooling offers a wide class of achievable steady-states as the cooling parameters are varied. It is therefore possible to reach situations more general than just the thermal equilibrium of the gas. Also, contrarily to evaporative cooling, this approach does not require any loss. It is not based on collisions and the quantum degeneracy could be reached for very low densities. This is particularly important for systems where inelastic collisions play a significant role and limit the efficiency of evaporative cooling.

We show in the first part of this paper that the radiative cooling of a Bose gas is closely related to the mechanism at the basis of a laser. Starting from atoms in an internal state a, we rely on the process

$$a \longrightarrow b + \text{photon} \qquad (1)$$

to accumulate bosonic atoms in another internal state b and in a given mode of an atomic resonator. The same decay is used in the laser, except that one then takes advantage of the bosonic nature of the photon, the atoms in b being removed as fast as possible in order to maintain population inversion. This parallel between radiative cooling and the laser operation has suggested names such as *atom laser*, *atomaser* or *Boser* for this new device[4,5,6,7,8].

This analogy leads us to reformulate in terms of matter waves questions familiar to quantum optics physicists. As an example, we address in the second part of this paper the problem of the relative phase of two "atom lasers". We consider two atomic samples, each containing N particles in the same internal+external quantum state and we show that they should appear coherent in an experiment that measures

---

[a]Unité de Recherche de l'Ecole Normale Supérieure et de l'Université Pierre et Marie Curie, associée au CNRS(URA 18).

the beatnote between them. The result of the beating experiment is therefore the same as if the two atomic samples are assumed to be in a coherent state with a well defined relative phase.

2 The principle of an "atom laser"

Laser cooling in standard optical molasses leads to temperatures such that $k_B T \geq 20\, E_R$, where $E_R = \hbar^2 k^2/2m$ is the recoil energy associated with a single photon emission or absorption [9]; $\hbar k$ denotes the photon momentum and m the atomic mass. In optical molasses, the atoms are constantly absorbing and emitting photons and the light-assisted collisions between the cold atoms limit the phase space density to $n_0 \Lambda_T^3 \sim 10^{-5}$, where n_0 is the spatial density and $\Lambda_T = h/\sqrt{2\pi m k_B T}$ is the thermal wavelength [10,11].

In order to increase by five orders of magnitude the phase space density and to reach quantum degeneracy, the basic idea is to accumulate the atoms in a state where they do not interact with light anymore. We model such a process using 3-level atoms (figure 1); a and b denote two internal states which are supposed to be stable in absence of radiation. The third relevant internal level e is unstable and it decays rapidly to a or b. Atoms in a and b are confined in a square well potential with a volume V, and the external states are labelled by their momentum $\vec{p}$. The cooling process can be modeled as the repetition of the following cycle. The N atoms are initially prepared in b. They may then be transferred to a using a velocity selective excitation, chosen such that the transfer probability is minimal for $\vec{p} = 0$ (figure 1a). Finally the atoms in a are put back in b using an incoherent pumping process (figure 1b); we assume for simplicity that the momentum distribution of the atoms once they are pumped back into b is uniform within a sphere of radius p_0, of the order of the recoil momentum $\hbar k$. It is clear that the repetition of these cycles favors the accumulation of atoms in the state $|b, \vec{p} = 0\rangle$.

Consider now the evolution of the mean occupation number $n_b(\vec{p})$ of a given state $|b, \vec{p}\rangle$; for simplicity, we use here rate equations, as it is commonly done for usual (photon) lasers. During the first phase of the cooling process, whose duration is noted δt, $n_b(\vec{p})$ decreases by:

$$\delta n_b(\vec{p}) = -\gamma_b(\vec{p})\, \delta t\, n_b(\vec{p}) \tag{2}$$

The departure rate $\gamma_b(\vec{p})$ is supposed to be chosen such as it is minimal for $\vec{p} = 0$, in order to favor the accumulation of atoms into the particular state $|b, \vec{0}\rangle$. In the following we use

$$\gamma_b(\vec{p}) = \gamma_{b0} + \alpha p^2 \tag{3}$$

which is easily implemented experimentally (see below), but other variations of $\gamma_b(\vec{p})$ with $|\vec{p}|$ can also be achieved.

After this first phase the number of atoms present in a is

$$N_a = \sum_{\vec{p}} \delta n_b(\vec{p}) \tag{4}$$

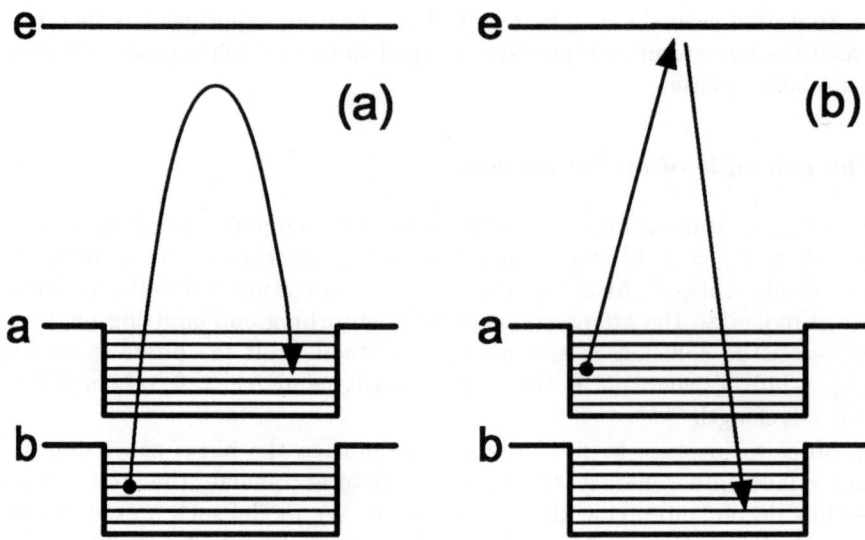

Figure 1: The atomic level scheme considered in this paper. Levels a and b are stable in absence of laser light. Level e is unstable and it decays to a or b. The atoms are initially in state b. They may be transferred to a using a velocity selective excitation (figure 1a). They are then put back in b using an incoherent optical pumping process (figure 1b).

We assume for simplicity in the following that only a small fraction of the atoms is actually transferred to a in a single pulse: $N_a \ll N$. These atoms are then pumped back into b and the population of a given state $|b, \vec{p}\rangle$ increases as:

$$\delta n_b(\vec{p}) = \frac{N_a}{N_{\text{lev}} + N}(1 + n_b(\vec{p})) \qquad (5)$$

N_{lev} represents the number of levels $|b, \vec{p}\rangle$ which can be reached through the spontaneous Raman process $a \to e \to b$ ($N_{\text{lev}} = (V/h^3)(4\pi p_0^3/3)$). In practice this number is much larger than the number of atoms N, otherwise quantum degeneracy would occur after a single cooling cycle. In (5), we take into account both the spontaneous and the stimulated emission [6,12] of a bosonic atom into the state $|b, \vec{p}\rangle$, leading to the factor $1 + n_b(\vec{p})$.

Strictly speaking one should also add to (5) a loss term describing the reabsorption of the photons scattered in the pumping process, by atoms already in b. This corresponds to the inverse process of (1). This reabsorption problem is quite serious and it may actually limit in a dramatic way the accumulation of atoms in the desired state. Several solutions have been proposed to it. One might use a broad band repumping laser for initiating the transition $a \to e \to b$; this should leads to a reduced reabsorption cross-section [7]. The reabsorption problem may also be circumvented by the use of a sufficiently slow pumping process from a to b [13]. In addition well designed confining potentials may reduce the problem: these optimized wells may correspond to a quasi-bidimensional geometry, in which the atoms are confined into a plane, whereas the scattered photons emerge most probably out of this plane [7,14]; the use of small wells may also lead to tiny 3D traps close to the so-called Lamb-Dicke limit; the size of the trap is then comparable to the light wavelength and one can show [15] that the gain due to the process (1) exceeds the losses due to reabsorption. Finally we note that it has been predicted that, once a macroscopic population has appeared in a given state, quantum interference phenomena can suppress the problem due to reabsorption [16]: this is the Boson Accumulation Regime (BAR).

We have also neglected here the short range atom-atom interactions, which correspond to a $1/r^3$ resonant dipole-dipole potential. They may play a significant role for high densities [17,18,19] since they lead to resonant light absorption by pairs of close atoms, even when the repumping laser inducing the $a \to e \to b$ is off resonant for the $a \to e$ transition for a single atom. The spurious effect of these short range interactions, which may empty the lasing state $|b, \vec{p} = 0\rangle$, could be reduced if one uses a red detuned repumping laser [18]. It has also been suggested [20] that those interactions may, for well chosen parameters, be benefic and induce the operation of the atom laser.

The steady state of (2,5) is

$$n_b(\vec{p}) = \frac{1}{\xi \gamma_b(\vec{p}) - 1} \quad \text{with} \quad \xi = \frac{N_{\text{lev}} + N}{N_a} \delta t \qquad (6)$$

The coefficient ξ is determined from the total number of atoms present in the system:

$$N = \sum_{\vec{p}} n_b(\vec{p}) \qquad (7)$$

The structure of this result is very similar to the Bose-Einstein distribution

$$n(\vec{p}) = \frac{1}{\xi \exp(E(\vec{p})/k_B T) - 1} \qquad (8)$$

where $\xi = \exp(-\mu/k_B T)$ is also determined from the total number of particles and where μ is the chemical potential. In particular a phenomenon similar to the Bose-Einstein condensation, i.e. a macroscopic population in the state $|b, \vec{p}\rangle = 0$, occurs when the rate of escape γ_{b0} out of this state is sufficiently low [7,21] (c.f. fig. 2). The condition for this "condensation" or "atom lasing" is:

$$4\pi \frac{\gamma_{b0}}{\alpha p_0^2} \leq n_0 \Lambda_0^3 \qquad (9)$$

where $n_0 = N/V$ denotes the atomic density and $\Lambda_0 = h/p_0$.

In order to implement experimentally the basic ideas presented above, which involve optical pumping into in a state which is decoupled from the light, several schemes are presently investigated. The velocity selective coherent population trapping (VSCPT) scheme consists in accumulating atoms in a coherent superposition of two or several atomic Zeeman substates, such that the photon absorption probability amplitudes from these substates interfere destructively [22]. The excitation rate of this "uncoupled state", which plays the role of the state labelled above as b, varies as p^2 as in (3). The VSCPT scheme has been demonstrated experimentally

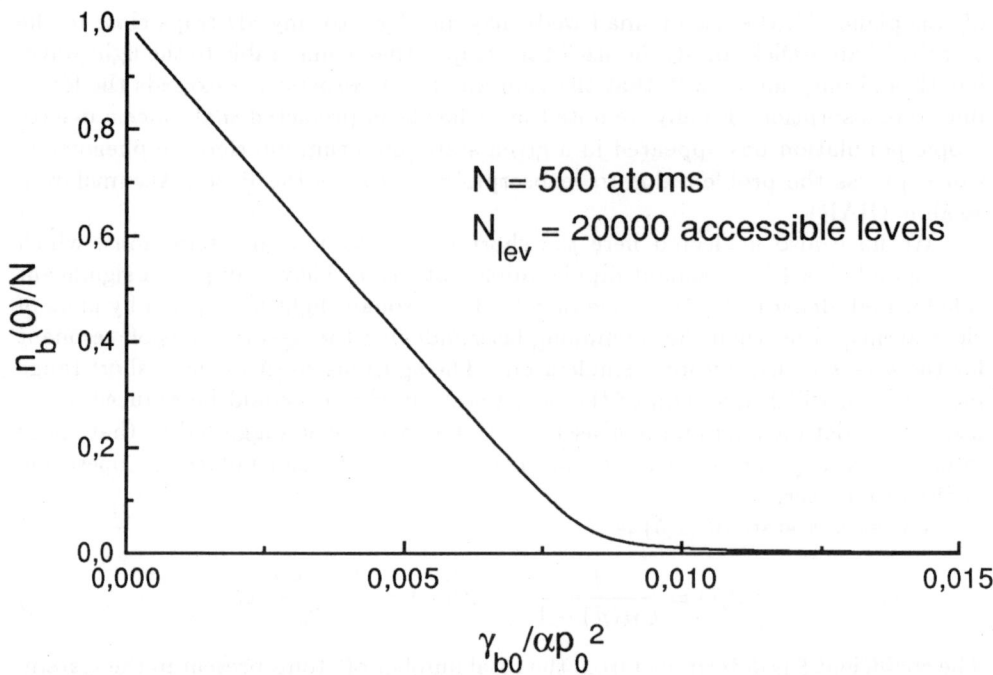

Figure 2: Population in the state $|b, \vec{p} = 0\rangle$ as a function of the ratio $\gamma_{b0}/(\alpha p_0^2)$. γ_{b0} represents the rate of transfer from b to a for $\vec{p} = 0$, while $\alpha p_0^2 (\gg \gamma_{b0})$ is the transfer rate for the fast atoms $|\vec{p}| = p_0$. This figure has been calculated for $N = 500$ atoms in a box such that $V/\Lambda_0^3 = 5000$, corresponding to $N_{\text{lev}} \simeq 20000$. The transition predicted from (9) occurs for $\gamma_{b0}/(\alpha p_0^2) = 0.0796$.

in 3D [23] for free atoms, and it has been studied in the context of an atom laser by several authors [4,18,21].

The Raman cooling method, which was first demonstrated experimentally for free atoms [24,25,26,27], has now been extended to trapped atoms [28,29]. The levels a and b correspond in this case to the two hyperfine ground levels of an alkali atom. This cooling scheme consists in transferring atoms from b to a using laser pulses whose frequency and temporal shape are carefully chosen. In this way, one achieves excitation rates $\gamma_b(\vec{p})$ scaling as p^β, where the exponent β depends on the precise temporal variation of the laser pulses [27]. The case $\beta = 2$ (eq. 3) corresponds to a square temporal profile. The atomic trap is provided using a far off resonant laser: Argon ion laser (488 and 514 nm) for sodium atoms [28], or Yag laser (1064 nm) for cesium atoms [29]. For a blue detuning [28], the laser forms a box with sheets of light which repel the atoms. For a red detuning [29], the atoms accumulate at the laser waist, where the light intensity is maximal. The method of Raman cooling of trapped atoms has recently led to the highest increase in phase space achieved by an all-optical cooling method [28].

There is also an experimental effort for going to geometries where the atoms are tightly confined. The guiding of atoms in a blue detuned doughnut mode, combined with transverse Sisyphus cooling, is currently studied in Hanover [30]. The

Konstanz group is presently investigating the optical pumping of rare gas atoms in a far detuned blue optical lattice [5,31]. The photon scattering by atoms trapped in the lattice then provides a simple mode selection, since this scattering is larger for hotter atoms, exploring regions with higher laser intensity. Finally experiments with reduced dimensionality have been proposed [33,14,32]; they consist in accumulating atoms in the immediate vicinity of a dielectric prism, in the potential well created by an evanescent wave propagating at the surface of the prism.

3 The relative phase of two condensates

The previous section indicates clearly the vitality of this field of research, aiming to reach a degenerate atomic gas using pure optical cooling. The ultimate goal is to accumulate nearly all atoms in a given mode of an atomic resonator. In the previous section we considered cooling parameters (in particular (3)) such that the lasing mode is $|b, \vec{p} = 0\rangle$; the analogy with BEC is then quite clear.

Once all, or nearly all atoms, are accumulated in a given state, the question of the phase coherence of the atomic gas rises, both for BEC and for an atom laser. Theoretically, this phase appears naturally as a result of a broken symmetry in the theory of BEC [34,35]. At zero temperature, the atomic sample is described by a coherent state, i.e. an eigenstate of the annihilation operator for a particular state of the one-atom Hilbert space. A classical field $|\psi_0| e^{i\phi}$ is associated to this coherent state, with a well defined amplitude $|\psi_0|$ and phase ϕ. This description in terms of atomic coherent states can be viewed also as a direct transposition to the atom laser problem of the standard laser theory which leads, at the output of a photonic laser, to an electromagnetic field described by a coherent state [36]. Experimentally however, one can in principle measure the exact number of trapped atoms. The gas is then described by a Fock state (or number state) with no definite phase. The question then arises of whether these two different descriptions lead to identical predictions for a given experimental setup.

To investigate this problem, we consider the following *Gedanken* experiment, using two atomic resonators with the same atomic species [37]. We assume that each resonator ends with an atomic mirror through which the atoms can tunnel (figure 3). The phase between the two emerging beams can be probed by "beating" them together, i.e. by mixing them onto a 50–50 atomic beam splitter [38].

If each resonator is in a coherent state with the same average number of atoms, the beams incoming from the left and right resonators onto the beam splitter are described by the two fields $|\psi_0| e^{i\phi_l}$ and $|\psi_0| e^{i\phi_r}$. The intensities in the two outports of the beam splitter are then:

$$I_+ = 2|\psi_0|^2 \cos^2 \phi \quad I_- = 2|\psi_0|^2 \sin^2 \phi \tag{10}$$

where $\phi = (\phi_l - \phi_r)/2$. From the ratio I_+/I_-, one then deduces the value of the relative phase $2|\phi|$. Note that ϕ is an unpredictable random variable, which takes a different value for any new realization of the experiment.

We now turn to the description of the system in terms of Fock states. We assume that the system is prepared in the Fock state $|N_l, N_r\rangle$ with $N_l = N_r = N$, neglecting for simplicity shot to shot variations of N_l and N_r. Our purpose is to

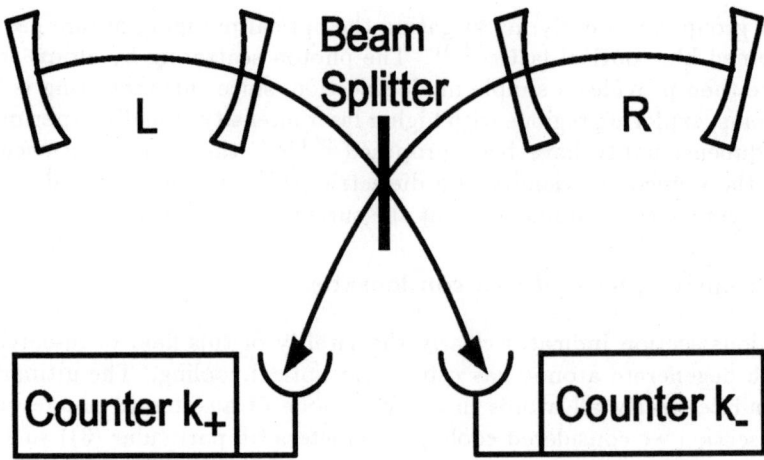

Figure 3: A *gedanken* experiment: atoms leaking from two atomic resonators L and R are detected in the output channels ($\pm$) of a 50-50 beam splitter.

show that, in the absence of interaction between the L and R gases, the predictions corresponding to the initial state $|N, N\rangle$ are identical to (10) if $N \gg 1$. The notion of phase broken symmetry is therefore not mandatory to understand the beating of two atom lasers or two condensates [39]. On the other hand it provides a simple way for the analysis of such an experiment while, as we see below, Fock states are more delicate to handle in such a situation.

The problem that we are facing here is analogous to the question raised by P.W. Anderson [40]: *Do two superfluids which have never "seen" one another possess a definite relative phase?* As pointed out by A.J. Leggett [34], the question is meaningless as long as no measurement is performed onto the system. One can also show [34] that for some class of interactions between the two condensates, the ground state of the total system has a definite relative phase between the two condensates. J. Javanainen addressed recently a similar question by considering the spatial interferences of two *ideal* condensates prepared in the state $|N, N\rangle$, and arriving on a given array of detectors [41]. He showed numerically that after the detection of all the atoms of the two condensates, the count distribution on the set of detectors was similar to the one predicted from a phase broken symmetry state. The effect of the atomic interactions onto the interference pattern has been included by Naraschewski *et al*[42].

We assume that $k \ll N$ atoms are detected on $D_\pm$, and we focus first on the case where all these detections occur in the $(+)$ channel. If the system is initially in a coherent state, the probability for such a sequence is $\cos^{2k} \phi$, which after averaging over the unknown relative phase 2ϕ gives:

$$W_k = \int_0^\pi \frac{d\phi}{\pi} \cos^{2k} \phi = \frac{(2k)!}{(2^k k!)^2} \sim_{k \gg 1} \frac{1}{\sqrt{\pi k}} \qquad (11)$$

For $k = 100$, this probability to get all counts in the (+) channel is $\sim 6\%$. Suppose now that the system is in the Fock state $|N, N\rangle$; a naive argument could consist in saying that since $k \ll N$, the probability for detecting the n-th atom ($n \leq k$) in the channel (+) is nearly independent of the $n - 1$ previous detection results. The probability for k detections in the channel (+) should then be 2^{-k}. This is obviously very different from the result W_k obtained from a coherent state ($2^{-k} < 10^{-30}$ for $k = 100$).

However the latter reasoning is wrong; the first detection of an atom in the (+) channel projects the atom in a state proportional to $(\hat{l} + \hat{r})|N, N\rangle \propto |\Psi\rangle = |N, N - 1\rangle + |N - 1, N\rangle$, where $\hat{l}$ and $\hat{r}$ annihilate a particle in the left and right condensates respectively. To calculate the probability to detect a second atom in the (+) channel, we have to compare the squared norm of the two vectors corresponding to a detection in the ($\pm$) channels:

$$\begin{aligned}(+) : (\hat{l} + \hat{r})|\Psi\rangle &= \sqrt{N-1}(|N-2, N\rangle + |N, N-2\rangle) \\ &\quad + 2\sqrt{N}|N-1, N-1\rangle \end{aligned} \quad (12)$$

$$(-) : (\hat{l} - \hat{r})|\Psi\rangle = \sqrt{N-1}(|N-2, N\rangle - |N, N-2\rangle) \quad (13)$$

For $N = 1$ we recover the well known interference effect leading to a bunching of the two Bosons in a single outport of the beam splitter [43]. For $N \gg 1$ the squared norms of these two vectors are in the ratio 3:1. This indicates that once a first atom has been detected in the (+) channel, the probability for detecting the second atom in the same channel is 3/4, while the probability for detecting this second atom in the (-) channel is only 1/4. This somehow counter-intuitive result shows clearly that the successive detection probabilities are strongly correlated in the case of an initial Fock state, even if the number of detected atoms is very small compared to the number of atoms present in the condensates. The reasoning can be extended to k detections and we find that the probability to detect respectively $k_+ = k$ and $k_- = 0$ atoms in the two channels is:

$$\mathcal{P}(k, 0) = \frac{1}{2} \frac{3}{4} \cdots \frac{2k-1}{2k} \quad (14)$$

which is equal to W_k for any k. The predictions for an initial Fock state and for an initial coherent state with random phase are therefore equivalent, but the result for the coherent state is obtained in a much more straightforward and intuitive manner than for the Fock state.

Consider now the general case of $k_\pm$ detected atoms in the ($\pm$) channels, for a fixed number of measurements $k = k_+ + k_-$. An initial coherent state is not modified by the measurements since it is an eigenstate of $\hat{l}$ and $\hat{r}$; the probability for the result (k_+, k_-) is:

$$\mathcal{P}(k_+, k_-, \phi) = \frac{k!}{k_+! \, k_-!} (\cos \phi)^{2k_+} (\sin \phi)^{2k_-} \quad (15)$$

In the limit $k_\pm \gg 1$, using $\ln n! \sim n \ln n - n$ for $n \gg 1$, we find that $\mathcal{P}(k_+, k_-, \phi)$ is maximal for $k_-/k_+ = \tan^2 \phi$, as expected from (10). In other words, for $k \gg 1$, mean and most probable intensities coincide, since the shot noise on the signal in the two channels (±) becomes negligible.

For an initial Fock state $|N, N\rangle$, the evolution due to the sequence of measurements is conveniently analyzed by expanding $|N, N\rangle$ onto the overcomplete set of *phase states*[34] $|\phi\rangle_{2N}$:

$$|\phi\rangle_{2N} = 2^{-N}((2N)!)^{-1/2} \left(\hat{l}^\dagger e^{i\phi} + \hat{r}^\dagger e^{-i\phi}\right)^{2N} |0\rangle. \quad (16)$$

If the system is in a given state $|\phi\rangle_{2N}$ there exists a well defined relative phase 2ϕ between the left and right gases. Any state $|\Psi\rangle$ with $2N$ particles can be expanded in the phase set:

$$|\Psi\rangle = \int_0^\pi d\phi\, c(\phi)\, |\phi\rangle_{2N} \quad (17)$$

where the phase amplitude $c(\phi)$ is obtained as:

$$c(\phi) = \frac{1}{\pi} \sum_{n=-N}^{N} 2^N \left(\frac{(N+n)!\,(N-n)!}{(2N)!}\right)^{1/2} e^{2in\phi} \langle N-n, N+n|\Psi\rangle. \quad (18)$$

In what follows we will use the quasi-orthogonality of the phase states valid for large N and for $0 \leq \phi, \phi' \leq \pi$:

$$_{2N}\langle\phi|\phi'\rangle_{2N} = \cos^{2N}(\phi - \phi') \simeq \sqrt{\pi/N}\, \delta(\phi - \phi') \quad (19)$$

According to (18) the initial state $|N, N\rangle$ has a flat phase amplitude:

$$|N, N\rangle \propto \int_0^\pi d\phi\, |\phi\rangle_{2N} \quad (20)$$

After a sequence of $(k_+, k_- = k - k_+)$ detections the state of the system is:

$$|\Psi(k_+, k_-)\rangle \propto (\hat{l} + \hat{r})^{k_+}(\hat{l} - \hat{r})^{k_-}|N, N\rangle \propto \int_0^\pi d\phi\, (\cos\phi)^{k_+}(\sin\phi)^{k_-}|\phi\rangle_{2N-k} \quad (21)$$

For $k_\pm \gg 1$, we find using the stationary phase method:

$$|\Psi(k_+, k_-)\rangle \propto \int_0^\pi d\phi\, \left(e^{-k(\phi-\phi_0)^2} + (-1)^{k_+}e^{-k(\phi-\pi+\phi_0)^2}\right)|\phi\rangle_{2N-k} \quad (22)$$

with $\tan\phi_0 = \sqrt{k_-/k_+}$, $0 \leq \phi_0 \leq \pi/2$. The interpretation of this result is quite clear: initially, the relative phase of the two condensates is indefinite since the vector state of the system projects equally onto the various phase states (20). After $k \gg 1$ detections, the system has evolved into a state where the phase ϕ is well defined; more precisely, the phase distribution is a double Gaussian, centered on ϕ_0 and $\pi - \phi_0$, with a standard deviation $1/\sqrt{2k}$. Note that this ambiguity between ϕ_0 and $\pi - \phi_0$ also arises in the determination of ϕ from (10).

In addition we can check that the probability for getting the result $(k_+, k_- = k - k_+)$ is the same for (i) a system prepared initially in a Fock state and (ii) a system prepared in a coherent state after average over the unknown phase ϕ. Using (21) and (19), and summing over the various possible ways to get $k_\pm$ counts in the ($\pm$) channels, we get for $N \gg 1$:

$$\mathcal{P}(k_+, k_-) = \alpha(k_+, k_-) \| (\hat{l} + \hat{r})^{k_+} (\hat{l} - \hat{r})^{k_-} |N, N\rangle \|^2 \sim \int_0^\pi \frac{d\phi}{\pi} \mathcal{P}(k_+, k_-, \phi) \quad (23)$$

with

$$\alpha(k_+, k_-) = \frac{(2N-k)!}{2^k (2N)!} \frac{k!}{k_+! \, k_-!} \quad (24)$$

Note that this equivalence is obtained for any, non necessarily large k.

In this model, it is also possible to include the effects of atomic interactions within each subsystem L and R [37]. As the system evolves, these interactions lead to collapses and revivals of the peaked phase distribution (22) [45].

To summarize we have two different points of view onto the system: for an initial coherent state, the measurement "reveals" the pre-existing phase through $\tan^2 \phi = k_-/k_+$; for an initial Fock state, the detection sequence "builds up" the phase [44]. There is no possibility to favor one particular point of view from experimental results if $N \gg 1$. If the same experimental sequence involving k detections is repeated with the same initial preparation – but a different phase varying randomly from shot to shot in the coherent state point of view, since those are different realizations – the predicted occurrence of a given result $k_+, k_- = k - k_+$ is identical in the two points of view.

This work has been done in collaboration with Y. Castin, R. Dum and M. Olshanii. Very helpful discussions with C. Cohen-Tannoudji, F. Laloë and C. Salomon are also acknowledged. This work is supported by the DRET.

1. M.H. Anderson, J.R. Ensher, M.R. Matthews, C.E. Wieman and E.A. Cornell, Science **269**, 198 (1995).
2. C.C. Bradley, C.A. Sackett, J.J. Tollett, and R.G. Hulet, Phys. Rev. Lett. **75**, 1687 (1995).
3. K. Davis, M.O. Mewes, M.R. Andrews, N.J. van Druten, D.S. Durfee, D.M. Kurn and W. Ketterle, Phys. Rev. Lett. **75**, 3969 (1995).
4. H.M. Wiseman and M.J. Collett, Phys. Lett. **A 202**, 246 (1995).
5. R. Spreeuw, T. Pfau, U. Janicke and M. Wilkens, Europhys. Lett. **32**, 469 (1995).
6. Ch.J. Bordé, Phys. Lett. **A 204**, 217 (1995).
7. M. Olshanii, Y. Castin and J. Dalibard, Proceedings of the XII Laser Spectroscopy International Conference (Capri, June 1995), M. Inguscio, M. Allegrini and A. Sasso Edts, World Scientific (1995).
8. We concentrate here on "atom lasers" based on interaction of atoms with light. Another scheme for achieving the same result, *i.e.* stimulated emission of atoms in a mode of an atomic resonator, is based on elastic collisions: M. Holland, K. Burnett, C. Gardiner, J.I. Cirac and P. Zoller, unpublished (1995); H. Wiseman, A. Martins and D. Walls, Quantum Semiclass. Opt. **8**, 737 (1996).

9. C. Cohen-Tannoudji and W.D. Phillips, Physics Today, October 1990, p. 33. H. Metcalf and P. van der Straten, Phys. Rep. **244**, 203 (1994).
10. M. Drewsen, P. Laurent, A. Nadir, G. Santarelli, A. Clairon, Y. Castin, D. Grison and C. Salomon, Appl. Phys. **B 59**, 283 (1994).
11. C.G. Townsend, N.H. Edwards, C.J. Cooper, K.P. Zetie, C.J. Foot, A.M. Steane, P. Szriftgiser, H. Perrin and J. Dalibard, Phys. Rev. **A 52**, 1423 (1995).
12. M. Naraschewski, H. Wallis and A. Schenzle, Phys. Rev. **A 54**, 677 (1996).
13. J.I. Cirac, M. Lewenstein and P. Zoller, preprint (May 1996).
14. T. Pfau and J. Mlynek, preprint (March 1996).
15. U. Janicke and W. Wilkens, to be published.
16. J.I. Cirac and M. Lewenstein, Phys. Rev. **A 53**, 2466 (1996).
17. E. Goldstein, P.Pax, K.J. Schernthanner, B. Taylor and P. Meystre, Appl. Phys. **B 60**, 161 (1995).
18. T.W. Hijmans, G.V. Shlyapnikov and A.L. Burin, preprint (April 1996).
19. K. Burnett, P. Julienne and K.A. Suominen, preprint (April 1996).
20. A.M. Guzmán, M. Moore and P. Meystre, Phys. Rev. **A 53**, 977 (1996).
21. M. Olshanii, Y. Castin and J. Dalibard, Proceedings of the CXXXI International School of Physics Enrico Fermi, *Coherent and Collective Interaction of Particles and Radiation* (Varenna, July 1995), A. Aspect, W. Barletta and R. Bonifacio, Edts. (1996).
22. A. Aspect, E. Arimondo, R. Kaiser, N. Vansteenkiste and C. Cohen-Tannoudji, Phys. Rev. Lett. **61**, 826 (1988).
23. J. Lawall, S. Kulin, B. Saubamea, N. Bigelow, M. Leduc and C. Cohen-Tannoudji, Phys. Rev. Lett. **75**, 4194 (1995).
24. M. Kasevich and S. Chu, Phys. Rev. Lett. **69**, 1741 (1992).
25. N. Davidson, H.J. Lee, M. Kasevich and S. Chu, Phys. Rev. Lett. **72**, 3158 (1994).
26. J. Reichel, O. Morice, G.M. Tino and C. Salomon, Europhys. Lett. **28**, 477 (1994).
27. J. Reichel, F. Bardou, M. Ben Dahan, E. Peik, S. Rand, C. Salomon and C. Cohen-Tannoudji, Phys. Rev. Lett. **75**, 4575 (1995).
28. H.J. Lee, C.S. Adam, M. Kasevich and S. Chu, Phys. Rev. Lett. **76**, 2658 (1996).
29. H. Perrin, A. Kuhn, W. Hänsel and C. Salomon, to be published.
30. W. Ertmer, communication in the Les Houches Workshop on Bose-Einstein Condensation, April 1996.
31. R. Spreeuw, communication in the Les Houches Workshop on Bose-Einstein Condensation, April 1996.
32. P. Desbiolles and J. Dalibard, to appear in Opt. Commun. (1996).
33. J. Söding, R. Grimm and Yu. B. Ovchinnikov, Opt. Commun. **119**, 652 (1995).
34. A. J. Leggett and F. Sols, Found. of Physics **21**, 353 (1991).
35. A. Griffin, Phys. Rev. B **53**, 9341 (1996) and references therein.
36. For a review see M. Lewenstein and L. You, to appear in *Advances of Atomic and Molecular Physics*, B. Bederson and H. Walther Edts (1996).

37. Y. Castin and J. Dalibard, submitted for publication (June 1996).
38. C.S. Adams, M. Sigel and J. Mlynek, Physics Reports **240**, 143-210 (1994).
39. The optical equivalent of this phenomenon has been recently discussed by K. Mølmer (submitted to Phys. Rev. A).
40. P.W. Anderson, in *The Lesson of Quantum Theory*, edited by J. de Boer, E. Dal and O. Ulfbeck, Elsevier Science Publishers B.V. (1986).
41. J. Javanainen, Phys. Rev. Lett. **76**, 161 (1996).
42. M. Naraschewski, H. Wallis, A. Schenzle, J.I. Cirac and P. Zoller, to be published in Phys. Rev. A. (1996).
43. R.P. Feynman, R.B. Leighton and M. Sands, *The Feynman lectures on Physics, Quantum Mechanics* (Addison Wesley, 1965).
44. A similar conclusion is obtained in the frame work of continuous measurement theory in J.I. Cirac, C.W. Gardiner, M. Naraschewski and P. Zoller, preprint (June 1996).
45. A similar result holds for superfluid helium and superconductors (F. Sols, Physica B **194-196**, 1389 (1994)), and for the order parameter $\langle \psi(\vec{r}) \rangle$ of a single trapped condensate: E.M. Wright and D.F. Walls, preprint (March 1996); M. Lewenstein and L. You, preprint (July 1996).

Bose-Einstein condensation of a weakly-interacting gas

C.G. Townsend, N. J. van Druten[a], M.R. Andrews, D.S. Durfee, D.M. Kurn, M.-O. Mewes and W. Ketterle

Research Laboratory of Electronics and Department of Physics, Massachusetts Institute of Technology, Cambridge, MA 02139, USA

Abstract

We review the recent achievements in observing Bose-Einstein condensation (BEC) in dilute atomic gases, and summarize our own studies of BEC in sodium. Thermal sodium atoms were optically trapped and cooled and then transferred to a cloverleaf magnetic trap. Radio-frequency induced evaporation increased the phase-space density by six orders of magnitude and condensates with up to ten million atoms were observed. Basic properties of the condensate such as condensate fraction and mean-field energy were found to be in agreement with theory. "Dark-ground" imaging has been used to observe the condensate directly and non-destructively. We have also investigated the low frequency oscillation modes of a condensate, and taken the first steps towards an "atom laser" by demonstrating an rf output coupler for a Bose condensate.

1 INTRODUCTION

Bose-Einstein condensation (BEC) is one of the most intriguing phenomena that is predicted by quantum statistical mechanics. As Einstein calculated in 1924, and published in 1925 [1], an ideal quantum gas will undergo a phase transition when the average interparticle spacing becomes comparable to the thermal de Broglie wavelength. To be specific, the phase transition occurs when

$$n\Lambda^3 = 2.612, \qquad (1)$$

where n is the peak number density of the sample, and $\Lambda = (2\pi\hbar^2/mk_BT)^{1/2}$ is the thermal de Broglie wavelength, with m the mass of the particle and T the gas temperature[2]. Below the transition temperature a macroscopic fraction of the particles "condenses" by occupying the single quantum-mechanical ground state of the system. The rest of the particles, the "normal" fraction, behaves like a saturated gas, with the saturated density given by Eq. (1), where n is now the peak number density of the normal fraction. The quantity $n\Lambda^3$ is generally used as a measure of proximity to BEC. (For high temperatures it is the occupancy of the ground state and is usually referred to as the "phase-space density").

The history of the theory of BEC is interesting and deserves a review on its own. For instance, Einstein made his predictions before quantum theory had been fully developed, and before the differences between bosons and fermions had been revealed. After Einstein, important contributions were made by, most notably, London, Landau, Bogoliubov, Penrose and Onsager, Feynman, Lee, Yang and Huang,

---

[a] Present address: Huygens Laboratory, Leiden University, P.O. Box 9504, 2300 RA Leiden, The Netherlands.

Goldstone and Anderson (see Refs.[2,3,4]). An important issue has always been the relationship between BEC and superfluidity in liquid helium.

An important aspect of BEC is the role of interactions between the particles. On the one hand BEC is unique in that it is a purely quantum-statistical phase transition, i.e. it occurs even in the absence of interactions (Einstein described the transition as "condensation without interactions"[1]). This makes BEC an important paradigm of statistical mechanics. On the other hand, real-life particles will always have some weak interaction, and the weakly-interacting Bose gas behaves qualitatively differently than the ideal Bose gas (see, e.g., Ref.[2]). Finally, although BEC has been invoked as an important process in such diverse fields as condensed matter, nuclear, elementary-particle and astrophysics, the only experimental systems that exhibit Bose-Einstein condensation were, until recently, liquid helium and excitons[4,5]. However, the strong interactions at the densities of a liquid or a solid considerably modify and complicate the nature of the phase transition. This explains why the experimental realization of BEC in a dilute atomic gas was a long-standing goal in atomic physics. It was realized quite early on that a gas of atomic hydrogen would be an excellent candidate for observing BEC[6]. After the stabilization of atomic hydrogen by Silvera and Walraven in 1979[7], experimental efforts first focussed on compressing hydrogen in cells, and then on evaporative cooling in magnetic traps (see, for example, Refs.[8,9,10] for reviews of the hydrogen work). The closest approach to BEC with $n\Lambda^3 = 0.4$ was achieved by Doyle, Greytak, Kleppner and collaborators in 1991[11].

In the last few years experimental progress in atomic hydrogen has been thwarted by the presence of decay processes (recombination and dipolar decay), and by difficulties in detection due to the lack of easily accessible optical transitions from the hydrogen ground state. In contrast, this same period witnessed tremendous progress for alkali atoms stemming from developments in laser cooling and trapping in combination with magnetic trapping and evaporative cooling. In 1995, within a few months, this culminated in two independent groups observing BEC[12,13] and a third one obtaining evidence for evaporative cooling into the quantum-degenerate regime[14]; see also Table 1.

Although this field is still in its infancy (at the time of writing BEC was observed about one year ago), experiments on BEC in atomic alkali gases have already provided the first quantitative verification of the theory for weakly-interacting Bose gases, and are posing new challenges for theorists. In this paper we summarize the experimental techniques that we have used to obtain BEC and the studies undertaken so far. The review extends and updates our previous summary[15], and a similar version will appear in the proceedings of LT 21.

2 EXPERIMENTAL TECHNIQUES

The development of laser cooling and trapping of atoms during the 80's has profoundly changed atomic physics. It has opened up a new route to low temperature physics that does not depend on cryogenic methods, and submillikelvin samples of laser-cooled atoms are now routinely used in a large variety of experiments. The techniques commonly employed are Doppler cooling[17,18], polarization-gradient

	JILA 95[12]	Rice 95[14]	MIT 95[13]	MIT 96[16]
Atom	^{87}Rb	^{7}Li	Na	Na
Scattering length [nm]	+6	−1.5	+3	+3
Trap	TOP	permanent magnetic trap	opt. plugged magnetic trap	cloverleaf magnetic trap
$B_x'' B_y'' B_z''$ [$(10^3$ Gcm$^{-2})^3$]	27	1	4×10^6	100
First BEC	June 95	(July 95)	September 95	March 96
Evidence	TOF	(Halo)	TOF	TOF, in-situ image
N_C	2×10^4	2×10^5	2×10^6	15×10^6
T_C [μK]	0.1	0.4	2	1.5
n_C [cm^{-3}]	2×10^{12}	2×10^{12}	1.5×10^{14}	10^{14}
N_0	2,000		5×10^5	5×10^6
Cooling time	6 min.	5 min.	9 s	30 s
BEC atoms/s	6		60,000	200,000
Lifetime [s]	≈ 15	≈ 20	≈ 1	≈ 20

Table 1: Comparison of BEC experiments reported thus far. The relevant figure of merit of a trap for evaporative cooling is the product of the three effective magnetic field curvatures, which are defined to be the curvatures of the trapping potential divided by the magnetic moment [16]. N_c, T_c and n_c are atom number, temperature and number density at the phase transition respectively. N_0 is the number of condensate atoms. Recently the Boulder group reported reaching $N_0 = 1.5 \times 10^6$ in a purely magnetic trap (see their review paper in this volume).

cooling [19,20,18] and the magneto-optical trap (MOT) [21]. However, in these now conventional cooling and trapping techniques temperatures are limited by heating due to spontaneous emission and densities by radiation trapping effects [22]. Despite recent advances [23,24,25,26] the closest approach to BEC that one can achieve with them is short in phase-space by five orders of magnitude. This prompted activity to find different techniques to increase phase-space density, including work on sub-recoil cooling [27,28,29], optical lattices [30,31,32] and far-off-resonant dipole traps [28,33]. The closest approach to BEC using only optical techniques reached a phase-space density 400 times lower than that required for BEC [28].

In the end the successful approach was to combine laser cooling and optical trapping with magnetic trapping and evaporative cooling. The technique of evaporative cooling was developed by Doyle, Greytak, Hess, Kleppner and collaborators at MIT as a method for cooling atomic hydrogen which had been precooled by cryogenic methods [34,35,11,8]. The challenge in applying this scheme to alkali atoms was to achieve simultaneously effective laser cooling and trapping, which work best at low number densities, and efficient evaporative cooling, which requires high densities. This shifted the emphasis of the optical techniques from attaining low temperatures and high phase-space density towards achieving high elastic collision rates in an ultra-high vacuum environment. Work in this direction started in our group in 1991 and became the sole focus in 1992 (until 1993 in collaboration with D. E. Pritchard). Similar techniques were developed in parallel by E. Cornell and C. Wieman at Boulder (see their review paper in this volume and Ref. [12]). Related work was done at Rice University [14] and at Stanford [36]. An overview of the development of evaporative cooling in alkali gases is shown in Fig. 1.

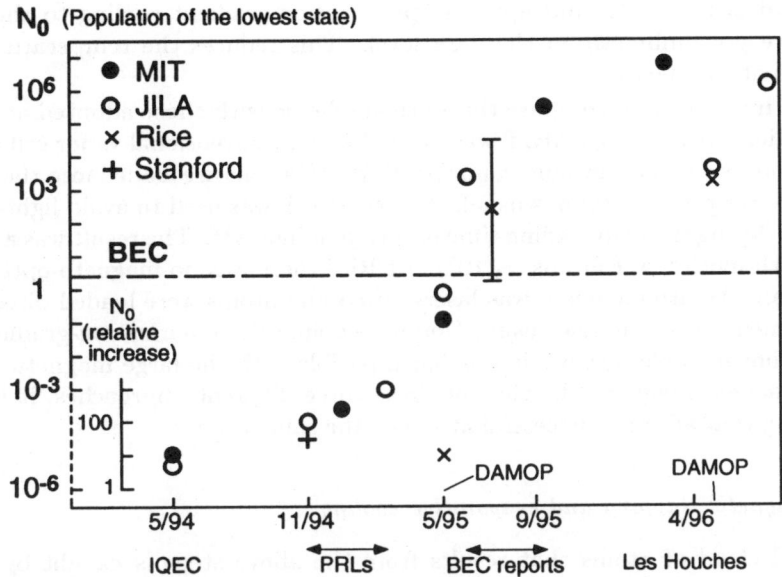

Figure 1: Progress in evaporative cooling of alkali atoms. This diagram shows all experiments reported thus far. The number of atoms in the lowest quantum state is proportional to the so-called phase-space density, and has to exceed a critical number of 2.612 to achieve Bose-Einstein condensation. For $N_0 < 10^{-3}$, the increase in phase-space density is plotted. The Rice result in 1995 was qualitative evidence for reaching quantum degeneracy.

2.1 Laser cooling and trapping

Laser light is used to slow atoms in a sequence of precooling steps, which is necessary because the trap depths are small, typically one Kelvin for the magneto-optical trap and in the millikelvin range for magnetic traps.

The three groups that have cooled into the quantum-degenerate regime (Table 1) have used laser cooling techniques in different ways. In our experiment the source of atoms is a thermal (600 K) beam of sodium atoms which effuses from an oven with an average velocity of $800\,\text{ms}^{-1}$ and with a phase-space density which is roughly twelve orders of magnitude away from BEC. The atoms are first slowed with a Zeeman slower [37] to a velocity of approximately 30 m/s, which is sufficient for the atoms to be captured by a MOT. The distinguishing feature of our Zeeman slower is the high flux of up to 10^{12} slow atoms per second [38], which enables more than 10^{10} atoms to be loaded into the MOT with a filling time of about 1 s. The MOT compresses the atoms to a small cloud about 2 mm in diameter and additionally cools them to about 1 mK. To produce higher densities we developed a "dark" version of the MOT in which the atoms are mainly confined to a hyperfine level that hardly interacts with the trapping light (the "dark" level). In this "dark SPOT" (dark SPontaneous-force Optical Trap) reabsorption of scattered photons is reduced, removing strong radiation pressure forces between atoms and allowing an improvement in density by one or two orders of magnitude [39]. Once cooled and compressed in the dark SPOT, the cloud is cooled further by switching off the magnetic field of the MOT and applying polarization-gradient cooling to the atoms, which are predominantly in the dark level. This reduces the temperature of the atoms to about $100\,\mu$K.

It is interesting to compare this optical scheme with those adopted at Boulder and at Rice. At Boulder a MOT was formed with a conventional vapor cell arrangement[40] but with a background vapor (from which atoms are loaded into the trap) at a much lower pressure than is usual. A dark SPOT was used to avoid light-induced trap loss during the trap loading time of several minutes[12]. The result was a trapped cloud with much fewer atoms ($\sim 10^6$). At Rice there was no magneto-optical trapping at all. An atomic beam was laser-slowed and atoms were loaded directly into a permanent magnetic trap using Doppler cooling[41]. Polarization-gradient cooling was not possible because it was incompatible with the large magnetic fields of the permanent magnets. In view of these three different approaches, it is rather surprising that all were successful at about the same time.

2.2 Magnetic trapping and evaporative cooling

The cold cloud of atoms that results from the above steps is caught by quickly switching off all the laser beams and switching on an inhomogeneous magnetic field **B** with a local minimum in its amplitude $|\mathbf{B}|$. Atoms in the "low-field-seeking" state ($F = 1$, $m_F = -1$) are trapped near this minimum, resulting in 2×10^9 magnetically-trapped atoms, at a phase-space density some six orders of magnitude away from BEC; this gap is bridged by evaporative cooling.

Evaporative cooling works by selectively removing atoms from the cloud with more than average energy; when the remaining atoms rethermalize via elastic collisions the cloud is then at a lower temperature. Moreover, the high-energy tail of the distribution is repopulated so the cooling process continues. The essential requirement for evaporative cooling is for the collisional rethermalization time to be much shorter than the lifetime of an atom in the trap (for reviews of evaporative cooling see Refs.[42,43]). This in turn requires a tightly confining trap (to enhance the collision rate) and an ultra-high vacuum (to increase the lifetime of the sample, typically limited by collisions with the background gas). In our experiment and at Boulder, the dark SPOT was instrumental in ensuring that the thermalization time was much shorter than the trap lifetime.

The tightest confinement by magnetic fields is obtained with a spherical quadrupole trap, which yields a linear potential. With such a trap our group and that at Boulder were able to combine for the first time laser cooling with evaporation[44,45]. However, the quadrupole trap suffers from loss at the centre of the trap, where the magnetic field is zero ("Majorana spinflips"). The small volume where the trap loss occurs was dubbed the "hole" in the trap. This loss process limited the increase in phase-space density to a factor of 5 at Boulder and 190 at MIT[46,47].

Both groups had to devise schemes to circumvent these losses. At Boulder a rotating linear magnetic field was added to the spherical quadrupole field, which continually moved the hole away from the atoms (the "TOP" trap[46]). We prevented losses by adding a beam of blue-detuned far-off-resonant light, focused at the point where the magnetic field was zero. The light acted as a highly localized, strongly

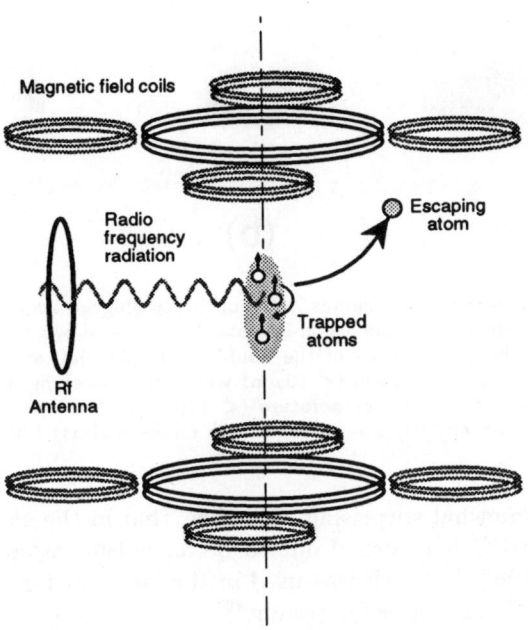

Figure 2: Cloverleaf configuration of trapping coils used in experiments at MIT. The central coils provide axial confinement, the outer coils (the "cloverleaves") provide tight radial confinement. The rf radiation induces spinflips of hot atoms, leading to evaporative cooling.

repulsive potential for the atoms, and thus "plugged" the hole in the trap [13]. The result was an extremely tight trap with a large trapping volume, and we were able to evaporatively cool to the BEC phase transition within seven seconds.

The magnetic traps used until recently in BEC experiments had major limitations: the time-dependent rotating field of the TOP trap [12], the complications of having two condensates in the optically-plugged trap [13], the inflexibility of a permanent magnetic trap [14], or the restrictions of a cryogenic environment necessary for superconducting coils [35,48,49]. In March 1996 we achieved BEC in a novel "cloverleaf" magnetic trap which overcame those limitations [16]. This trap is a variation of the Ioffe-Pritchard configuration [50,51]. The bottom of this trap forms an anisotropic harmonic potential, with a nonzero minimum magnetic field to prevent Majorana losses. The confinement is tight in two ("radial") dimensions, and relatively weak in the third ("axial") dimension direction, resulting in cigar-shaped trapped clouds. The novelty of our trap lies in the "cloverleaf" winding pattern we have devised (see Fig. 2). It allows us to obtain tight confinement using d.c. electromagnets, combined with excellent optical access and independent and fast control over bias field and axial and radial confinements. The variable aspect ratio is important to enable transfer of an almost spherical cloud from the dark SPOT and then adiabatically compress it into a cigar shape. The cloverleaf trap has proven to be reliable and versatile, and most of the results discussed here were obtained with this trap.

In the last few years many efforts have gone into the development of novel

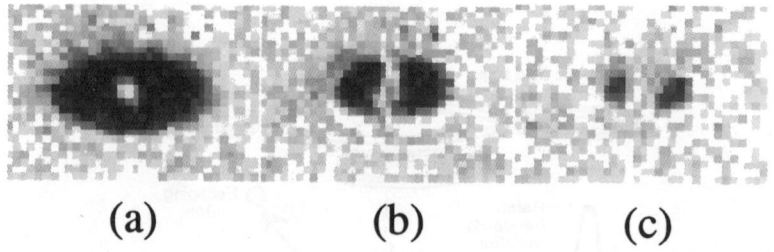

(a) (b) (c)

Figure 3: Absorption images of atom clouds in the optically-plugged trap [13]. Cloud (a) is already colder than was attainable without the argon-ion laser beam, (b) shows the break-up of the cloud into two "pockets" in the two minima of the double-well potential, and (c) the cloud reaches the optical limit of the imaging system ($< 10\,\mu$m) while still absorbing 90 % of the probe light, which sets an upper bound on the temperature ($< 10\,\mu$K) and a lower bound on the density ($5 \times 10^{12}\,\text{cm}^{-3}$). The width of each image is about $120\,\mu$m.

atom traps. It is somewhat surprising, therefore, that in the end the most suitable magnetic trap for BEC has turned out to be an optimization of a configuration suggested back in 1983 [51], which was used in the late 80's for trapping sodium [52], atomic hydrogen [53,48], and later for cesium [40].

For evaporation we use an rf field that selectively flips the spin of relatively energetic atoms that sample high magnetic fields, thus ejecting them from the trap. The effective trap depth is controlled by the frequency of the rf field, and can thus be conveniently reduced as the atomic cloud cools by evaporation [42]. This technique, first proposed by Pritchard [54] and Walraven [55], was first demonstrated in our group [56], and has been used by all experiments that have cooled into the quantum-degenerate regime.

Table 1 compares the reported observations of BEC to-date.

3 EXPERIMENTAL STUDIES OF BOSE-EINSTEIN CONDENSATES

3.1 Initial observations

Progress in the cooling process can be observed by imaging the shrinking cloud. This is done by shining a probe laser beam onto the atoms and recording the "shadow" of the cloud (absorption imaging).

Figure 3 shows the shape of the cloud in the optically-plugged trap. For high temperatures the cloud had an aspect ratio of 2:1 due to the anisotropy of the spherical quadrupole field. For temperatures below $20\,\mu$K the cloud broke up into two parts confined to the two minima of the trapping potential (the total potential has a double well due to the anisotropy of the magnetic field). Close to BEC the size of the cloud reached the limit of our optical resolution. This limit was avoided by suddenly switching off the trapping potential and imaging the expanded cloud after a delay time of 6 ms, thus observing the velocity distribution of the cloud. Figure 4 shows such time-of-flight images above and below the transition point. The striking signature of condensation was the sudden appearance of a bimodal velocity distribution when the sample was cooled below the critical temperature

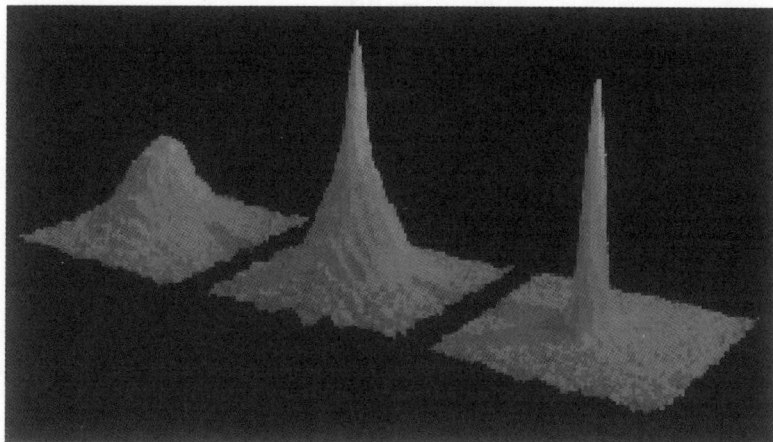

Figure 4: Two-dimensional probe absorption images after 6 ms time of flight, showing evidence for BEC [13]. The leftmost image is the velocity distribution of a cloud cooled to just above the transition point. The middle image was taken just after the condensate appeared, and the rightmost figure shows the velocity distribution after further evaporative cooling has left an almost pure condensate. The data were taken using the optically-plugged spherical quadrupole trap. The width of the images is 0.9 mm.

$T_C \sim 2\,\mu$K. The distribution consisted of an isotropic normal part and an elliptical core attributed to the expansion of a dense condensate [13].

From time-of-flight images such as Fig. 4 one can obtain the total number of atoms, the condensed fraction N_0/N and the temperature T. Below T_C the condensed fraction should vary as $N_0/N = 1 - (T/T_C)^3$ in agreement with our results [16] (Fig. 5). Note that in liquid helium the condensed fraction is difficult to measure and does not exceed 10 % [57], whereas in a dilute atomic gas, the condensate fraction is 100 % for sufficiently low temperatures, which can be obtained simply by lowering the final rf frequency.

For small numbers of atoms the condensate wavefunction is simply the single-particle ground state of the harmonic oscillator, and the density is proportional to N_0. As the number increases the effect of repulsive interactions is to cause the cloud to expand in size [58,59]. The density thus increases more slowly with N_0. When the mean-field energy becomes large compared with the kinetic energy (Thomas-Fermi regime) the confining forces due to the trapping potential are balanced by the repulsive interactions. For a harmonic potential, this implies that the cloud size is proportional to $N_0^{1/5}$, and that the density scales as $N_0^{2/5}$. We verified this dependence by measuring the kinetic energy in time-of-flight as a function of N_0 [16] (Fig. 6). The dominant contribution to this energy is the mean-field energy, which is proportional to density. The results in Fig. 6 are therefore a direct measurement of the condensate density.

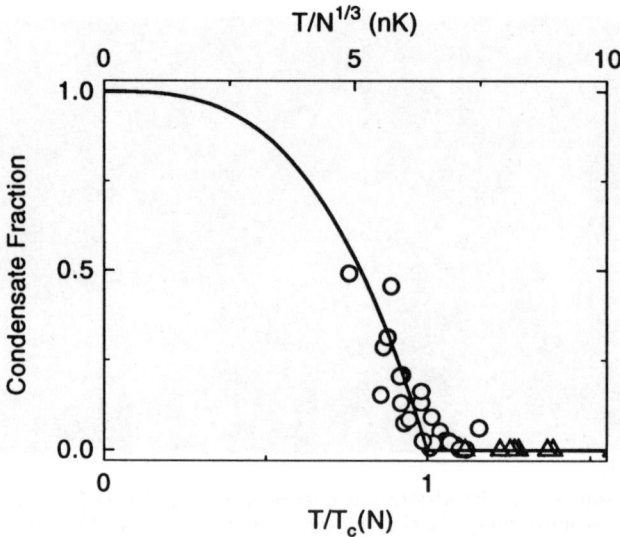

Figure 5: Condensate fraction versus normalized temperature T/T_C. Solid line: theoretical curve. The experimental data was determined from fits to time-of-flight images. Figure taken from Ref.[16].

3.2 Direct and non-destructive observation

The initial observations of Bose condensation were done destructively by switching off the trap and letting the cloud expand. Early in 1996 we observed a Bose condensate non-destructively using "dark-ground" imaging, a technique that uses the dispersive, rather than the absorptive, properties of the atom cloud[60].

The interaction of a probe laser beam with a cloud of atoms can be described by a complex index of refraction. Either photons are absorbed and then incoherently scattered through large angles (the imaginary part of the refractive index), or they are coherently scattered in the forward direction which produces a phase-shift of the laser beam (dispersion, the real part of the refractive index). For a cloud that is finite-sized, coherent scattering occurs only in an angular region that is determined by diffraction, and for our experiments this was typically ~ 0.01 radians near the BEC transition temperature. Thus the recoil heating due to coherent scattering is small. If conditions are chosen such that many more photons are scattered dispersively than absorptively, a condensate can be imaged with orders of magnitude less heating than in absorption imaging. At large detunings from atomic resonance the phase-shift from dispersive scattering falls off less rapidly with detuning than the absorption rate, and an atomic cloud looks like a phase object. It may then be observed with a phase-contrast method, which is a technique common in microscopy[61].

There is a second reason why dispersive imaging is preferable to absorption for the imaging of small, dense clouds. If the optical density is large (in our experiment it is typically 300 below T_C), then to obtain information from absorption the probe

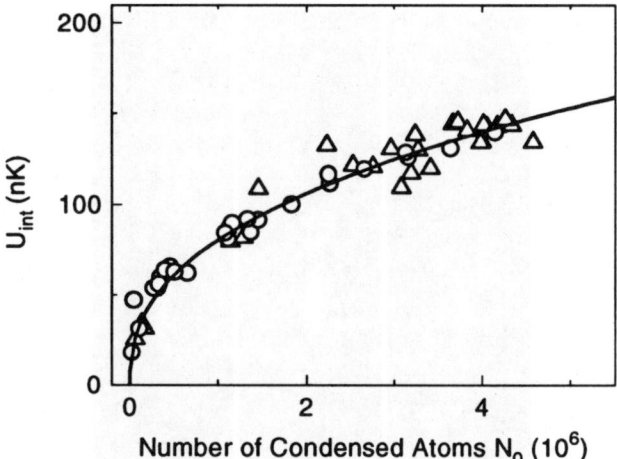

Figure 6: Mean-field energy per condensed atom versus number of atoms in the condensate. The solid line is the theoretical prediction proportional to $N_0^{2/5}$, with only the scattering length as an adjustable parameter. Figure taken from Ref.[16].

laser must be detuned. However, since the atom cloud has a non-vanishing index of refraction, the probe beam is refracted (i.e. the cloud acts like a lens). If the refraction angle exceeds the diffraction angle of the cloud, the resolution of absorption imaging is degraded. This can only be avoided by using large detunings, where absorption is small and is exceeded by the signal obtained from coherently scattered photons by roughly the resonant optical thickness, and thus dispersive imaging is preferable.

In phase-contrast imaging the phase modulation introduced by a phase object is converted into amplitude modulation by the use of spatial filtering[61]. The transmitted part of the probe is shifted in phase by $\pi/2$ with respect to the phase-modulated part enabling the two to interfere constructively. The phase-shifting is done in the Fourier plane of the imaging system where the transmitted and modulated parts are spatially separated. We have used dark-ground imaging, a technique where the transmitted light is completely blocked. This has the advantage that the phase modulation is observed against a zero-signal background, but the disadvantage of a signal quadratic in line-of-sight integrated density.

To image the condensate dispersively we used light red-detuned by about 350 linewidths. A thin wire (0.2 mm) blocked the focus of the probe beam in the Fourier plane, leaving only those photons scattered by the condensate and the surrounding normal cloud to be imaged by the optics. Figure 7 shows the growth of the condensed cloud at the phase transition[60]. To demonstrate that the imaging was non-destructive we took two images of a condensate and varied the probe intensity. When the signal in the second image deteriorated, the number of photons scattered was sufficient to take 100 consecutive images of the same condensate.

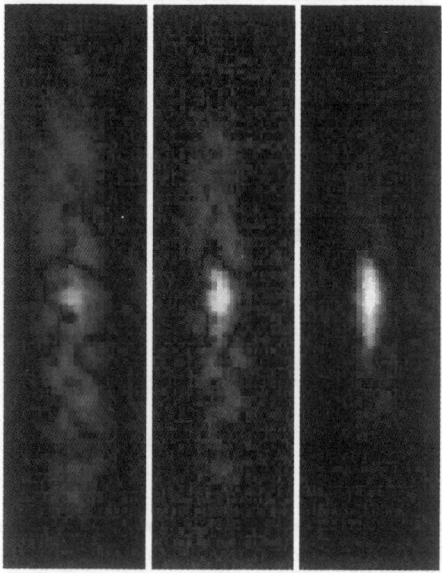

Figure 7: Direct observation of Bose-Einstein condensation of magnetically-trapped atoms by dispersive light scattering[60]. The figures show clouds with a condensate fraction that is increasing from close to 0% (left) to almost 100% (right). The signal for the normal component is rather weak and interferes with the speckle pattern of stray laser light, giving it a patchy appearance. The horizontal width of each figure is about 150 μm.

This work has revealed that dispersive imaging will be an important tool for probing condensates. It can be used to observe real-time dynamics of a single condensate, such as formation and decay, and collective excitations (see below).

3.3 Collective excitations of a dilute Bose gas

Collective quantum excitations played a key role in determining the superfluid behavior of liquid helium. In 1941 Landau described the excited states of liquid helium in terms of the elementary excitations of a quantum liquid [62]. In so doing he accounted for various thermodynamic and transport properties of He II, including the propagation of "second" sound which had been predicted earlier by Tisza using the two-fluid model[63]. In 1947 Bogoliubov reemphasized the link between superfluidity and Bose statistics [64], which had been originally suggested by London in 1938 [65]. Bogoliubov found that for a weakly-interacting Bose system the single-particle energy spectrum is modified by a macroscopic population of the ground state in such a way as to give the low-lying phonon states predicted by Landau. In the years 1953–57 the nature of superfluidity in a Bose liquid was thoroughly addressed by Feynman. With a series of physical arguments concerning the nature of the macroscopic wavefunction, Feynman obtained a form for the excitation spectrum which agreed reasonably with data obtained from neutron scattering[66]. He used this form of the spectrum to consider the flow of a superfluid, eventually linking the critical velocity for superfluidity with the creation of quantized vortex lines [67].

The theoretical description of a dilute Bose gas is much simpler than that of a Bose liquid because the interactions are binary and can be represented by a single parameter, the scattering length a. The excited states of a trapped Bose gas are understood qualitatively as follows. Consider first the simplified case of a spherically symmetric harmonic potential of frequency ν_0. In the absence of any interatomic interactions both a condensed and an uncondensed cloud will have excitations whose frequencies are integer multiples of ν_0. In particular, the two lowest frequency modes are a rigid-body sloshing of the centre-of-mass at ν_0, and a quadrupole shape oscillation at $2\nu_0$. The effect of weak interactions between the atoms in both condensed and uncondensed clouds is to give rise to sound waves, but with the mechanism for propagation being different in the two situations because of the different role of the interactions. In a classical gas elastic collisions enable local equilibrium to be established and pressure waves will propagate if the collision time is much less than the oscillation period in the trap; however, the speed of sound is independent of the interactions. In a dilute Bose gas, when the mean-field energy $n_0\tilde{U} = 4\pi\hbar^2 a n_0/m$ (where n_0 is the peak density) is much larger than the kinetic energy (Thomas-Fermi regime), sound waves propagate as oscillations in the mean-field energy, and the speed of these waves depends upon the strength of the interactions. The condensate wavefunction at zero temperature is described by the Gross-Pitaevskii (or non-linear Schrödinger) equation [68], and its solution in a harmonic potential has been the subject of several recent studies [69,70,71,72,73,74,75,76,77]. A quantitative study of the collective modes of a dilute Bose gas thus provides a critical test of the mean-field theory and is a first step towards investigating superfluid effects in such a system.

We have made a first investigation of the low energy collective oscillations of a trapped Bose gas[78], contemporaneous with a similar study undertaken at Boulder[79]. We ensured the zero temperature regime by evaporating below the BEC phase transition until the background of normal atoms was no longer visible. Collective modes were then excited by introducing a small time-dependent modulation in the trapping potential for a short period of time. When the perturbation was removed the condensate was allowed to oscillate for a chosen time and was then released into ballistic expansion. The oscillatory behavior of the condensate was thus detected in velocity-space as shape oscillations in time-of-flight pictures. By varying the time between excitation and release of the cloud, the free evolution of the oscillating condensate was strobed.

To identify the significant oscillation modes we "kicked" the condensate by reducing the radial gradient by a small percentage for a few milliseconds. A sequence of time-of-flight images similar to the one shown in Fig. 8 was obtained, with each image representing a different condensate. The dominant mode was a shape oscillation analogous to the quadrupole oscillation for an isotropic cloud, with a frequency of roughly $1.5\nu_0$ in the axial direction. We measured the frequency of this mode accurately and compared it directly to ν_0 to obtain a quantitative test of mean-field theory. The ratio of the two frequencies agreed to 1.5 % with a recent analytical solution of the Gross-Pitaevskii equation in the Thomas-Fermi regime[71].

It is interesting to consider the damping of collective excitations. If a coherent oscillation were found to persist much longer than that of the equivalent classical

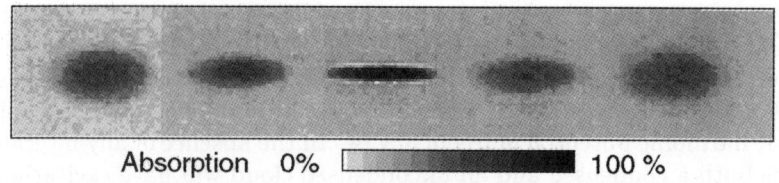

Figure 8: Shape oscillations of a Bose condensate. After excitation the cloud was allowed to freely oscillate for a variable time ranging from 16 ms (left) to 48 ms (right). The absorption images were taken after 40 ms of ballistic expansion, and the horizontal width of each cloud is 1.2 mm. Figure taken from Ref.[78].

oscillation then this would be evidence for superfluid behavior. A normal mode may be thought of as a coherent excitation of quasi-particles, and thus the lifetime of these quasi-particles is given by the damping time of the shape oscillations in our time-of-flight series. Damping may be caused either by interactions between quasi-particles and thermal atoms, or by non-linear coupling between normal modes in a process analogous to phonon decay in crystals. So far there is no theoretical prediction for the damping of excitations in an inhomogeneous condensate. In our case of a nearly pure condensate damping due to collisions with thermal atoms should be negligible.

We observed a damping time of about 0.25 s for the collective "quadrupole" oscillation. A similar excitation in a thermal cloud ($T/T_c \approx 2$) had half this damping time. The density of the thermal cloud was high enough that the observed excitation corresponded to a sound wave. The measured damping time agreed quantitatively with a classical estimate of sound wave damping due to thermal conduction and viscosity. Theoretical discussions have tended to emphasize the frequency shifts of condensate oscillations compared to integer multiples of the trapping frequency, but it is clear from this discussion that the normal modes of classical clouds can also show frequency shifts, due to sound wave propagation. A detailed study of oscillations and damping at a different temperature would be very interesting and could reveal analogous modes to "first" and "second" sound propagation in superfluid liquid helium.

For small oscillation amplitudes, the frequency spectrum of a weakly-interacting condensate is the normal mode spectrum derived by Bogoliubov[64]. In the above experiments we excited condensates with drive amplitudes of only a few percent of the d.c. field strength. At larger drives we observed striations in the time-of-flight pictures parallel to the radial direction, as shown in Fig. 9. We speculate that these patterns are due to the self-interference of an expanding matter-wave, reflecting the nodal structure of a strongly-driven condensate.

3.4 The phase of a condensate

An intriguing property of a Bose condensate is the existence of a macroscopic wavefunction, i.e. the existence of a coherent phase for the whole cloud. The gradient of the phase is proportional to the superfluid velocity. Considerable theoretical effort

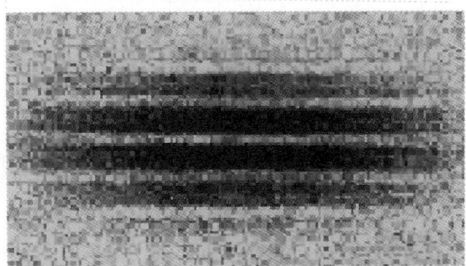

Figure 9: Ballistic expansion of a strongly-driven condensate. The time-of-flight image shows high-contrast striations. The image was taken after 75 ms of ballistic expansion and has a horizontal width of 3 mm.

has been directed at interference experiments that could directly probe the phase of the condensate.

In the theory of BEC a breaking of symmetry naturally allows a condensate to be described by a coherent state [2]: the boson field is classical with a well-defined macroscopic wavefunction and a well-defined (but arbitrary) phase. This assumption of a phase is very convenient for the interpretation of interference experiments. However, a coherent state is made up of a linear combination of number states which implies that there is no definite value for the number of atoms in the condensate. Yet in principle one can measure this number and attribute the wavefunction to the appropriate number state. This apparent conflict in descriptions is, in fact, only superficial. Recent theoretical studies have shown that during a measurement which is sensitive to the phase of a condensate (such as observing the interference between two condensates), the process of detection itself causes the condensate to evolve from a number state into a coherent state with a definite phase, without violation of number conservation [80,81,82,83,84,85]. Thus for the purposes of describing an interference experiment one may simply adopt the picture of a macroscopic wavefunction with an arbitrary but well-defined phase.

The interference pattern of overlapping condensates is analogous to interfering two completely independent laser beams. It is different than division of the wavefront or amplitude of a single wavetrain, where the phase of the interference pattern depends only on geometry. Two independent condensates are expected to have a relative phase that is random, and thus the phase of the pattern will differ for each experimental realization. Two studies have modelled the ballistic expansion of the two condensates formed in our earlier optically-plugged trap [84,83]. High contrast fringes are predicted with a wavelength that corresponds to the relative momentum of the expanding clouds.

We have taken the first steps towards observing the interference pattern of two overlapping condensates. With a procedure similar to that for the optically-plugged trap, we partitioned the potential of the cloverleaf trap with a sheet of far blue-detuned laser light directed at the trap centre along a radial direction. As a result, evaporation produced two independent condensates at the separated minima of the potential.

The large aspect ratio of the cloverleaf trap should be beneficial for observing

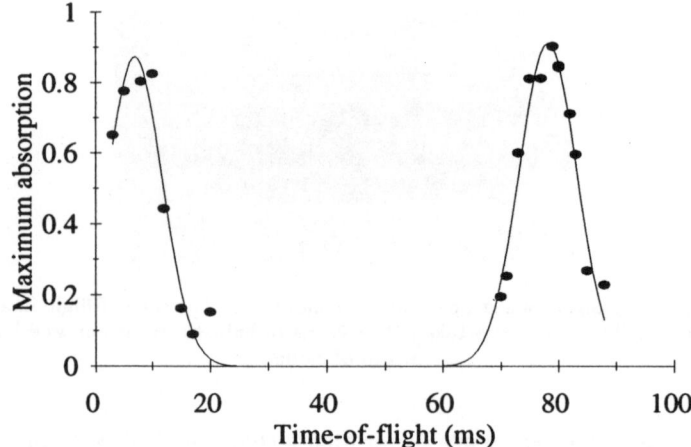

Figure 10: An atomic fountain for Bose-condensed atoms. The absorption signal shown was obtained from a horizontal probe place just above the trap centre, and each point corresponds to a different set of atoms. The atoms return with a roundtrip time of 80 ms which corresponds to a fountain height of about 9 mm.

matter-wave interference. Very little mean-field energy is released in the weakly-confined direction in time-of-flight, resulting in a larger fringe spacing in this direction. With 10^6 condensed atoms the mean-field energy is of order 100 nanokelvin, which corresponds to a de Broglie wavelength for the radial motion of about 2.5 μm. The aspect ratio in the time-of-flight expansion is about 10:1, so we should expect fringe spacings of roughly 10 μm due to the relative axial motion of the two condensates. This estimate is in agreement with a recent simulation of interference from a divided cloverleaf trap [86].

Because the expansion in the axial direction is slow it is necessary to wait a long time for the two clouds to overlap, typically around 150 ms. During this time the condensates fall 10 cm under gravity, making imaging difficult. To avoid this problem we have launched the condensates vertically against gravity and probed the atoms as they passed back through their starting position. The launch was made by temporarily shifting the bottom of the trap upwards by a millimetre with a uniform magnetic field, and allowing the atoms to accelerate towards the new equilibrium position. Just before the atoms reached the bottom (after a few milliseconds) all magnetic fields were switched off, slinging the atoms into an atomic fountain. Figure 10 confirms the fountain action. A horizontal probe initially detects the atoms on their vertical rise and then subsequently on their return. In contrast to the conventional optical launching technique [87,88], the magnetic slinging preserves sub-recoil temperatures. So far, we have reached fountain heights of 3.5 cm, limited by the strength of the applied magnetic fields.

3.5 An output coupler for a Bose condensate

The observations of BEC have stimulated interest in an "atom laser," a coherent source of atomic matter waves. The build-up of atoms in the ground state of a magnetic trap is analogous to stimulated emission into a single mode of an optical laser. An important feature of a laser is a means of emitting a fraction of the coherent light, and we have been able to demonstrate a scheme for doing this with Bose condensed atoms. A variable fraction of atoms was extracted coherently from the condensate by applying rf radiation to the cloud, thereby coupling atoms to the $m_F = 0, +1$ ($F = 1$) untrapped states. This was done in two ways: either by sweeping the rf frequency rapidly through the condensate, or by applying a short rf pulse. The fraction of atoms coupled out was varied by controlling the amplitude of the rf radiation. The propagation of atoms in the untrapped ($m_F = 0$) and antitrapped ($m_F = +1$) states was observed in time-of-flight images. When the rf amplitude was increased, the rf sweep approached 100% coupling efficiency (rapid adiabatic passage), whereas the rf pulse showed Rabi oscillations in the population of the remaining condensate. To investigate the coherence properties of the pulsed atoms remains a challenge for the future; however, the spin dynamics in an rf field is a unitary evolution without any dissipation and should preserve coherence.

4 OUTLOOK

The realization of BEC has not only provided us with a new form of quantum matter, but also with a unique source of ultracold atoms. Evaporative cooling to the transition temperature has been used to increase the phase space density by six orders of magnitude (Table 1). A small price to pay for this increase was a reduction in the number of trapped atoms by about one hundred. The next seven orders of magnitude, i.e., the condensation of 10^7 atoms into a Bose condensate, reduced the number by only four — thus BEC provides free cooling!

Evaporative cooling has extended the temperature range of atoms from microkelvin obtained with laser cooling to nanokelvin, and the densities from 10^{12} cm^{-3} to 10^{14} cm^{-3}. We have observed pure condensates of 10^7 atoms in the cloverleaf trap, with a cooling cycle of typically 20 s. With further optimization we hope to obtain 10^8 Bose condensed atoms with cycle times of five to ten seconds. This production rate is comparable to the performance of a standard light trap and indicates that Bose condensation has the potential to replace the MOT as a bright source of ultracold atoms for precision experiments and matter-wave interferometry.

Just over a year has passed since the first observations of BEC in dilute gases, and some of the basic properties of condensation have been studied and found to be in good agreement with theory. However, there remain a number of fundamental questions that can be addressed by experiments, in particular, the relationship between BEC and superfluidity in dilute atomic gases.

This work was supported by ONR, NSF, JSEP and the Sloan Foundation. C.G.T. acknowledges support from a NATO Science Fellowship, N.J.v.D. from "Nederlandse Organisatie voor Wetenschappelijk Onderzoek (NWO)" (Talent fellowship) and NACEE (Fulbright fellowship), M.-O.M. from Studienstiftung des

Deutschen Volkes and D.M.K from an NSF Graduate Research Fellowship.

1. A. Einstein, Sitz. Preuss. Akad. Wiss. 3 (1925).
2. K. Huang, *Statistical Mechanics* (John Wiley and Sons, Inc., New York, 1987).
3. K. Huang, in Ref.[4].
4. *Bose-Einstein Condensation*, edited by A. Griffin, D. W. Snoke, and S. Stringari (Cambridge University Press, Cambridge, UK, 1995).
5. J. L. Lin and J. P. Wolfe, Phys. Rev. Lett. **71**, 1222 (1993).
6. W. C. Stwalley and L. H. Nosanow, Phys. Rev. Lett. **36**, 910 (1976).
7. I. F. Silvera and J. T. M. Walraven, Phys. Rev. Lett. **44**, 164 (1980).
8. T. J. Greytak, in Ref.[4].
9. I. F. Silvera, in Ref.[4].
10. I. D. Setija, O. J. Luiten, M. W. Reynolds, H. G. C. Werij, T. W. Hijmans, and J. T. M. Walraven, in *Atomic Physics 14*, edited by D. J. Wineland, C. E. Wieman, and S. J. Smith (AIP Press, New York, 1995), p. 389.
11. J. M. Doyle, J. C. Sandberg, I. A. Yu, C. L. Cesar, D. Kleppner, and T. J. Greytak, Phys. Rev. Lett. **67**, 603 (1991).
12. M. H. Anderson, J. R. Ensher, M. R. Matthews, C. E. Wieman, and E. A. Cornell, Science **269**, 198 (1995).
13. K. B. Davis, M. O. Mewes, M. R. Andrews, N. J. van Druten, D. S. Durfee, D. M. Kurn, and W. Ketterle, Phys. Rev. Lett. **75**, 3969 (1995).
14. C. C. Bradley, C. A. Sackett, J. J. Tollett, and R. G. Hulet, Phys. Rev. Lett. **75**, 1687 (1995).
15. W. Ketterle, M. R. Andrews, K. B. Davis, D. S. Durfee, D. M. Kurn, M.-O. Mewes, and N. J. van Druten, to be published in Physica Scripta, Proceedings of the 15th General Conference of the Condensed Matter Division of the European Physical Society, Baveno-Stresa, April 1996.
16. M.-O. Mewes, M. R. Andrews, N. J. van Druten, D. M. Kurn, D. S. Durfee, and W. Ketterle, Phys. Rev. Lett. **77**, 416 (1996).
17. S. Chu, L. Hollberg, J. Bjorkholm, A. Cable, and A. Ashkin, Phys. Rev. Lett. **58**, 48 (1985).
18. P. D. Lett, W. D. Phillips, S. L. Rolston, C. E. Tanner, R. N. Watts, and C. I. Westbrook, J. Opt. Soc. Am. B **6**, 2084 (1989).
19. J. Dalibard and C. Cohen-Tannoudji, J. Opt. Soc. Am. B **6**, 2023 (1989).
20. P. Ungar, D. Weiss, E. Riis, and S. Chu, J. Opt. Soc. Am. B **6**, 2058 (1989).
21. E. L. Raab, M. Prentiss, A. Cable, S. Chu, and D. Pritchard, Phys. Rev. Lett. **59**, 2631 (1987).
22. *Laser Manipulation of Atoms and Ions*, edited by E. Arimondo, W. D. Phillips, and F. Strumia (North-Holland, Amsterdam, 1992), proceedings of "Enrico Fermi" Summer School, Course CXVIII, Varenna, Italy.
23. A. M. Steane and C. J. Foot, Europhys. Lett. **14**, 231 (1991).
24. M. Drewsen, P. Laurent, A. Nadir, G. Santarelli, A. Clairon, Y. Castin, D. Grison, and C. Salomon, Appl. Phys. B **59**, 283 (1994).
25. W. Petrich, M. H. Anderson, J. R. Ensher, and E. A. Cornell, J. Opt. Soc. Am. B **11**, 1332 (1994).
26. C. G. Townsend, N. H. Edwards, C. J. Cooper, K. P. Zetie, C. J. Foot, A. M. Steane, P. Szriftgiser, H. Perrin, and J. Dalibard, Phys. Rev. A **52**, 1423

(1995).
27. J. Lawall, S. Kulin, B. Saubamea, N. Bigelow, M. Leduc, and C. Cohen-Tannoudji, Phys. Rev. Lett. **75**, 4194 (1995).
28. H. J. Lee, C. S. Adams, M. Kasevich, and S. Chu, Phys. Rev. Lett. **76**, 2658 (1996).
29. T. Esslinger, F. Sander, M. Weidemüller, A. Hemmerich, and T. W. Hänsch, Phys. Rev. Lett. **76**, 2432 (1996).
30. A. Kastberg, W. D. Phillips, S. L. Rolston, R. J. C. Spreeuw, and P. S. Jessen, Phys. Rev. Lett. **74**, 1542 (1995).
31. A. Hemmerich, M. Weidemüller, T. Esslinger, C. Zimmermann, and T. W. Hänsch, Phys. Rev. Lett. **75**, 37 (1995).
32. B. P. Anderson, T. L. Gustavson, and M. A. Kasevich, Phys. Rev. A **53**, R3727 (1996).
33. J. D. Miller, R. A. Cline, and D. J. Heinzen, Phys. Rev. A **47**, R4567 (1993).
34. H. F. Hess, Phys. Rev. B **34**, 3476 (1986).
35. N. Masuhara, J. M. Doyle, J. C. Sandberg, D. Kleppner, T. J. Greytak, H. F. Hess, and G. P. Kochanski, Phys. Rev. Lett. **61**, 935 (1988).
36. C. S. Adams, H. J. Lee, N. Davidson, M. Kasevich, and S. Chu, Phys. Rev. Lett. **74**, 3577 (1995).
37. W. D. Phillips and H. Metcalf, Phys. Rev. Lett. **48**, 596 (1982).
38. M. A. Joffe, W. Ketterle, A. Martin, and D. E. Pritchard, J. Opt. Soc. Am. B **10**, 2257 (1993).
39. W. Ketterle, K. B. Davis, M. A. Joffe, A. Martin, and D. E. Pritchard, Phys. Rev. Lett. **70**, 2253 (1993).
40. C. Monroe, W. Swann, H. Robinson, and C. Wieman, Phys. Rev. Lett. **65**, 1571 (1990).
41. J. J. Tollet, C. C. Bradley, C. A. Sackett, and R. G. Hulet, Phys. Rev. A **51**, R22 (1995).
42. W. Ketterle and N. J. van Druten, in *Advances in Atomic, Molecular and Optical Physics*, edited by B. Bederson and H. Walther (Academic Press, San Diego, 1996), Vol. 37, in press.
43. J. T. M. Walraven, in *Quantum Dynamics of Simple Systems*, edited by G. Oppo, S. Barnett, E. Riis, and M. Wilkinson (Institute of Physics, London, 1996), p. 315.
44. K. B. Davis, M.-O. Mewes, M. A. Joffe, and W. Ketterle, in *Fourteenth International Conference on Atomic Physics* (Book of Abstracts, 1M-3, Boulder, Colorado, 1994).
45. W. Petrich, M. H. Anderson, J. R. Ensher, and E. A. Cornell, in *Fourteenth International Conference on Atomic Physics* (Book of Abstracts, 1M-7, Boulder, Colorado, 1994).
46. W. Petrich, M. H. Anderson, J. R. Ensher, and E. A. Cornell, Phys. Rev. Lett. **74**, 3352 (1995).
47. K. B. Davis, M.-O. Mewes, M. A. Joffe, M. R. Andrews, and W. Ketterle, Phys. Rev. Lett. **74**, 5202 (1995).
48. R. van Roijen, J. J. Berkhout, S. Jaakkola, and J. T. M. Walraven, Phys. Rev. Lett. **61**, 931 (1988).

49. P. A. Willems and K. G. Libbrecht, Phys. Rev. A **51**, 1403 (1995).
50. Y. V. Gott, M. S. Ioffe, and V. G. Telkovsky, in *Nuclear Fusion, 1962 Suppl., Pt. 3* (International Atomic Energy Agency, Vienna, 1962), p. 1045.
51. D. E. Pritchard, Phys. Rev. Lett. **51**, 1336 (1983).
52. V. S. Bagnato, G. P. Lafyatis, A. G. Martin, E. L. Raab, R. N. Ahmad-Bitar, and D. E. Pritchard, Phys. Rev. Lett. **58**, 2194 (1983).
53. H. F. Hess, G. P. Kochanski, J. M. Doyle, N. Masuhara, D. Kleppner, and T. J. Greytak, Phys. Rev. Lett. **59**, 672 (1987).
54. D. E. Pritchard, K. Helmerson, and A. G. Martin, in *Atomic Physics 11*, edited by S. Haroche, J. C. Gay, and G. Grynberg (World Scientific, Singapore, 1989), p. 179.
55. T. W. Hijmans, O. J. Luiten, I. D. Setija, and J. T. M. Walraven, J. Opt. Soc. Am. B **6**, 2235 (1989).
56. W. Ketterle, K. B. Davis, M. A. Joffe, A. Martin, and D. E. Pritchard, invited oral presentation at OSA Annual meeting, Toronto, Canada, October 3–8, 1995.
57. P. E. Sokol, in Ref.[4].
58. V. V. Goldman, I. F. Silvera, and A. Legget, Phys. Rev. B **24**, 2870 (1981).
59. D. A. Huse and E. D. Siggia, J. Low Temp. Phys. **46**, 137 (1982).
60. M. R. Andrews, M.-O. Mewes, N. J. van Druten, D. S. Durfee, D. M. Kurn, and W. Ketterle, Science **273**, 84 (1996).
61. E. Hecht, *Optics* (Addison-Wesley, Reading, Massachusetts, 1989).
62. L. D. Landau, J. Phys. (USSR) **5**, 71 (1941).
63. L. Tisza, Nature **141**, 913 (1938).
64. C. N. Bogoliubov, J. Phys. (USSR) **11**, 23 (1947).
65. F. London, Nature **141**, 643 (1938).
66. P. Nozières and D. Pines, *The Theory of Quantum Liquids* (Addison-Wesley, Redwood City, California, 1990), Vol. 2.
67. R. P. Feynman, in *Progress in Low Temperature Physics*, edited by C. J. Gorter (North-Holland, New York, 1955), Vol. 1, p. 17.
68. A. L. Fetter and J. D. Walecka, *Quantum Theory of Many-Particle Systems* (McGraw-Hill, New York, 1971).
69. M. Edwards, P. A. Ruprecht, K. Burnett, R. J. Dodd, and C. W. Clark, Phys. Rev. Lett. **77**, 1671 (1996).
70. P. A. Ruprecht, M. Edwards, K. Burnett, and C. W. Clark, to be published in Phys. Rev. A.
71. S. Stringari, Phys. Rev. Lett. **77**, 2360 (1996).
72. L. You, W. Hoston, and M. Lewenstein, preprint.
73. Y. Castin and R. Dum, preprint.
74. V. M. Pérez-García, H. Michinel, J. Cirac, M. Lewenstein, and P. Zoller, preprint.
75. A. L. Fetter, Phys. Rev. A **53**, 4345 (1996).
76. M. Marinescu and A. Starace, preprint.
77. K. G. Singh and D. S. Rokshar, Phys. Rev. Lett. **77**, 1667 (1996).
78. M.-O. Mewes, M. R. Andrews, N. J. van Druten, D. M. Kurn, D. S. Durfee, C. G. Townsend, and W. Ketterle, Phys. Rev. Lett. **77**, 988 (1996).

79. D. S. Jin, J. R. Ensher, M. R. Matthews, C. E. Wieman, and E. A. Cornell, Phys. Rev. Lett. **77**, 420 (1996).
80. J. Javanainen and S. M. Yoo, Phys. Rev. Lett. **76**, 161 (1996).
81. J. I. Cirac, C. W. Gardiner, M. Naraschewski, and P. Zoller, preprint.
82. Y. Castin and J. Dalibard, preprint.
83. M. Naraschewski, H. Wallis, A. Schenzle, J. I. Cirac, and P. Zoller, preprint.
84. W. Hoston and L. You, preprint.
85. T. Wong, M. J. Collett, and D. F. Walls, preprint.
86. H. Wallis, A. Rohrl, M. Naraschewski, and A. Schenzle, preprint.
87. A. Clairon, C. Salomon, S. Guellati, and W. D. Phillips, Europhys. Lett. **16**, 165 (1991).
88. K. Gibble and S. Chu, Phys. Rev. Lett. **70**, 1771 (1993).

ZEEMAN AND HIS CONTEMPORARIES: DUTCH PHYSICS AROUND 1900

A.J. KOX

Institute of Theoretical Physics, University of Amsterdam, Valckenierstraat 65, 1018 XE Amsterdam, The Netherlands

An overview is given of physics in the Netherlands around the turn of the century. In addition, some of the factors that contributed to the quantitative and qualitative expansion of Dutch science in the last decades of the nineteenth century are analyzed.

1 Introduction

On May 31, 1941, at a meeting of the Section of Sciences of the Royal Dutch Academy of Sciences, the chairman of the Section read an obituary of the recently deceased mathematician Diderik Johannes Korteweg. It contains the following remarkable passage:

> At the end of his life Korteweg had become an almost legendary figure, being the last survivor of the heroic age of the Amsterdam natural sciences faculty, an age during which to the amazement of the rest of the world a few almost blindingly bright lights illuminated the dark labyrinth of the laws of nature.[1]

These words, though somewhat tortured and exaggerated, refer to the period that started around 1875, in which Dutch science showed a remarkable and unexpected expansion, both quantitatively and qualitatively. It is the period of Hendrik Antoon Lorentz, Johannes Diderik van der Waals, Heike Kamerlingh Onnes, Pieter Zeeman and Jacobus van 't Hoff, to mention only the Dutch Nobel Prize winners. When one looks back on this period, the question arises whether specific factors or developments can be isolated that have influenced this expansion or have contributed in some way to the historical development. In this paper I will provide some historical background on this remarkable period in Dutch science, discussing the people involved in it and the specific topics that interested them.

2 The "Second Golden Age" of Dutch Science

In a lecture delivered in 1908, Pieter Zeeman summarized the situation in the Netherlands around 1850 in the following way:

> Although in the first half of the last century physics was diligently cultivated here, the truly new and great discoveries were expected from Paris, Berlin, London or Göttingen. There, one felt, would the results have to be published, and a big discovery, it was thought, could not be made in our country.[2]

But, Zeeman continued, "Nowadays, the results of Dutch physics can compare with the best other people produce." Zeeman did not exaggerate. Within a few decades after 1850, the situation had changed completely and the Netherlands had become a world-power in science.

The status of Dutch science around 1900 becomes even more interesting if we look at the size of the scientific community in those days.[a] In Amsterdam there were three full professors and two assistants in 1900; in the same year, Leiden had two professors, four assistants and one "privaatdocent" (a person who was not officially employed by the university, but had the right to teach courses). That made the Amsterdam and Leiden departments the largest nationally: Utrecht had two professors and two assistants and Groningen only one professor and one single assistant. (I have not counted technical personnel.)

These are modest figures: Germany, for instance, counted around 100 professors and staff members, approximately ten times as much as the Netherlands. If we relate these figures to the number of inhabitants, however, a different picture emerges: 4.1 university position per million inhabitants in the Netherlands, against 2.9 in Germany. The German number is characteristic for all of Europe (with the exception of Switzerland, with 8.1).[3]

It is interesting to investigate whether the rise of Dutch science was pure coincidence, or whether it was connected to external factors and developments. Of course, it can be maintained that a fortunate concentration of talented people caused this rise to fame of Dutch science. But talent needs the opportunity to develop. According to recent research it appears that two important developments played a role in changing the position of Dutch science. The first one was the creation of a new type of secondary school; the second one was a reform of the university system.[b]

In 1862, a new type of secondary school was introduced in the Netherlands. It was called HBS, an abbreviation for "Hoogere Burger School." The school was meant to provide a secondary school education to middle and upper middle class students who could not or would not attend the already existing elitist classical gymnasium. In contrast to the gymnasium, which prepared for a subsequent university education, the HBS trained for positions in commerce and industry, as well as for the civil service. The curriculum was very practically oriented, with much emphasis on modern languages and the sciences. In the science subjects, much time was devoted to practical exercises and laboratory classes. Owing to the creation of the HBS, the number of secondary school students and thus the number of secondary-school graduates with a good science background increased markedly. Many of these enrolled in the university, in particular in one of the sciences. This was a remarkable and totally unexpected side-effect of the creation of the HBS since the HBS was not meant to prepare for university. In fact, one had to take an additional exam in Latin and Greek to be admitted to the university.

A few figures make clear how many HBS-graduates entered a university.[c] In 1863 15 HBS schools were created. In the period 1864–1876 already more than half of the HBS-graduates entered university. In the period between 1879 and 1884, six

---

[a] I will limit myself to the physics departments at the various universities, but the situation in the other disciplines was comparable.

[b] For a more detailed discussion than can be given here, see refs. 4 and 5.

[c] See ref. 5 for the sources of these numbers.

of the 12 new members of the Academy of Sciences were former HBS students.[d] And, finally, four of the five Nobel prize winners listed in the Introduction were HBS-graduates. Clearly, the establishment of this new type of school was a factor in the rise of Dutch science.

The second factor, university reform, was no less important. In 1876, university education in the Netherlands was reformed by the passage of a new law. As was the case in other countries, in particular Germany, independent research was explicitly stated to be a task of the universities, in addition to their already existing educational task. In order to achieve this change, the university system was expanded considerably. The number of professors in the science faculties increased from 20 in 1877 to 28 in 1879, and continued to increase in later years. Salaries were raised considerably, and, in addition, financial means were provided to expand existing facilities and construct new buildings. In this way, university science was rapidly professionalized and increased in quality. At the same time, thanks to increased funding, experimental physics became more prominent than it had been in earlier days. Whereas around 1850 most Dutch dissertations discussed experimental work by others without adding new experimental facts and data, experimental work became an integral part of the education of a student and of dissertation research.[e] In addition, theoretical physics was developing as a separate discipline.

One of the characteristics of Dutch science in the latter part of the nineteenth century is the close interaction between experimental and theoretical physics. In Leiden, Lorentz occupied one of the first chairs of theoretical physics in Europe and had as his colleague the eminent experimentalist Kamerlingh Onnes; in Amsterdam Van der Waals, though formally in charge of experimental as well as theoretical physics, focused on theoretical issues, both in his teaching and his research, and left the carrying out of experiments to assistants and later to his famous colleague Zeeman. The emergence of theoretical physics also forged a strong link between physics and mathematics (which was made easier by the fact that, at the time, mathematics and the sciences were organized in one single faculty at the universities).

In short, a new climate arose in which scientific interest was stimulated in a large group of secondary school students, who were given a first class education and who could pursue their interests at rapidly expanding universities of increasing quality. And for the majority who, after their university education, did not pursue a scientific career, there were many teaching positions available at the secondary schools, which profited from their professionalism and good education. The influence secondary school teachers had (and still have) on talented students cannot be overestimated.

3 Physics and the physicists around 1900

At the university of Amsterdam, the 63-year old Van der Waals was the leading figure. He was a self-made man, who had become famous with his dissertation of 1873, on the continuity of the gaseous and liquid states.[8] The work on the equation

---

[d]Sixty years later, in 1937–38, of the professors in the science departments of the universities, 202 were HBS students, against 85 gymnasium and 31 others.

[e]See refs. 6 and 7 for more details on these developments.

of state, begun in the dissertation, had in the meantime been extended by Van der Waals and a number of collaborators (including the mathematician Korteweg, who was Van der Waals's first doctoral student in 1878). Three main achievements should be mentioned: the law of corresponding states, the theory of binary mixtures and the theory of capillarity.

Although the theoretician Van der Waals was the most famous man in the Amsterdam Faculty of Natural Sciences, he shared the laboratory with a much younger colleague, who had already established a reputation and who would acquire much fame: Pieter Zeeman, the discoverer of the Zeeman effect. The staff was further completed by Remmelt Sissingh.

Research in Amsterdam focused on two subjects. The first was thermodynamics in the widest sense of the word, including molecular theory and kinetic theory; the second was spectroscopy, pioneered by Zeeman and in the forefront of current research thanks to the Zeeman effect and related magneto-optic phenomena, that were of crucial importance in the investigation of the internal structure of atoms. Both of these lines of research have been pursued in Amsterdam for many decades.[f]

Strong personal ties existed between the Amsterdam laboratory and its counterpart in Leiden. The two important figures in Leiden in 1900 were the theoretician Hendrik Antoon Lorentz and the experimentalist Heike Kamerlingh Onnes. The latter was director of the laboratory and inspirator and gifted manager of an impressive research program. In the years around the turn of the century, Lorentz was rapidly acquiring fame as the leading theoretical physicist in Europe, a position that he would hold for several decades.

Lorentz's work, though extremely broad and many-faceted, was focused in those years on his electron theory, his dualistic theory of electromagnetism. The dualistic nature of the theory lies in the fact that it is based on the existence of small charged particles (electrons) on the one hand, and the existence of an electromagnetic ether on the other hand. The particles were the sources of the fields and of the electromagnetic interaction; the ether was the medium through which the charged particles interacted. After the publication of an influential monograph on this subject in 1895,[10] Lorentz worked on the completion of his theory, a goal that he would essentially achieve in 1904.[g]

In the Leiden laboratory much energy was spent on research in the field of electromagnetism and magneto-optics. The topic of Zeeman's dissertation, for example, was the Kerr effect, while Faraday rotation was another subject of experimental investigation.[h] Lorentz's theoretical work on the Zeeman effect should of course be mentioned as well (see section 4 for more details). This work resulted in a long-lasting collaboration between Lorentz and Zeeman, not just on spectroscopy, but also on other investigations by Zeeman, such as the experimental determination of the speed of light in moving transparent media, a topic closely related to special relativity.

But the collaboration between Leiden and Amsterdam was not limited to Zeeman and Lorentz. In close collaboration with Van der Waals in Amsterdam, Kamer-

---

[f] See ref. 9 for more details on the history of physics in Amsterdam.

[g] See ref. 11 for a comprehensive review of Lorentz's electron theory.

[h] See ref. 12 for a more detailed review of the experimental work done in Leiden.

lingh Onnes had started an ambitious program of research in low-temperature physics, aimed at the liquefaction of the permanent gases. This program reached its greatest success with the liquefaction of helium in 1908.

In the other two university cities, Utrecht and Groningen, the activity was somewhat less lively. Physics was a bit of a family affair in Utrecht: the two professors there were V.A. Julius and his nephew W.H. Julius. The latter was an interesting physicist, who was much involved in solar physics, a field in which he developed a somewhat controversial theory, explaining phenomena like the shifting and widening of solar spectral lines on the basis of refraction and dispersion of light in the solar atmosphere. The theory was forgotten after Julius died, but it was sufficiently intriguing to interest Albert Einstein, as becomes clear from the lively correspondence between Einstein and Julius in the years 1910–1911.

The one professor in Groningen, Herman Haga, was an experimentalist, who had acquired some fame with his experiments (together with Cornelis Wind) on the nature of X-rays. They had in fact established that X-rays gave rise to diffraction phenomena, which supported a wave-theory for X-rays.

4 Electromagnetism

Among the topics of research at the physics departments of the various universities, the theory of electromagnetism took a special position. Its development during the nineteenth century had been so successful that the years around 1900 were characterized by a great confidence in the theory, a confidence that even inspired the idea that electromagnetism might take the place of mechanics as the basic science, to which all physical phenomena could eventually be reduced.

This point can be illustrated by two lectures, both held in the year 1900, by prominent physicists from different universities.

In the academic year 1899–1900, Lorentz was Rector Magnificus of the university of Leiden, and in that capacity he gave a lecture on a topic "from the realm of my daily research and thinking," as Lorentz put it himself, on February 8th, the 325th anniversary of the university. The title was "Electromagnetic theories of physical phenomena."[13] After a review of some of the achievements of nineteenth-century physics, such as the law of conservation of energy, the second law of thermodynamics, and the molecular theories of matter, Lorentz reached the real topic of the lecture:

> Among the theories of contemporary physics there is one group—one may call them electromagnetic theories of physical phenomena—that seem so promising for the near future that a more detailed discussion seems justified.

Next, Lorentz discussed the work of Faraday and Maxwell and Hertz's experimental confirmation of Maxwell's prediction of the existence of electromagnetic waves. He continued with a discussion of recent advances in the microscopic theory of the emission of light and the production of spectral lines, the theory in which it is postulated that inside the atoms and molecules small electrically charged particles are present that can vibrate and thus emit electromagnetic radiation. It is characteristic for Lorentz's style that nowhere he even hinted at the fundamental contributions he himself had made in this particular field. His next topic is spectroscopy and its importance for the study of the internal structure of atoms. Lorentz introduced this topic with the following words:

> After these considerations an experimental confirmation will not be unwelcome. In the laboratory of Professor Kamerlingh Onnes, this jewel of our university, not because of its external appearance, but because of the internal organization, we encounter, a little more than three years ago, Dr. Zeeman, working with a sodium flame placed between the poles of a strong electromagnet. The yellow light is analyzed in a spectrum...
>
> Zeeman tried, with the help of modern techniques, to find a phenomenon that Faraday apparently had already looked for, without success. Would not the forces due to the magnetic poles, the effect of which on certain light phenomena was known, bring about some change in the emission of light as well?
>
> Zeeman found a slight widening of the spectral lines, and careful considerations and control experiments taught him, that these were indeed due to a direct influence of the magnetic forces.
>
> The electromagnetic theory clarified the phenomenon and predicted its particulars...[i]

The importance of the Zeeman effect for the study of the internal structure of atoms was enormous. For the first time a technique was available to probe the inside of atoms: because spectra were due to processes that take place inside atoms, a change in the emitted spectrum (as brought about by the Zeeman effect) had to correspond to a change in those internal processes. The quantitative electromagnetic theory for the Zeeman effect, a theory proposed by Lorentz, made this connection even more explicit. That this theory was electromagnetic in nature was very important for Lorentz, as becomes clear from the following quotation:

> The regularities found and the strong resemblance of the spectra of some elements, give hope that all this will once be cleared up, that perhaps much will be achieved through one happy insight. And without doubt, in sofar as it can be judged right now, electromagnetic theory provides the best possibility for a solution of the problem.

Lorentz concluded his lecture with a few remarks on the possible electromagnetic origin of molecular and even gravitational forces. The latter possibility was also the topic of a paper that appeared later that year.

In Lorentz's mind, electromagnetism was a serious candidate to replace mechanics as the basic physical science and research in this field was extremely important for the advancement of our knowledge. That Lorentz was not alone in that view becomes clear from a lecture by Zeeman.

A month after Lorentz's lecture, on 12 March 1900, Pieter Zeeman delivered his inaugural lecture as extraordinary professor of physics at the university of Amsterdam. Its title was: "Experimental investigations on entities, smaller than atoms."[14]

---

[i]This account by Lorentz, almost an eyewitness account, has been recently supplemented with new material from the Zeeman archive (see the Appendix). See also the paper by P F A Klinkenberg in these Proceedings for a detailed account of the discovery of the Zeeman effect.

In his lecture, Zeeman focused on the new insights in the structure of atoms and molecules that had led to the assumption that within atoms even smaller particles are present. In Zeeman's own words:

> We know that atoms can only be called indivisible in sofar as we cannot divide them up any further. But the complicated spectra of even the simplest gases make us conclude with certainty that the structure of atoms must be very complicated and that certain parts must be distinguished in them. Very recent investigations, however, have proven the existence of entities smaller than atoms and have even opened the possibility that the dream of the alchemists, the transmutation of elements, may contain some truth.

Zeeman's lecture gave an overview of the experimental results that support the existence of small charged particles, ranging from gas discharge phenomena to the photo-electric effect. But a place of honor is given to the Zeeman effect. After a clear exposé on the effect and its theoretical explanation, Zeeman concluded:

> We cannot doubt that this negative ion must play a fundamental role in all electrical theories. Perhaps it is the fundamental quantity in which all electrical processes can be expressed, for its mass and charge seem to be constant and also independent of the electric processes that create it and the substances it is created from....
>
> The experimental study of radiation phenomena, in a variety of situations, seems to me to furnish in more than one respect important building-blocks for our knowledge of nature.
>
> It will be my aim to stimulate the Physics Laboratory to these investigations, that are so closely related to the final foundations on which this world is built.

Again, there is a strong emphasis on the importance of electromagnetism to obtain fundamental insights in the structure of matter, in this case supplemented with a research program to reach that goal. In spite of the important Dutch contributions to thermodynamics and gas theory, there was clearly a strong focus on electromagnetism around 1900 and much was expected from the further development of this theory. We now know that these expectations were exaggerated. Nevertheless, the successes of electromagnetism ensured its continuing central role in physics.

Appendix: Pieter Zeeman and the Zeeman Archive

Since 1989, much more information about Pieter Zeeman's life and work has been available, because in that year his personal and scientific archive was discovered. The discovery came as a great surprise, because Zeeman's papers were assumed to have been lost. The surprise was even greater when the extent of the discovery became clear. Few archives of scientists are as complete as this one: it contains not only a large amount of scientific notes, lecture notes, and laboratory notebooks, but also more than 10,000 correspondence items (including many drafts of outgoing letters) and many more personal papers such as diaries and appointment-books.

On the basis of the archival material we can now develop a detailed and balanced view of Zeeman's life, work, and personality.[j]

Pieter Zeeman was born in 1865 in a small village in the Dutch province of Zeeland. He was the son of a protestant minister. After concluding his HBS secondary school education he continued his studies at the University of Leiden, where he specialized in physics and served as an assistant to the theoretician Hendrik Antoon Lorentz and the experimenter Heike Kamerlingh Onnes. The latter was Zeeman's dissertation advisor; the dissertation dealt with the experimental investigation of the Kerr effect.[16] It was the start of Zeeman's interest in magneto-optical phenomena and fitted in with an already existing line of research in Leiden.

Not long after the discovery of the Zeeman effect in the fall of 1896, Zeeman was appointed as lecturer at the University of Amsterdam. Soon afterwards he became extraordinary and later ordinary professor. In spite of several offers from other places, he remained in Amsterdam until his retirement in 1935. He used the offers he received to improve his position and to work towards the goal of having his own laboratory and being completely independent. That goal was reached in 1923. Zeeman received many honors, of which the Nobel prize in 1902 was the most important. He shared the prize with Lorentz for their work in magneto-optics. Zeeman died in 1943, in the middle of the Second World War.

Zeeman was an experimenter par excellence, whose experimental work is characterized first and foremost by extraordinary precision. He had a unique talent for designing precision experiments and performing them with great persistence. After his initial work on the Zeeman effect, he had to shift to different fields of research because of the unfavorable conditions in the Amsterdam laboratory. After having worked on related topics, such as magnetic double refraction and the rotation of the plane of polarization in a magnetic field, he applied his talents to experimental relativity: first he repeated Fizeau's experiment with unprecedented precision, confirming the prediction of special relativity for the speed of light in moving media (in particular the occurrence of a dispersion term in the expression for the speed of light). A few years later he investigated the equality of gravitational and inertial mass for anisotropic and radioactive materials, using a very sensitive torsion-balance. Toward the end of his career he tried to observe the transverse Doppler effect and became involved in nuclear physics through his investigation of the hyperfine structure of spectral lines. From the material in the archive, Zeeman emerges as a dedicated scientist who was constantly looking for ways to apply his skills to new problems and who kept a lively interest in current developments in physics.

References

1. *Versl. Kon. Ak. Wet.* **50**, 35 (1941)
2. P Zeeman, *Ned. Tijdschr. Geneesk.* **53-I**, 522 (1909).
3. P Forman, J L Heilbron, S Weart *Hist. Stud. Phys. Sci.* **5**, 1 (1975).

---

[j]See ref. 15 for a biographical sketch of Zeeman based on much new material. The Zeeman Archive is kept in the Rijksarchief in Noord-Holland, Haarlem, The Netherlands.

4. B Willink, *Burgerlijk sciëntisme en wetenschappelijk toponderzoek* (diss. Amsterdam, 1988).
5. B Willink, *Social Studies of Science* **21**, 503 (1991).
6. K van Berkel, *In het voetspoor van Stevin* (Boom, Meppel, 1985).
7. F van Lunteren, *Gewina* **18**, 102 (1995).
8. J D van der Waals, *Over de continuïteit van den gas- en vloeistoftoestand* (Sijthoff, Leiden, 1873).
9. A J Kox, *Physics in Amsterdam: A brief history* (North-Holland, Amsterdam, 1990).
10. H A Lorentz, *Versuch einer Theorie der electrischen und optischen Erscheinungen in bewegten Körpern* (Teubner, Leipzig, 1895).
11. H A Lorentz, *The theory of electrons* (Teubner, Leipzig, 1909).
12. *Het Natuurkundig Laboratorium der Rijks-Universiteit te Leiden in de jaren 1882–1904* (IJdo, Leiden, 1904).
13. H A Lorentz in *Jaarboek der Rijks-Universiteit te Leiden 1899–1900* (Brill, Leiden, 1900).
14. P Zeeman, *Experimenteele onderzoekingen over deelen kleiner dan atomen* (Scheltema en Holkema, Amsterdam, 1900).
15. A J Kox in *Een brandpunt van geleerdheid in de hoofdstad*, ed. J C H Blom et al. (Verloren, Amsterdam, 1992).
16. P Zeeman, *Metingen aan het verschijnsel van Kerr* (Van Doesburgh, Leiden, 1893).

ZEEMAN'S GREAT DISCOVERY

P.F.A. KLINKENBERG
Bosweg 16, NL-7314 AP, Apeldoorn, The Netherlands

Considerations and experiments leading to the discovery of the Zeeman effect in 1896, and reactions of the scientific community to the announcement are sketched. Lorentz's model, its application to obtain a value of e/m for the "luminiferous ion" and the introduction of photographic observations is discussed. The importance of Voigt's phenomenological theory on Zeeman's views and his interest in searching for asymmetries and other magneto-optical effects are mentioned. The significance of the Zeeman splitting for measuring magnetic fields in extraterrestrial objects is pointed out, in connection with Hale's sunspot observations. A discussion of the anomalous Zeeman effect, its application to the problem of atomic structure and spectra as well as the effect on hyperfine structure (Paschen-Back and Back-Goudsmit effect included) follows. A short survey of magnetic resonance experiments concludes the report, whereby the main developments of the Zeeman effect during its first half century are covered.

1 The Leiden years

According to Zeeman himself the magnetic splitting of spectral lines was discovered in August 1896 [1]. What he saw through an eyepiece in a rather primitive setup was a symmetrical broadening of the sodium D_1 and D_2-lines in the radiation from a flame, obtained by letting a Bunsen burner – or a gas/oxygen burner – heat a piece of asbestos soaked in a solution of kitchen salt, when the flame was subjected to a homogeneous magnetic field. It was viewed perpendicular to the field lines and the light was dispersed by means of a concave grating of 3 m radius acquired shortly before by the Laboratory of Physics in the University of Leiden, where he was an assistant to Professor Kamerlingh Onnes.

August was the regular vacation month for the laboratory, but for Zeeman it was the ideal period to do research because normally he was burdened with teaching obligations. At the same time he was engaged in a series of routine measurements on the absorption of electromagnetic waves in solutions of various salts, a continuation and extension of the work he had been doing during a 6 months stay with Professor E. Cohn in the University of Strasbourg. A first entry on the magneto-optical observation in his research notebook is dated September 2, 1896. Of course, Zeeman was aware of the profound influence of magnetic fields on vapors and gaseous discharges which contain electric charges. So he performed a series of experiments to check whether the broadening could be caused by changes in temperature, pressure or other macroscopic parameters before concluding that only a direct action of the magnetic field on the individual atoms could be responsible for the phenomenon that apparently was caused directly by the magnetic field. Among those experiments there were tests in which the D-lines were observed in absorption and others in which different materials than sodium were used. So, for instance, he found the red line of lithium to behave just like the Na D-lines and molecular band lines to be unaffected by the field.

When Zeeman had convinced himself of the reality of the new phenomenon he wrote a report on the observations to be handed over to Kamerlingh Onnes on his return from the vacation resort Bad Reichenhall (Bavaria), with an eye to have it published in the Proceedings of the Royal Academy of Sciences, which Zeeman had been patronizing many times before. Onnes, who was to become famous for his cryogenic work, had little experience with optics and with the observation of very faint spectral lines from an unsteady source in a dark room, but Zeeman managed to show him the phenomenon just one day before the monthly meeting of the Section for Natural Sciences of the Academy in Amsterdam, on October 31, where Kamerlingh Onnes presented the (Dutch) text. It was in this session that Lorentz, the great Leiden theoretician, heard of the event for the first time. That was the more remarkable because Zeeman had been his assistant for some time and in that capacity had been stimulated to investigate the Kerr effect, – the change of polarization of light by reflection on magnetized mirrors. The subject had been reported as a prize essay of the Dutch Society of Sciences at Haarlem and he had been awarded the gold medal. Subsequently the essay became his dissertation for the doctor's degree (1893) under supervision of Kamerlingh Onnes, Professor in Experimental Physics. It was during this investigation that the idea came to Zeeman's mind that if magnetism influences the state of polarization of light – as in the Kerr effect, and earlier in the Faraday effect – it should perhaps influence the production of light too. Reading a sketch on Faraday by Maxwell he was struck by the fact that Faraday in 1862 had tried in vain to detect such an influence, and he was not impressed by statements like Maxwell's, in his address to the Mathematical and Physical Sections of the British Association for the Advancement of Science, Liverpool, September 15, 1870: *"No power in nature can now alter in the least either the mass or the period of any one of them"*[2] with reference to atoms and molecules. Maxwell was eager to find anything in nature that could produce indestructible and unalterable standards of high precision. In the last quarter of the 19th century, however, there was a growing insight that, in order to understand spectral lines, one had to assume that inside the atoms something must "rattle". But there were also many physicists clinging to the vortex model who passionately defended the idea of the undivisible atom. This was not Zeeman's primary concern. If anything he regarded Maxwell's statement as a challenge and furthermore he reasoned that if a Faraday had thought of the possibility of the magnetic interaction with the atom it would be worthwhile repeating the experiment using the more advanced techniques at his disposal. Apart from these considerations he had more fundamental reasons to assume that a splitting might occur. In the first place it was known from the theories of W. Thomson (the later Lord Kelvin) and Maxwell that a homogeneous magnetic field acting on a charge can be described as a rotational motion "of the aether" around the lines of force. We know this as Larmor's precession, but this was proven and published by Larmor only in 1897. Zeeman concluded that the orbital motions of luminiferous particles with a certain natural frequency would be distorted and give rise to sort of combination frequencies on both sides of the natural frequency. In a somewhat analogous mechanical system, a double "pendulum" consisting of a vibrating system coupled to a similar vibration participating in a rapid rotation, treated by Thomson in 1856, leads to circular motions of different periods depending on the sense of the rotation. Lorentz, who had listened to Onnes's presentation of Zeeman's work, decided to invite Zeeman for a conversation which took place on Monday, November 2, in his office. In 1895 Lorentz had published the theory which we now call the electron theory, but it should be realized that the electron was unknown yet. He had introduced "luminiferous ions" to account for the interaction between light and matter, without giving a reason for the choice of a word already standard in electrochemistry and electrolysis. (The name electron had been invented by Stoney in 1891, but merely to denote the elementary charge, without connecting it with any massive particle). The luminiferous ion was supposed to be elastically bound in matter, so that it is capable of performing harmonic oscillations, and Lorentz showed Zeeman how the influence of a magnetic field on such a vibrating ion could be described in this "Lorentz" model.

Zeeman picked this up quickly and the next day informed Lorentz about the value of e/m that would then follow from his observations. His precision can be appreciated as one derives e/m from his data: At a field strength of 10^4 gauss the displacement of the outer components is 1/40th of the distance between the Na D-lines. We then obtain $e/m = 1.28 \times 10^7$ e.m.u./gram on the present-day theory while we know it to be 1.758882×10^7. Lorentz, however, considered it to be incompatible with all current views on the constitution of atoms, from which one may infer that he had expected the luminiferous particles to be much more massive, more "ion"-like. From Lorentz's model an important consequence was the polarization of the light. In the perpendicular direction this could be checked easily, but for longitudinal observation where circular polarization was predicted, it was necessary to pierce the poles of the magnet and to install equipment for detecting the circularity. After the meeting with Lorentz, Zeeman worked frantically to get everything ready in time for the next Academy session, on November 28, where the second half of his article was to be presented. He succeeded in observing the opposite-circularly polarized components by converting them to linearly polarized ones by means of a $\lambda/4$ crystal plate and looking through a nicol that would suppress one of the two σ-components and pass the other one, or just the other way around, depending on the position of the nicol. As the model demanded and observation verified, the π-component was invisible so that the separation of the σ-components was easier than in transversal observation, but the most distinct effect is that the observer can produce at will each of the σ-components, while the other one is suppressed, by simply rotating the nicol which causes a periodic jumping of the intensity in the spectrum, easily detected by the eye. Reversing the magnet current produces the same phenomenon in principle. Zeeman was elated: In his research notebook one reads (translated from the Dutch): *"Finally established that the influence of magnetisation on light vibrations is real."* The entry was dated November 23, and Zeeman managed to demonstrate the longitudinal effect as described above to Lorentz the next day. Lorentz was impressed and heartily congratulated Zeeman on this "slice of unexpected fortune", showing that there were indeed particles in the sodium flame which obey the laws of the model. He graciously authorized Zeeman to have his contribution acknowledged in the second part of the article which was to be presented at the Academy session of November 28, - not by Onnes who was absent, but by the secretary of the section, J.D. van der Waals from Amsterdam. So there were two papers by

Zeeman in the proceedings, with continuous numbering of paragraphs, under the title (translated) *"On the influence of a magnetization on the nature of the light emitted by a body"*, which really formed one publication, as he had wished all the time [3]. Notwithstanding the Dutch language the transactions did not go unnoticed abroad. For there were foreign correspondents at the Academy sessions, and the British Journal *Nature* carried reports on Academic meetings in Paris, Berlin, Amsterdam, Copenhagen and several others, in short in each place where important progress could be expected. Already in its issue of December 25, 1896 there was a report on the session of October 31 [4]. The session of November 28, however, was mentioned rather late and it missed the most important information [5]. From those columns one can draw the conclusion that Zeeman's results were discussed in scientific circles already and he indeed received many congratulations, mostly through the hands of Kamerlingh Onnes, since he had moved to Amsterdam where he had accepted a lectureship starting January 1, 1897. Several of them also had critical questions or expressed a great surprise on the value of e/m, which Zeeman had given as *"of the order of* 10^7*"* in part II. Of course the extravagance of the experimental value of e/m had not escaped the attention of the Leiden scientists. Certainly Lorentz must have been pondering on the question whether the detected particle really was the one he had postulated in his corpuscular theory of electricity in condensed matter. In my opinion it is unreasonable to assume that Lorentz, within a period of 4 weeks, converted from a state of distress to one of happiness over essentially the same numerical result. His scepticism led Zeeman to present the result without comment on the meaning it might have for the constitution of atoms. This permitted Lorentz to stay away from the disputes following the publication.

Of his Dutch text Zeeman made translations in English, French and German. The English text was accepted by The Philosophical Magazine [6], the French one was taken by Archives Néerlandaises des Sciences Exactes et Naturelles, a journal (now extinct) of the Dutch Society of Sciences at Haarlem, and the German text appeared in Verhandlungen der Physikalischen Gesellschaft zu Berlin.

To modern physicists educated with electrons, electronics and electric devices everywhere, it may seem very strange that the electron, instead of being welcomed as the missing entity in nature, was so slow in being accepted as a constituent of matter. Well into the 20th century a number of scientists opposed. One has to plunge oneself into the 19th century science situation to understand the culture shock connected with a change by a factor of more than a thousand in the lowest mass known.

2 The electron

In Amsterdam the first investigation that Zeeman undertook concerned the magnetic splitting of a blue cadmium line which is not susceptible to self-reversal. He succeeded in resolving the two σ-components completely in the longitudinal effect, and, when a nicol was used to suppress the π-polarization, also in the transversal direction [7]. Two months later the full triplet was observed so that the predicted polarizations could be experimentally verified [8]. In the mean time Zeeman had performed an accurate check on the $\lambda/4$ mica plates he had been using since the

Leiden time and found the marks indicating the directions of the axes to be wrong. A corresponding rectification regarding the sign of e could be added to the text of the former paper [7] as a footnote.

As the Amsterdam grating was inferior to the one in Leiden, he ordered a new grating [a] having a radius of 6 1/2 m which, however, was not received until 1900. But in the University of Groningen Professor Haga allowed him to use a beautiful grating mounting with which in a field of 22400 Gauss triplets for the D lines of sodium were observed and the best value of e/m was established at 1.6×10^7, again working visually, but making long series of measurements [8]. Back in Amsterdam, he was able to get the Leiden grating on loan between October 1897 and December 1898, and started photographic observations, mainly because photography allows to compare many spectral lines simultaneously in exactly the same situation, with regard to conditions in the light source as well as the strength and homogeneity of the magnetic field, that was never quite reproducible after having been switched off. But, of course, he also realized that at lower wavelength, including the ultraviolet, the photographic emulsion is more sensitive than the human eye, and that it has the advantage to yield a faithful document that can be used afterwards, again and again. Finally the photographic method allows compensating the weakness of certain lines by prolonged exposures. But again he was frustrated, now because of the fact that the building's (upper) floor was not stable enough and that vibrations caused by persons, machines, traffic and so on had the consequence that of 30 photographs only one was useful for measurements. What he really was after, was to detect whether the value of e/m was stable, or whether there were significant differences pointing to the existence of different kinds of luminiferous ions. He had an unshakable confidence in the Lorentz-model, about which Lorentz had written in a German article [9] (translated) *"Mr. Zeeman has in his communications on the spectral lines from a light source acted on by a magnetic field developed the simple theory, through which the phenomena he discovered can be explained and partly predicted."* The formulation illustrates Lorentz's courteousness, although one might also sense from it a certain reservation, an indication that he did not (yet) want to be associated with the puzzling value of e/m derived from the theory for his "luminiferous ions and molecules". In his article [3] (part II) Zeeman had given the theory using simultaneous differential equations, which makes it less transparent than the treatment he chose in his monograph of 1913 [10] which we loosely will follow. An electric particle with charge e moving in a homogeneous magnetic field H experiences a Lorentz force $Hev \sin \alpha$ with v the velocity and α the angle between H and v, perpendicular to the field lines and the velocity. If it moves in a circular path normal to the field the force is directed along the radius r, and if it moves parallel to the field H the force is zero. Now an elastically bound charge performing a harmonic vibration in an arbitrary axis can be thought of as having three components, i.e. a vibration along the magnetic field lines ω_0 and two circular motions in opposite senses, ω_1 and ω_2. The vibration (π-component) will not be affected by the field and stay at the field-free position ω_0 in the spectrum (ω, the angular frequency, being $2\pi\nu$, ν being the frequency). As an antenna it will not radiate in the directions parallel to

---

[a]This grating bears an engraved inscription saying that it was ruled by Professor Rowland himself especially for Zeeman.

H, so in longitudinal observation it will disappear. The other two components (σ-components) will be accelerated by the Lorentz force (σ_1) respectively decelerated (σ_2), the circle radius staying the same. Now for a circular motion one needs a centripetal force $mr\omega^2$, which is the resultant of the natural elastic force $mr\omega_0^2$ and the Lorentz force $Hev = +Her\omega_1$, respectively $-Her\omega_2$. Thus for σ_1 we get $mr\omega_1^2 = mr\omega_0^2 + Her\omega_1$ and for σ_2 $mr\omega_2^2 = mr\omega_0^2 - Her\omega_2$. Hence $mr(\omega_1^2 - \omega_2^2) = Her(\omega_1 - \omega_2)$ or $m(\omega_1 + \omega_2) = He$. Since $\omega_1 - \omega_2 = 2\delta\omega$, $\delta\omega$ being the shift from the unperturbed position, we obtain $m \times 2\delta\omega = He$ or $\delta\omega = He/2m$ or $\delta\nu = \frac{He}{4\pi m}$ in which expression one recognizes the Larmor precession frequency. In wavenumber units (cm^{-1}) one gets $\delta\sigma = \frac{He}{4\pi mc}$ which expression is called the Lorentz unit and has the value $4.6686 \times 10^{-5} H$ (in Gauss).

It would appear that for each oscillator one obtains a different intensity distribution over the 3 components, but with an isotropic distribution of velocities over all directions it is easy to see that the triplet in transversal observation must have relative intensities (in %) 25 : 50 : 25. The middle component, π, linearly polarized parallel to the magnetic field lines, the two outer components, σ, linearly polarized normal to the magnetic field lines, as the circular motions are viewed "edge-on".

Around 1897/98 there were observations being published by various workers in the field (Cornu, Preston, Michelson) which were not in harmony with the theory, since they showed either more than three components or they were triplets whose splittings strongly depart from the Lorentz width. It is very probable that Zeeman had seen the anomalous patterns of the sodium D-lines, but he thought they were faked as a result of self-reversals. The situation became serious when reports with improved resolution continued to appear, so that Lorentz began to work on *systems* of charged "ions", trying to explain splittings different from those predicted by the "single particle" model [9]. Zeeman, in his hunt for e/m, chose to ignore the anomalous patterns, since without an adequate theory there seemed no point in trying to extract a value of e/m from them, but probably also in the opinion that the details would be due to some external or internal perturbation. This intuitive idea, we know, is correct in principle, since the Lorentz triplet is ultimately always found, when the magnetic field is made high enough: In the complete Paschen-Back effect, discovered only in 1912, the effect of spin magnetism is eliminated and a complete line multiplet degenerates into the normal triplet! In 1898 Lorentz remarked casually that the value of e^2/m determined from the disperson of light in solid transparent materials approximately agrees with Zeeman's e/m [11]. Meanwhile another great discovery had come to the rescue. It was J.J. Thomson who determined e/m for the negative corpuscles in cathode rays. On April 30, 1897, he had made this known in his address to the Royal Institution of Great Britain [12], which ended with the following note: *"It is interesting to note that the value of e/m which we have found from the cathode rays, is of the same order as the value 10^7 deduced by Zeeman from his experiments on the effect of a magnetic field on the period of the sodium light."* It was also published in Phil.Mag. [12]. Thomson's method was not the later famous technique of combined electric and magnetic deflection in vacuo using a Lenard window. On the contrary the cathode rays were studied inside the discharge tube in which a calorimetric measurement was needed to determine the velocity, assuming that all kinetic energy was converted into heat. The magnetic deflection took place on the way passing an external magnet. From Thomson's data we derive $e/m = 0.74 \times 10^7$ emu/g. It may be remarked that Thomson started his investigation with the idea that the corpuscles did not penetrate an Al window as Lenard had used in 1893. With Lenard he supposed the "Lenard rays" rather to be ether waves, such as Röntgen had discovered, and that the ionization found behind the Lenard window was due to these waves. There was no notion about the emptiness of solid materials yet. The technique with two deflecting fields in a high vacuum, suggested by Thomson, that led to high precision data for e/m at later times was inflicted with a systematic error for several years, stemming from the fact, that researchers did not realize the influence of space charges on the electrodes, which weaken the external electric field strength. So the results were systematically low, but gradually increased to an asymptotic value as the vacuum pumping technique improved. However, the agreement was good enough to conclude that Lorentz's hypothetic luminiferous ions, Zeeman's particles and Thomson's corpuscles were identical, so that they unified under the name electron around 1900. In 1902 the Nobel prize for Physics was awarded to Lorentz and Zeeman, with the common motivation: *"in recognition of the extraordinary service they rendered by their researches into the influence of magnetism upon radiation phenomena"* while Lenard received it in 1905 with the motivation: *"for his work on cathode rays"* and Thomson in 1906 with the motivation: *"in recognition of the great merits of his theoretical and experimental investigations on the conduction of electricity by gases."* Most remarkably the electron is not mentioned in any of these citations. It is known that the Nobel committee is always very cautious in its statements, but were there still members holding doubts about the reality of the electron? As far as Lorentz and Zeeman are concerned, the joint award may have contributed to the impression that they closely collaborated, or to the statement in certain textbooks that Lorentz predicted the effect and Zeeman verified it experimentally. This is totally mistaken, as Lorentz has stressed for example in his acceptance speech at receiving the honoris causa doctor's degree of the Delft technical University (March 7, 1918). His final words on this matter, namely *"he could not have succeeded, if the mass of the electron had been ten times greater with the identical charge"*, sound like an echo from the time, that Lorentz was still thinking in terms of ions. The translation is from Zeeman himself[13]. The fact is, however, that the two men held one another in high esteem always and that they closely cooperated in preparing their Nobel lectures. They became close friends till Lorentz's death in 1928; Zeeman then was 62, at the zenith of his fame. The first honoris causa doctorate, Göttingen 1905, was connected with a change in his interests. Seeing, that further efforts to improve the value of e/m by optical means would be futile and considering that his technical means in Amsterdam were still inadequate, he decided to study phenomena, connected to the magnetic splitting, which did not require the extreme resolution and stability, needed for studies of the anomalous Zeeman effect.

3 Exploring triplets and doublets

This was stimulated by Professor Woldemar Voigt at Göttingen who had an approach different from that of Lorentz, in that he treated absorption rather than emission, and then used the Kirchhoff law to describe emission features. He started out already in 1898 with a thorough study of the Kirchhoff relation in spectra [14] and treated the longitudinal Zeeman effect in connection with the Faraday rotation. Other magneto-optical phenomena were described in detail, especially magnetic double refraction, later to be called the Voigt effect. This differs from the Cotton–Mouton effect, observed with liquids having anisotropic molecules which are lined up by a magnetic field, and is usually smaller by orders of magnitude. The many predictions from Voigt's theory were investigated meticulously under Zeeman's guidance. Only certain slight intensity dissymmetries predicted by Voigt could not be confirmed. The 178 pieces of correspondence in the Zeeman archive, between 1898 and 1919, give testimony of the close personal relationship between the two men and their intense collaboration. Lorentz, who also highly appreciated Voigt's approach, commented that this theory could be so powerful just because it refrains from going into the mechanism of light emission [15]. This is especially true with respect to the anomalous Zeeman effect [16]. When Voigt, on one of his frequent visits, watched the splitting in a demonstration experiment he exclaimed: *"Das allein ist schon die Reise nach Amsterdam wert!"* [This alone would make it worthwhile traveling to Amsterdam.] The most impressive achievement of the phenomenological theory was the explanation of the behaviour of the D-lines in varying magnetic fields, from the 10 components (four π- polarized and six σ-polarized) in the low field pattern to the final Paschen–Back stage: A Lorentz triplet. In this case he assumed three electrons interacting, namely one with the natural D_1 frequency and two swinging at the natural D_2 frequency (thus accounting for the 1:2 intensity ratio at zero field). The magnetic field then coupled the three motions in such a way that the transition could be quantitatively described[17]. The classical formula, again derived from the absorption case first, was later quantized in terms of the old quantum theory by Sommerfeld (for the emission case) to read $W = \overline{W} + (h/2)(m \pm \sqrt{1 + \frac{2}{3}mv + v^2})\Delta\nu_L$, where $\Delta\nu_L$ is the Lorentz unit, $v = \Delta\nu_0/\Delta_L$, $\Delta\nu_0$ being the natural doublet interval. It now became clear what Zeeman had seen already in 1896, and later photographed: the red Li line always behaving as a perfect Lorentz triplet, because the D-lines nearly coincide and they are in the Paschen–Back stage even at low field intensity. In 1907 it was not yet known that the red line is a doublet, but Zeeman detected it and found the interval to be only 0.13 Å, equivalent to 0.29 cm^{-1} in wavenumber units (0.34 cm^{-1} accepted today, of the order of a hyperfine structure). This discovery also removed the last doubts about Li being a genuine alkali metal [19].

The search for dissymmetries instigated by Voigt had led to the detection of an unexpected asymmetry in the Lorentz triplet of the mercury line at 579.1 nm which turned out to consist of a shift of the π-component from its zero position towards the red [20]. This was in flat contradiction with the Lorentz model and therefore intrigued Zeeman very much. In the middle of his studies, in which he also made use of a Michelson echelon and an etalon (he wanted to exclude possible instrumental defects), he was surprised by the work of P. Gmelin from Tübingen who prepared for the doctor's thesis. He also had observed the phenomenon and moreover had found a quadratic dependence of the shift on the magnetic field strength [21], which Zeeman could verify. Both of them had the correct suspicion that the weak satellite line at 579.0 nm might have something to do with the shift, but the matter was not resolved until 1934, when it was quantitatively explained as the result of Paschen–Back perturbation in jj-coupling, the lower levels of the two lines being the same, 6s6p $^1P_1$ [22]. Hence again Zeeman had run into a Paschen–Back phenomenon by studying magnetic triplets, this time an intermediate stage. The real discovery of course took place in 1912 [23]. An accompanying phenomenon which they called "magnetic completion of line groups" was treated by the same physicists and published as a tribute to Zeeman on the occasion of the 25th anniversary of his great discovery [24].

Around the same period that he had been struggling with the yellow mercury line and other dissymmetries, Zeeman had another and most pleasant experience with (pseudo)normal triplets. There came a letter from G.E. Hale, the founder of Mt. Wilson Observatory. It was accompanied by a manuscript for a note on *Solar vortices and the Zeeman effect*, intended to be submitted to *Nature*. It was followed by two glass photographs showing the spectrum of sunspots in the wavelength region 6250–6360 Å, taken through a Fresnel rhomb and a nicol in front of the slit of the grating mounting down the "Tower" telescope. The picture showed doublets for a certain position of the nicol and Hale asked whether Zeeman could agree that these were magnetic splittings, and if so, whether he would be inclined to write a note in support of Hale's view, also in *Nature*.

Spectroheliographic pictures of the sun had already revealed the existence of extended cyclonic motions in the sunspots. If charged masses are involved in such vortices a magnetic field is bound to occur which for spots not too far from the centre would be directed towards the observer, so that the doublets could be explained by the longitudinal Zeeman effect. In agreement herewith Hale reported circular polarization to be seen. Although Zeeman, after inspecting the material sent, was nearly convinced, he asked Hale to try to observe spots near the sun's limb which should be expected to show the transversal effect, and moreover to check whether in the central spots the sign of the circular polarization reverses with the direction of rotation. Both question were answered in the affirmative in a wire from Hale on September 21, 1908, reading: "Vortices rotating opposite direction show opposite polarities; spot lines near limb plane polarized."

Zeeman sent the note which was printed in conjunction with Hale's communication [25], and Hale gave details in an exhaustive publication [26]. The magnetic fields in the spots turn out to be at most 4500 gauss, but lighter elements give lower values than heavier ones owing to the different altitudes where they have their highest concentrations. The determination of extra-terrestrial magnetic fields is one of the finest applications of the Zeeman effect. It has given rise to a vast field of research in astrophysics. A class of magnetic stars has been detected in which completely resolved Zeeman splittings have been observed in metal lines. Magnetic fields up to 35000 gauss have been found according to a report given at the Zeeman Centennial Conference in 1965 [27]. This is the general field, since in stars local fields cannot be studied. The general magnetic field of the sun is of the order of 50 gauss and can only be measured by refined optical techniques combined with photo-electric detection (G. Thiessen, Forschungsbericht Fraunhofer-Institut 1945).

Because all sunspot observations show the inverse effect, Zeeman decided to start a series of absorption experiments in the laboratory, in order to simulate as much as possible the situation in the spots. His collaborator in this was a Polish guest. A new wing to the laboratory had just been built, and on the stable ground floor they could install the 6 1/2 m Rowland grating. The so-called Wadsworth mounting had been modified by replacing the collimator lens by a concave mirror so that a more compact spectrograph could be constructed. On the unstable upper floor this Wadsworth–Zeeman mounting had never given satisfactory results. Light from the positive crater of an arc source passed the sodium flame subjected to the magnetic field. The D-doublet was photographed under the widest variation of external parameters, i.e. temperature and sodium density, intensity and direction of the magnetic field, and state of polarization of the incoming light. Predictions from the theories of Voigt and Lorentz (which coincided) were experimentally checked and satisfactory resemblance with the astronomical pictures was obtained. A number of publications appeared in the Academy Proceedings. This attracted the attention of the Editor of The Astrophysical Journal who invited Zeeman to write a more exhaustive paper for his journal in order to serve the astronomical community better. So there appeared a comprehensive article entitled: *The magnetic separation of absorption lines in connection with sunspot spectra*[28].

4 Anomalous effect and optical spectroscopy

The complete failure of existing theories to account for the anomalous patterns, – with the exception of the alkali resonance doublets, so successfully treated by Voigt –, combined with the relative inferiority of his apparatus, had not encouraged Zeeman to pay much attention to those anomalous features. But he was not blind to the fact that other spectroscopists were making progress and busy to establish certain rules from the wealth of data that were being produced in more complicated spectra (Preston, Runge, Landé). Bohr's theory (1913) which finally incorporated the electron in a consistent atomic model, – albeit a one-electron system, applicable to the hydrogen atom and hydrogenlike ions only – , could not explain the anomalous effect either. However, it gave a new outlook on the origin of the magnetic splitting, because this was now ascribed to the stationary energy levels, while the optical effect in the lines became the difference of the level splittings. Due to the selection and polarization rules, derived from the correspondence principle, all allowed transitions coincided in the Lorentz triplet, which thus was explained in a much more complex way than before. The magnetic levels were henceforth denoted as Zeeman levels. To spectroscopists this development was not a radical change, for since the days of Rydberg, Ritz and Balmer they had been accustomed to see spectral wavenumbers as differences between terms and the principal quantum number was a well-established concept. They had a sense for quantum effects long before the quantum principle was enunciated by Planck (1900). Of course Lenard and Rutherford had paved the way by showing that atoms are practically empty space. Nevertheless Bohr's postulates constituted a bold step, the more so because they were not really understood. The extension by elliptic orbits, requiring two new quantum numbers (Sommerfeld 1916), the radial and the azimuthal one, did not contribute anything to the understanding of the anomalous Zeeman effect directly, but it fostered the idea that the anomalous patterns had something to do with unequal splitting factors in the two combining levels. The new quantum numbers stood for mechanical momenta, which are vectorial quantities so that their orientation with respect to an external field could not be neglected. The interaction with a magnetic field occurs through the magnetic dipole momenta which are connected with the motions of the electrons. However, the existence of space quantization was experimentally proven only by the famous atomic beam experiment of Stern and Gerlach (1921). This was done with silver atoms and the result was rather puzzling at first, because the beam was split in two, not in three. It was not yet known that the magnetic moment of the silver atom does not originate in an orbital motion, but is the intrinsic property of the electron only. However, the experiment gave a decisive support to the view that the anomalous Zeeman effect also is a manifestation of space quantization: Without the field the Zeeman substates of a level coincide and in the field they split up, provided the atom has a magnetic moment in that particular state.

It will be clear that a formal trick for describing anomalous Zeeman patterns lies in the introduction of a correction factor g to the Lorentz unit such that $g = \mu/j$. In the following years theorists, among whom particularly Sommerfeld, Hund, Landé, Heisenberg and Pauli, worked on the development of the vector model, obeying the quantum conditions inclusive space quantization. A key role was attributed to the concept of multiplicity, originating from Catalán's analysis of manganese spectra, 1922. He found rather close line groups showing characteristic spacings and intensities which he called multiplets which connected groups of levels with regular spacings, – terms –, denoted as doublets, quartets, sextets, and so on, or triplets, quintets, and so on. This multiplicity turned out to be a general feature in many spectra, though not always as conspicuous as in Mn, and is closely connected with the chemical periodicity as well as the stage of ionization. In one and the same spectrum it is either even (Mn I) or odd (Mn II). In 1923 Landé succeeded in finding an expression for g as a function of the quantum numbers, including one for multiplicity, which satisfied nearly all available experimental data on anomalous splittings [29]. Space quantization being taken care of by the magnetic quantum number m,– projection of $\mathbf{j}$ on the axis of the magnetic field –, the transition rules $\Delta m = 0$ for π-components and $\Delta m = \pm 1$ for σ-components, with the interdiction of the 0–0 transition in case $\Delta j = 0$, then permit to derive the anomalous Zeeman pattern. The formula of Landé is not yet unambiguous, since the quantum number j representing the total momentum has to be replaced by $j - 1/2$ for odd multiplicity, but this ambiguity was removed when the true nature of the multiplicity was unveiled as a consequence of electron spin magnetism.

The solution proposed in 1925 by Goudsmit and Uhlenbeck consisted in allotting a spin 1/2 and magnetic moment of one Bohr magneton to the electron [30], which made $g_s = (-)2$. This spin anomaly agrees with quantitative Stern–Gerlach experiments and the agreed structure of the silver atom as an inert core (Pd) plus one electron in an s-state, so that the magnetic moment is entirely due to the electron's intrinsic magnetism. Landé's g-formula can then be simplified and Pauli's exclusion principle formulated more precisely, whereby the notion of closed shells obtains its modern shape. Finally, the level scheme of hydrogen is reinterpreted so that an observed fine structure component can be explained. However, the doublet splitting in the alkali's came out twice as large as observed and a classical picture of a rotating electron made no sense, because the peripheral velocity would have to be some 50 times the velocity of light in vacuo in order to produce the correct magnetic moment with the mechanical moment $h/4\pi$. When the two young men approached Professor Ehrenfest in Leiden, successor of Lorentz since 1912, he was cooperative. But Pauli immediately declared the idea to be nonsensical. Lorentz, consulted by Uhlenbeck personally, raised severe objections in connection with the mass equivalent of the magnetic energy. Thereupon Uhlenbeck wanted to withdraw the paper, meanwhile sent off by Ehrenfest who is reported to have said that the authors as yet had no reputation to lose. The hypothesis, suspected by Ehrenfest and Lorentz, ridiculed by Pauli, became as we know, a resounding success. The objection regarding the doublet splitting was soon removed in a relativistic treatment of the spin–orbit interaction, introducing the Thomas precession [31]. Relativistic quantum mechanics (Dirac 1928) produced spin and magnetic moment of the electron as vital intrinsic properties of that particle.

The Goudsmit–Uhlenbeck hypothesis became the keystone of the vector model, which has been of tremendous importance for understanding the level structures of atomic and ionic spectra in the following decades. In the modified Landé formula, $g = 1 + \frac{J(J+1)+S(S+1)-L(L+1)}{2J(J+1)}$, capitals are used to denote, that the vectors are resultants for all electrons outside closed shells, and it applies to the case of Russell–Saunders coupling which is approximated in the lighter elements. Many other coupling schemes have been developed and g-formulae can be derived for any pure coupling scheme. But cases with pure coupling are rare. Whatever the deviations from the Landé-formula, be it as a result of intermediate coupling conditions, or of an incidental local perturbation, the sum of g-factors belonging to one particular value of J is a constant. This g-sum rule, formulated by Pauli[32], was originally proposed for pure electron configurations. Only quantum mechanics is able to cope with the intermediate coupling situations where the many interactions in a complex atomic system are of the same order of magnitude. Even the concept of electron configuration has lost its sacrosanct status of early times. Configuration interaction is quite common and sometimes unexpectedly strong so that multi-configurational approaches are needed to explain, or describe, the properties of stationary states (including g-values and lifetimes) and those of spectral lines (including transition probabilities and oscillator strengths). In such cases the g-sum rule has to be stretched over all configurations involved. The g-formulae, obtained by trial and error, has the flavour of quantum mechanics in that squares of quantum numbers q are replaced by expressions $q(q+1)$ which are foreign to the vector model theory.

In the so-called *extended* vector model the interaction of electrons with nuclei having a nuclear spin I and magnetic moment μ_I can be accounted for. The resulting magnetic hyperfine structure is roughly on a scale 10^3 times smaller than the spin-orbit interaction, since $\mu_I = g_I \times I$ is roughly a thousand times smaller than a Bohr magneton. This implies that Russell–Saunders coupling between **J** and **I** is a better approximation than between **L** and **S**. In many cases, I has been determined by the application of the Landé interval rule to the hyperfine levels $\mathbf{F} = \mathbf{J} + \mathbf{I}$. Departures can occur for strongly deformed nuclei in the presence of an electric quadrupole interaction, but this is relatively rare. The narrowness of the magnetic hyperfine structure makes it difficult to observe the anomalous Zeeman effect in the optical hyperfine transitions. Even in very weak magnetic fields the coupling between **J** and **I** is broken and the two vectors will be precessing around the field lines independently, i.e. one sees the Paschen–Back effect of the hyperfine structure. The residual interaction is then given by $A_F M_I M_J$, where A_F is the interval constant of the field-free pattern. For a fixed value of M_J this gives rise to $2I + 1$ Zeeman sublevels and for suitable combinations a splitting of the Zeeman components into $2I + 1$ subcomponents can be observed. This phenomenon, the so-called Back–Goudsmit effect, offers the most reliable way to determine I, because the subcomponents are equidistant and equally strong. It was demonstrated on ^{209}Bi in a classic paper [33], that the Zeeman components indeed split in 10 subcomponents, corresponding to $I(^{209}\text{Bi}) = 9/2$. The investigation was made in Back's laboratory in Tübingen which had the best installation for such work. Goudsmit went there for a stay while he held an assistentship with Zeeman in Amsterdam. During his high school period (HBS) at The Hague his physics teacher was T. van Lohuizen, who was Zeeman's pupil (doctor's thesis in 1912) and recommended him to go into spectroscopy. So he did, but after concluding his university studies in Leiden. His academic carrier nicely illustrates the development of Dutch physics sketched by Kox in these proceedings. In several letters to Zeeman, Goudsmit described life in Germany, and later also in the USA, with his signature usually preceded by "yours devoted pupil" or similar expressions of affection. Around 1938, when he was working in Ann Arbor, he seriously considered an offer to come to Amsterdam as Zeeman's successor.

In the thirties progress was made in the field of the magnetic effects on hyperfine structure by the introduction of high resolution interference instruments combined with atomic beams viewed perpendicularly in order to minimize Doppler broadening. Especially the study of resonance lines in absorption was successful in exploring the intermediate region between the natural hyperfine structure and the Back–Goudsmit effect, e.g. in the Na D-lines (for sodium $I = 3/2$). A comprehensive report was presented to the Zeeman congress [34], held in 1946, to celebrate the 50 year Jubilee of the Zeeman effect. Also in emission experiments a few cases had been successfully resolved (Kopfermann - Cs; Rasmussen - Cd; Back - Tl) [35]. Summarizing it may be stated that the theory of the intermediate region has been tested in hyperfine structure to a higher degree of accuracy than in the fine structure case. At the same time these experimental investigations represent about the best one can do in optical measurements.

5 Zeeman effect at radiofrequencies

The real drawback of the optical method lies in the fact, that in optical spectral lines, Zeeman effect and hyperfine structure are observed as very small differences between huge frequencies. An enormous gain can be expected by *directly* measuring the distance between e.g. adjacent Zeeman levels, but the corresponding transitions have magnetic dipole character (normally "forbidden" transitions), and secondly the natural population of such levels is practically the same at normal temperature. Observation of spontaneous radiation therefore is out of the question in situations of thermal equilibrium. One of the first scientists in the Netherlands to realize that the development of radiofrequency methods could yield new information by absorption experiments at very low temperature, was C.J. Gorter, who studied paramagnetic relaxation, absorption, dispersion and resonance. In particular he had tried to detect nuclear magnetic resonance by subjecting suitable crystals at low temperature to a steady homogeneous magnetic field and a superposed a.c. field. A sudden rise in the temperature of the sample at slowly varying the field strength then would indicate resonance, but the absence of such a caloric effect was ascribed to the lack of a coupling between the nuclei and a lattice. In 1937 Gorter, then lecturer at Groningen University, visited Columbia University and saw Rabi's installation intended to study nuclei in molecular beams. It was obvious that the molecular beam method is superior in sensitivity because any change in spin orientation contributes to the signal, i.e. the signal will be proportional to the *sum* of the populations of the two levels involved, whereas in calorimetric measurements the signal depends on the population *difference*. However, Gorter also saw that the method for inducing the spin flip, by a magnetic field whose direction would be rotated in space, is inefficient in comparison with his device, in which even a small oscillator would create a much stronger depolarization. So he proposed Rabi to introduce that method and in the first announcement of the discovery of nuclear magnetic resonance [36] Gorter's contribution was acknowledged in a note: *"We are very much indebted to Dr. Gorter who, when visiting our laboratory in September 1937, drew our attention to his stimulating experiments in which he attempted to measure nuclear moments by observing the rise in temperature of solids placed in a constant magnetic field on which an oscillating field was superimposed"*. The MBMR technique was very successful in yielding high precision data for quite a number of nuclei although the diamagnetism of the electrons imposes a limit on the accuracy. It should be remarked that this technique represents the pinnacle of a long series of investigations on beams in inhomogeneous fields at Columbia, aimed at precision determination of magnetic properties of atomic states. In Europe this work became widely known through Kopfermann's book [35]. It was the Columbia group that initiated the Brookhaven Conferences (1955-1964) and from these ICAP sprang in 1968 [37]. Rabi's work was awarded the Nobel prize 1944, with the citation *"for his resonance method for recording the magnetic properties of atomic nuclei"*.

Gorter was to become Zeeman's successor in 1940 and he was the director of the Zeeman Laboratory up to the fall of 1946. In this period he undertook a new attempt at observing NMR in the condensed state, without success [38]. F. Bloch, and independently, E.M. Purcell, succeeded finally in reaching that goal in 1946 and this discovery opened a vast area of research in which g_I values of practically all stable nuclei and several unstable ones could be determined. The universality of the method is the greatest advantage over the MBMR technique, but apart from the diamagnetic shift, which can be calculated, the results are afflicted with an uncertainty due to the more erratic "chemical shift". For instance, in ammonium nitrate one finds two resonance peaks for ^{14}N, one belonging to the NH_4 group, the other one to the nitrate, owing to the different environment of the nitrogen atoms. While the resonance can be determined with a precision of 10^{-5} easily, the g_I value is not much better than 10^{-3}, - which of course is a big improvement over optical determinations. As is known, NMR in the condensed state has developed into a mighty diagnostic tool in the form of MRI (Magnetic Resonance Imaging) which concentrates on the protons, so abundantly present in organic materials. That this application of the Zeeman effect would conquer the medical world was not even foreseen in September 1950 at the *"International Conference on Spectroscopy at Radiofrequencies"*, organized by Gorter in Amsterdam, and recorded in *Physica* **17** 169–484 (1951). It saw an unusual concentration of future Nobel prize winners, namely Bloch and Purcell who were to share the prize in 1952 and further P. Kusch (1955), Ch. Townes (1964), A. Kastler (1966), A. Bohr (1975), J.H. van Vleck (1977), and N.F. Ramsey (1989). But also other physicists of great fame participated, among which H.B.G. Casimir, A. Fokker, H. Kopfermann, H.A. Kramers, R. Kronig, F.M. Penning, G. Racah and L. Rosenfeld.

Kastler presented a paper on *"Méthodes optiques d'étude de la résonance magnétique"*[39] in which he discussed the optical pumping technique and the double resonance experiments on Hg by Brossel, Sagalyn and Bitter, and launched a proposal for an optical analogon of the Rabi MBMR apparatus. Optical pumping is based on the selection of Zeeman states; the polarization, used for detection is a consequence of the non-equilibrium population of Zeeman states reached by irradiation of properly polarized resonance light. In passing, Kastler also mentioned the Lamb–Retherford experiment and the role played by the Zeeman effect in establishing the resonance.

An inkling on the subject had been discernible already at the Zeeman Congress four years earlier, when Kastler talked about *"The Zeeman effect and the intensity and polarization of resonance and fluorescence radiation"*. Making use of simple old-style diagrams, but with an emphasis on the intensities of the Zeeman transitions he discussed some interesting cases, including stepwise excitation, to conclude that the Zeeman components of spectral lines have a physical significance even in the absence of a magnetic field. Kastler's work can be considered as the start of the period of manipulating atoms by means of the Zeeman effect which has developed into an established art nowadays, as witnessed by many contributions to recent ICAP meetings.

Acknowledgment – The author wishes to express his gratitude to Drs B. Lightfoot and J.E. Hansen for critically reading the manuscript and suggesting several improvements and corrections.

References

1. e.g. P. Zeeman, *Journ. Franklin Inst.* **200**, 305–311 (1925).
2. *The scientific Papers of James Clerk Maxwell* Vol. II, ed. W.D. Niven (Dover, New York, 1965), p 225.
3. P. Zeeman, *Over den invloed eener magnetisatie op den aard van het door een stof uitgezonden licht* in *Verslagen Koninklijke Akademie van Wetenschappen (Afdeling Natuurkunde)* **5**, I:181–4 and II:242–8 (1896).
4. *Nature* **55**, 192 (1896).
5. *Nature* **55**, 431–2 (1897).

6. P. Zeeman, *Phil. Mag.* **43**, 226-37 (1897) (March issue).
7. P. Zeeman, *Phil. Mag* **44**, 55-60 (1897).
8. P. Zeeman, *Phil. Mag.* **44**, 255–9 (1897).
9. H.A. Lorentz, *Ann. d. Physik* **63**, 278–84 (1897).
10. P. Zeeman, *Researches in Magneto-Optics* (McMillan, London, 1913) pp30–4.
11. H.A. Kramers, *Nederl. Tijdschr. v. Natuurkunde* **8**, 137–141 (1941), p 140.
12. J.J. Thomson, *Proc. Roy. Institution* **15**, 419–32 (1897)
    and *Phil. Mag.* **44**, 293–316 (1897) (October issue).
13. P. Zeeman, *Nature* **128**, 365–8 (1931) (Faraday Memorial).
14. W. Voigt, *Ann. d. Physik* **67**, 366–87 (1898).
15. H.A. Lorentz, *Physica* **1**, 228-41 (1921) (p 240, in Dutch).
16. W. Voigt, *Ann. d. Physik* **24**, 193–224 (1907) *ibid.* **36**, 866–70; 873–906 (1911).
17. W. Voigt, *Ann. d. Physik* **41**, 403–40 (1913) *ibid.* **42**, 210–30 (1913).
18. A. Sommerfeld, *Atombau und Spektrallinien* (Vieweg, Braunschweig, 3rd Ed. 1922), p 492.
19. P. Zeeman, *Physik. Zeitschr.* **14**, 405–6 (1913)
    and *Proc. Roy. Acad. Amsterdam* **15**, 1130–1 (1913).
20. P. Zeeman, *Proc. Roy. Acad. Amsterdam* **10**, 351–9; 566–74; 574–8; 862–4 (1908) *ibid.* **11**, 473–7 (1909).
21. P. Gmelin, *Physik. Zeitschr.* **9**, 212–4 (1908) *ibid.* **11**, 1193–5 (1910).
22. J.B. Green and R.A. Loring, *Phys. Rev.* **46**, 888–93 (1934), p 890.
23. F. Paschen and E. Back, *Ann. d. Physik* **39**, 897–932 (1912) *ibid.* **40**, 960–70 (1913).
24. F. Paschen and E. Back, *Physica* **1**, 261–73 (1921).
25. G.E. Hale, *Nature* **78**, 369–70 (1908)
26. G.E. Hale, *Astroph. Journ.* **28**, 315–43 (1908).
27. H.W. Babcock in *Proc. Zeeman Centennial Conf., Amsterdam, September 1965, Physica* **33**, 102–21 (1967).
28. P. Zeeman and B. Winawer, *Astroph. Journ.* **32**, 329–62 (1910).
29. A. Landé, *Zeitschr. f. Physik* **15**, 189–205 (1923).
30. G.E. Uhlenbeck and S.A. Goudsmit, *Naturwissenschaften* **13**, 953-4 (1925)
    also *Nature* **117**, 264–5 (1926) and *Physica* **6**, 273–90 (1926).
31. L.H. Thomas, *Nature* **117**, 514(L) (1926).
32. W. Pauli, *Zeitschr. f. Physik* **16**, 155–64 (1923).
33. P. Zeeman, E. Back and S.A. Goudsmit, *Zeitschr. f. Physik* **66**, 1–12 (1930).
34. D.A. Jackson, *Physica* **12**, 568–80 (1946).
35. H. Kopfermann, *Kernmomente* (Akad. Verlagsgesellschaft, Leipzig, 1940; revised and extended 1956). Translated into *Nuclear Moments* by E.E. Schneider (Acad. Press, New York, 1958).
36. I.I. Rabi, J.R. Zacharias, S. Millman and P. Kusch, *Phys. Rev.* **53**, 318(L) (1938).
37. See *Atomic Physics* **12** (Proc. ICAP 12, Ann Arbor 1990) pp 585–607.
38. C.J. Gorter and L.J.F. Broer, *Physica* **9**, 591–6 (1942).
39. A. Kastler, *Physica* **17**, 191–204 (1951).
40. A. Kastler, *Physica* **12**, 619–26 (1946).

The Zeeman Effect : a Tool for Atom Manipulation

Claude Cohen-Tannoudji
Collège de France et Laboratoire Kastler Brossel *
24 rue Lhomond, 75005 Paris, France

Abstract

We review in this paper experiments which have been carried out during the last fifty years and which use the Zeeman effect, in conjunction with other effects, for manipulating the various degrees of freedom of an atom. We consider first the internal degrees of freedom and we show how the polarization selection rules of the Zeeman effect have played an essential role in the development of optical methods, such as double resonance or optical pumping, allowing one to control and to detect the polarization of atomic states. The importance of linear superpositions of Zeeman sublevels (Zeeman coherences) is emphasized as well as the possibility to change the Zeeman splittings by non resonant optical or RF fields. The second part of the paper will review more recent experiments where spatially dependent Zeeman shifts are used to control the position and the velocity of a neutral atom. Various schemes will be described, such as Zeeman slowers, Sisyphus cooling, magneto-optical traps, magnetostatic traps, which have been developed recently and which have just culminated with the observation of Bose-Einstein condensation.

1 Introduction

The initial purpose of this paper was to review the applications of the Zeeman effect in modern atomic physics. In fact, the scope of such a paper would have been too broad. There are practically no experiments in atomic physics where Zeeman sublevels, Zeeman shifts or magnetic couplings are not involved! I have thus thought that it would be more appropriate here to try to find a simple guideline along which I could organize this paper and which would allow me to put in perspective several important developments which have occurred during the last fifty years. In this respect, atom manipulation is a good guideline because the Zeeman effect turns out to play an essential role in the different methods which have been developed for controlling the various degrees of freedom of an atom.

Consider first the internal degrees of freedom of an atom, i.e. its angular momentum and its energy. By playing with the polarization selection rules of the Zeeman effect, which result from the conservation of the total angular momentum of the atom–photon system in absorption or emission processes, it is possible to prepare an atom or to detect its presence in a given Zeeman sublevel $|M\rangle$, or in a linear superposition of such sublevels $\sum_M c_M |M\rangle$. The first part of this paper will be devoted to a review of several developments based on these ideas, such as double resonance, optical pumping, Hanle effect, quantum beats, etc. Another interesting topic is the possibility to use non-resonant optical or RF fields for perturbing the energy of Zeeman sublevels. Non-resonant light produces light shifts which can vary from one Zeeman sublevel to another, and which thus change the Zeeman splittings. A high frequency non-resonant RF field can modify, and even cancel the g-factor of an atomic state, giving rise to dressed magnetic moments. All these developments, which have taken place from the early fifties to the middle seventies, do not rely on the monochromaticity of the exciting light, but only on its polarization. This explains why they predate the use of lasers in atomic physics.

The second part of the paper will deal with the external degrees of freedom of atoms, i.e. their position and their velocity. Then, the monochromaticity of laser sources can be combined with spatially varying Zeeman shifts for controlling the exchanges of linear momentum between atoms and photons. A lot of new developments have occurred since the early eighties, concerning the possibility to slow down, to cool and to trap atoms. A few of them will be briefly reviewed, such as Zeeman slowers, Sisyphus cooling, optical lattices, magneto-optical traps, magneto-static traps. These developments have culminated recently with the observation of Bose-Einstein condensation on alkali atoms. Several contributions in this volume are devoted to these problems. It is clear that a new research field is being opened by these new states of matter, and that the Zeeman effect will continue to find applications in this domain.

2 Internal degrees of freedom

2.1 *Preparing or detecting an atom in a given Zeeman sublevel*

Recall first the well known polarization selection rules for the various Zeeman components $|g, M_g\rangle \longleftrightarrow |e, M_e\rangle$ of an optical line connecting a ground state g to an excited state e, M_g and M_e being the magnetic quantum numbers labelling the eingenvalues of the total atomic angular momentum along the quantization axis. For electric dipole transitions, $\Delta M = M_e - M_g = -1, 0,$ or $+1$. Fig. 1 gives the polarization corresponding to each value of ΔM: σ^+ for $\Delta M = +1$, π for $\Delta M = 0$, σ^- for $\Delta M = -1$. These results are a direct consequence of the conservation of the total angular momentum[1]. Photons corresponding to σ^+-polarized light (resp. σ^-) have an angular momentum along the axis of quantization equal to $+\hbar$ (resp. $-\hbar$), whereas π-polarized photons have an angular momentum equal to 0. This is precisely the angular momentum which is gained by the atom when it absorbs such a photon and is excited from M_g to M_e.

Consider an atom with a transition $J_g = 0 \longleftrightarrow J_e = 1$. By exciting it with resonant light having a well defined polarization, σ^+, σ^- or π, it is thus possible to prepare it in a well defined excited Zeeman sublevel. Similarly, by monitoring the fluorescence light reemitted by such an atom with a well defined polarization, one can infer from what Zeeman sublevel the photon has been emitted. This is the principle of the double resonance method [2,3], which is recalled in Fig. 2 and which is an optical method for studying magnetic resonance in atomic excited states.

Optical methods also apply to atomic ground states having several Zeeman

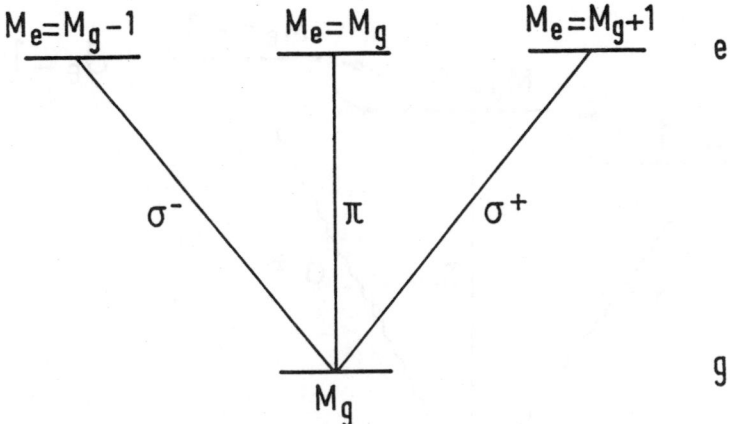

Figure 1: Polarization of the various Zeeman components of an optical line

sublevels. Angular momentum can be transferred from polarized photons to atoms in absorption–spontaneous emission cycles. The principle of such a method, called optical pumping [4], is recalled in Fig. 3 for a transition $J_g = 1/2 \longleftrightarrow J_e = 1/2$. It allows one to achieve high degrees of spin polarization in the ground state. Since the amount of absorbed light depends on the relative populations of the ground state Zeeman sublevels, it is also possible to detect optically any variation of these populations due to resonant radiofrequency transitions or to relaxation processes.

2.2 Linear superpositions of Zeeman sublevels–Zeeman coherences

When the polarization of the exciting light (or of the detected fluorescence light) is a linear superposition of the basic polarizations σ^+, σ^- and π, the atom is prepared (or detected) in a linear superposition of excited Zeeman sublevels. The atomic density matrix σ has then off–diagonal elements $\sigma_{12} = \langle 1|\sigma|2\rangle$ (where 1 and 2 are shorter notations for M_e and M_e'), which are called "Zeeman coherences"[5]. They are at the origin of several interesting effects which can be observed using the fluorescence light which is emitted by the atom and which depends not only on the populations σ_{11} and σ_{22} of the two sublevels $|1\rangle$ and $|2\rangle$, but also on the Zeeman coherences σ_{12} and σ_{21}. These effects result from quantum interferences between two emission amplitudes starting from the two sublevels $|1\rangle$ and $|2\rangle$ and ending into the same ground state sublevel. We review now a few of them.

It is convenient for that to start from the equation of motion of σ_{12}, which may be shown to have, in several important cases [5,6], the following form

$$\dot{\sigma}_{12} = R - i\frac{E_1 - E_2}{\hbar}\sigma_{12} - \Gamma\sigma_{12} \qquad (1)$$

The various factors which determine the rate of variation of σ_{12} are: the optical excitation, which prepares σ_{12} at a rate R; the free evolution in the magnetic field **B** at the Larmor frequency $(E_1 - E_2)/\hbar$; the damping with a rate Γ due to spontaneous emission. We can now look for the solution of equation (1) in a certain number of cases.

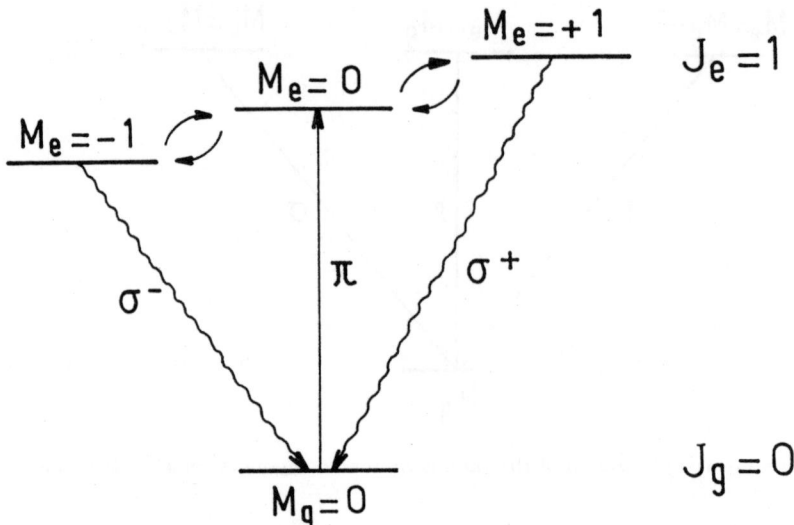

Figure 2: Principle of the double resonance method for a transition $J_g = 0 \longleftrightarrow J_e = 1$. Atoms are selectively prepared in the sublevel $M_e = 0$ by excitation with π-polarized light. During the time spent in the excited state, resonant radiofrequency transitions transfer them from $M_e = 0$ to $M_e = +1$ and $M_e = -1$. Such transfers are detected by monitoring the σ^+ or σ^- fluorescence light reemitted by the atom.

Suppose first that the intensity of the exciting light beam is constant. R is then constant, and equation (1) has a steady-state solution.

$$\sigma_{12} = \frac{\hbar R}{i(E_1 - E_2) + \hbar \Gamma} \tag{2}$$

which clearly exhibits resonant variations when the magnetic field **B** is scanned around zero, in an interval determined by $\mid E_1 - E_2 \mid \leq \hbar \Gamma$. This provides a quantitative interpretation of the zero field level crossing resonance, which is called also the Hanle effect[7]. The same equation (1) also explains why level crossing resonances can be observed near values of the magnetic field where two Zeeman sublevels $|1\rangle$ and $|2\rangle$ belonging to two different hyperfine levels cross. This is the Franken effect[8] and $\langle 1|\sigma|2\rangle$ is then a hyperfine coherence.

Suppose now that one uses a modulated excitation :

$$R = R_0 e^{-i\Omega t} \tag{3}$$

Equation (1) then admits a solution of the form :

$$\sigma_{12}(t) = \frac{\hbar R_0}{i(E_1 - E_2 - \hbar \Omega) + \hbar \Gamma} e^{-i\Omega t} \tag{4}$$

which shows that σ_{12} is modulated at the same frequency Ω as the exciting light and exhibits resonant variations when $E_1 - E_2$ is scanned around $\hbar \Omega$. Such resonant modulations of the fluorescence light have been first observed on cadmium atoms[9].

Another interesting situation is found when one uses a percussional excitation :

$$R = R_0 \delta(t) \tag{5}$$

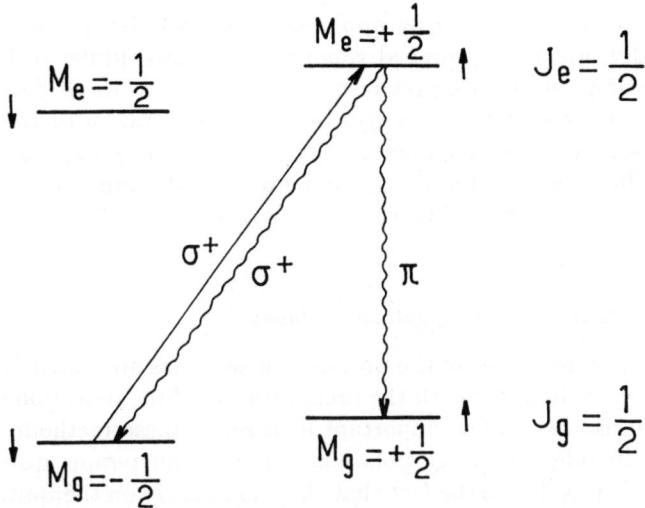

Figure 3: Principle of the optical pumping method for a transition $J_g = 1/2 \longleftrightarrow J_e = 1/2$. Atoms are selectively excited from $M_g = -1/2$ to $M_e = +1/2$ by excitation with σ^+-polarized light. From there, they fall back in the ground state by spontaneous emission of a photon which can be, either σ^+-polarized, in which case the same cycle can be repeated, or π-polarized, in which case the atom remains trapped in $M_g = +1/2$. After such an optical pumping cycle, the ground state becomes fully polarized in the $M_g = +1/2$ sublevel. Note that the absorption of the incoming σ^+ light is directly proportional to the population of the $M_g = -1/2$ sublevel, which provides an optical detection signal for monitoring the variations of this population.

$\delta(t)$ being the delta fonction (more physically, one uses a pulse of exciting light, with a duration much smaller than $\hbar/|E_1 - E_2|$ and $1/\Gamma$). For $t > 0$, the solution of equation (1) reads

$$\sigma_{12}(t) = R_0 e^{-i(E_1-E_2)t/\hbar} e^{-\Gamma t} \tag{6}$$

Such damped oscillations at the frequency $(E_1 - E_2)/\hbar$ are nothing but the so-called "quantum beats" and they have been first observed in 1964 on cadmium atoms [10].

Note also that, even if it is not prepared directly by the optical excitation ($R = 0$), σ_{12} can build up from $\sigma_{11} - \sigma_{22}$ under the effect of a resonant radiofrequency field $\mathbf{B}_1 e^{-i\Omega t}$, perpendicular to $\mathbf{B}$. Modulations then appear in σ_{12} which are resonant when the frequency Ω of the RF field is close to $(E_1 - E_2)/\hbar$. The corresponding modulations at frequency Ω of the fluorescence light have been called "light beats" [11]. In fact, one of the first demonstrations of the importance of Zeeman coherences was the observation of a narrowing of the double resonance curves in the excited state of mercury atoms when the density of the atomic vapour increases [12]. The Zeeman coherence induced in the excited state by the RF field is partially transferred from this atom to another one by multiple scattering of resonance radiation and this explains why the effective lifetime of Zeeman coherences becomes longer at higher vapour pressures leading to more efficient imprisonment of resonance radiation.

All the previous considerations can be easily extended to atomic ground states. Zeeman coherences are associated with the existence of an anisotropy of the atomic orientation or alignment in the plane perpendicular to $\mathbf{B}$. Such a transverse orientation or alignment can be prepared in the ground state by a transverse optical pumping, perpendicular to **B**, or by applying a resonant RF (or microwave) field to a longitudinally oriented or aligned vapour. Equations similar to (1) can be established [5], allowing one to interpret resonances similar to those described above. Because relaxation times are much longer in the ground state g than in the excited state e, these resonances are much narrower. For example, zero field level crossing resonances have been observed in the ground state of rubidium atoms, which are so narrow that they can be used to detect very weak magnetic fields, on the order of 3.10^{-10} Gauss [13,14].

2.3 A few important features of optical methods

The various schemes described in the previous subsections are called "optical methods" because they use light for both the preparation and the detection of the atomic state. We summarize here a few important features of these methods.

First, they provide very large polarizations at room temperature and in low magnetic fields. This is due to the fact that they do not rely on the Boltzmann factor $\exp(-H_{\text{Zeeman}}/k_B T)$. For the same reason, they can be applied to states having a purely nuclear paramagnetism ($J = 0$ and $I \neq 0$). Optical pumping methods are thus very efficient for polarizing nuclear spins, which can lead to interesting applications, such as the magnetic resonance imaging of human organs (see for example the contribution of E. Otten in this volume and the references therein).

Optical methods have also a very high sensitivity. The magnetic resonance is detected, not by measuring the absorption of the RF or microwave power, but by monitoring a modification of the light absorbed or emitted by the atoms. One can thus study very dilute media, such as atomic vapours.

Finally, optical methods are not sensitive to the optical Doppler effect. Zeeman splittings, fine or hyperfine structures are not determined from a difference between two optical frequencies. They are measured directly from the frequency of the resonant RF or microwave field, or from the evolution frequency of a Zeeman or hyperfine coherence. This explains why it has been possible to develop a high resolution spectroscopy before the advent of monochromatic laser sources.

2.4 Perturbing Zeeman splittings with nonresonant optical or RF fields

If the exciting light is detuned from resonance, one can show [5,6] that it produces energy shifts of the ground state Zeeman sublevels, called "light shifts" or "ac Stark shifts". The magnitude of such light shifts is proportional to the light intensity and inversely proportional to the detuning $\delta = \omega_L - \omega_A$ between the laser frequency ω_L and the atomic frequency ω_A (in the limit when the Rabi frequency Ω_1 describing the light-atom interaction is small compared to $|\delta|$). Because of the polarization selection rules, light shifts depend on the polarization of the exciting light and vary from one Zeeman sublevel to another.

For example, in the case of the transition $J_g = 1/2 \longleftrightarrow J_e = 1/2$ of Fig. 3, a σ^+-polarized exciting light shifts only the sublevel $M_g = -1/2$ (if $\omega_L \neq \omega_A$), whereas a σ^--polarized light shifts only the sublevel $M_g = +1/2$. The Zeeman splitting between the two sublevels can thus be changed by a nonresonant light (see Fig. 4), which produces a shift of the magnetic resonance curve in the ground state.

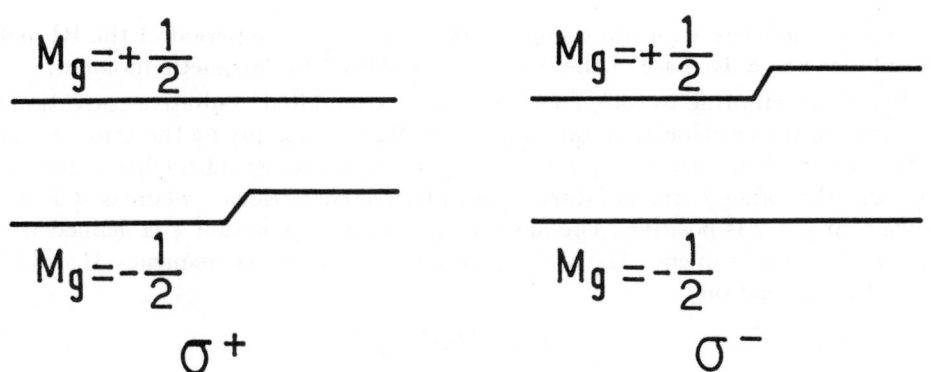

Figure 4: Light shifts of the ground state Zeeman sublevels for the transition $J_g = 1/2 \longleftrightarrow J_e = 1/2$ of Fig. 3. The Zeeman degeneracy is removed by a static magnetic field. The light beam exciting the transition is slightly detuned from resonance. The detuning δ is positive, so that light shifts are positive. Depending whether the light polarization is σ^+ or σ^-, only Zeeman sublevel $M_g = -1/2$ or $M_g = +1/2$ is light-shifted.

The sign of this shift changes when the polarization of the exciting light changes from σ^+ to σ^-. Because magnetic resonance curves are very narrow in the ground state, it is possible in this way to detect very small light shifts, on the order of one Hertz, produced by the light emitted by a discharge lamp [15]. Now, with laser sources, light shifts on the order of one Gigahertz can be easily produced.

In the absence of external magnetic fields, light shifts can remove the Zeeman degeneracy and their effect is equivalent with the one which would be produced by dc "fictitious" magnetic or electric fields [16,17].

Light shifts can be considered from different points of view. First, they are "stimulated" radiative corrections, which can be interpreted as resulting from virtual absorptions and reemissions of photons by the atom. In this respect, they are the equivalent, for the absorption-stimulated emission process, of the Lamb shift for spontaneous emission. Secondly, they introduce perturbations to high precision measurements using optical methods, which must be taken into account before extracting from these measurements spectroscopic data. Finally, they are now more and more frequently used for manipulating the energy of Zeeman sublevels. For example, it is easy to produce a light field whose polarization changes from σ^+ to σ^- every quarter of wavelength (see Fig. 6). From Fig. 4, we then deduce that one can produce in this way spatial modulations of the Zeeman splittings on an optical wavelength scale, which would not be easily achieved with real magnetic fields. We will see in the next section interesting applications of such a situation.

Zeeman splittings can be also modified by nonresonant RF fields. In particular, it can be shown that the g-factor of an atomic state can be reduced, and even cancelled by a high frequency nonresonant RF irradiation [18,19]. Such an effect has been calculated in a nonperturbative way with the dressed atom approach. One can also interpret semiclassically why the effective magnetic moment of the atom is reduced by the interaction with the RF field. The motion of the magnetic moment in the RF field consists of an angular vibration of the direction of this magnetic moment which keeps a constant length. Averaging over one period of the RF field can only lead to a decrease of the static component of the magnetic moment.

It is then tempting to consider that such a "stimulated" radiative correction is analogous to the electron spin anomaly $g-2$. However, applying the same picture to the motion of the electron spin in vacuum fluctuations would predict a decrease of g from the value 2 (in the absence of radiative corrections), whereas it is well known that $g-2$ is positive. The answer to this paradox is that g is defined from both the Larmor frequency Ω_L of the spin and the cyclotron frequency Ω_C of the charge by the relation

$$g/2 = \Omega_L/\Omega_C \qquad (7)$$

It is not enough to consider the radiative corrections to the Larmor frequency of the spin. One must also consider the modifications of the cyclotron motion. One then finds [20] that both Ω_L and Ω_C are reduced, Ω_C being more reduced than Ω_L, so that g becomes larger than 2 according to equation (7). The physical interpretation of such a result, in the nonrelativistic domain, is that a charge is more coupled to its self field than a magnetic moment, which results in a more efficient slowing down of the cyclotron motion. A full relativistic calculation, to all orders in $1/c$, but to order 1 in the fine structure constant α, confirms this interpretation [21]. Similar conclusions have been obtained from a different approach [22]. Finally, such a discussion shows that, for understanding $g-2$, it is necessary to consider the modification of the motion of both the charge and the spin of the electron. There is here a certain analogy with the situation encountered when one tries to interpret the "anomalous" Zeeman effect. Such an effect cannot be understood by considering only the motion of the charge in the applied magnetic field. One must also take into account the magnetic coupling of the spin.

3 External degrees of freedom

3.1 Zeeman slowers

We review now a few mechanisms using spatially dependent Zeeman shifts for controlling the position and the velocity of a neutral atom, and we begin by describing the so-called "Zeeman slowers" which are used for decelerating and stopping an atomic beam [23,24].

Consider an atomic beam which is irradiated by a counterpropagating resonant laser beam (see Fig. 5). Photons are absorbed from the laser beam and spontaneously reemitted in all possible directions. In an elementary absorption-spontaneous emission cycle (fluorescence cycle), the average momentum transferred to the atom is equal to the momentum $\hbar \mathbf{k}$ of a laser photon, because spontaneously emitted photons have equal probabilities to be emitted in opposite directions and their mean momentum is equal to zero. When the atomic transition is saturated, the mean number of fluorescence cycles per unit time is equal to $\Gamma/2$, where Γ is the spontaneous emission rate (the atom spends half of its time in the upper state). It follows that the mean radiation pressure force experienced by the atom is equal

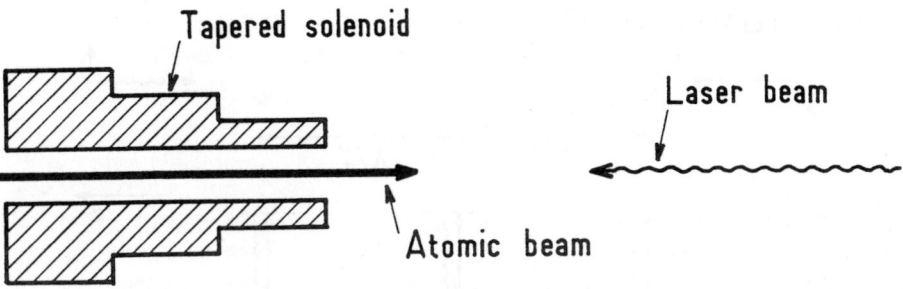

Figure 5: Principle of a Zeeman slower. The radiation pressure force exerted on an atomic beam by a counterpropagating resonant laser beam decelerates the atoms. The Doppler shift due to such a deceleration is compensated for by a spatially dependent Zeeman shift associated with an inhomogeneous magnetic field produced by a tapered solenoid. This allows the laser beam to remain in resonance with the atoms during the whole deceleration process.

to $\hbar k \Gamma / 2$, leading to a mean acceleration (or deceleration) given by

$$a_{\text{Max}} = \frac{\hbar \mathbf{k}}{M} \frac{\Gamma}{2} = \mathbf{v}_R \frac{\Gamma}{2} \tag{8}$$

where $\mathbf{v}_R = \hbar \mathbf{k}/M$ is the recoil velocity of an atom absorbing a laser photon. Such a recoil veloity is usually very small, on the order of $10^{-2}\,\text{m.s}^{-1}$. But Γ can be very large, on the order of $10^8\,\text{s}^{-1}$, so that a_{Max} can reach values on the order of $10^6\,\text{m.s}^{-2}$, *i.e.* 10^5 times the acceleration of gravity.

There is however a difficulty due to the Doppler shift associated with the deceleration. Such an effect shifts the atoms out of resonance and the mean radiation pressure force decreases. This is precisely where Zeeman shifts can be useful. The Doppler shift due to the deceleration process can be compensated for by a spatially dependent Zeeman shift associated with the inhomogeneous magnetic field produced by a tapered solenoid (see Fig. 5). Such a scheme, now called Zeeman slower, has been first demonstrated with sodium atoms [23,24]. It is quite general and it allows the deceleration to remain at its maximum value during the whole deceleration process. Continuous beams of slow atoms can be easily obtained in this way. Atomic beams can even be completely stopped over distances of the order of one meter.

3.2 Spatially modulated Zeeman splittings. Sisyphus cooling and optical lattices.

We describe now a laser cooling mechanism using spatially modulated Zeeman splittings due to light shifts produced by a laser light whose polarization is spatially modulated. Consider for example the laser configuration of Fig. 6.a, consisting of two counterpropagating plane waves along the z-axis, with orthogonal linear polarizations and with the same frequency and the same intensity. At a certain position z_0 along the z-axis, the phase difference between the electric fields of the two waves

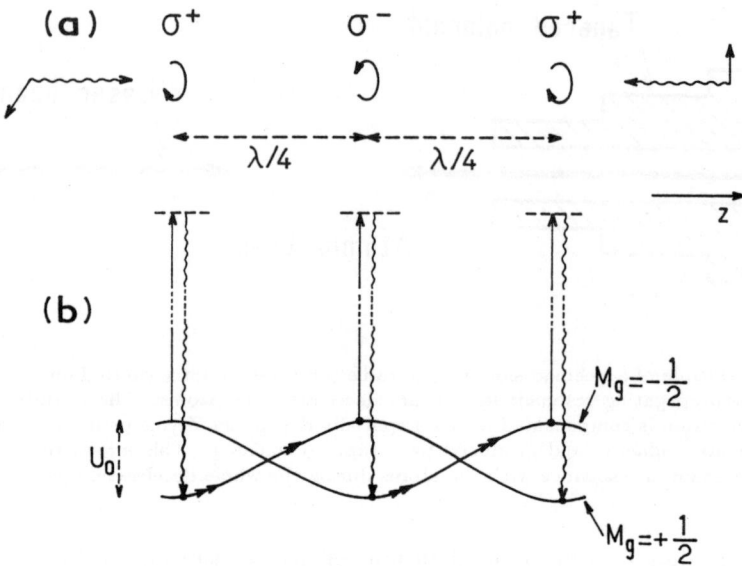

Figure 6: a–Laser configuration formed by two counterpropagating plane waves along the z-axis, with orthogonal linear polarizations. The polarization of the resulting total field is spatially modulated with a period $\lambda/2$. Every $\lambda/4$, it changes from σ^+ to σ^-. In between, it is elliptical or linear. b–Light shifts and optical pumping transfers (vertical arrows) for an atom having two Zeeman sublevels $M_g = \pm 1/2$ in the ground state and put in such a laser configuration. The spatial modulation of the laser polarization results in correlated spatial modulations of the light shifts of the two sublevels and of the optical pumping rates between them. Because of these correlations, a moving atom can run up potential hills more frequently than down (double arrows).

is equal to $\pi/2$, so that the total field is σ^+-polarized. A distance $\lambda/4$ farther, at $z = z_0 + \lambda/4$, the phase difference between the two fields has increased by π and becomes equal to $3\pi/2$, so that the total field is σ^--polarized, and so on. Every $\lambda/4$, the light polarization changes from σ^+ to σ^- and vice versa. In between, it is elliptical or linear.

Consider now the simple case where the atomic ground state has an angular momentum $J_g = 1/2$. As shown in subsection (2.4), the two Zeeman sublevels $M_g = \pm 1/2$ undergo different light shifts, depending on the laser polarization, so that the Zeeman degeneracy in zero magnetic field is removed. One can always choose the detuning δ between the laser frequency and the atomic frequency so that, in the places where the polarization is σ^+ (resp. σ^-), the sublevel $M_g = -1/2$ (resp. $M_g = +1/2$) is above the sublevel $M_g = +1/2$ (resp. $M_g = -1/2$). We get in this way the energy diagram of Fig. 6.b showing spatial modulations of the Zeeman splitting between the two sublevels with a period $\lambda/2$.

If the detuning δ is not too large, there are also real absorptions of photons by the atom followed by spontaneous emission, which give rise to optical pumping transfers between the two sublevels, whose direction depends on the polarization: $M_g = -1/2 \longrightarrow M_g = +1/2$ for a σ^+ polarization, $M_g = +1/2 \longrightarrow M_g = -1/2$ for a σ^- polarization. Here also, the spatial modulation of the laser polarization results in a spatial modulation of the optical pumping rates with a period $\lambda/2$ (vertical arrows of Fig. 6.b).

The two spatial modulations of light shifts and optical pumping rates are of course correlated because they are due to the same cause, the spatial modulation of the light polarization. These correlations clearly appear in Fig. 6.b. With the sign chosen for the detuning, optical pumping always transfers atoms from the higher Zeeman sublevel to the lower one. This can lead to a very efficient cooling mechanism, called "Sisyphus cooling" or "polarization gradient cooling"[26,27] (see also [25]). Consider an atom moving to the right and starting from the bottom of a valley, for example in the state $M_g = +1/2$ at a place where the polarization is σ^+. The atom can climb up the potential hill and reach the top of the hill where it has the maximum probability to be optically pumped in the other sublevel, i.e. in the bottom of a valley, and so on (double arrows of Fig. 6.b). Like Sisyphus in the Greek mythology, the atom is running up potential hills more frequently than down. When it climbs a potential hill, its kinetic energy is transformed into potential one which is then dissipated by light, since the spontaneously emitted photon has an energy higher than the absorbed laser photon (anti–Stokes Raman processes of Fig. 6.b). Such a cooling mechanism is very efficient and can lead to temperatures T on the order of a few microkelvins, given by $k_B T \simeq U_0$, where U_0 is the depth of the optical potential wells of Fig. 6.b, i.e. the maximum differential light shift. Equation $k_B T \simeq U_0$ cannot remain valid when U_0 tends to zero, because we have neglected the recoil due to the spontaneously emitted photons. There is a threshold for U_0, on the order of a few recoil energies $E_R = \hbar^2 k^2 / 2M$, below which Sisyphus cooling can no longer work.

Note finally that, for the optimal conditions of Sisyphus cooling, atoms become so cold that they get trapped in the quantum vibrational levels of the potential wells of Fig. 6.b. More precisely, one must consider energy bands in this periodic structure [28]. Experimental observation of such a quantization of atomic motion in an optical potential has been first achieved at one dimension [29,30]. Atoms then form a spatial periodic array, called "1D–optical lattice", with an antiferromagnetic order, since two adjacent potential wells correspond to opposite spin polarizations. 2D and 3D optical lattices have been achieved subsequently (see the review papers [31,32]; see also the contribution of A.Hemmerich in this volume).

3.3 The magneto-optical trap (MOT)

We describe now an example of a trap for neutral atoms which uses the radiation pressure force **F** already mentioned in subsection (3.1) and resulting from the exchanges of linear momentum between atoms and photons in resonant absorption-spontaneous emission cycles. Other types of radiative forces can be used for trapping atoms, the so-called dipole or gradient forces which result from position dependent light shifts or dressed-state energies [25,33]. But they require in general higher intensities. Using radiation pressure, which is a resonant process, one can hope to build deeper and larger traps.

In most cases, the radiation pressure force **F** is simply proportional to the Poynting vector **G** of the laser field. This is the case when the induced dipole moment **d** is proportional to the laser electric field $\mathbf{E}_L$. Condition $\nabla.\mathbf{G} = 0$ then results in $\nabla.\mathbf{F} = 0$: the radiation pressure force is divergence-free. This means

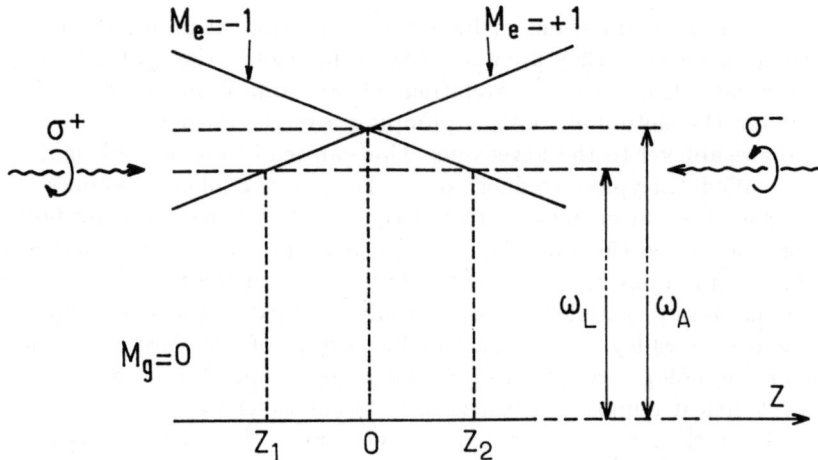

Figure 7: Principle of a one-dimensional magneto-optical trap. An atom with a transition $J_g = 0 \longrightarrow J_e = 1$ is put in a magnetic field gradient along the z-axis and is irradiated with two couterpropagating waves, with a red detuning ($\omega_L < \omega_A$) and with opposite circular polarizations σ^+ and σ^-. The two waves are resonant in different places $z = z_1$ and $z = z_2$, so that the two radiation pressure forces are not balanced, giving rise to a restoring force.

that **F** cannot be a restoring force in all directions and that stable traps cannot be achieved with radiation pressure forces. Such a result is known as the optical Earnshaw theorem [34].

In fact, it is possible to overcome such a limitation. Suggestions have been made to change the proportionality between **d** and $\mathbf{E_L}$ in a position-dependent way using external fields or optical pumping, so that $\nabla.\mathbf{F}$ no longer vanishes [35].

It is here that position dependent Zeeman shifts can be very useful, as suggested first by Jean Dalibard in 1986 with the following one-dimensional scheme (see Fig. 7). Consider an atom with a transition $J_g = 0 \longrightarrow J_e = 1$, put in a magnetic field gradient along the z-axis and irradiated with two couterpropagating waves, with a red detuning ($\omega_L < \omega_A$) and with opposite circular polarizations σ^+ and σ^-. Because of the polarization selection rules, the σ^+ wave excites only the transition $M_g = 0 \longrightarrow M_e = +1$, whereas the σ^- wave excites only the transition $M_g = 0 \longrightarrow M_e = -1$. The spatial variation of the energy of the sublevels $M_g = \pm 1$ and the non zero value of the detuning ($\omega_L \neq \omega_A$) result in the fact that the two waves cannot be resonant at the same place : the σ^+ wave is resonant with the transition $M_g = 0 \longrightarrow M_e = +1$ at $z = z_1$, whereas the σ^- wave is resonant with the transition $M_g = 0 \longrightarrow M_e = -1$ at $z = z_2$ (see Fig. 7). It follows that the radiation pressure forces of the two waves are not balanced. The radiation pressure force of the σ^+ wave predominates at $z = z_1$, whereas the radiation pressure force of the σ^- wave predominates at $z = z_2$. This results in a restoring force towards the point $z = 0$ where the two sublevels $M_e = \pm 1$ cross. Atomic motion is thus confined in a zone $z_1 \leq z \leq z_2$ whose width $z_2 - z_1$ can be adjusted by varying the detuning $\delta = \omega_L - \omega_A$. Furthermore, the non zero value of the detuning provides a Doppler cooling [36,37].

In fact, such a scheme can be extended to three dimensions and leads to robust,

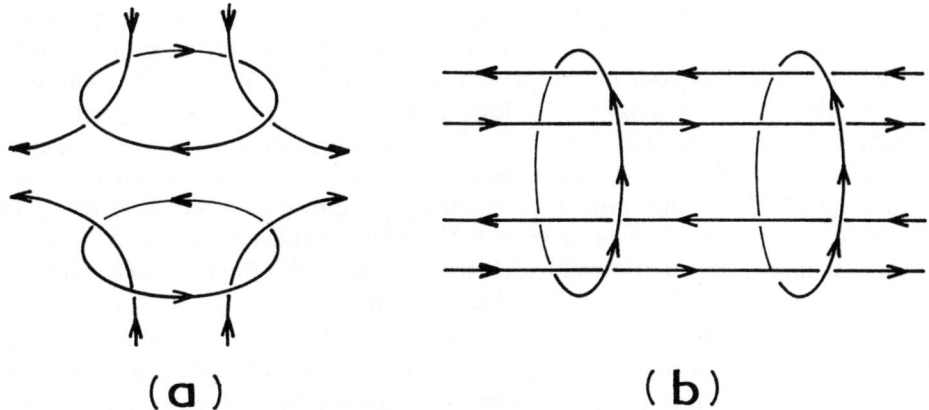

Figure 8: Examples of magnetostatic traps : the quadrupole trap (Fig.a) and the Ioffé-Pritchard trap (Fig.b).

large and deep traps [38]. It combines trapping and cooling, it has a large velocity capture range and it can be used for trapping atoms in a cell [39]. With all these advantages, the MOT has become the "workhorse" trap in laser cooling.

3.4 Magnetostatic traps

Magnetostatic traps use purely magnetic forces for trapping neutral atoms. These magnetic forces are those which are responsible for the Stern-Gerlach effect. They are due to spatially dependent shifts $E_M(\mathbf{r})$ of the ground state Zeeman sublevels, giving rise to M-dependent forces $\mathbf{F}_M(\mathbf{r}) = -\nabla E_M(\mathbf{r})$. Using these forces for controlling the motion of neutral particles has been suggested and used by several authors [40,41,42,43].

To make a trap with purely magnetic forces, one must achieve a magnetic field configuration exhibiting a local extremum of the modulus $|\mathbf{B}|$ of the magnetic field $\mathbf{B}$. In fact, it can be shown that a local maximum of $|\mathbf{B}|$ in a source-free region cannot exist [44]. Only local minima can be achieved, giving rise to trapping of "low field seekers" atoms.

The depth of magnetostatic traps is rather small. For a magnetic moment of one Bohr magneton μ_B, and for a field depth B of 200 Gauss, equation $\mu_B B = k_B T$ gives $T = 13$ mK. This is why magnetostatic traps can work only with precooled atoms. The first magnetostatic trap for atoms to be demonstrated was for laser precooled sodium atoms [45]. Magnetostatic trapping has been also achieved for polarized hydrogen atoms precooled by cryogenic techniques [46,47] (for a review of atom traps, see [48]).

Fig. 8 gives two examples of magnetostatic traps. The quadrupole trap (Fig.8a) consists of two identical coils with the same axis and with opposite currents. The magnetic field $\mathbf{B}$ vanishes at the center of symmetry $\mathbf{r} = 0$ and its modulus increases linearly with the distance from this point along the three principal axes. Near $\mathbf{r} = 0$, the moving spin cannot follow adiabatically the spatial changes of $\mathbf{B}$ and there are leaks due to Majorana transitions to non trapping spin states. Two methods have been used for overcoming this difficulty. In the first one, a rotating RF field is added so that the "hole" of the trap is rotating sufficiently rapidly for preventing atoms from moving into it [49]. In the second method, the hole of the trap is plugged by a detuned focussed laser beam introducing a repulsive potential [50]. The Ioffé-Pritchard trap (Fig.8b) consists of a four parallel wire configuration producing a confining transverse quadrupole field in the plane perpendicular to the wires, and of two identical coils with the same axis parallel to the wires and with identical currents. These two coils provide a confining longitudinal field, parallel to the wires, near the center of the trap $\mathbf{r} = \mathbf{0}$ [43]. The modulus of the field no longer vanishes. It increases quadratically with the distance from the center of symmetry along the three principal axes and the losses are reduced.

Spectacular developments have occurred recently when magnetostatic trapping of laser precooled alkali atoms was combined with evaporative cooling, leading to the observation of Bose-Einstein condensation [51,50] and quantum degeneracy effects [52]. We refer the reader to the contributions of C.Wieman and W.Ketterle in this volume for a review of the most recent developments in this field. With the MOT used for capturing first the alkali atoms and for precooling them, with the magnetostatic trap then replacing the MOT, the Zeeman effect plays an essential role in these achievements. One can hope that magnetic couplings will continue to find new applications in this domain. For example, it has been suggested that the scattering length for alkali atoms in the lower hyperfine state could be tuned by an external magnetic field around a Feshbach resonance [53,54]. In view of the importance for BEC of the scattering length, and in particular of its sign, such a possibility would be very attractive.

4 Conclusion

From the various examples discussed in this paper, one can try to point out a few general trends in the evolution of the modern researches using the Zeeman effect.

Rather than being considered only as a source of informations on the structure of atoms, the Zeeman effect has now become a very useful tool for manipulating them. Extensive studies have been first devoted to the control of the internal degrees of freedom : spin polarization and energy of the Zeeman sublevels. More and more attention is paid now to the evolution of the translational degrees of freedom. In fact, there is a certain continuity between these two types of studies, since many effects dealing with internal variables, such as optical pumping and light shifts, turn out to play an important role in new cooling and trapping schemes, such as Sisyphus cooling. With the recent observation of BEC and the production of condensates of atoms, a new research field is being opened, where magnetic couplings will certainly play an important role.

We finally mention a few important spectroscopic applications of the Zeeman effect which are discussed in other contributions in this volume : the investigation of the energy diagram of Rydberg states in high magnetic fields, when the magnetic energy becomes on the order of the Coulomb energy or larger (contribution of J.Delos); the magnetic resonance imaging using optically pumped nuclei in magnetic field gradients (contribution of E.Otten).

I would like to thank J. Dalibard and J. Brossel for very helpful discussions.

* Unité de Recherche de l'Ecole Normale Supérieure et de l'Université Pierre et Marie Curie, associée au CNRS (URA 18).

1. A.Rubinowicz, Phys.Z. **19**, 442, 465 (1919).
2. J.Brossel and A.Kastler, C.R.Acad.Sci.(Fr) **229**, 1213 (1949).
3. J.Brossel and F.Bitter, Phys.Rev. **86**, 311 (1952).
4. A.Kastler, J. Phys. Rad. **11**, 255 (1950).
5. C.Cohen-Tannoudji, Ann. Phys. Paris **7**, 423 and 469 (1962).
6. J-P.Barrat and C.Cohen-Tannoudji, J. Phys. Rad. **22**, 329 and 443 (1961).
7. W.Hanle, Z. Phys. **30**, 93 (1924); **35**, 346 (1926).
8. F.D.Colegrove, P.A.Franken, R.R.Lewis and R.H.Sands, Phys. Rev. Lett. **3**, 420 (1959).
9. A.Corney and G.W.Series, Proc. Phys. Soc. **83**, 207 (1964).
10. J.N.Dodd, R.D.Kaul and D.M. Warrington Proc. Phys. Soc. **84**, 176 (1964); J.N.Dodd, W.J.Sandle and D.Zisserman, Proc. Phys. Soc. **92**, 497 (1967).
11. J.N.Dodd, W.N.Fox, G.W.Series and M.J. Taylor, Proc. Phys. Soc.**74**, 789 (1959); J.N.Dodd and G.W.Series, Proc. Roy. Soc. **A263**, 353 (1961).
12. M-A.Guiochon-Bouchiat, J-E.Blamont and J.Brossel, C.R.Acad.Sci.(Fr) **243**, 1859 (1956).
13. J.Dupont-Roc, S.Haroche and C.Cohen-Tannoudji, Phys.Lett. **28A**, 638 (1969).
14. C.Cohen-Tannoudji, J.Dupont-Roc, S.Haroche and F.Laloë, Phys. Rev. Lett. **22**, 758 (1969).
15. C.Cohen-Tannoudji, C.R.Acad.Sci.(Fr) **252**, 394 (1961).
16. W.Happer and B.S.Mathur, Phys.Rev. **163**, 12 (1967).
17. C.Cohen-Tannoudji and J.Dupont-Roc, Phys.Rev. **A5**, 968 (1972).
18. C.Cohen-Tannoudji and S.Haroche, C.R.Acad.Sci.(Fr) **262**, 268 (1966).
19. C.Landré, C.Cohen-Tannoudji, J.Dupont-Roc and S.Haroche, J. Phys. Rad. **31**, 971 (1970).
20. J.Dupont-Roc, C.Fabre and C.Cohen-Tannoudji, J.Phys.**B11**, 563 (1978).
21. J.Dupont-Roc and C.Cohen-Tannoudji, in New Trends in Atomic Physics, Les Houches, Session XXXVIII 1982 (Edited by G.Grynberg and R.Stora), p.156, Elsevier Science Publishers B.V. (1984).
22. H.Grotch and E.Kazes, Am.J.Phys. **45**, 618 (1977).
23. W.D.Phillips and H.Metcalf, Phys.Rev.Lett. **48**, 596 (1982).
24. J.V.Prodan, W.D.Phillips and H.Metcalf, Phys.Rev.Lett. **49**, 1149 (1982).
25. J.Dalibard and C.Cohen-Tannoudji, J.Opt.Soc.Am. **B2**, 1707 (1985).
26. J.Dalibard and C.Cohen-Tannoudji, J.Opt.Soc.Am. **B6**, 2032 (1989).
27. P.J.Ungar, D.S.Weiss, E.Riis and S.Chu, J.Opt.Soc.Am. **B6**, 2058 (1989).
28. Y.Castin and J.Dalibard, Europhys.Lett. **14**, 761 (1991).
29. P. Verkerk, B. Lounis, C. Salomon, C. Cohen-Tannoudji, J-Y. Courtois and G. Grynberg, Phys. Rev. Lett. **68**, 3861 (1992).
30. P.S.Jessen, C.Gerz, P.D.Lett, W.D.Phillips, S.L.Rolston, R.J.C.Spreeuw and C.I.Westbrook, Phys.Rev.Lett. **69**, 49 (1992).
31. G.P.Collins, Physics Today, June 1993, p.17.

32. J-Y.Courtois and G.Grynberg, Europhysics News, **27**, (January/February 1996).
33. S.Chu, J.E.Bjorkholm, A.Ashkin and A.Cable, Phys.Rev.Lett. **57**, 314 (1986).
34. A.Ashkin and J.P.Gordon, Opt.Lett. **8**, 511 (1983).
35. D.E. Prithard, E.L. Raab, V.S. Bagnato, C.E. Wieman and R.N. Watts, Phys. Rev. Lett. **57**, 310 (1986).
36. T.Hansch and A.L.Schawlow, Opt.Commun. **13**, 68 (1975).
37. D.Wineland and H.Dehmelt, Bull.Am.Phys.Soc. **20**, 637 (1975).
38. E.L.Raab, M.Prentiss, A.Cable, S.Chu and D.E.Pritchard, Phys.Rev.Lett. **59**, 2631 (1987).
39. C.Monroe, W.Swann, H.Robinson and C.E.Wieman, Phys.Rev.Lett. **65**, 1571 (1990).
40. C.V.Heer, Rev.Sci.Instrum. **34**, 532 (1963).
41. K.J.Kugler, W.Paul and U.Trinks, Phys.Lett. **72B**, 422 (1978).
42. V.S.Letokhov and V.G.Minogin, Opt.Commun. **35**, 199 (1980).
43. D.E.Pritchard, Phys.Rev.Lett. **51**, 1336 (1983).
44. W.H.Wing, Prog.Quantum Electron. **8**, 181 (1984).
45. A.L.Migdall, J.V.Prodan, W.D.Phillips, T.H.Bergeman and H.J.Metcalf, Phys.Rev.Lett. **54**, 2596 (1985).
46. H.F. Hess, G.P. Kochansky, J.M. Doyle, N. Masuhara, D. Kleppner and T.J. Greytak, Phys. Rev. Lett. **59**, 672 (1987).
47. R.van Roijen, J.J.Berkhout, S.Jaakkola and J.T.M.Walraven, Phys.Rev.Lett. **61**, 931 (1988).
48. D.E.Pritchard, K.Helmerson and A.G.Martin, in Atomic Physics **11**, ed. by S.Haroche, J-C.Gay and G.Grynberg (World Scientific, Singapore 1989).
49. W. Petrich, M.H. Anderson, J.R. Ensher and E.A. Cornell, Phys. Rev. Lett. **74**, 3352 (1995).
50. K.B. Davis, M.O. Mewes, M.R. Andrews, N.J. van Druten, D.S. Durfee, D.M. Kurn and W. Ketterle, Phys. Rev. Lett. **75**, 3969 (1995).
51. M.H.Anderson, J.R.Ensher, M.R.Matthews, C.E.Wieman and E.A.Cornell, Science, **269**, 198 (1995).
52. C.C.Bradley, C.A.Sackett, J.J.Tollett and R.G.Hulet, Phys.Rev.Lett. **75**, 1687 (1995).
53. A.J.Moerdijk, B.J.Verhaar and A.Axelsson, Phys.Rev. **A51**, 4852 (1995).
54. N.R.Newbury, C.J.Myatt and C.E.Wieman, Phys.Rev. **A51**, R2680 (1995).

THE ZEEMAN EFFECT A CENTURY LATER: NEW INSIGHTS INTO CLASSICAL PHYSICS

JOHN B. DELOS

Department of Physics, College of William and Mary,
Williamsburg, VA 23187, U.S.A.

Abstract

Zeeman's measurements helped guide the revolution in thought in which classical mechanics was replaced by quantum mechanics. Today the same kinds of measurements are providing new insights into classical dynamics--we are learning about periodic orbits, bifurcations, and the transition from order to chaos.

1 The Zeeman Effect and the Development of Quantum Theory

It was just one century ago, at a meeting of the Academy of Sciences here in Amsterdam, Pieter Zeeman described the observations that now bear his name. Over twenty years earlier, Faraday had searched for an effect of magnetic fields on spectral lines, but the effect was too small to observe. Zeeman used what must have been the best technology then available: using a grating with 15,000 lines per inch, and a 3 Tesla electromagnet running at 27 amps, he showed that each of the two sodium D-lines split into doublets or triplets, depending on the polarization of light. The tiny effect was a major discovery, because H.A. Lorentz provided an interpretation based on his new "electron-theory" of matter. Presuming that charged particles were bound in atoms as harmonic oscillators, he showed theoretically that a magnetic field should induce a frequency shift. From the observed magnitude of the frequency shift, Lorentz and Zeeman could calculate the charge/mass ratio, $|e/m|$ of the particles, and the sense of circular polarization of absorbed or emitted light showed that the charge of the oscillators was negative. Within a year, J.J. Thomson had shown that cathode rays also behave like negatively charged particles with the very same charge/mass ratio. By means of these two discoveries, the existence of electrons as free particles and as particles bound in atoms was firmly established.

All physicists know about the fundamental difficulties of atomic theory that followed. As ideas about atomic structure slowly developed, the Zeeman effect continued to play a central role. One story is worth retelling, because we can easily forget the depth of the problems physicists were confronting. Soon after Lorentz's successful explanation of the Zeeman effect, it was observed that in many atoms the effect is far more complex, and after thirty years of failed explanations by a large number of physicists, Samuel Goudsmit and George Uhlenbeck postulated that the electron has its own rotational degree of freedom. They wrote their ideas up in a note that their advisor, Paul Ehrenfest, submitted for publication, and Uhlenbeck went to consult with Lorentz. "Lorentz was very kind and interested, although I got the idea that he was rather skeptical," Uhlenbeck related. Lorentz apparently recognized that a wide variety of phenomena could be explained by this new spin-hypothesis, but he pointed out that such a large angular momentum in such a small particle would involve an extraordinarily rapid motion: the speed of the surface of this spinning ball of charge would be at least ten times the speed of light (today we would say 137 times the speed of light).

Jagdish Mehra continues the story [1]: "Uhlenbeck and Goudsmit immediately went to Ehrenfest, and told him that [their] idea...was nonsense, and it would be better not to publish it. But to their surprise, Ehrenfest answered that he had sent off the paper quite a while ago, and that it would appear in print pretty soon. Ehrenfest comforted the perplexed authors by remarking: 'You are both young enough to be able to afford a stupidity like that.'"

Needless to say, despite the objections of many physicists [including Pauli, who, less gently than Lorentz, called the spin-hypothesis "Irrlehre" ("false doctrine")], the idea eventually became an integral part of the subsequently developed quantum theory of matter.

Today observations of the Zeeman effect are playing a different role. In the same years that Zeeman was making measurements that helped to bring an end to classical physics, Poincaré was writing his treatises on celestial mechanics, works that represented the highest development of classical mechanics for much of the next century [2]. In our time, with the availability of computers and effective graphical visualization techniques, classical mechanics is enjoying a renaissance. Concepts anticipated by Poincaré -- order and chaos, quasiperiodicity, mixing, bifurcations of periodic orbits, homoclinic oscillations and fractals -- are now part of the common language of physicists. Science is full of surprises -- while the new generation of computational physicists were examining classical orbits of stars or planets or coupled pendulums, no one could have guessed that among the experimental measurements that would assume a central role, we would find again the Zeeman effect. If we examine the very highly excited states of an atom, the electron behaves somewhat like a classical particle. And if we put that atom in a magnetic field, we find in measurements manifestations of classical periodic orbits, of bifurcations, and of a transition from orderly to chaotic motion. New observations of the atomic Zeeman effect are stimulating further development of these new concepts in classical mechanics, and they are providing some of the most detailed experimental tests of those concepts.

In this article, we briefly review measurements and theoretical developments made in the past few years.

2 Closed Orbits and Recurrences

The critical experiment was the measurement of the absorption spectrum of a Hydrogen atom in a magnetic field (Fig. 1). Welge's group in Bielefeld showed experimentally that absorption near the ionization threshold consists of a smooth background plus a superposition of sinusoidal oscillations [3]. Furthermore, they showed by calculations that the oscillations were correlated with closed orbits or "recurrences" of the excited electron: classical orbits in which the electron travels a long distance away from the atom and later returns.

The observations reminded us of the work of Gutzwiller [4], who showed a relationship between periodic classical orbits and the quantum density-of-states. He examined the density of states as a function of energy, and he showed that this quantity could be written as a smooth term $\rho_0(E)$ plus a sum of sinusoidally oscillating terms

$$\rho(E) = \sum_n \delta(E - E_n) = \rho_0(E) + \sum_n D_n(E) \sin[S_n(E)/\hbar + \text{phases}]$$

The smooth term is related to the volume in phase space (each quantum state occupies a volume of order $(2\pi\hbar)^n$ where n is the number of degrees of freedom of the system). Each oscillatory term arises from a periodic orbit (PO) of the system. $S_n(E)$ is the action of the periodic orbit, and $D_n(E)$ is a coefficient related to the stability of the orbit. Short PO's give the large-scale structure of the spectrum, and long PO's give higher-resolution detail.

In his Ph.D. thesis, Meng Li Du showed that quantum properties of any system show sinusoidal fluctuations as a function of energy [5]. Those fluctuations are apparent if the property is measured at finite resolution, averaging over a band of eigenstates of the system. In the case of an atomic absorption spectrum, the fluctuations are associated not only with periodic orbits but with closed orbits. We developed a quantitative formula to describe these oscillations. Let $Df(E,B)$ represent the oscillator strength density for electron energy E in the presence of a magnetic field of strength B. Then

$$Df(E,B) = Df_0(E) + \sum_{k,n} C_k^n(E,B) \sin \Delta_k^n(E,B)$$

$Df_0(E)$ represents a slowly varying "background" absorption rate. The sum is over all repetitions (n) of all closed orbits (k) up to some maximum return time T_{max}. This determines the absorption spectrum to a resolution $\Delta E = 2\pi\hbar/T_{max}$. The phase $\Delta_k^n(E,B)$ is the classical action on the closed orbit, plus small corrections associated with Maslov indices. For each orbit k, the phase $\Delta_k^n(E,B)$ depends on energy approximately as

$$\Delta_k^n(E,B) = \Delta_k^n(E=0,B) + n T_k E$$

where T_n is the closure time of the orbit. It follows that the Fourier transform of $Df(E,B)$ with respect to E will show peaks at times $T = T_k$ corresponding to the classical recurrences. Therefore we call the Fourier-transform of such an absorption spectrum the "Recurrence Spectrum." The height of the peak associated with each recurrence is proportional to the coefficient C_k^n, so we call that coefficient the "Recurrence Amplitude," and its square the "Recurrence Strength." This quantity is related to the current of returning electrons associated with the closed orbits and its neighbors.

We derived formulas for the recurrence amplitude C_k^n and phase Δ_k^n. Good agreement was found between this "Closed Orbit Theory" and the measurements. (Closely related work was done independently by Wintgen and Friedrich and by Bogomolnyi [6].)

3 Scaled-Variable Measurements

The elaboration of these ideas and these measurements has led to the development of "Recurrence Spectroscopy." An important advance for recurrence spectroscopy came with the invention of the scaled-variables method [3]. The idea arose in a conversation between Dieter Wintgen and Jörg Main in Bielefeld. It was known that the classical Hamiltonian for a Hydrogen atom in a magnetic field obeys a scaling law. This scaling law asserts that the shapes of trajectories do not depend upon the energy and magnetic field separately, but only upon the scaled energy $\epsilon = EB^{-2/3}$. Furthermore, at each fixed ϵ, the size of any orbit changes with B such that the classical action is proportional to $B^{-1/3}$. With this in mind, the Bielefeld group measured absorption vs. $B^{-1/3}$, varying the photon energy and magnetic field simultaneously to keep $EB^{-2/3}$ fixed. Now the Fourier transform gives peaks at locations corresponding to the classical action of the closed orbit. One result is shown in Fig. 2, and it is compared to the results of closed-orbit theory.

In Fig. 4, recurrence spectra at many different scaled energies are drawn in a single picture. At low scaled energies, only a few recurrences are present. As the scaled energy increases, recurrences proliferate, and individual peaks split into mountain ranges. Since the time of Poincaré and Birkhoff it has been known that regular systems possess a small number of periodic orbits, while chaotic systems admit many. (The number of periodic orbits having period less than T_{max} increases as T_{max}^2 for a regular system with two degrees of freedom, but approximately as $e^{T_{max}}/T_{max}$ in a chaotic system.)

Calculations had already shown that classical orbits of the electron change from orderly to chaotic with increasing energy [7]. Now the measured proliferation of recurrences could provide an experimental manifestation of this change.

A "bifurcation" is the creation of a new periodic orbit or the splitting of one periodic orbit into several. Bifurcation theory has been an active branch of mathematics, and, happily, theorems that describe bifurcations of periodic orbits in classical Hamiltonian systems have been established by K. Meyer [8]. He showed that there are precisely five "generic" forms of bifurcation in Hamiltonian systems. We call them saddle-node bifurcations, period-doublings, 3-touch-and-go bifurcations, 4-touch-and-go or 4-island-chain bifurcations, and 5-and-higher-island chain bifurcations (Fig. 5). "Generic" means that other types can exist, but they would be atypical -- they would happen only "accidentally," or in the presence of some special symmetry. In fact, our system has a number of symmetries: time reversal invariance, and reflection symmetries about the ρ- and z-axes. We examined the consequences of these symmetries, and we showed that in certain cases the structure of bifurcations is modified by the presence of these symmetries.

This general framework provides a logical structure for interpretation of the observations. We have shown that: (1) the "exotic" orbits discovered in ref. 3 are produced by saddle-center bifurcations; (2) a sequence of pitchfork and period-doubling bifurcations produces the "main sequence"; (3) both generic (touch-and-go) and special (island chain) period-triplings produce some of the peaks that are visible in the experiment; (4) a focusing effect associated with a 4-island-chain bifurcation is also visible.[9] At present, this measurement of the diamagnetic Zeeman effect on an atom constitutes the most detailed experimental study of bifurcations of classical orbits in a conservative system.

4 Recurrence Spectroscopy

The above ideas lead us to think about additional closely-related measurements: one can study atoms in electric fields, or with both electric and magnetic fields together, parallel or crossed. Non-hydrogenic atoms should give new effects. Different energy ranges, different initial states of the atom and different polarization states of the laser could show other phenomena. Also we can look at electron detachment from negative ions in electric or magnetic fields.

a. Polarization of the Laser

The first Bielefeld experiments also examined excitation to states having $|m_\ell| = 1$. Classical orbits can, to reasonable approximation, be replaced by the orbits having $L_z = m_\ell \hbar = 0$. The most important change is in the angular distribution of outgoing waves. The changed angular distribution causes certain recurrences not to be excited; in this case the orbit perpendicular to the field was not present [9b].

Another phenomenon occurs in this case which has been examined theoretically [9c]. If $L_z = 0$, there is an orbit lying parallel to the magnetic field, and as mentioned above, it undergoes a sequence of "pitchfork" bifurcations producing the "main sequence" of orbits, the first of which has the shape of a balloon. However, if L_z is small but non-zero, the "parallel" orbit is displaced off-axis, and as the energy is increased it smoothly evolves into the balloon. The other orbits in the main sequence are produced in saddle-node bifurcations. Pitchfork bifurcations of periodic orbits occur only in the presence of symmetries, and the change from a pitchfork to a saddle-node bifurcation is an example of symmetry-breaking. A non-generic structure is changed into a generic structure.

b. Different Initial States

Another interesting phenomenon connected with bifurcations was observed by van der Veldt et al. [10]. They excited an atom by a sequence of steps that produced an outgoing p-wave. This has a node at 90°, and so again it does not excite the recurrence on the orbit perpendicular to the magnetic field. Nevertheless, a small recurrence associated with the fifth return of this orbit was visible in the measurements. Quantum calculations of the absorption spectrum confirmed the existence of this recurrence.

Kus et al. [10] suggested that this peak was a sign of a bifurcation that was about to occur at a slightly higher energy. Just before a new closed-orbit is created, there are unclosed orbits that nevertheless return close enough to the atom that they produce a detectable recurrence. The resulting signal is called a "ghost" of the orbit that is yet to come.

Further calculations indicated that even at energies that are not close to a bifurcation, an orbit that lies in a node of a quantum wave function produces a small but non-zero recurrence.[10] The neighbors of the nodal orbit carry waves of small amplitude, and these waves produce recurrences that are visible in calculations (though not yet in measurements).

c. Ionization in Parallel Electric and Magnetic Fields

In Berlin, the group led by H. Rinneberg measured the absorption spectrum of Ba in parallel electric and magnetic fields. Again many bifurcations were visible as the electric field was varied [11]. (In this case the locations of recurrences also provided the best available measure of the electric field in the system.)

d. Electron Detachment in Electric and Magnetic Fields

Electron detachment from a negative ion $h\nu + H^- \rightarrow H + e^-$ is in some ways simpler than excitation or ionization of a neutral atom because there is no long-range Coulomb force affecting the active electron. Fabrikant [12a] developed the theory in a purely quantum framework, and he showed that the photodetachment cross-section shows oscillations, but he could not give any simple interpretation of these oscillations. Closed-orbit theory gives a simple and transparent description [12b]. The classical orbits have the shape of trochoids (Fig. 7) circular cyclotron motion combined with linear E x B drift; as the energy changes the radius of the cyclotron motion increases, and at certain bifurcation energies new closed orbits are created. These bifurcations are analogous to saddle-node bifurcations.

If detachment takes place in parallel electric and magnetic fields, then we have uniform acceleration down the z-axis and circular cyclotron motion in (xy), which is equivalent to harmonic oscillation in ρ. This is, therefore, a very simple system, but it has keys that unlock solutions to a number of problems. (1) The relevant parts of the model are exactly solvable, and everything in the system can be understood by elementary methods. (2) It admits a simple structure of closed orbits and their associated recurrences, and it possesses an orderly sequence of bifurcations (Figs. 8,9). (3) At each bifurcation a certain geometrical structure -- a cylindrically focused cusp -- passes through the origin. This causes the semiclassical approximation to fail. (4) The failure is repaired by a simple diffraction function, a Fresnel integral. The integral provides a uniform approximation which is always finite and which behaves correctly in all limiting cases. (5) The focused cusp is similar to the structures found in excitation of neutral atoms [12c,14].

e. Excitation and Ionization of Alkali Atoms in Electric Fields

The scaled-variable method was also used to study spectra of atoms in electric fields, and again a number of closed orbits were identified [13]. Systematic study of this case was begun by J. Gao [14]. Stimulated by experiments of groups led by Gallagher, by Metcalf, and by Welge [15], she used closed orbit theory to describe ionization and excitation of sodium in electric fields. Electric fields are easy because (excluding core effects) the Schröedinger equation is separable in parabolic coordinates. Furthermore, above the zero-field ionization threshold, there is only one closed orbit of the electron; it goes straight up against the electric field and then returns to the atom. The more challenging aspect involved incorporation of spin-orbit coupling and core effects. The quantum defect produces a small phase shift to the outgoing and returning waves, and spin-orbit coupling in the initial state (3p $^2P_{j=3/2}$) modifies the angular distribution of the outgoing waves.

More interesting phenomena occur below the ionization threshold. The parallel orbit undergoes a sequence of bifurcations, sending out additional closed orbits. Near a bifurcation, the recurrence-strength of the nth-repetition of the parallel orbit should get very large; indeed, semiclassical formulas for the recurrence-strength again diverge at a bifurcation, and need to be repaired. The divergence again results from the passage of a focused-cusp through the origin, and the repair of the divergence is similar to the theory developed for detachment. While this theory was being developed, experiments in the Kleppner group came online, and we were able to test
the formulas against those measurements. Again good agreement was obtained between

theory and experiment.

f. Scattering of Electrons by the Ion Core

When the active electron travels around a closed orbit and returns to an alkali ion, it feels both the Coulomb field and a short range field within the core of the residual ion. This short-range field produces phase-shifts in the lower partial-waves (related to quantum defects in the energy levels). The result is that the electron wave-function splits into a superposition of a Coulomb-scattered part (which scatters backwards to retrace the original orbit) and a core-scattered part, which is approximately spherical, and which produces outgoing waves on every other orbit.

Gao sketched the theory of core-scattering (Fig. 10) and in her calculations showed that these effects were too small to be visible in the experiments we were then analyzing. Other cases, and subsequent measurements and calculations showed clear evidence of core-scattering. Dando et al. [16] put Gao's formulas into more explicit form, and showed that these semiclassical formulas give an accurate description of core-scattering. The core produces "shadows" on the original orbit, reducing the recurrence-strength associated with subsequent repetitions of the orbit, and it produces combination-recurrences, in which the electron is scattered from one closed orbit to another.

g. Excitation and Ionization in Crossed Fields

Essentially all of the above work refers to systems with just two degrees of freedom. If we measure the spectrum of an electron in an atom in crossed fields, then the system has three degrees of freedom. Many recurrences have been seen, and it is time for a systematic theory to be developed.

h. Real-time Recurrences

All of the above experiments detected recurrences by a kind of interferometry. The laser frequency is sharp, and light shines on the atom for a time that is long compared to return-times of classical orbits. This steady excitation of the atom produces a steady stream of outgoing waves, which propagate along classical paths. Returning orbits carry a steady stream of returning waves, which interfere with the outgoing waves and with each other to produce oscillations in the large-scale-structure of the absorption spectrum.

Recurrences can also be observed in real time. A short laser pulse produces a wave packet, which also travels along classical paths and returns to the atom some time later. The return can be detected by a second laser pulse that ionizes the atom. Theory has been developed by Alber and Zoller, and a variety of observations have been made [18].

References

1. *The Historical Development of Quantum Theory* **Vol. I**, pts. 1 and 2, p. 172 ff, p. 445 ff, p. 684 ff and 697 ff, J. Mehra and H. Rechenberg, Springer-Verlag, NY (1982). See also The Early History of the Theory of the Electron, in Aspects of Quantum Theory, A. Pais (A. Salam and E. P. Wigner, eds), Cambridge U. Press 1972, p. 79.
2. *Les Méthodes Nouvelles de la Méchanique Céleste* **Tome I-III**, H. Poincaré, Gauthier-Villars, Paris 1892, 1899, 1907.
3a. A. Holle, G. Weibusch, J. Main, B. Hager, H. Rottke and K. H. Welge, *Phys. Rev. Lett.* **56**, 2594 (1986)
3b. J. Main, G. Weibusch, A. Holle and K. Welge, *Phys. Rev. Lett.* **57**, 2789 (1986).
4. M. C. Gutzwiller, *J. Math. Phys.* **8**, 1979 (1967); **10**, 1004 (1969); **11**, 1791 (1970); **12**, 343 (1971). A broad exposition is given in M. C. Gutzwiller, Chaos in Classical and Quantum Mechanics, Springer-Verlag (1990).
5. M. L. Du and J. B. Delos, *Phys. Rev. Lett.* **58**, 1731 (1987); *Phys. Rev.* **A38**, 1896, 1913 (1988); J. B. Delos and Meng Li Du, *IEEE Journal of Quantum Electronics* **24**, 1445 (1988); J. B. Delos and M. L. Du in *Atomic Spectra and Collisions in External Fields*, K. T. Taylor, M. H. Nayfeh and C. W. Clark, eds, Plenum, NY (1989).
6. E. G. Bogomolnyi, *Sov. Phys. JETP* **69**, 275 (1989), and D. Wintgen

and H. Friedrich, *Phys. Rev.* **A36**, 131 (1987).

7. J. B. Delos, S. K. Knudson and D. W. Noid, *Phys. Rev.* **A30**, 1208 (1984).

8. K. R. Meyer, *Transaction of the American Mathematical Society* **149**, 95 (1970). See also, M. A. M. de Aguiar, C. P. Malta, M. Baranger and K. T. R. Davies, *Ann. Phys.* (NY) **180**, 167 (1987); M. A. M. de Aguiar and C. P. Malta, *Physica* (Amsterdam) **30D**, 413 (1988), and K. Meyer, *Introduction to Hamiltonian Dynamical Systems and the n-Body Problem*, Springer (1992).

9a. J. M. Mao and J. B. Delos, *Phys. Rev.* **A45**, 1746 (1992); J. Main, G. Weibusch, K. Welge, J. Shaw and J. B. Delos, *Phys. Rev.* **A49**, 847 (1994); J. M. Mao, J. Shaw and J. B. Delos, *J. Stat. Phys.* **68**, 51 (1992).

9b. J. Goetz and J. B. Delos, Quantum dynamics of chaotic systems; proceedings of the Third Drexel Symposium on Quantum Nonintegrability, Philadelphia, PA. Gordon and Breach Science Publishers, Langhorne, PA, 1993.

9c. W. Schweizer, R. Niemeier, G. Wunner and H. Ruder, *Z. Phys.* **D25**, 95 (1993). This paper was misquoted in ref. 19.

10. T. van der Veldt, W. Vassen and W. Hogervorst, *Europhys. Lett.* **21**, 903 (1993); M. Kus, F. Haake and D. Delande, *Phys. Rev. Lett.* **71**, 2167 (1993); D. Delande, K. T. Taylor, M. H. Halley, T. van der Veldt, W. Vassen and W. Hogervorst, *J. Phys.* **B27**, 2771 (1994); John A. Shaw, J. B. Delos, Michael Courtney and Daniel Kleppner, *Phys. Rev.* **A51**, 3695 (1995).

11. A. König, J. Neukammer, K. Vietzke, M. Mohl, H.-J. Grabka, H. Heironymus and H. Rinneberg, *Phys. Rev.* **A38**, 547 (1988); J. M. Mao, K. A. Rapelje, S. J. Blodgett-Ford, J. B. Delos, A. König and H. Rinneberg, *Phys. Rev.* **A48**, 2117 (1993).

12a. I. I. Fabrikant, *Phys. Rev.* **A43**, 258 (1991).

12b. Aaron D. Peters and John B. Delos, *Phys. Rev.* **A47**, 3020, 3036 (1993). See also M. L. Du and J. B. Delos, *Phys. Rev.* **A38**, 5609 (1988); H. C. Bryant, A. Mohagheghi, J. E. Stewart, J. B. Donahue, C. P. Quick, R. A. Reeder, V. Yuan, C. E. Hummer, W. W. Smith,

S. Cohen, W. P. Reinhardt and L. Overman, *Phys. Rev. Lett.* **58**, 2412 (1987).

12c. A. D. Peters, C. Jaffé and J. B. Delos, *Phys. Rev. Lett.* **73**, 2825 (1994).

13. U. Eichmann, K. Richter, D. Wintgen and W. Sandner, *Phys. Rev. Lett.* **61**, 2438 (1988).

14. J. Gao, J. B. Delos and M. Baruch, *Phys. Rev.* **A46**, 1449 (1992); J. Gao and J. B. Delos, *Phys. Rev.* **A49**, 869 (1994); Michael Courtney, Hong Jiao, Neal Spellmeyer, Daniel Kleppner, J. Gao and J. B. Delos, *Phys. Rev. Lett.* **74**, 1538 (1995).

15. H. Rottke and K. H. Welge, *Phys. Rev.* **A33**, 301 (1986); T. S. Luk, L. Dimauro, T. Bergeman and H. Metcalf, *Phys. Rev. Lett.* **47**, 83 (1981); W. Sandner, K. A. Safinya and T. F. Gallagher, *Phys. Rev.* **A23**, 2448 (1981).

16. P. A. Dando, T. S. Montiero, D. Delande and K. T. Taylor, *Phys. Rev. Lett.* **74**, 5321 (1995).

17. G. Raithel, M. Fauth and H. Walther, *Phys. Rev.* **A44**, 1898 (1991); <u>47</u>, 419 (1993); G. Raithel et al., *J. Phys.* **B27**, 2849 (1994); G. Raithel and M. Fauth, *J. Phys.* **B28**, 1687 (1995). Also dissertations of E. Flöthmann and R. Ubert, University of Bielefeld (1995). Interesting theoretical work is by J. von Milczewski, G. H. F. Diercksen and T. Uzer, *Int. J. Bifurcations and Chaos* **4**, 905 (1994).

18. G. Alber and P. Zoller, *Phys. Rep.* **199**, 231 (1991); M. W. Beims and G. Alber, *Phys. Rev.* **A48**, 3123 (1993); B. Broers, J. F. Christian, J. H. Hoogenraad, W. J. van der Zande, H. B. van Linden van den Heuvell and L. D. Noordam, *Phys. Rev. Lett.* **71**, 344 (1993); J. A. Yeazell, M. Mallalieu, J. Parker and C. R. Stroud, *Phys. Rev.* **A40**, 5040 (1989); J. A. Yeazell, G. Raithel, L. Marmet, H. Held and H. Walther, *Phys. Rev. Lett.* **70**, 2884 (1993).

19. K. R. Meyer, J. B. Delos and J. M. Mao, *Fields Institute Communications* **8**, 93 (1996).

266

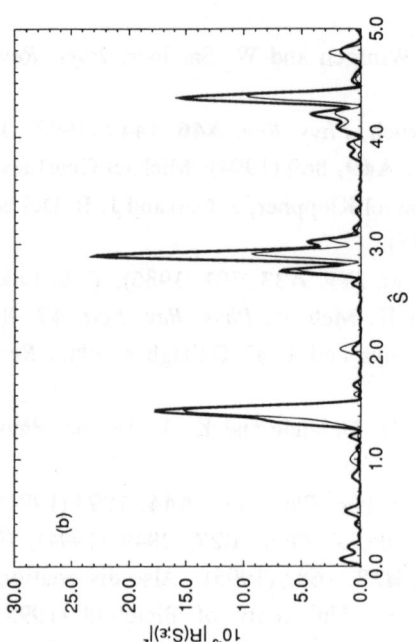

FIG. 2. A scaled-variables Recurrence Spectrum. Heavy line: Experiment. Light line: Theory. Each peak occurs at an action corresponding to a closed orbit, and the height of each peak is the Recurrence Strength.

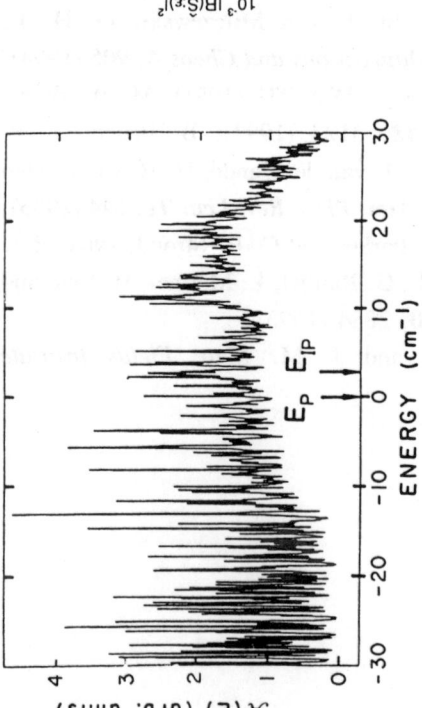

FIG. 1 Absorption spectrum of the hydrogen atom in a magnetic field $B=5.96$ T. The zero-field and actual ionization thresholds are indicated. The initial state was $2p_z$, and the final states have $L_z = m\hbar = 0$.

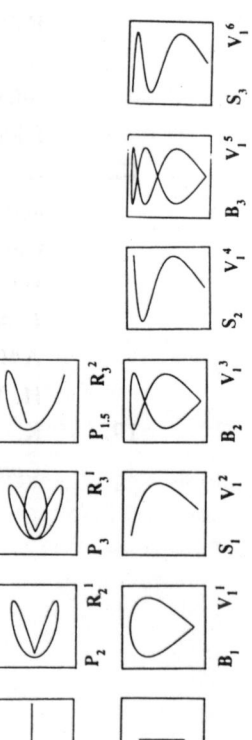

Fig. 3 Some of the closed orbits of an electron in an atom in a magnetic field drawn in (ρ, z) coordinates. The second row is the "Main Sequence."

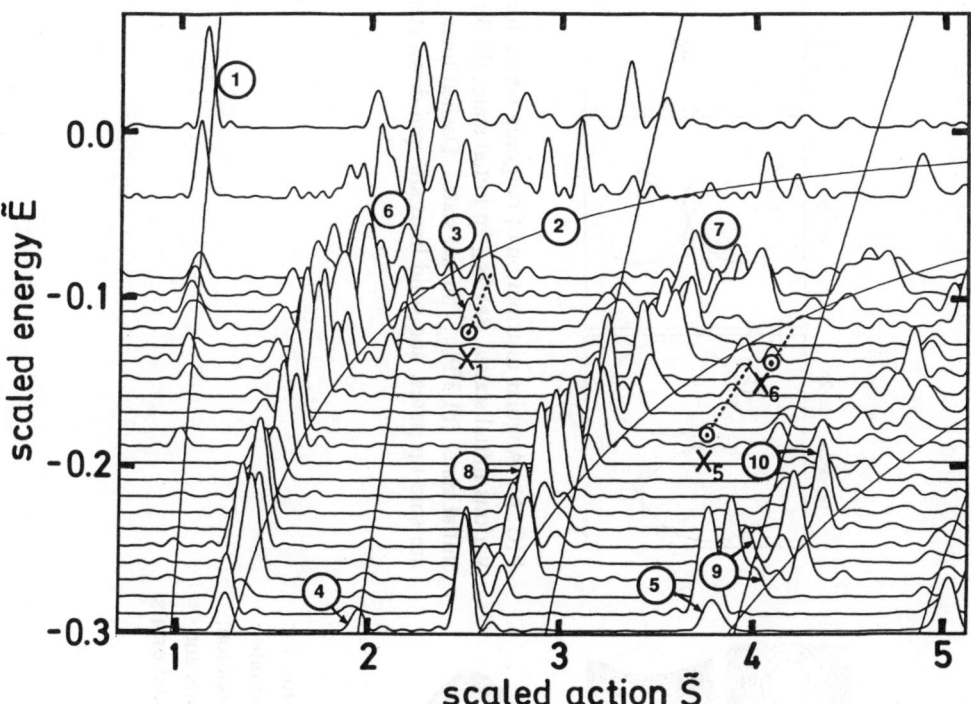

FIG. 4 Experimental |(Fourier transforms)|² (recurrence strengths) at each scaled energy are drawn in a single graph. The following structures were explained. (1) The lowest-action peak results from an orbit that lies perpendicular to the magnetic field. The nearly vertical lines represent the action vs. energy for the first return and subsequent returns of this orbit. (2) The curved line represents the action vs. energy of the orbit that lies parallel to the magnetic field. (3) A new peak appearing "out of nowhere" is caused by a saddle-node bifurcation that produces an "exotic" orbit. (4) A focusing effect gives a strong recurrence at the second return of the perpendicular orbit and again, (5) at the fourth return. (6) The first "mountain range" is produced by the "main sequence" of orbits that bifurcate from the parallel orbit.

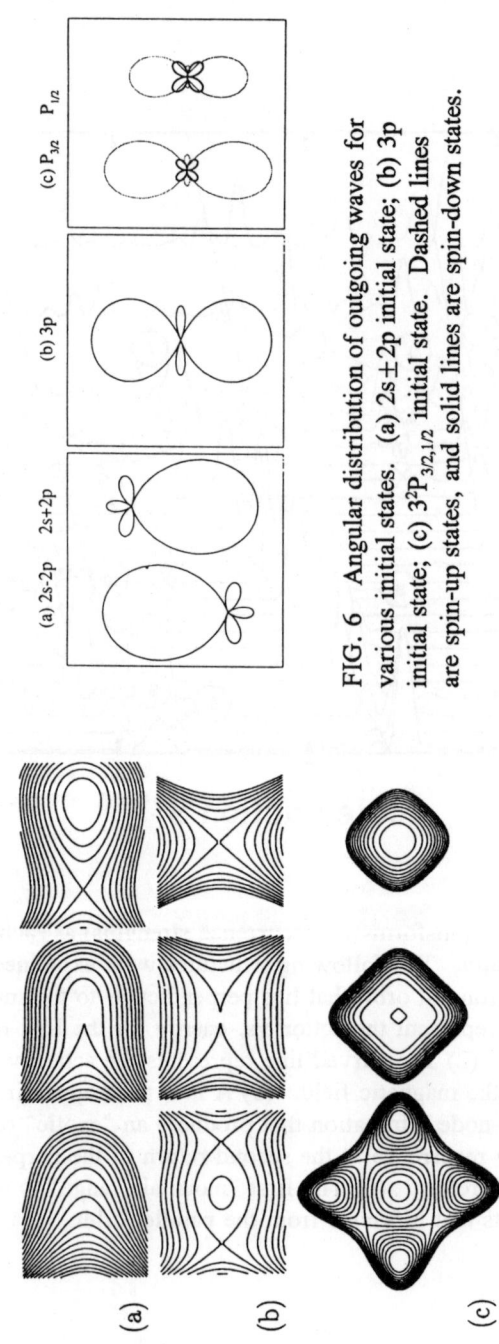

FIG. 6 Angular distribution of outgoing waves for various initial states. (a) $2s\pm 2p$ initial state; (b) $3p$ initial state; (c) $3^2P_{3/2,1/2}$ initial state. Dashed lines are spin-up states, and solid lines are spin-down states.

FIG. 5 Patterns in the Poincaré surfaces of section for three of the five generic bifurcations. (a) Saddle-node bifurcation; (b) period-doubling or pitchfork bifurcation; (c) period-4 island-chain bifurcation. The O-points and X-points are respectively stable or unstable periodic orbits.

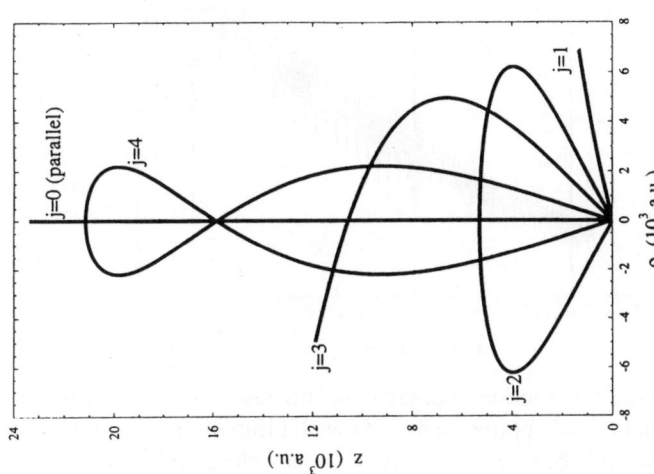

FIG. 8 Some orbits closed at the nucleus for parallel electric and magnetic fields

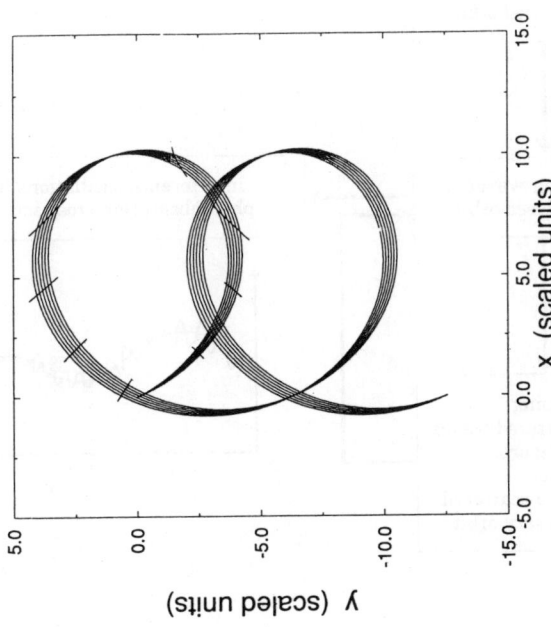

FIG. 7 A pencil of trajectories representing electrons propagating away from a Hydrogen atom in crossed electric and magnetic fields. The magnetic field is out of the paper and the electric field is to the right.

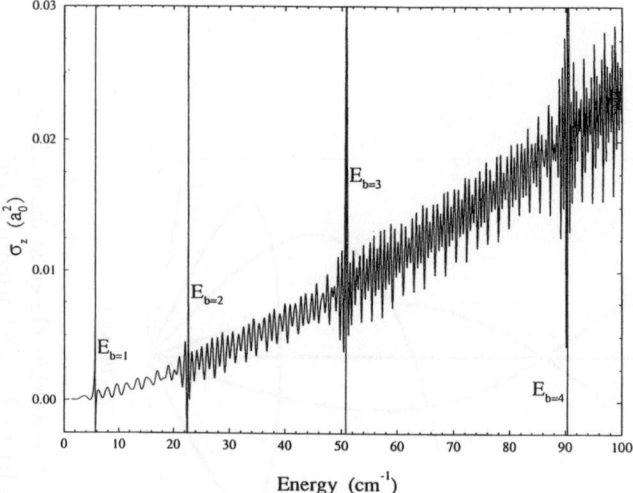

FIG. 9 Photodetachment cross section in parallel electric and magnetic fields as calculated using a semiclassical approximation. At each bifurcation the semiclassical approximation diverges, and above each bifurcation a new closed orbit produces another series of sinusoidal oscillations.

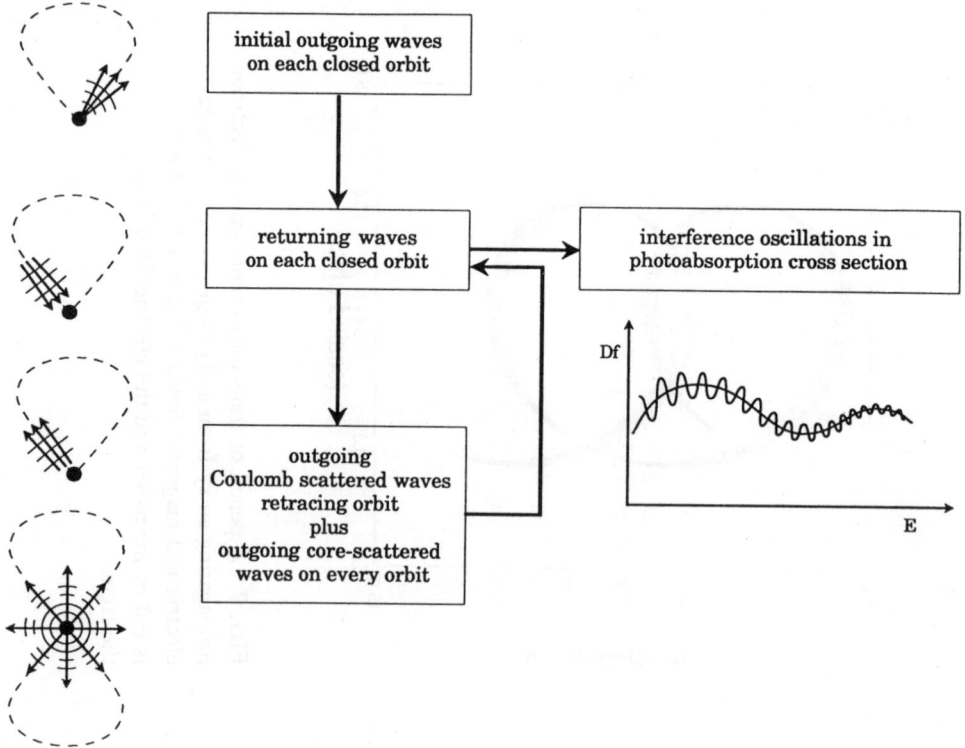

FIG. 10 Schematic of effect of core-scattering.

QED EFFECTS IN FEW-ELECTRON HIGH-Z SYSTEMS

I. LINDGREN, H. PERSSON, S. SALOMONSON, P. SUNNERGREN
*Department of Physics, Chalmers University of Technology and
Göteborg University, S-412 96 Göteborg, Sweden*

QED effects in heavy, highly charged ions are reviewed, particularly the energy-level shift (Lamb shift) for one-, two-, and three-electron ions. The results of numerical calculations (to all orders in $Z\alpha$) are compared with those obtained by the $Z\alpha$ expansion as well as with recent experimental results. Also numerical calculations of the hyperfine structure and Zeeman effect in heavy hydrogenlike ions are discussed. These calculations can be performed also for low Z and compete in accuracy with those based on the $Z\alpha$ expansion.

1 Introduction

Very accurate tests of QED have been performed on light atomic systems, and impressive agreement between theory and experiment has been obtained for a number of properties, particularly for the free-electron g-value, but also for the hyperfine structure and Lamb shift of positronium, muonium and neutral hydrogen. Impressively accurate calculations have for a long time been performed by Kinoshita *et al.* and more recently also by Eides, Grotch, Karshenboim, Pachucki, and others [1]. Lately, considerable progress has been made - experimentally as well as theoretically - also in studying few-electron systems with high Z, which will make it possible, for the first time, to perform accurate tests of QED also at strong fields, where the effects are much more pronounced. The Lamb shift in the ground state of hydrogen-like uranium, for instance, has recently been measured at GSI by Beyer *et al.* [2] to be 470 eV (including a nuclear-size effect of about 200 eV) with an experimental uncertainty of 16 eV. This could be compared to the corresponding shift in neutral hydrogen of 30 μeV. If the experimental accuracy could be somewhat improved, which is anticipated, the Lamb shift of heavy hydrogen-like ions will constitute important test objects for strong-field QED. Also systems with more than one electron can be used for this purpose. The splitting between the $2p_{1/2}$ and $2s$ states in Li-like uranium has been measured by Schweppe *et al.* [3] to be 280,59 eV with an uncertainty of only 0,09 eV, which has challenged several theoretical groups to improve their computational techniques. At Livermore Marrs *et al.* [4] have measured the binding energies of some He-like ions, and by comparing with the corresponding data for H-like ions the two-electron contribution to the level shift can be extracted. These data can be used for direct test of the *screening* of the first-order Lamb shift.

Also other quantities than the level shift can be useful for testing the strong-field QED. The hyperfine structure can now be measured with high accuracy for heavy H-like ions [5] and experiments of the Zeeman effect (g-factor) of such ions are in progress [6]. Such data will serve as important complement to the level-shift data.

For light systems, the standard theoretical technique is to treat the nuclear field as a perturbation ($Z\alpha$ expansion), starting from plane-wave solutions of the Schrödinger or the Dirac equation. For very heavy systems, on the other hand, where $Z\alpha$ approaches unity, such an expansion is no longer meaningful. Instead, the calculations have to be performed *non-perturbatively*, starting from electronic states generated in the external field. This technique has been developed for Coulomb potentials particularly by Peter Mohr [7], following pioneering work of Brown et al. [8] and Desiderio and Johnson [9]. In recent years, new techniques have been developed, which are applicable also for non-coulombic potentials [10-13].

In the present talk a review will be given of the application of QED to heavy ions with few electrons. We will begin by defining a many-body perturbation scheme (*No-Virtual-Pair Approximation*), which will form the starting point for the QED calculations. Then we shall discuss the evaluation of the Lamb shift for hydrogen-, helium- and lithium-like ions and make comparison with experimental results. Finally, we shall analyze QED effects on the hyperfine structure and the Zeeman effect and the possibilities of making significant comparison with the corresponding experimental data.

2 Relativistic Many-body Perturbation Theory

2.1 No-Virtual-Pair Approximation

Non-relativistic many-body perturbation theory (MBPT) is based upon the Hamiltonian

$$H = \sum_i h_S(i) + \sum_{i<j} \frac{e^2}{4\pi\varepsilon_o r_{ij}}, \qquad (2.1)$$

where h_S is the single-electron Schrödinger hamiltonian

$$h_S = -\tfrac{1}{2}\nabla^2 - \frac{e^2 Z}{4\pi\varepsilon_o r}. \qquad (2.2)$$

Relativistic MBPT has to be based on the Dirac equation, rather than the Schrödinger equation, and a reasonable starting point might then be a Hamiltonian of the type

$$H = \sum h_D + \sum V_{ij}, \qquad (2.3)$$

where h_D is the single-electron Dirac hamiltonian

$$h_D = c\boldsymbol{\alpha}\cdot\boldsymbol{p} + \beta mc^2 - \frac{e^2 Z}{4\pi\varepsilon_o r} \qquad (2.4)$$

($\boldsymbol{\alpha}$ and β being the Dirac operators) and V_{ij} is the interelectronic interaction. This Hamiltonian, however, suffers from two serious problems. Firstly, the eigenvalues have no lower bound, due to the existence of the negative-energy solutions to the Dirac equation (Brown-Ravenhall disease [14]), and secondly the interelectronic potential V_{ij} is not uniquely determined (gauge dependent).

The first problem can be remedied by introducing *projection operators*, Λ_+, which eliminate the negative-energy states [15]

$$H = \Lambda_+ \left[\sum_i h_D(i) + \sum_{i<j} V_{ij} \right] \Lambda_+ . \quad (2.5)$$

This is the so-called *No-Virtual-Pair Approximation* (NVPA), which is a sound starting point for relativistic many-body calculations. One problem remains, though, namely to determine the interelectronic potential V_{ij}. For that purpose we have to analyse the interelectronic interaction by means of QED.

2.2 Bound-state QED

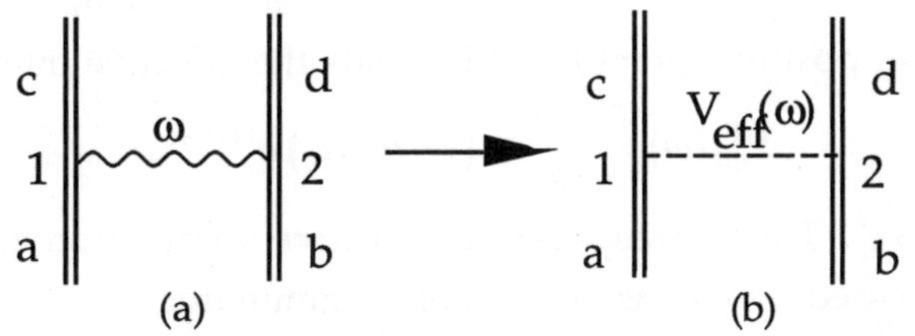

Figure 1. The Feynman diagram of single-photon exchange (a) is compared with that of potential scattering (b).

In QED the interaction between the electrons is represented by the exchange of virtual photons, as illustrated in Fig. 1(a). The second-order S-matrix for single-photon exchange is given by [16]

$$S^{(2)} = -e^2 \iint d^4x_1 d^4x_2 \ T\left[\left(\Psi^\dagger \alpha^\nu A_\nu \Psi\right)_2 \left(\Psi^\dagger \alpha^\mu A_\mu \Psi\right)_1\right]. \quad (2.6)$$

(We use here relativistic units, $\hbar = m = c = \varepsilon_0 = 1$. This implies that $e^2 = 4\pi\alpha$, α being the fine-structure constant, but for the time being we shall keep e in the equations.) In the S-matrix, T represents the Wick time-ordering operator, Ψ, $\Psi^\dagger$ the field operators, A_μ the electromagnetic field and α^μ the Dirac operator in covariant form (related to the gamma matrices by $\alpha^\mu = \beta \gamma^\mu$). In bound-state QED we use the *Furry interaction picture*, where the field operators are composed of orbitals generated in the field, V, of the nucleus (and possibly the other electrons),

$$\Psi = \sum_i a_i \phi_i \ ; \quad \Psi^\dagger = \sum_i a_i^\dagger \phi_i^* \quad (2.7)$$

$$h_D \phi_i = \varepsilon_i \phi_i; \quad h_D = \boldsymbol{\alpha} \cdot \boldsymbol{p} + \beta m + V. \quad (2.8)$$

Here, $a_i^\dagger$ and a_i are creation and destruction operators, respectively.

We find it here convenient to work in a *mixed energy-space representation*, obtained by integrating over time. This leads to the S-matrix element for the diagram shown in Fig. 1(a)

$$\langle cd|S^{(2)}|ab\rangle = -2\pi i \delta(\varepsilon_a + \varepsilon_b - \varepsilon_c - \varepsilon_d)\langle cd|\alpha_1^\mu \alpha_2^\nu e^2 D_{F\nu\mu}(x_2 - x_1, \omega_{ac})|ab\rangle. \quad (2.9)$$

$D_{F\nu\mu}(x_2 - x_1, \omega_{ac})$ is here the photon propagator in the mixed representation and $\omega_{ac} = \varepsilon_a - \varepsilon_c$ the energy parameter. The delta factor indicates that energy is conserved at the interaction. The expression above can be compared with the S-matrix of *potential scattering*, represented by the Feynman diagram (b) in Fig. 1. This yields an *effective interaction potential*

$$V_{\text{eff}}(\omega) = \alpha_1^\mu \alpha_2^\nu e^2 D_{F\nu\mu}(x_2 - x_1, \omega). \quad (2.10)$$

This potential is *energy-dependent* and, in addition, *gauge dependent*, due to the appearance of the photon propagator.

In the *Feynman gauge* the photon propagator is

$$D_{F\nu\mu}(x_2 - x_1, \omega) = -g_{\nu\mu}\int \frac{d^3k}{(2\pi)^3} \frac{e^{i\mathbf{k}\cdot(x_2-x_1)}}{\omega^2 - k^2 + i\eta}, \quad (2.11)$$

where η is a small positive quantity. This yields the effective interaction

$$V_{\text{eff}}^F(\omega) = \frac{e^2}{4\pi r_{12}}(1 - \boldsymbol{\alpha}_1\cdot\boldsymbol{\alpha}_2)e^{i\omega r_{12}}. \quad (2.12)$$

where $r_{12} = |x_1 - x_2|$. The corresponding effective interaction in the *Coulomb gauge* can be expressed by means of a double commutator,

$$V_{\text{eff}}^C(\omega) = \frac{e^2}{4\pi}\left\{\frac{1}{r_{12}} - \boldsymbol{\alpha}_1\cdot\boldsymbol{\alpha}_2\frac{e^{i\omega r_{12}}}{r_{12}} + \left[\boldsymbol{\alpha}_1\cdot\nabla_1, \left[\boldsymbol{\alpha}_2\cdot\nabla_2, \frac{e^{i\omega r_{12}} - 1}{\omega^2 r_{12}}\right]\right]\right\}. \quad (2.13)$$

The ω dependence of the interactions represents the *retardation*, which is a relativistic effect. The *unretarded* (frequency independent) *limits* of these interactions are

$$V_{\text{eff}}^F(\omega \Rightarrow 0) = \frac{e^2}{4\pi r_{12}}(1 - \boldsymbol{\alpha}_1\cdot\boldsymbol{\alpha}_2) \quad (2.14a)$$

$$V_{\text{eff}}^C(\omega \Rightarrow 0) = \frac{e^2}{4\pi r_{12}}\left(1 - \tfrac{1}{2}\boldsymbol{\alpha}_1\cdot\boldsymbol{\alpha}_2 - \frac{(\boldsymbol{\alpha}_1\cdot r_{12})(\boldsymbol{\alpha}_2\cdot r_{12})}{2r_{12}^2}\right), \quad (2.14b)$$

respectively. Here, we recognize the (instantaneous) Coulomb interaction as the first part of these expressions and as the remaining parts the *Gaunt* (2.14a) and the *Breit interactions* (2.14b), respectively.

The potentials given above for the Feynman and Coulomb gauges give significantly different results when applied in iterative schemes, such as multi-configuration Dirac-Fock (MCDF) or MBPT [17, 18]. The question is then: Which is the best interaction to use in many-body calculations? In order to answer that question it is necessary to analyse the *two-photon interaction* between the electrons (see Fig. 2).

In the single-photon interaction treated above *energy is conserved*, as indicated by Eq. (2.9). This implies that the interactions derived are strictly speaking valid only in first-order. In higher orders, energy is *not* conserved in the intermediate states. This is the reason for the gauge dependence when the effective single-photon interactions are used iteratively.

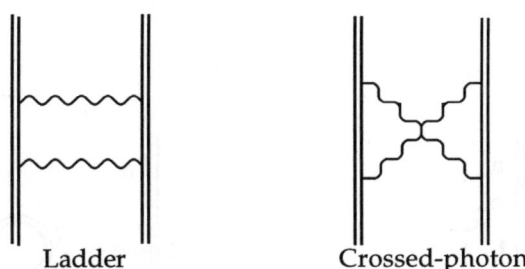

Figure 2. The Feynman diagrams for two-photon exchange between the electrons, the ladder and the crossed-photon diagrams.

The reason for the relatively large gauge-dependence, when the single-photon interactions are used iteratively, can be understood by considering the retardation dependence of the two interactions. It is found that in the Feynman gauge (2.12) the leading Coulomb interaction is retarded (ω dependent), while in the Coulomb gauge (2.13) only the much weaker second (magnetic) part is retarded. Since retardation is not described correctly in higher orders by these interactions, it follows that the error in the Feynman gauge is considerably larger (in fact of order α^2 Hartree) than in the Coulomb gauge (of order α^3 Hartree). In addition, the effect of negative-energy states (as well as all radiative effects) are left out in these schemes, but these contribute first in the order α^3 Hartree. This explains the large gauge-dependence observed. It can be shown that the two gauges yield identical results to order α^2 Hartree, when the two second-order diagrams in Fig. 2 are included [19, 20]. Recently, the two-photon contribution has been evaluated numerically (to all orders of $Z\alpha$) for the ground state of He-like systems by Blundell *et al.* [21] and by Lindgren *et al.* [22]. The results confirm that the two gauges give numerically identically results, when the two-photon contribution is included.

The results of the analysis of the two-photon exchange shows that the Coulomb + Breit interaction, derived in the Coulomb gauge (in the limit of no retardation), leads to results correct to order α^2 Hartree, when used iteratively. Therefore, this constitutes a good approximation for relativistic MBPT. This is the NVPA based on the *Dirac-Coulomb-Breit Hamiltonian*

$$H = \Lambda_+ \left[\sum h_D + \sum \left(\frac{e^2}{4\pi\varepsilon_0 r_{ij}} + B_{ij}^o \right) \right] \Lambda_+ \qquad (2.15)$$

where
$$B_{12}^o = -\frac{e^2}{4\pi\varepsilon_0} \left(\frac{\alpha_1 \cdot \alpha_2}{2r_{12}} + \frac{(\alpha_1 \cdot r_{12})(\alpha_2 \cdot r_{12})}{2r_{12}^3} \right) \qquad (2.16)$$

is the unretarded Breit interaction.

Many-body calculations based on the Dirac-Coulomb-Breit Hamiltonian will form the starting point for our analysis. Effects beyond that approximation are defined as "QED-effects". These are of two kinds, *radiative effects* (self energy and vacuum polarization), and *non-radiative QED effects*, due to retardation and negative-energy states in the non-radiative diagrams.

3 The Lamb shift

3.1 General

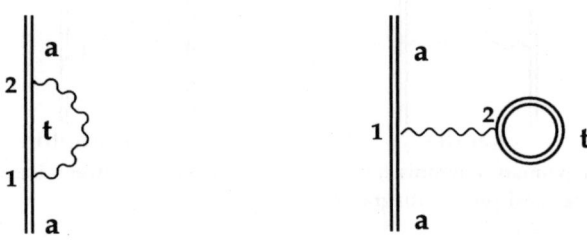

Figure 3. The Feynman diagrams for the first-order Lamb shift of a bound electronic state, a. The first diagram represents the electron self energy and the second diagram the vacuum polarization.

The first-order Lamb shift is caused by the effects represented by the two diagrams in Fig. 3, the *electron self energy* and the *vacuum polarization*. We shall start with the self energy, which is somewhat more complicated to handle. The Feynman amplitude for the first-order self energy is in the mixed energy-space representation

$$M = \iint d^3x_1 \, d^3x_2 \int \frac{d^3k}{(2\pi)^3} \phi_a^*(x_2) i e \alpha^\nu$$

$$\times \int \frac{d\omega}{2\pi} i S_F(x_2, x_1, \varepsilon_a - \omega) i e \alpha^\mu \phi_a(x_1) i D_{F\nu\mu}(x_2 - x_1, \omega) \quad (3.1)$$

Here, the *electron propagator* is

$$S_F(x_2, x_1, \omega) = \sum_t \frac{\phi_t(x_2) \phi_t^*(x_1)}{\omega - \varepsilon_t(1 - i\eta)}, \quad (3.2)$$

and the photon propagator in the Feynman gauge is given by (2.11). This leads to the first-order bound-state self energy

$$\Delta E_{bau} = e^2 \sum_t \left\langle at \left| (1 - \boldsymbol{\alpha}_1 \cdot \boldsymbol{\alpha}_2) \int \frac{d^3k}{(2\pi)^3} \frac{e^{i\mathbf{k}\cdot(\mathbf{x}_2-\mathbf{x}_1)}}{\varepsilon_a - \varepsilon_t - k\,\text{sign}(\varepsilon_t)} \right| ta \right\rangle. \quad (3.3)$$

For the numerical treatment it is convenient to make an expansion in *spherical waves* (L),

$$\Delta E_{bau} = -\frac{e^2}{4\pi^2} \sum_L (2L+1) \int k\,dk \sum_t \frac{\left\langle at \left| (1 - \boldsymbol{\alpha}_1 \cdot \boldsymbol{\alpha}_2) j_L(kr_1) j_L(kr_2) \, \mathbf{C}^L(1) \cdot \mathbf{C}^L(2) \right| ta \right\rangle}{\varepsilon_a - \varepsilon_t - k\,\text{sign}(\varepsilon_t)}. \quad (3.4)$$

Here, $j_L(kr)$ is a spherical Bessel function and $\mathbf{C}^L$ is a spherical tensor operator, closely related to the spherical harmonics.

For the numerical evaluation of an expression of the type (3.4), some kind of "complete" single-electron spectrum is required. This can be generated by solving the Dirac equation (2.8), using numerical basis set of spline [23] or space discretization type [24]. In the expression (3.4) the summation over the intermediate states t has to be performed over the entire spectrum, i.e. over positive-energy (particle) as well as negative-energy (hole) states. Each term in the partial-wave expansion is finite, but the L sum diverges, and the expression has to be *renormalized*.

For a free electron the self energy constitutes a part of the "physical" electron mass. This part is also present in the bound-state energy and has to be removed, before the physically significant effect can be extracted (the "*mass counter term*"). This is the *mass renormalization*. The mass counter term is the average of the free-electron self energy for the bound state considered, evaluated *on the mass shell*,

$$-e^2 \sum_{p,p',q} \langle a|p'\rangle \left\langle p'q \left|(1-\boldsymbol{\alpha}_1\cdot\boldsymbol{\alpha}_2)\int \frac{d^3k}{(2\pi)^3} \frac{e^{ik\cdot(x_2-x_1)}}{\varepsilon_p - \varepsilon_q - k\,\text{sign}(\varepsilon_q)}\right|qp\right\rangle \langle p|a\rangle, \quad (3.5)$$

as illustrated in Fig. 4. Here, p, p' and q are free-electron states. It should be noted that the energy parameter in the denominator is the free-electron energy (ε_p) in contrast to that of the bound expression (3.4).

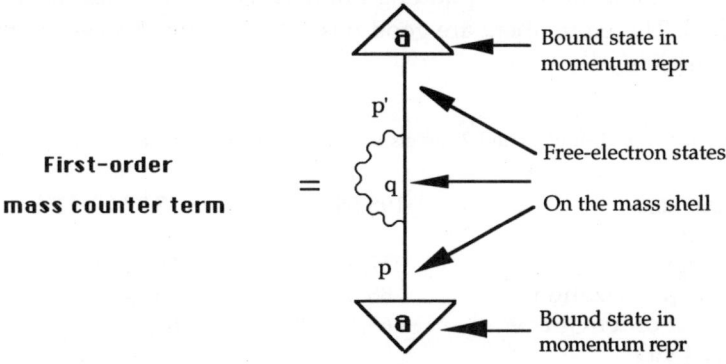

Figure 4. Illustration of the first-order mass renormalization. The bound state is expressed in the momentum representation and the matrix elements of the free-electron self energy is evaluated "on the mass shell".

One possible renormalization procedure is to expand also the mass counter term in partial waves, and to perform the renormalization for each partial wave, the so-called *partial-wave renormalization* (PWR) [11, 12]. This works well in first order without any further regularization, but some precaution is needed in higher orders [25].

The approximate effect of the first-order *vacuum polarization* can be obtained in a simple way by means of the *Uehling potential* [26]. The remaining part, the so-called Wichmann-Kroll effect [27], can be evaluated numerically with high accuracy, as shown by Mohr and Soff [28] and Persson et al. [29].

3.2 The Lamb shift of H-like uranium

The Lamb shift of hydrogen-like heavy ions can now be measured with good accuracy, and such systems therefore constitute a good testing ground for QED at strong fields. Numerical calculations of the first-order self energy on such systems were pioneered by Desiderio and Johnson [9], based on a procedure introduced by Brown, Langer and Schaefer [8]. Later the numerical technique has been developed to a high degree of sophistication, particularly by Peter Mohr [7]. The first-order Lamb shift for hydrogen-like systems can now be calculated with such an accuracy that the uncertainty is negligible for all practical purposes [30].

The Lamb shift of the 1s level of H-like uranium - compared to the Dirac value for a point nucleus - has recently been measured by Beyer et al. [2] to be 470±16 eV, and higher accuracy is anticipated. A major uncertainty in the corresponding theoretical evaluation is due to the partly unknown nuclear structure. As can be seen from the results in Table 1, the effect of the finite nucleus for H-like uranium is about 200 eV. The size and shape of the uranium nucleus is quite well known, however, and the corresponding uncertainty can be reduced to a few tenths of an eV [25]. Therefore, there are good possibilities with this system to test also higher-order effects.

Table 1. Lamb shift of 1s and 2s levels in H-like Uranium (in eV)

	1s	2s
Finite nuclear size	198,68 (32)	37,77 (8) [25]
First-order QED		
Self energy	355,05	65,42 [7]
Vacuum polarization	-88,60	-15,64 [29]
Nucl size + first-order QED	465,13 (32)	87,55 (8)
Second-order QED		
Second-order vacuum pol	-0,94	-0,16 [29, 25, 33]
Comb self-energy-vac pol.	1,27	0,23 [25, 34]
Second-order self energy	NOT CALCULATED	
Second-order QED (calculated so far)	0,33	0,07
Nuclear polarization	-0,18	-0,03 [31]
Nuclear recoil	0,51*	0,13* [32]
TOTAL THEORY	465,8 (4)	87,72
EXPERIMENTAL	470, (16) [2]	

* This includes the reduced-mass effect.

The effects of nuclear polarization and nuclear recoil, which appear on the level of a few tenths of an eV, have recently been evaluated with good accuracy by Plunien et al. [31] and Artemyev et al. [32], respectively. More uncertain for the moment is the second-order (two-photon) Lamb shift, represented by the diagrams shown in Fig. 5. These are of three kinds, second-order vacuum polarization, second-order self energy and combined vacuum-polarization--self-energy. The effects of second-order vacuum polarization and the combined vacuum-polarization--self-energy, which are of the order of 1 eV, have recently been calculated by Soff et al. [33], Persson et al. [25, 29] and Lindgren et al. [34]. The second-order self-energy diagrams, on the other hand, which can be expected to be at least of the same order, have not yet been evaluated [35, 36].

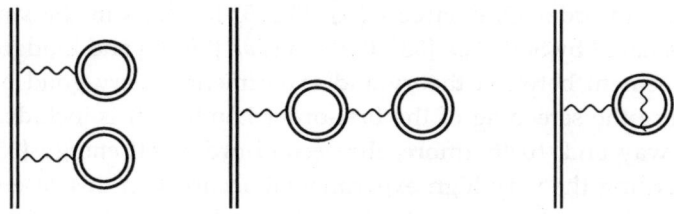

Second-order vacuum polarization

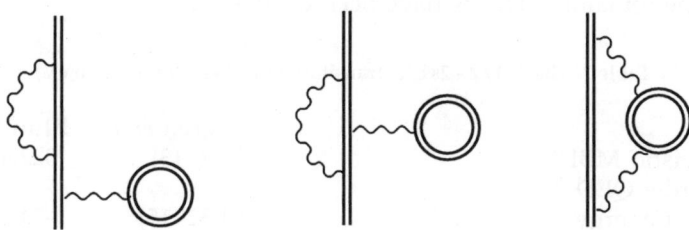

Combined vacuum polarization and self energy

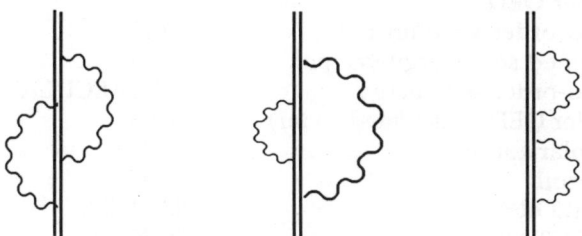

Second-order self energy

Figure 5. Feynman diagrams for the second-order Lamb shift for single-electron systems.

3.3 The Lamb shift of Li-like uranium

The energy separation between the $2p_{1/2}$ and $2s$ states of Li-like uranium was measured very accurately a few years ago at the Super-Highlac at Berkeley by Schweppe et al. [3]. Although this system has three electrons, it can mainly be treated as a single-electron system by starting from Dirac-Fock functions generated in the $1s^2$ core. The many-body effects (beyond Dirac-Fock), though, are significant but can be evaluated quite accurately. The remaining uncertainty indicated in Table 2 is mainly due to the nuclear-size effect [37]. The difference between the many-body results obtained by Lindgren et al. [11] and by Blundell [10] is mainly due to the fact that the former contains also some higher-order

Breit interactions. The first-order Lamb shift given in the table includes also - in an approximate way - the effects of screening, due to the fact that the electron orbitals were generated in the potential of the nucleus *and* the core electrons. The nuclear polarization and nuclear recoil contributions have, as in the previous case, been obtained by Plunien *et al.* [31] and Artemyev *et al.* [32], respectively. Also some second-order QED effects (see Fig. 5) have, as in the single-electron case, been evaluated by Soff *et al.* [33], Persson *et al.* [25, 29] and Lindgren *et al.* [34]. The final agreement between theory and experiment is very good but might be fortuitous, since the screening of the first-order Lamb shift is included only in an approximate way and, furthermore, the second-order self energy [35, 36] is still missing. Regarding the very high experimental accuracy in this case (0,09 eV), as well as the small uncertainty due to the finite nuclear size, this system is a very good candidate for a serious test of second-order QED effects at strong nuclear field, once the remaining effects have been evaluated.

Table 2. The 2p1/2 - 2s1/2 transition in Li-like Uranium (in eV)

	Lindgren *et al.*	Blundell
Relativistic MBPT	322,33 (3)	322,41
First-order QED		
Self energy	-54,32 (15)	-54,24
Vacuum polarization	12,55 (4)	12,56
First-order QED Total	-41,77 (15)	-41,68
Second-order QED		
Second-order vacuum pol	0,13	
Combined self-energy-vac pol.	-0,19	
Second-order self energy	NOT CALCULATED	
Second-order QED (calculated so far)	-0.08	
Nuclear polarization	0,03	0,03
Nuclear recoil	-0,07	-0,07
TOTAL THEORY	280,44 (20)	280,84 (10)
EXPERIMENTAL	280,59 (9)	

3.4 Two-electron Lamb shift

Recently, the two-electron contribution to the binding energy of the ground state of some He-like ions has been measured at the Super-EBIT facility at Livermore by Marrs *et al.* [4]. In this experiment the binding energies of He- and H-like ions of the same element have been compared, which makes it possible to eliminate very accurately all single-particle effects and to extract the pure two-particle contribution. In this way, most of the nuclear effect as well as the single-electron Lamb shift is eliminated. The remaining two-electron effect in second order is represented by the diagrams in Fig. 6. The first two diagrams represent the MBPT effect as well as the non-radiative QED effects, discussed above, and the remaining ones the radiative effects, i.e. the *screening* of the first-order Lamb shift. A complete QED calculation of the two-electron contribution to second order has recently been performed by Persson et al. [38], and the results are compared with the corresponding experimental results in Table 3. The agreement between the theory and experiment is good, although the experimental accuracy is for the time being not sufficient for testing the QED contributions. However, only a moderate increase of the accuracy is needed for this purpose. The uncertainty due to the finite nuclear size is very small in this case, and therefore these systems constitute potentially good objects for testing second-order QED effects at strong fields.

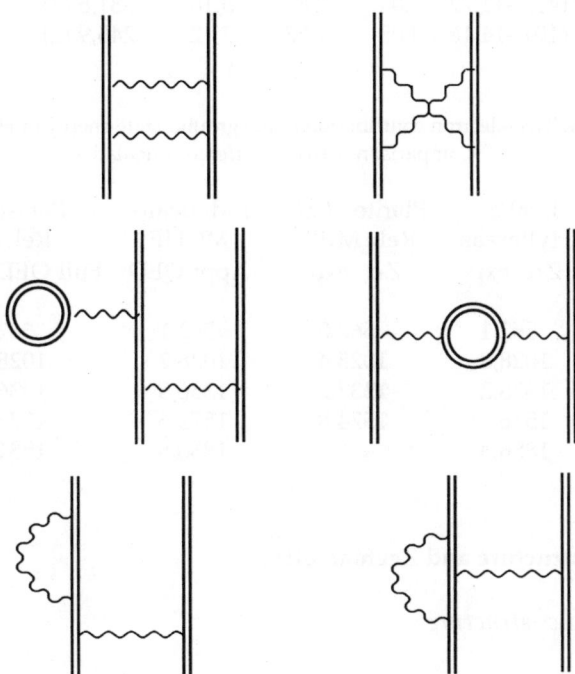

Figure 6. Feynman diagrams of the two-electron contribution in second order to the binding energy of He-like systems. The first two diagrams represent the many-body part and the non-radiative QED part, and the remaining ones the radiative contribution (screening of the first-order Lamb shift).

In Table 4 we have compared the results of various theoretical evaluations of the two-electron contribution to the binding energy of some He-like ions. The results of Drake [39] are obtained using very accurately correlated wave functions of Hylleraas type together with the QED results to order $(Z\alpha)^3$ Hartree, derived by Araki [40] and Sucher [41], and subtracting the hydrogenic binding energies of Johnson and Soff [30a]. The results of Plante et al. [42] are obtained in a similar way, using relativistic MBPT, while the results of Indelicato [43] are obtained with MCDF functions and some approximate scheme for evaluating the QED effects.

Table 3. Two-electron contribution to the ground-state energy of He-like ions
Comparison between theory and experiment (in eV)

Nuclear charge	M B P T First order	2nd	3rd	Non-radiative	Lamb shift	Total theory	Experimental Marrs et al.
32	567,61	-5,22	0,02	0,03	-0,42	562,02 (10)	562,6 ±1,6
54	1036,56	-7,04	0,03	0,16	-1,56	1028,15 (10)	1027,2 ±3,5
66	1347,45 (1)	-8,59	0,03	0,36	-2,66	1336,59 (10)	1341,6 ±4,3
74	1586,93 (2)	-9,91	0,04	0,55	-3,68	1573,93 (10)	1568,9 ±15,
83	1897,56 (4)	-11,77	0,04	0,86	-5,16	1881,5 (2)	1876, ±14,
92	2265,87 (10)	-14,16	0,05	1,28	-7,12	2245,9 (2)	

Table 4. Two-electron contribution to the ground-state energy of He-like ions
Comparison between different calculations.

Nuclear charge	Drake Hylleraas $Z\alpha$ exp	Plante et al. Rel. MBPT $Z\alpha$ exp	Indelicato MCHF Appr QED	Persson et al. Rel. MBPT Full QED, all order $Z\alpha$
32	562,1	562,0	562,1	562,0
54	1028,8	1028,4	1028,2	1028,2
66	1338,2	1337,2	1336,5	1336,6
74	1576,6	1574,8	1573,6	1573,9
83	1886,3		1880,8	1881,5

4 Hyperfine structure and Zeeman effect

4.1 The hyperfine structure

The hyperfine structure of the ground state of H-like bismuth has recently been measured with high accuracy at GSI by Klaft et al. [5]. The corresponding numerical calculations have been performed [44], and the results are compared in Table 5. Here, the effect of the nuclear charge distribution has been evaluated, using available experimental data [47], and the effect of the magnetic distribution (the Bohr-Weisskopf effect) has been taken from Tomaselli et al. [48]. An additional uncertainty in the theoretical evaluation is due to the experimental nuclear magnetic moment, which is based on an old nmr measurement [45], and thus possibly subject to a significant but largely unknown chemical shift [46].

Table 5. Hyperfine structure in H-like Bi (in eV)

Point-nucleus value	5,8249 *
Finite nuclear size**	
Charge distribution	−0,6335 (4)
Magnetic distribution (Bohr-Weisskopf)	−0,107 (7)
Non-QED value	5,0844 (8)
First-order QED corrections	
Self energy	−0,0614
Vacuum polarization	0,0346
Sum QED corrections	−0,0268
TOTAL THEORY	5,058 (7)
EXPERIMENTAL	5,084 (1)

* Based on magnetic moment of 4,1106 nuclear magnetons. Uncertainty not considered.
** Based on nuclear rms 5,519 fm.

Numerical calculations (to all orders of $Z\alpha$) can now be performed with high accuracy also for low Z, down to Z=1, which is demonstrated for the one-loop self energy by Persson et al. [44]. This makes it interesting to compare the numerical results with the corresponding results of the $Z\alpha$ expansion.

Conventionally, the hyperfine splitting (for an infinitely heavy point nucleus) is expressed in the form

$$\Delta E = \frac{\alpha}{\pi} \Delta E^{(1)} \times F(Z\alpha),$$

where $\Delta E^{(1)}$ is the (non-relativistic) first-order splitting and $F(Z\alpha)$ is a general function of $Z\alpha$. The coefficients for $Z\alpha$ and $(Z\alpha)^2$ for the one-loop self energy (including $\log(Z\alpha)$ terms) were calculated some time ago [1a], but recently a new value for the coefficient of the quadratic term, originally calculated by Sapirstein [49], have been obtained independently by Pachucki and Nio [50]. In addition, a significant term of the order $(Z\alpha)^3 \ln(Z\alpha)$ has recently been evaluated by Karshenboim [51]. The original value of Sapirstein leads for Z=1 to a contribution to $F(Z\alpha)$ due to the first-order self energy of 0,43805, which with the new term of Karshenboim is reduced to 0,43800. The corresponding values obtained with the new value of Pachucki and Nio (which agree to the accuracy considered here) become 0,43816 and 0,43811, respectively. These values can be compared with the result of the numerical calculations of Persson et al. [44] of 0,4380 with an uncertainty of one unit in the last decimal place.

This example demonstrates that the numerical calculations have now reached to such a degree of accuracy that they can well compete with the most accurate $Z\alpha$-expansion results also for low Z. This will provide an additional test of the complicated analytical calculations as well as a check of the significance of uncalculated terms. Evidently, this will be of great importance in future tests of QED and in the determination of the fundamental constants [52].

4.2 The Zeeman effect

The Zeeman effect (g-factor) of singly charged ions has for some time been accurately studied in ion-trap experiments [53], and similar experiments on highly charged ions are now being prepared by the Mainz group [6]. To start with, experiments up to Z=20 are being planned, and the anticipated accuracy is $1:10^7$. For a free electron the g-factor is accurately known to be 2x1,001.159.652.1884(43), where the dominating deviation from the Dirac value of 2 is the Schwinger correction, α/π. For a bound electron there are additional corrections, a relativistic correction, first evaluated by Breit [54], and additional radiative and recoil corrections. The leading radiative correction beyond the Schwinger correction has been calculated by Grotch and Hegstrom to be $\alpha(Z\alpha)^2/6\pi$ [55].

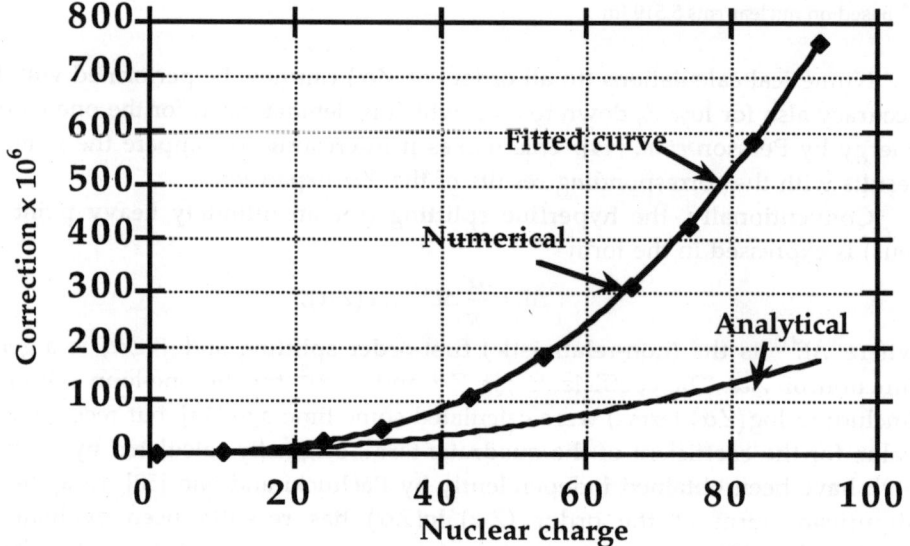

Figure 7. Radiative correction to the g-values of H-like ions. The analytic result is that of Grotch and Hegstrom [55], and the numerical result is obtained by Persson et al. [56].

Persson et al. [56] have recently calculated the radiative corrections numerically for a number of H-like ions to all orders of $Z\alpha$, and the results are displayed in Fig. 7 together with the analytical results of Grotch and Hegstrom. For low Z the numerical results agree well with the earlier predictions by Grotch and Hegstrom, but for high Z there is a substantial deviation. Furthermore, for high Z the calculations show that the uncertainty due to nuclear structure is small and thus strongly motivate the bound g-factor experiment in progress.

5 Summary and Conclusions

In this review we have concentrated on a comparison between some recent experimental and theoretical results for few-electron, high-Z systems, which can be used for testing QED at strong nuclear fields. For such systems, the conventional $Z\alpha$ expansion is no longer applicable, and new all-order techniques have to be applied.

The binding energy in the ground state of H-like uranium, relative to the Dirac value for a point nucleus, has been measured to be 470±16 eV, and this can be well explained by considering the finite nuclear size and the first-order Lamb shift. With somewhat improved experimental accuracy also the second-order (two-photon) Lamb shift could be detected. These effects are not yet fully evaluated but expected to be of the order of a few eV.

The energy separation between the $2p_{1/2}$ and $2s$ states of Li-like uranium has been measured very accurately, and good agreement between theory and experiment is obtained by considering relativistic many-body effects and the first-order Lamb shift. Here, the experimental accuracy is already sufficient for verifying the second-order effects, once they are fully evaluated.

The binding energies of He-like ions have been compared experimentally with the corresponding H-like ions, yielding experimental values of the two-electron contribution to the binding energy of the He-like ions. Recent calculations give good agreement with experiments, although the experimental accuracy is not yet sufficient for testing the QED contributions.

The hyperfine structure of H-like bismuth has recently been measured with high accuracy. Recent calculations including QED effects give good agreement with the experimental result. However, more accurate value for the nuclear magnetic moment is here needed, before the QED contributions can be tested.

Very accurate measurements of the g-factor of highly charged H-like ions are now in preparation. Recent calculations show that the QED effects beyond leading order could be easily detected, also for quite low Z. The effect of the finite nuclear size is here extremely small, and therefore these systems may constitute very good objects for QED test of tightly bound electrons.

The numerical calculations (to all orders of $Z\alpha$) have now reached such a degree of accuracy that they can compete favourably with the results of the $Z\alpha$ expansion also for low Z, particularly for the hyperfine structure and the Zeeman effect. This may have future implications for the determinations of the fundamental constants.

Acknowledgments

Much of the work reported here has been done in collaboration with Walter Greiner and Stefan Schneider, Frankfurt, Gerhard Soff, Günter Plunien and Thomas Beier, Dresden, Leonti Labzowsky and Alexander Mitrushenkov, St. Petersburg. In addition, the authors want to acknowledge the fruitful collaboration with the experimental Atomic Physics group at GSI, Darmstadt, under the leadership of Jürgen Kluge. Furthermore, the financial support of the Swedish Natural Research Council, the Knut & Alice Wallenberg Foundation and the von Humboldt Stiftung is acknowledged.

References

1. For a review, see, for instance, (a) J. Sapirstein and D.R. Yennie, in *Quantum Electrodynamics*, edited by T. Kinoshita (World Scientific, Singapore, 1990) and (b) M. Boshier, in the proceedings of this conference. See also (c) M.I. Eides, S.G. Karshenboim, V.A. Shelyuto, *Physics of Atomic Nuclei*, 57, 1240 (1994); (d) M.I.Eides and V.A. Shelyuto, *Phys. Rev.* A52, 954 (1995); (e) M.I. Eides and H. Grotch, *Phys. Rev.* A52, 3360 (1995); (f) T. Kinoshita and M. Nio, *Phys. Rev.* D53, 4909 (1996); (g) K. Pachucki, *Phys. Rev. Lett.* 72, 3154 (1994); (h) S.G. Karshenboim, *JETP* 82, 403 (1996), *J. Phys.* B29, L29 (1996).
2. H.F. Beyer et al., *Z. Phys.* D35, 169 (1995).
3. J. Schweppe et al., *Phys. Rev. Lett.* 66, 1434 (1991).
4. R.E. Marrs, S.R. Elliott and Th. Stöhlker, *Phys. Rev.* A52, 3599 (1995) and Th. Stöhlker, in the proceedings of this conference.
5. I. Klaft et al., *Phys. Rev. Lett.* 73, 2425 (1994).
6. K. Hermanspahn et al., *Acta Physica Polonica* B27, 357 (1996).
7. P. Mohr, *Phys. Rev.* 88, 26 (1974); *Phys. Rev. Lett.* 34, 1050 (1982); P.J. Mohr and G. Soff, *Phys. Rev. Lett.* 70, 158 (1993).
8. G.E. Brown, J.S. Langer and G.W. Schaefer, *Proc. Roy. Soc.* A251, 92 (1959).
9. A.M. Desiderio and W.R. Johnson, *Phys. Rev.* A3, 1267 (1971).
10. N.J. Snyderman, *Ann. Phys.* A44, 1427 (1991); S.A. Blundell and N.J. Snyderman, *Phys. Rev.* A44, 1427 (1991). S. A. Blundell, *Phys. Rev.* A46, 3762 (1992).
11. H. Persson, I. Lindgren and S. Salomonon, *Physica Scripta* T46, 125 (1993); I.Lindgren, H. Persson, S, Salomonson and A. Ynnerman, *Phys. Rev.* A47, R4555 (1993).
12. H.M. Quiney and I.P. Grant, *J. Phys.* B27, L199 (1994).

13. P. Indelicato and P.J. Mohr, *Theor. Chem. Acta* **80**, 207 (1991).
14. G.E. Brown and D.G. Ravenhall, *Proc. Roy. Soc. London* **A251**, 92 (1951).
15. J. Sucher, *Phys. Rev.* **A22**, 348 (1980).
16. See, for instance, F. Mandl and G. Shaw, *Quantum Field Theory*, Wiley & Sons (1984).
17. P. Indelicato, O. Gorciex and J.P. Desclaux, *J. Phys.* **B20**, 651 (1987); O. Gorciex and P. Indelicato, *Phys. Rev.* **A37**, 1087 (1988), O. Gorciex, P. Indelicato and J.P. Desclaux, *J. Phys.* **B20**, 639 (1987).
18. E. Lindroth and A.-M. Mårtensson-Pendrill, *Phys. Rev.* **A39**, 3794 (1989).
19. J. Sucher, *J. Phys.* **B21**, L585 (1988).
20. I. Lindgren, *J. Phys.* **B23**, 1985 (1990).
21. S. Blundell, P.J. Mohr, W.R. Johnson and J. Sapirstein, *Phys. Rev.* **A48**, 2615 (1993).
22. I. Lindgren, H, Persson, S. Salomonson and L. Labzowsky, *Phys. Rev.* **A51**, 1167 (1995).
23. W.R. Johnson, S.A. Blundell and J. Sapirstein, *Phys. Rev.* **A41**, 4670 (1989).
24. S. Salomonson and P. Öster, *Phys. Rev.* **A40**, 5548, 5559 (1989).
25. H. Persson, I. Lindgren, L. Labzowsky, G. Plünien, Th. Beier and G. Soff, *Phys. Rev.* A 1996 (to appear); H. Persson, S. Salomonson, P. Sunnergren, I. Lindgren and M.H.G. Gustavsson, *Proceedings of the Euro-conference*, June 1996, Baltzer Journals.
26. E.A. Uehling, *Phys. Rev.* **48**, 55 (1935).
27. E.H. Wichmann and H.M. Kroll, *Phys. Rev.* **101**, 843 (1956)
28. G. Soff and P.J. Mohr, *Phys. Rev.* **A38**, 5066 (1988).
29. H. Persson, I. Lindgren, S. Salomonson and P. Sunnergren, *Phys. Rev.* **A48**, 2772 (1993).
30. (a) W.R. Johnson and G. Soff, *Atomic and Nuclear Data Tables* **33**, 405 (1985); (b) P.J. Mohr and G. Soff, *Phys. Rev. Lett.* **70**, 158 (1993).
31. G. Plünien *et al.*, *Phys. Rev.* **A43**, 5853 (1991) and unpublished.
32. A.N. Artemyev, V.M. Shabev and V.A. Yerokhin, *Phys. Rev.* **A52**, 1884 (1995).
33. Th. Beier and G. Soff, *Z. Phys.* **D8**, 129 (1988); S.M. Schneider, W. Greiner and G. Soff, *J. Phys.* **B26**, L529 (1993).
34. I. Lindgren, H. Persson, S. Salomonson, V. Karasiev, L. Labzowsky, A. Mirtushenkov and M. Tokman *J. Phys.* **B26**, L503 (1993).
35. L. Labzowsky and A. Mitrushenkov, *Phys. Lett.* **A198**, 333 (1995).
36. I. Lindgren, H. Persson, S. Salomonson and P. Sunnergren, Phys. Rev A (submitted).
37. A. Ynnerman, J. James, I. Lindgren, H. Persson and S. Salomonson, *Phys. Rev.* **A50**, 4671 (1994).
38. H. Persson, S. Salomonson, P. Sunnergren, and I. Lindgren, Phys. Rev. Lett. **76**, 204 (1966)
39. G. Drake, *Can. J. Phys.* **66**, 586 (1988).

40. H. Araki, *Prog. Theor. Phys.* **17**, 619 (1957).
41. J. Sucher, *Phys. Rev.* **109**, 1010 (1957) and Ph.D thesis, Columbia University, 1957 (unpublished).
42. D.R. Plante, W.R. Johnson and J. Sapirstein, *Phys. Rev.* **A49**, 3519 (1994).
43. P. Indelicato, taken from ref. [4]. See also ref. [17].
44. H. Persson, S.M. Schneider, W. Greiner, G. Soff and I. Lindgren, *Phys. Rev. Lett.* **76**, 1433 (1966).
45. Y. Ting and D. Williams, *Phys. Rev.* **89**, 595 (1989).
46. O. Lutz and G. Stricker, *Phys. Lett.* **35A**, 397 (1971).
47. H. de Vries, C.W. de Jager and C. de Vries, *Atomic and Nuclear Data Tables* **36**, 495 (1987).
48. M. Tomaselli, S.M. Schneider, E. Kankeleit and T. Kühl, *Phys. Rev.* **C51**, 2989 (1995).
49. J. Sapirstein, *Phys. Rev. Lett.* **51**, 985 (1983)
50. K. Pachucki, presented in a poster at this conference and private communication.
51. S.G. Karshenboim, *Z. Phys.* **D36**, 11 (1996)
52. E.R. Cohen and B.N.Taylor, *Rev. Mod. Phys.* **59**, 1121 (1987).
53. A. Hubrich, H. Knab, K.H. Knöll and G. Werth, *Z. Phys.* **D18**, 113 (1991).
54. G. Breit, *Nature* **122**, 649 (1928).
55. H. Grotch, *Phys. Rev. Lett.* **24**, 39 (1970); H. Grotch and R.A. Hegstrom, *Phys. Rev.* **A4**, 59 (1971).
56. H. Persson, S. Salomonson, P. Sunnergren and I. Lindgren, *Phys. Rev. Lett.* (submitted).
57. P.F. Winkler, D. Kleppner, T. Myint and F.G. Walther, *Phys. Rev.* **A5**, 83 (1972).

LAMB SHIFT EXPERIMENTS ON HIGH-Z ONE- AND TWO-ELECTRON SYSTEMS

TH. STÖHLKER

*Institut für Kernphysik, Universität Frankfurt, August-Euler-Straße 6,
D-60486 Frankfurt, and GSI-Darmstadt, Postfach 110552, D-64220 Darmstadt,
Germany*

The heavy-ion storage ring ESR and the trap for highly-charged ions Super-EBIT open up unique possibilities for the study of the effects of quantum electrodynamics (QED) in the domain of strong electric fields. Here, higher-order corrections can be investigated which are not accessible for low-Z ions. In this review special emphasis is given to x-ray spectroscopic investigations of the ground-state transition energies in high-Z hydrogen-like ions up to H-like uranium. By using the intense beams of cooled heavy ions provided by the storage ring ESR, these experiments allow for a precise study of the ground-state binding-energies in high-Z H-like ions where the effects of QED are strongest. Also, the complementary experiments conducted at Super-EBIT are discussed which are aiming at a measurement of the small electron-electron QED corrections, present in the ground state of high-Z helium-like systems.

1. Introduction

Since the early days of quantum mechanics the simple level scheme of atomic hydrogen has provided a stringent testing ground for atomic structure theory. For hydrogen, Lamb and Retherford discovered in 1947 the tiny level spacing between the $2s_{1/2}$ and $2p_{1/2}$ states [1], the so-called Lamb shift, which cannot be explained within the relativistically correct formulation of quantum mechanics, the Dirac theory. This led to the development of modern Quantum Electrodynamics which explains the slightly lowered binding energy of the $2s_{1/2}$ level with respect to the $2p_{1/2}$ state by the interaction of the electron with its own radiation field. The Lamb shift is defined as the difference between the real binding energy and the Dirac-Coulomb energy of an atomic state which for one electron systems is commonly expressed by [2,3]

$$L = \frac{\alpha}{\pi} \frac{(Z\alpha)^4}{n^3} F(Z\alpha) m_0 c^2, \qquad (1)$$

where α is the fine-structure constant, n is the principal quantum number, $m_0 c^2$ is the electron rest mass, and $F(Z\alpha)$ is a slowly varying function of Z. The latter function comprises all the QED corrections and includes in addition the shift in binding energy caused by the finite size of the nucleus. As the Lamb shift scales approximately with Z^4/n^3, these corrections are largest for the ground state and for the strong fields of high-Z ions. The main radiative contributions for the Lamb shift arise from the self energy and the vacuum polarization of the electron bound in the external field of the nucleus, which are represented by the Feynman diagrams depicted in Fig. 1. At low-Z the self energy, i.e. the emission and reabsorption of a virtual photon, gives the most important Lamb shift correction. With increasing nuclear charge, however, the influence of the vacuum polarization, i.e. the virtual creation and annihilation of an electron-positron pair, increases continuously. In Fig. 2 the contributions of the self energy, vacuum polarization,

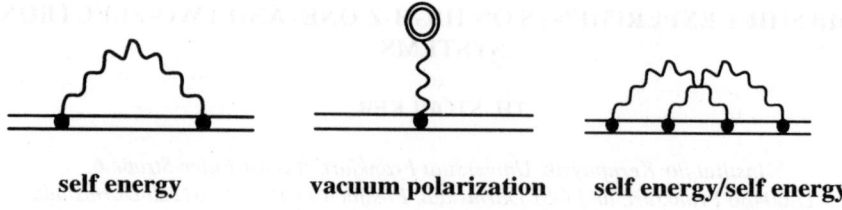

self energy   **vacuum polarization**   **self energy/self energy**

Figure 1: Feynman diagrams for self energy and vacuum polarization. For the case of the higher-order radiative corrections one two-loop diagram is shown in addition.

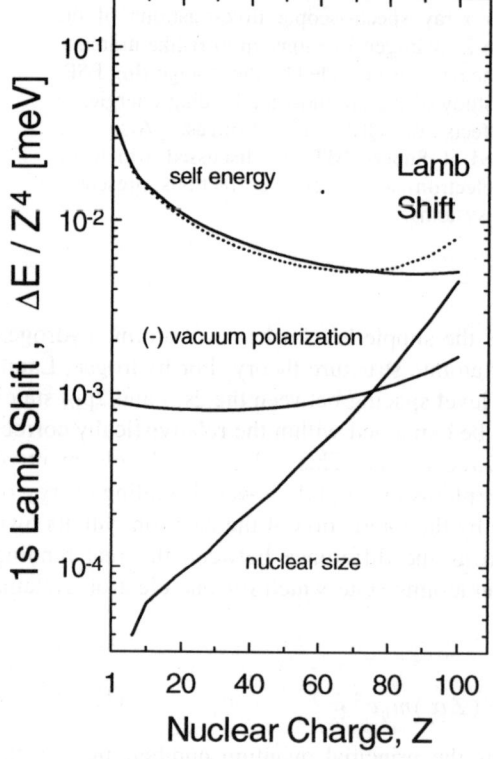
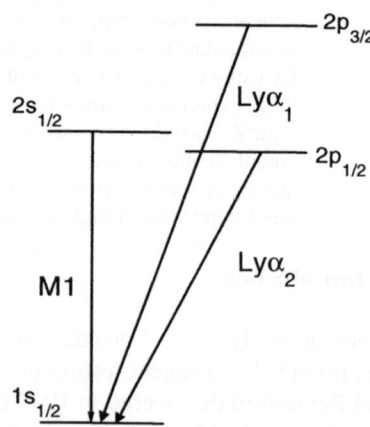

Figure 2: The various contributions to the ground state Lamb shift in hydrogen-like ions as a function of the nuclear charge, according to Ref. [3]. In addition a schematic presentation of the origin of the Lyman-α transitions is given.

and of the finite nuclear size to the Lamb Shift in hydrogen-like atoms are given separately as a function of the nuclear charge. For light one-electron systems such as atomic hydrogen, the theory of QED is now well confirmed with extraordinary precision [4]. Here, the experiments are sensitive to the lower orders of the function $F(Z\alpha)$ which, for low-Z, can be treated by an αZ expansion method. However, for a test of the higher order terms, which are not accessible using low-Z ions, the heaviest species such as H-like uranium are required. At high-Z the influence of the higher-order contributions becomes so important that the radiative corrections can no longer be treated by the αZ expansion method but must be calculated to all orders of αZ [5,6,7].

For high-Z H-like ions, the most direct experimental approach for the investigation of the effects of quantum electrodynamics in strong Coulomb fields is a precise determination of the x-ray energies emitted by transitions from bound (continuum) states into the ground state of the ion. In particular, the Lyman transitions are used in this kind of experiment as they appear most intense and well resolved in the x-ray spectra (the origin of the Lyman transitions is shown in Fig. 2). The goal of the experiments is to achieve a precision which not only tests the higher order αZ contributions but also probes QED corrections which are beyond the one-photon exchange corrections. These correspond to Feynman diagrams such as the two-photon exchange diagrams (see Fig. 1). For the case of uranium, where the total 1s Lamb shift contributes 465.5 eV to the total ground state binding energy of 131.814 keV [8], such a stringent test of QED in strong electric fields requires an absolute experimental accuracy of about 2 eV, which represents the accuracy theoreticans claim presently [6]. For such studies the ESR storage ring at GSI provides favorable experimental conditions. This will be shown in this review, by discussing the series of experiments performed at the ESR gasjet target [9,10] as well as at the electron cooler device [11,12,13] (for a review see also Ref. [14] and Ref. [15]).

Besides the one-electron systems, the two-electron ions are of particular interest as they represent the simplest multi-electron system. Investigations of these ions along the isoelectronic sequence probe uniquely our understanding of correlation, relativistic, and quantum electrodynamical effects. Very recently, the theoretical and experimental investigations of these fundamental systems achieved a considerable improvement in accuracy [16,17,18,19,20,21]. For the ground state the progress is in particular impressive, since even the two-electron QED effects can presently be calculated completely to second order. Here, all the two-photon QED contributions to the electron-electron interaction are considered in this new type of calculation [18,19].

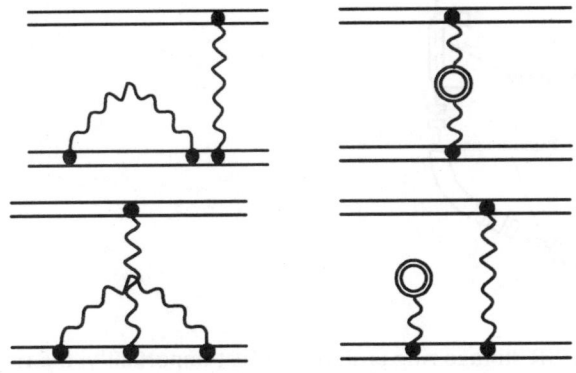

Figure 3: Feynman diagrams for the two-electron Self energy (left side) and Vacuum polarization (right side).

It comprises the so called non-radiative QED part of the electron-electron interaction as well as the two-electron Lamb shift, i.e. the two-electron self energy and the two-electron vacuum polarization (see the Feynman diagrams displayed in Fig. 3). Experimentally, the progress achieved manifests itself by a novel approach where the two-electron contributions to the binding energies in He-like ions can be experimentally isolated [20,21]. This technique exploits the x-ray transitions from the continuum into the vacant K-shell of H- and He-like high-Z ions in order to measure the ionization potentials of He-like species with respect to that of the H-like ions. Here, the potential of this technique will be discussed and the first experiments conducted at the Super-Electron Beam-Ion-Trap (Super-EBIT) in Livermore will be presented.

2. Ground state Lamb shift investigations at the ESR storage ring

The efficient production of bare or H-like high-Z ions is a prerequisite for Lamb shift experiments dealing with high-Z one-electron systems. The most common experimental approach is to use highly charged ions accelerated to high energies e.g. 300 MeV/u, corresponding to $\beta=0.6$, where β denotes the ion velocity in units of the speed of light. At such high energies the ions penetrate through thick solid target foils, such as Cu foils, where all electrons can be stripped off with a high probability. This technique is also exploited for the experiments at the heavy-ion storage ring ESR. Here the bare high-Z ions delivered from the heavy ion synchrotron SIS are injected into the ESR at typical beam energies between 200 and 400 MeV/u. Usually, many beam pulses are stacked and accumulated in the ring. For high-Z ions, e.g. uranium, up to 10^8 ions can be stored by this technique.

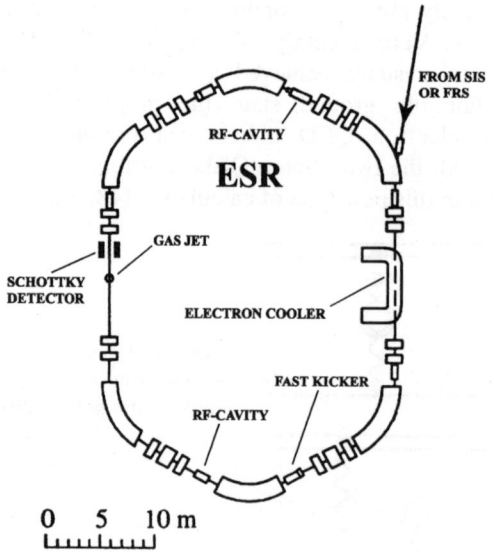

Figure 4: The experimental storage ring ESR.

In Fig. 4 a schematic sketch of the ESR storage ring is shown. Most importantly, the ring with its circumference of 108 m and a magnetic rigidity of 10 Tm is equipped with an electron cooler device. The stored ions are very efficiently cooled by Coulomb interaction with the cold co-moving electrons of the cooler. For this purpose electron currents of typically 100 to 300 mA are applied. This cooling technique guarantees a well defined constant beam velocity, generally of the order of $\Delta\beta/\beta \approx 10^{-4}$. Also it reduces the relative longitudinal momentum spread of the injected ion beam of $\Delta p/p \approx 10^{-3}$ to about 10^{-5}. Moreover, the electron cooling provides a small beam size with a typical diameter of 5 mm and an emittance of the beam of less than $0.1\ \pi$ mm mrad. Besides the cooler device, the storage ring is in addition equipped with a gasjet target (see Fig. 4). Here, various gas targets such as CH_4, N_2, Ar or even heavier targets can be used with densities of about

10^{+13} particles/cm^3 and a diameter of about 5 mm. Both the electron cooler and the gasjet can be applied for an intense production of characteristic Lyman radiation of the circulating high-Z ions. At the electron cooler, the free electrons might be captured by radiative recombination (i.e. the time reversed photo ionization process) into the bare ions, populating excited levels of the H-like species formed by the capture process. By cascades, all such events will lead to the emission of Lyman photons. A similar process occurs at the gasjet target. Here, capture of bound target electrons into the fast moving, bare projectiles populates excited levels of H-like ions which will finally cause the emission of Lyman photons

2.1 X-ray experiments at the electron cooler device and the gasjet target of the ESR

Although the brilliant, monochromatic beams of the ESR provide unique experimental conditions for spectroscopy of the Lyman-α transition energies, the main problem encountered is still the uncertainty introduced by the Doppler shift corrections, because the x-rays are emitted by ions moving with velocities of up to 65% of the speed of light. Due to the relativistic Doppler shift, given by the formula

$$E = E_{lab} \times \gamma \times (1 - \beta \cos \theta_{lab}) \quad (2)$$

(where E and E_{lab} are the x-ray energies in the emitter system and the laboratory frame, respectively, θ_{lab} is the laboratory observation angle, and γ the relativistic factor), the final uncertainty of the x-ray energy in the emitter frame is determined by the uncertainties in the absolute value of β and of the observation angle θ_{lab}. The influence of the latter uncertainties on the final result, however, depend crucially on the beam velocity and the observation angle chosen. This can easily be seen from the derivative of Eq. 2 given by

$$\left(\frac{\Delta E}{E}\right)^2 = \left(\frac{\beta \sin \theta_{lab}}{1 - \beta \cos \theta_{lab}} \Delta \theta_{lab}\right)^2 + \left(\gamma^2 \frac{\cos \theta_{lab} - \beta}{1 - \beta \cos \theta_{lab}} \Delta \beta\right)^2 + \left(\frac{\Delta E_{lab}}{E_{lab}}\right)^2. \quad (3)$$

For instance, due to the sinθ_{lab} dependence, the influence of the uncertainty $\Delta\theta_{lab}$ is completely eliminated at 0° and 180°, but the error due to Δβ is largest. Also, by choosing β = cosθ_{lab} the uncertainty introduced by Δβ can be minimized, however, here the uncertainty introduced by $\Delta\theta_{lab}$ is maximal.

2.2 X-Ray spectroscopy at the electron cooler

The experimental set-up for the measurements of x-ray radiation at the electron cooler device is shown in Fig. 5 [12,13]. At the electron cooler, the ion-beam/electron-beam interaction region is viewed by a solid state Ge(i) detector at an observation angle close to 0.55°. The detector is mounted 4.2 m downstream of the midpoint of 2.5 m long straight electron cooler section which results in a solid angle of $\Delta\Omega/\Omega = 4\times10^{-5}$. The x-rays produced by electron capture into the bare projectiles are recorded in coincidence with the down-charged ions. For this purpose a multi-wire proportional counter (MWPC) is installed behind the first dipole magnet, located downstream from the cooler section. Both, the Ge(i) and the MWPC detector are installed on moveable devices. This allows one to bring the detectors into the final position after the end of the beam stacking procedure.

By using this experimental set-up experiments have been performed for H-like Au^{78+} and U^{91+} at specific beam energies of 298 MeV/u and 321 MeV/u, respectively [11,12,13].

For the experiments, the number of stored bare ions was typically of the order of 5×10^7. As an example, the x-ray spectrum for initially bare uranium ions undergoing electron capture in the cooler is shown in Fig. 6, recorded in coincidence with down charged U^{91+} ions. Due to the observation angle of approximately $0°$, the characteristic Lyman-α transitions, with an x-ray energy of about 100 keV in the emitter frame, are strongly blue shifted and appear at energies close to 220 keV. Note, that the radiative recombinaton process at low relative velocities populates predominantly high-n,l states. Therefore, the $2s_{1/2} \rightarrow 1s_{1/2}$ ground state transitions contribute by only about 5% to the observed Lyα_2 intensity.

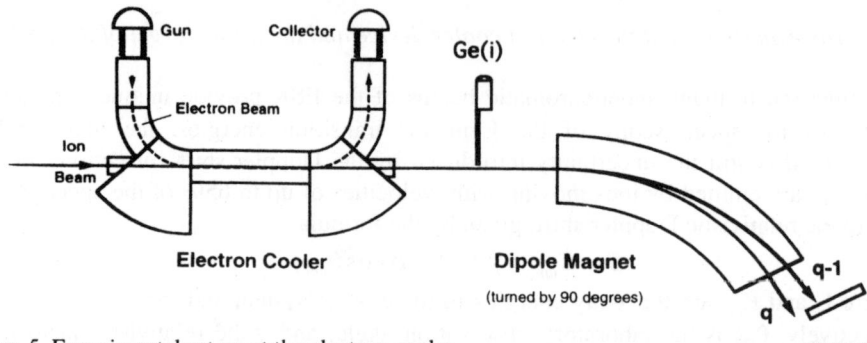

Figure 5: Experimental set-up at the electron cooler.

Besides the Lyman transitions, an additional x-ray line close to 300 keV is observed in the spectrum which is caused by radiative recombination transitions from the continuum state of the free cooler electrons into the vacant 1s-shell of bare uranium. Moreover, in contrast to the prompt RR transitions, distinctive tails on the low-energy side of the Lyman transitions are observed in the x-ray spectrum. They can be explained by the fact that the Lyman lines are induced by cascade transitions from high-n,l states. Such cascades may lead to a delayed Lyman emission, which then takes place within the 3m long distance between the end of the electron cooler and the Ge(i) detector. Such events are measured at observation angles between $0.8°$ and $9°$ which gives rise to an appreciable Doppler shift towards lower x-ray energies [11,12]. However, as these line profiles can be modeled precisely by cascade feeding calculations, no reduction in the final accuracy is caused by these delayed transitions. For the determination of the centroid energies of the Lyman-α transitions and the RR line, respectively, all lines were calibrated carefully by using ^{182}Ta and ^{192}Ir γ-sources. They provide γ-lines which are separated by less than 3 keV from the characteristic projectile transitions. Both the calibration lines and the projectile transitions were fitted by Gaussian functions of the same width, as the Doppler broadening of only about 30 eV can be neglected with respect to the intrinsic detector resolution of about 900 eV. The investigations at the cooler profit from the almost $0°$ observation angle which leads to the extremely small Doppler broadening (note that the uncertainties described by Eq. 3 can also be interpreted as widths) and makes the experiment insensitive to slight uncertainties in the observation angle. However, the experiments are extremely sensitive to $\Delta\beta$, as discussed above. In order to determine precisely the absolute beam velocity three different methods were employed: (1) the measurement of the revolution frequency, (2) the measurement of the high-voltage of the electron cooler, (3) the determination of energy differences between the lines appearing in the x-ray spectrum. All three methods led to a consistent result. For the case of the uranium experiment a β-value of β=0.66884(3) was deduced by the high voltage which turned out to be almost one order of magnitude more precise than the values obtained from the other two methods [11,12,13]. It is important to note that this result agreed within one standard deviation with the value measured by laser-induced recombination for Ar^{18+} ions [22].

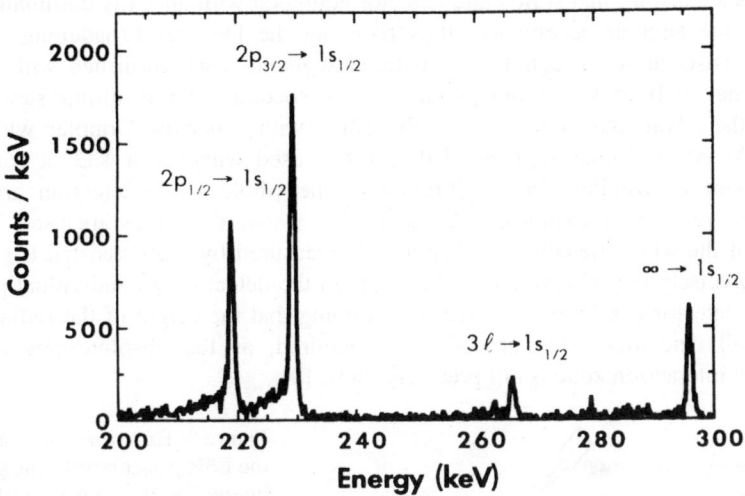

Figure 6: Coincident x-ray spectrum of U^{91+} measured at the electron cooler using a beam energy of 321 MeV/u.

Table 1: Experimental results obtained at the electron cooler device [11,12,13].

	1s binding energy (eV)	1s Lamb shift (eV)	uncertainty in beam velocity (eV)	statistical uncertainty (eV)
Z=79	93257.4 ± 7.9	202.3 ± 7.9	± 4.5	± 3.4
Z=92	131810 ± 16	470 ± 16	± 6.9	± 9.1

The results obtained at the electron cooler for the Lamb shift of Au^{78+} and U^{91+} are up to now the most precise experimental 1s Lamb shift results for high-Z ions. They are summarized in Table 1. In addition the error contributions due to the uncertainties in beam velocity and counting statistics are given separately in the table. Note that both error contributions are summed up linearly for the final uncertainty

2.3 X-ray spectroscopy experiments at the gasjet target

Due to the different geometrical constraints compared to the electron cooler device, the experimental technique applied at the gas target area of the ESR is based on a highly redundant x-ray detection set up. This allows for an intrinsic control of the beam/target/x-ray detector geometry by the simultaneous use of various x-ray detector devices at different observation angles [9,10,23,24]. The experimental set-up for the measurements of the Lyman-α radiation of high-Z ions at the gasjet target is shown in Fig. 7. The figure shows the reaction area which is surrounded by four Ge(i) detectors mounted at observation angles of 48°, 90°, and 132° with respect to the predicted ion-optical beam axis. The two detectors at 48° are installed symmetrically on opposite sides of the reaction chamber. This observation angle is, at high β-values, close to the magic angle $\theta_{lab}= \cos^{-1}(\beta)$ where the experiment is completely insensitive to uncertainties in the beam velocity. However, here the Doppler broadening and the uncertainties introduced by the uncertainty in observation angle are largest (see above). Therefore, one of these 48° detectors is a conventional solid state detector equipped with an x-ray collimator in order to confine the angular acceptance, thus reducing the Doppler broadening. The other detector consists of seven equidistant, parallel segments each furnished with a separate readout. They deliver seven independent x-ray spectra. The resulting sum spectrum combines the advantage of the large solid angle with a narrow Doppler width of one segment. At 90° a similar segmented detector is used whereas at 132° a conventional Ge(i) detector is installed. For calibration of the whole x-ray detection arrangement including its readout electronics, ^{179}Yb and ^{182}Ta standard sources are used. The exact geometry of the whole detector arrangement is measured by laser assisted trigonometry. Knowing precisely the relative angles between all the detectors the individual position of each x-ray detector can be determined by assuming that the origin of the radiation is the same for all detectors. This procedure is required, as the absolute position of the gasjet/beam interaction zone is not precisely know [9].

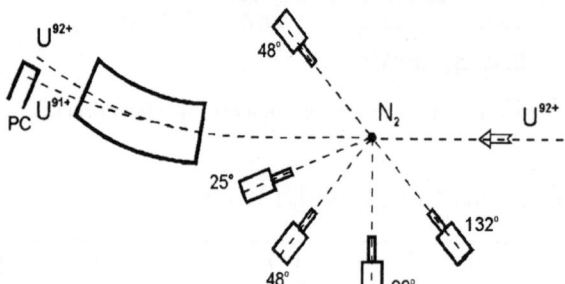

Figure 7: Experimental arrangement at the ESR gasjet target. The gasjet/beam interaction region is viewed by several Ge(i) x-ray detectors. X-rays are measured in coincidence with the down-charged U^{91+} projectiles detected in the particle detector (PC) [23,24].

The x-ray emission produced via electron pickup from the gasjet particles (usually a N_2 gasjet target is used) into the fast moving bare projectiles is registered by the x-ray detector array in coincidence with projectiles having captured one electron. For the latter purpose a fast plastic scintillator for detection of the down-charged U^{91+} ions is used which is located behind the dipole magnet. A typical coincident x-ray spectrum as observed for $U^{92+} \rightarrow N_2$ collisions at 358 MeV/u is shown in Fig. 8 (the spectrum is already transformed into the emitter frame). Beside the well resolved Lyman-α transitions observed, the spectrum is entirely dominated by radiative electron capture (REC) transitions into the projectile, a process very similar to radiative recombination except that bound target electrons are captured instead of free electrons.

In first investigations conducted at the gasjet target, the discussed experimental technique was applied for U^{92+} ions colliding with the N_2 gas-jet particles at 294.7 MeV/u [9]. At the time these experiments were performed the uncertainties in the determination of the beam velocity were rather large ($\Delta\beta/\beta \approx 4\times 10^{-4}$). Moreover, only up to 10^7 ions could be stored in the ring and only one segmented detector at 48° and the conventional detector at 132° were used in these experiments. The accuracy of the final result obtained was determined essentially by counting statistics. The final result of this first Lamb shift experiment performed at the storage ring ESR for U^{91+} is given in table 2 [9]. In addition the error contributions due to the uncertainties in beam velocity, counting statistics, and calibration are given separately in the table. Note, that the individual error contributions are summed up quadratically for the final uncertainty.

The results from the early measurements at the gasjet target suffered essentially from the low counting statistics. Meanwhile the situation has changed drastically, as demonstrated by the experiments performed at the electron cooler and a very recent experiment conducted at the gasjet target. In order to overcome the drawbacks associated with fast moving sources the latter experiment was carried out at various beam energies [23,24].

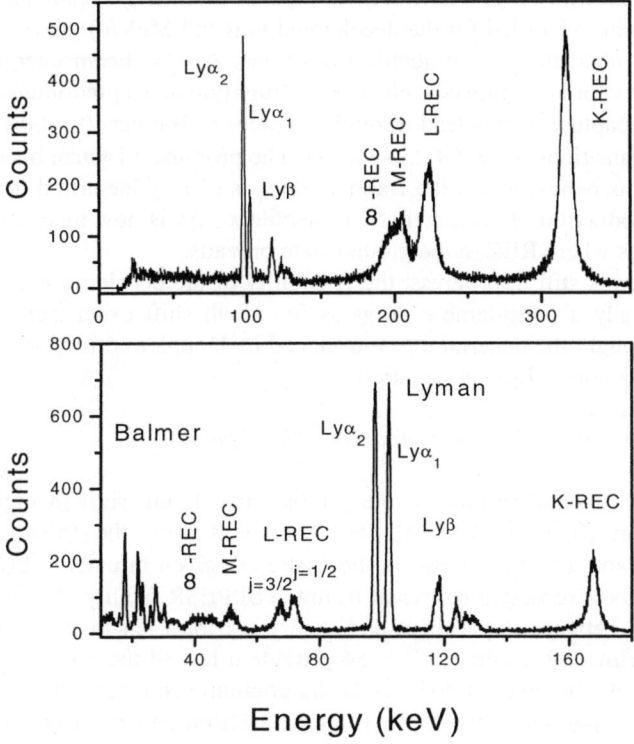

Figure 8: X-ray spectrum (projectile system) of H-like uranium measured at 358 MeV/u (top) in comparison with the one recorded at 68 MeV/u (bottom) [23,24].

Table 2: The result obtained for H-like U^{92+} at the gasjet target [9].

	1s binding energy (eV)	1s Lamb shift (eV)	uncertainty in beam velocity (eV)	statistical uncertainty (eV)	calibration (eV)
Z=92	131851 ± 63	429 ± 63	± 32	± 46	± 30

In particular, the deceleration mode of the ESR storage ring could be applied for the first time. It provides bare uranium ions at moderate energies as low as 49 MeV/u which reduces strongly the uncertainties associated with Lorentz transformation to the emitter system.

Bare uranium ions with an energy of 360 MeV/u delivered from the SIS were injected into the ESR storage ring. After finishing the stacking procedure, which was still performed at the energy of 360 MeV/u, the coasting DC-beam was rebunched and decelerated by simultaneously ramping down the magnetic fields. Subsequently, electron cooling was switched on in order to balance the beam energy loss in the gas target and to fix the ion velocity to the chosen final beam energies of 220, 68, and 49 MeV/u. By applying this procedure up to 2×10^7 ions could be decelerated with losses below 20%. The large reaction cross-sections of the gaseous target reduced drastically the beam lifetime to 5 min at 68 MeV/u and to about 1 min at 49 MeV/u.

The potential of the deceleration capabilities of the ESR is illustrated by the spectra of hydrogen-like uranium also shown in Fig. 8 (bottom). Compared to the high beam energy of 358 MeV/u, the x-ray spectra recorded for the decelerated ions (68 MeV/u) provide an abundant yield of different characteristic projectile transitions. At low beam energies, where non-radiative electron capture dominates, electron capture populates predominantly highly excited levels. Such capture events lead through cascades to Balmer ($n=3,4,... \rightarrow n=2$) as well as to Lyman transitions ($n=2,3,4,... \rightarrow n=1$). The prominent Lyman lines in the spectrum demonstrate this behavior and the Balmer series is clearly identified in the spectrum. Obviously, the production of characteristic projectile x-rays is now much more efficient than at high energies where REC to the ground state prevails.

Although the data evaluation is still in progress, the feasibility of the deceleration mode of the ESR constitutes already a considerable progress for Lamb shift experiments on high-Z ions as it reduces strongly the uncertainties introduced by Doppler corrections and lead to a very efficient production of Lyman radiation.

2.4 Summary of the experimental results and comparison with theory

In Fig. 9, all available experimental results for the ground state Lamb shift in high-Z systems (see Ref. given in [9,10,11,12,13,14]) are compared with the theoretical predictions [3]. For comparison, the data shown in the figure are given in units of $F(Z\alpha)$ (Eq. 1), where the solid symbols represent the result from the SIS/ESR facility. A general excellent overall agreement between experiment and theory can be seen in the figure. However, most of the experimental results for Z > 54 provide a test of the ground state Lamb shift contribution at only the level of 30%. Only the uranium result from the gasjet target (15% sensitivity) and in particular the results from the electron cooler for gold (4% sensitivity) and for uranium (3% sensitivity) have a considerably higher accuracy. The latter data provide already an accurate test of the QED corrections in lowest order α. Since most of the Lamb shift experiments for high-Z ions were performed for hydrogen-like uranium, these data are given in addition separately in the inset of Fig. 9 in comparison with the theoretical prediction (the data collected at the ESR are represented by solid symbols). The figure demonstrates the substantial improvement by almost one order of magnitude achieved at the storage ring compared to the former experiments conducted at the BEVALAC accelerator [25,26]. For the particular case of H-like uranium the various theoretical contributions to the 1s Lamb shift are quoted separately in table 3. The resulting 1s Lamb shift of 465.5 eV ± 2eV [6,8] has to be compared with the experimental value from the electron cooler of 470 eV ± 16 eV [11,12]. The largest theoretical uncertainty arises from higher-order QED contributions which are not yet calculated. However, an experimental test of these contributions still requires a further improvement of the experimental accuracy by almost one-order of magnitude in order to approach an absolute precision of 1eV. Note, that the finite nuclear size effect contributes more than 40% to the total Lamb shift correction. Although the uncertainties introduced by the latter effect are much smaller than the present experimental accuracy, they may prevent a direct test of QED in high-Z systems below 1 eV.

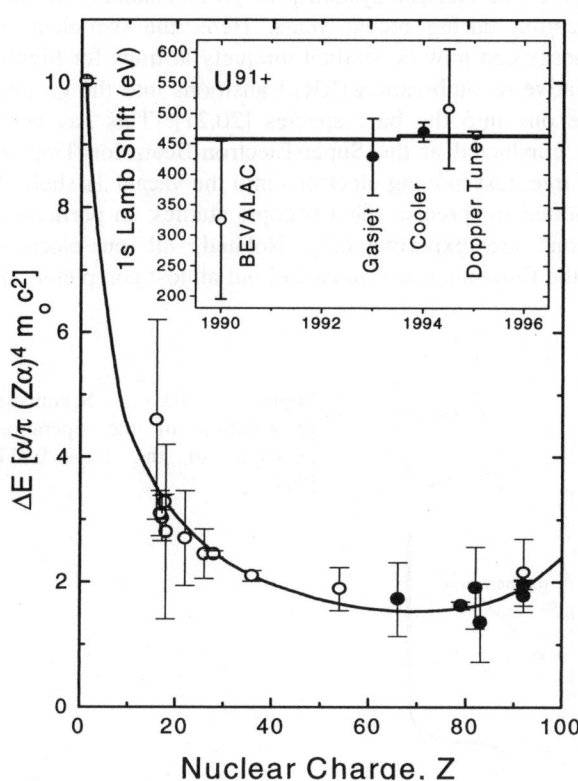

Figure 9: All available experimental results for the 1s Lamb shift in high-Z ions in comparison with theoretical predictions [3]. In the inset the available data for U^{91+} are depicted, where the solid points refer to the experiments performed at the ESR. The solid line shows the theoretical predictions of Ref. [3] and Ref. [6], respectively.

Table 3: The various theoretical contributions to the 1s-binding energy in H-like ^{238}U which lead to deviations from the Dirac eigenvalue for the point like nucleus of -132279.98 eV. All values are taken from Ref. [6]. Note, the reduced mass correction is by convention a non Lamb shift correction. The total theoretical binding energy is -131814 ± 2 eV.

Effect	Theoretical value (eV)
Nuclear size effect	$+ 198.7 \pm 0.3$
reduced mass correction	$+ 0.3$
relativistic recoil	$+ 0.2 \pm 0.1$
Nuclear polarization	$- 0.2 \pm 0.1$
Self energy	$+ 355.0$
Vacuum polarization	$- 88.6$
Calculated higher order QED	$+ 0.3$
Uncalculated higher order QED	± 2
Lamb shift (sum of all effects except of the reduced mass)	465.5 ± 2

3. Experiment on high-Z He-like ions conducted at the Super-EBIT

Besides the 1s binding energy in high-Z one-electron systems, the ground state of He-like high-Z ions gained increasing attention during recent years. Here, the two-electron contributions to the ground state energy can now be studied uniquely at traps for highly charged ions by measuring the radiative recombination (RR) transitions into the ground state of H-like ions relative to the one into the bare species [20,21]. This has been demonstrated in a first experiment conducted at the Super-Electron-Beam-Ion-Trap in Livermore, where RR transitions of free, fast moving electrons into the vacant 1s shell of bare and H-like ions has been exploited for precise spectroscopic studies. In particular, since the two-electron contributions are experimentally isolated, all one-electron contributions such as the effects of the finite nuclear size cancel out almost completely in this type of experiment [19].

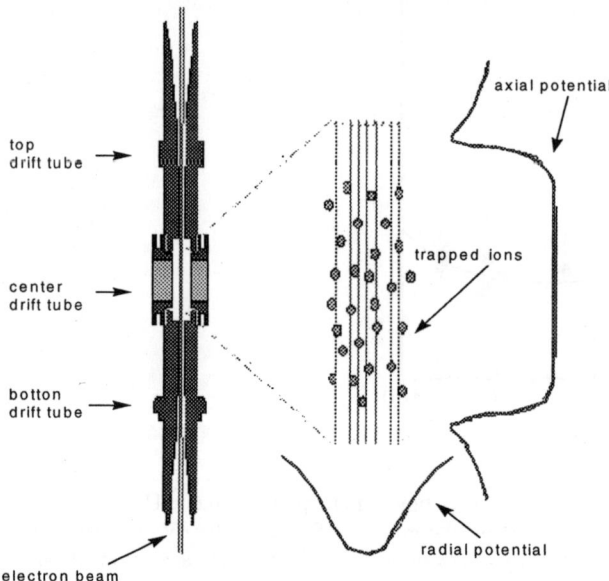

Figure 10: Schematic presentation of the operation principle of the Super-EBIT [27].

A schematic presentation of the operation principle of Super-EBIT is depicted in Fig. 10 [27]. At Super-EBIT bare and H-like ions of almost any element can be produced in an electron beam of arbitrary energy up to 200 keV and currents up to 200 mA [28]. Here, the ions are injected into the trap either by a metal vapor vacuum source or by injection of neutral gas. Simultaneously to the production of the highly-charged ions by electron impact, the ions are trapped radially by the space charge potential of the electron beam. For the latter purpose also a magnetic field of three Tesla is applied, which in addition compresses the electron beam to a typical diameter of 50 μm. In the axial direction the trapping is achieved by electrostatic potentials applied to the end drift tubes of the EBIT device. Moreover, to avoid a too strong heating of the stored ions by the electron-ion collisions, evaporative cooling is applied in the trap by injection of neutral Ne gas. In contrast to a storage ring, in the S-EBIT the ions are stationary in the laboratory system and normally several charge states are confined simultaneously in the trap. With such collision conditions, the fast moving free electrons may undergo a direct radiative recombination transition into the vacant K-shell of the bare and H-like species. Since RR is the time reversal of the photoelectric effect, the energy carried away by the photon is just given by

$$\hbar\omega = E_{kin} + V \qquad (4)$$

where E_{kin} denotes the kinetic energy of the electron captured and V is the ionization potential of the ionic system after undergoing radiative capture. Since both the bare and H-like ions are simultaneously trapped, i.e. both ion species are interacting with the same electron beam, the difference in the photon energies between radiative transitions into the bare and H-like ions is independent of the electron beam energy. It corresponds just to the difference in the ionization potentials between the H- and He-like species formed by the RR process which gives exactly the two-electron contribution to the ground state binding energy in He-like ions. A schematic presentation of this experimental situation at the EBIT is shown in Fig. 11. By applying the experimental method described above, data were obtained for Ge (Z=32), Xe (Z=54), Dy (Z=66), W (Z=74), Os (Z=76) and Bi (Z=83). For the experiments, x-ray spectra were taken by a solid state Ge(i) detector viewing the electron-beam/ion interaction zone through Be windows. They were saved in intervals of two hours in order to exclude the influence of possible electronic instabilities and drifts of the electron beam energy (for a detailed description of the data evaluation applied see Ref. [20]). In Fig. 12, sample spectra for the x-ray regime of RR into the vacant K-shell of germanium, xenon and bismuth are given separately. In all cases, the RR line splitting between RR into the bare and H-like ions appears well resolved.

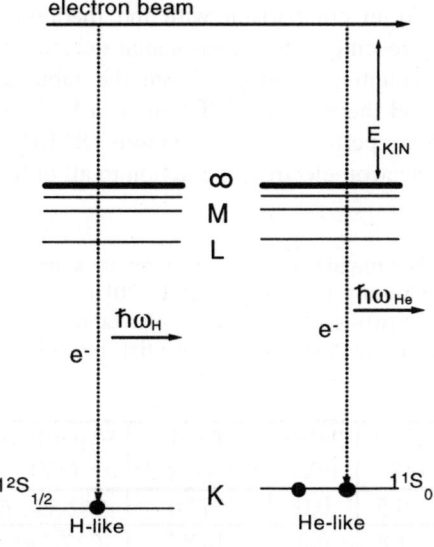

Figure 11: Schematic presentation of the RR process of free electrons into the initally bare and H-like ions. The energy difference $\hbar\omega_H - \hbar\omega_{He}$ gives exactly the two-electron contribution to the ionization potential in He-like ions.

Note that in contrast to fast ion-beam experiments the Doppler shift as well as the Doppler broadening are completely absent in this type of experiment. Here the width of the transition lines is almost completely determined by the intrinsic resolution of the Ge(i) detectors used. As observed in the spectra, the relative intensity of the peak from bare ions changes drastically between the two elements as the K-shell ionization cross section varies rapidly, as $1/Z^4$. Up to now, this has prevented the extension of the experiments to elements beyond Z=83.

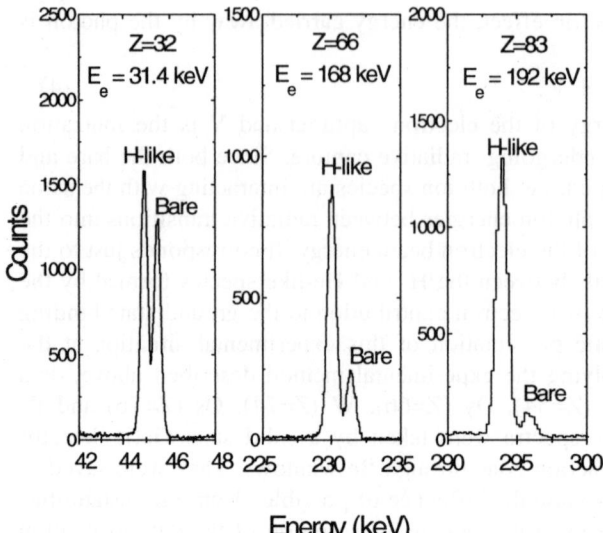

Figure 12: Sample K-shell RR line recorded for germanium (Z=32), dysprosium (Z=66), and bismuth (Z=83) at the Super-EBIT [20].

3.1 Comparison with Theory

The experimental results obtained for the two-electron contributions to the binding energy of some He-like ions are given in Table 4 [20] in comparison with the theoretical calculations of Persson et al. [19] performed very recently. The experimental uncertainty quoted in the table is entirely determined by counting statistics. From the table an excellent agreement between experimental data and theoretical predictions can be seen. The predictions are based on relativistic many-body perturbation calculations (RMBPT) which take into account the non-QED part of the electron-electron interaction to all orders [19].

Table 4: The individual two-electron contributions to the ground state binding energy in some He-like ions [19] in comparison with the experimental results from Super-EBIT [20] (NR: non-radiative QED as defined by Persson et al. [19]; 2eVP: two-electron vacuum polarization; 2eSE: two-electron self energy; Total theory: predicted difference in the ionization potentials between the H- and the He-like system).

Nuclear Charge	1st order RMBPT(eV)	$\geq 2^{nd}$ order RMBPT(eV)	NR (eV)	2eSE (eV)	2eVP (eV)	Total theory (eV)	Experiment (eV)
32	567.61	- 5.20	0.0	- 0.5	0.0	562.0	562.5±1.6
54	1036.56	- 7.01	0.2	- 1.8	0.2	1028.2	1027.2±3.5
66	1347.45	-8.56	0.4	- 3.2	0.6	1336.6	1341.6±4.3
74	1586.93	- 9.87	0.6	- 4.6	0.9	1573.9	1568±15
83	1897.56	- 11.73	0.9	-6.7	1.6	1881.5	1876±14

In particular, the two-electron QED contributions are considered for the first time complete to second order in α. These include the non-radiative QED part as defined by Persson et al. [19] as well as the two-electron Lamb shift, i.e. the two-electron self energy and the two-electron vacuum polarization (see Fig. 3). Note, that compared to the QED calculations for high-Z H-like systems, where some higher-order QED effects are still uncalculated, the claimed theoretical uncertainty for the two-electron QED contributions is very small and, for the particular case of He-like uranium, estimated to be of the order of only 0.1 eV. Most importantly, as has been shown in detail by Persson et al. [19], the two-electron QED effects are almost completely unaffected by the uncertainties of the nuclear charge radius, one of the most serious limitations for the QED tests in high-Z one-electron systems. As can be deduced from the experimental and theoretical results presented in Table 4 the experimental data provide already a meaningful test of the many-body non-QED part of the electron-electron interaction. Moreover, the data are already at the threshold of a sensitive test of the two-electron QED contributions. For this purpose only a moderate increase in accuracy of about half an order of magnitude is required.

4 Summary and outlook

In conclusion, the favorable experimental conditions for ground state QED studies of high-Z H-like ions were demonstrated within the first series of experiments performed at the ESR gasjet target as well as at the electron cooler device. For the case of the 1s-Lamb shift in hydrogen-like uranium, the achieved accuracy of ±16 eV is already a substantial improvement by almost one order of magnitude compared with former experiments conducted at the BEVALAC accelerator [25,26], and the available results from the ESR are now at the threshold of a real test of higher-order QED contributions. By using decelerated beams, further progress towards an absolute accuracy of 1 eV may be anticipated. As has been demonstrated in the first experiment with decelerated ions, the deceleration mode not only reduces the uncertainties in the Doppler corrections but it provides in particular a very efficient production of characteristic projectile radiation. Along with a strongly enhanced injection efficiency into the ESR, which is expected in the near future, it will allow one to implement high-resolution x-ray detection devices such as crystal spectrometers [29] or bolometers [30] which presently are under construction.

For the investigation of the ground state binding energy in He-like high-Z ions a novel technique has been introduced at the Super Electron Beam Ion Trap. The achieved experimental accuracy, which is only limited by counting statistics, already provides a meaningful test of the many-body part of the theory. In particular, only an improvement of half an order of magnitude is required for a test of the two-electron QED effects which, in contrast to the H-like systems, are almost unaffected by the finite nuclear size. For the latter purpose a more intense EBIT is required which is presently under construction [31] and also the ESR storage seems to be well suited for such investigations in the future[21].

Acknowledgment

The work at the ESR was done in collaboration with H.F. Beyer, F. Bosch, R.D. Deslattes, R.W. Dunford, A. Gallus, P. Indelicato, C. Kozhuharov, D. Liesen, G. Menzel, P.H. Mokler, H.T. Prinz and with the members of the ESR team, H. Eickhoff, B. Franzke, F. Nolden, P. Spädtke, and M. Steck. The experiments at LLNL Super-EBIT were done in collaboration with S.R. Elliott and R.E. Marrs. The author would like to thank H.-J. Kluge, L.N. Labzowsky, I. Lindgren, H. Persson, D. Schneider and G. Soff for stimulating discussions and A.E. Livingston and V.M. Shabaev for their helpful comments on the manuscript.

References

[1] W.E. Lamb and R.C. Retherford, Phys. Rev. **72**, 241 (1947).
[2] P.J. Mohr, At. Data and Nucl. Data Tables **29,** 453 (1983).
[3] W.R. Johnson and G. Soff, At. Data and Nucl. Data Tables **33**, 405 (1985).
[4] S. Bourzeix, B. de Beauvoir, F. Nez, M.D. Plimmer, F. de Tomasi, L. Julien, F. Biraden, D.N. Stacey, Phys. Rev. Lett. **76**, 384 (1996).
[5] P.J. Mohr, Phys. Rev. A **46,** 4421 (1993).
[6] H. Persson , S. Salomonson, P. Sunnergren, I. Lindgren, M.G.H. Gustavsson, submitted to Hyperfine Interaction (1996).
[7] I. Lindgren, see contribution to this conference.
[8] H. Persson, I. Lindgren, L. Labzowsky, G. Plunien, T. Beier, G. Soff, Phys. Rev. A, in print (1996).
[9] Th. Stöhlker, P.H. Mokler, K. Beckert, F. Bosch, H. Eickhoff, B. Franzke, M. Jung, T. Kandler, O. Klepper, C. Kozhuharov, R. Moshammer, F. Nolden, H. Reich, P. Rymuza, P. Spädtke, M. Steck, Phys. Rev. Lett. **71**, 2184 (1993).
[10] P.H. Mokler, Th. Stöhlker, C. Kozhuharov, R. Moshammer, P. Rymuza, F. Bosch, T. Kandler, Physica Scripta Vol. T51, **28** (1994).
[11] H.F. Beyer, D. Liesen, F. Bosch, K.D. Finlayson, M. Jung, O. Klepper, R. Moshammer, K. Beckert, H. Eickhoff, B. Franzke, F. Nolden, P. Spädtke, M. Steck, G. Menzel, R.D. Deslattes, Phys. Lett. **A184**, 435 (1994);
[12] H.F. Beyer, IEEE Trans. Intrs. Meas. **44**, 510 (1995).
[13] H.F. Beyer, G. Menzel, D. Liesen, A. Gallus, F. Bosch, R. Deslattes, P. Indelicato, Th. Stöhlker, O. Klepper, R. Moshammer, F. Nolden, H. Eickhoff, B. Franzke, M. Steck, Z.Phys. **D35**, 169 (1995)
[14] D. Liesen, H.F. Beyer, and G. Menzel, Comments At. Mol. Phys. **32**, 23 (1995).
[15] H.F. Beyer, H.-J. Kluge, V.P. Shevelko, 'X-ray Radiation of Highly Charged Ions' (Springer, Berlin, Heidelberg, 1996, in print).
[16] D.R. Plante, W.R. Johnson, J. Sapirstein, Phys. Rev. **A49**, 3519 (1994).
[17] I. Lindgren, Phys. Scripta **T59**, 179 (1995).
[18] V.A Yerkhin, V.M. Shabaev, Phys. Lett. **A207**, 274 (1995), **A210**, 437 (1996).
[19] H. Persson, S. Salomonson, P. Sunnergren, and I. Lindgren, Phys. Rev. Lett. **76**, 204 (1996).
[20] R.E. Marrs, S.R. Elliott, and Th. Stöhlker, Phys. Rev. **A52**, 3577 (1995).
[21] Th. Stöhlker, S.R. Elliott, R.E. Marrs, Hyperfine Interactions **99**, 217 (1996).
[22] S. Borneis, St. Becker, T. Engel, I. Klaft, O. Klepper, A. Kohl, T. Kühl, D. Marx, K. Meier, R.Neumann, F. Schmitt, P. Seelig, L. Völker, Proc. Resonance Ionization Spectroscopy 1994, H.J. Kluge, J.E. Parks, K. Wendt (eds.). NY AIP (1995).
[23] P.H. Mokler, Th. Stöhlker, R.W. Dunford, A. Gallus, T. Kandler, G. Menzel, T. Prinz, P. Rymuza, Z. Stachura, P. Swiat, A.Warczak, Z. Phys. D **35**, 77 (1995).
[24] Th. Stöhlker, GSI-Nachrichten, 05-95, (1995).
[25] J.P. Briand, P. Chevallier, P. Indelicato, K.P. Ziock, and D. Dietrich, Phys. Rev. Lett. **65**, 2761 (1990).
[26] J.H. Lupton, D.D. Dietrich, C.J. Hailey, R.E. Stewart, K.P. Ziock, Phys. Rev. **A50**, 2150 (1994).
[27] R.E Marrs, Comments At. Mol. Phys., Vol. **27** No. 2, 57 (1991).
[28] R.E. Marrs, S.R. Elliott, D. Knapp, Phys. Rev. Lett. **72**, 4082 (1994).
[29] H.F. Beyer, *Physics with Multiply Charged Ions*, Edited by D. Liesen, Plenum Press, New York (1995).
[30] P. Egelhof, H.F. Beyer, D. McCammon, F. v. Feilitzsch, A. v. Kienlin, H.-J. Kluge, D. Liesen, J. Meier, S.H. Moseley, Th. Stöhlker, Nucl. Instr. Meth. **A370**, 26(1996).
[31] R.E. Marrs, private communication (1996).

FUNDAMENTAL CONSTANTS OF NATURE

L.B. OKUN
ITEP, Moscow, 117218, Russia

A brief review, from basic atomic constants to "Mendeleev Table" of leptons, quarks, fundamental bosons, and then to superunification of all forces and particles.

1 Constants of atomic physics

The discovery in 1896 of Pieter Zeeman, which we are celebrating today, was a great step in unveiling the structure of atoms. At the same time it was a great step in measuring the fundamental constants of Physics. As was shown by H. Lorentz, the Zeeman splitting was determined by the ratio e/mc, where e and m, the charge and the mass of electron, and c, the velocity of light, are three out of the four fundamental constants of atomic physics. The fourth constant, $\hbar$, was introduced by Max Planck in 1900. (I am using the modern notations and terminology.) The fundamental constants $\hbar, e, m$ are the natural units for atomic physics. They determine the size and the energy levels of the hydrogen atom (but not its mass, which, as for any other atom, is determined by the mass of the nucleus). Three decades later this led to Nonrelativistic Quantum Mechanics. (An additional important ingredient was spin, the Pauli principle that explained the Mendeleev Table.)

Already in the original interpretation of the Zeeman effect by Lorentz an important role belonged to the velocity of light, which enters the expression of the Lorentz force. Were $\hbar, e, m$ the same as they are, but the velocity of light were infinite, the atoms would not emit and absorb light, and there would be no Zeeman splitting. In this sense atomic physics cannot be considered to be non-relativistic. Note that fine structure constant involves c:

$$\alpha = e^2/4\pi\hbar c$$

During the XX century the constants $\hbar, c$ took deep roots in Physics and have fundamentally changed its very basis. The electric charge e has been joined by the weak and strong charges. As for the mass of the electron, m, it turned out to be one of a whole constellation of fundamental masses.

2 QED, leptons and hadrons

The Dirac equation combining electron and positron opened a new chapter of Physics – the Relativistic Quantum Mechanics, which dealt with what we now call Feynman tree diagrams. Twenty years later, in the middle of the century, the Feynman loops became manageable and the QED (the Quantum Electrodynamics) arose, beautiful, as the Venus of Botticelli.

But it was clear that this beauty was not alone on the painting. Since early 1920's protons and since early 1930's neutrons and neutrinos were known, After the World War II new particles have been discovered. They belonged to two different groups: leptons and hadrons. Leptons are: electron, its neutrino and their relatives. The first leptons, identified after the war were muons. Hadrons are: proton, neutron and their relatives. The first hadrons discovered after the war were pions. Soon they were joined by a crowd of other strange creatures: strange mesons, hyperons, resonances. Botticelli was impetuously transforming into a Bosch.

A great relief and order was brought by three ideas:

1. that all hadrons are particles composed of a few building blocks (sakatons – in the 1950's, quarks – after 1964);

2. that in addition to the electromagnetic interaction, there are only two other interactions behind all this Boschian chaos: the strong and the weak one;

3. that the source of strong interaction are three basic, so called colour charges, whilst the source of weak interaction are two basic weak charges.

At present the "Mendeleev table" of basic elements consists of 16 particles, not counting antiparticles and colour degrees of freedom (colour charges).

3 The "Mendeleev table" of fundamental particles

The 16 basic elements are subdivided into two groups: 4 basic bosons with spin one, and 12 basic fermions with spin 1/2.

The four bosons are carriers of four forces:

γ	(photon)	electromagnetic force	$\alpha = e^2/4\pi\hbar c$
W		weak force for charged currents	$\alpha_W = f_W^2/4\pi\hbar c$
Z		weak force for neutral currents	$\alpha_Z = f_Z^2/4\pi\hbar c$
g	(gluon)	strong force	$\alpha_s = g^2/4\pi\hbar c$

The main difference between photon and Z and W bosons is that photon is massless, while $m_Z = 91$ GeV, $m_W = 80$ GeV.

The main difference between photon and gluon is that photon is single and electrically neutral, while there exist eight gluons carrying eight different combinations of colour charges, and emitting and absorbing themselves. The result of this self-interaction is the phenomenon of confinement of coloured gluons and quarks inside snow-white hadrons. The forces between hadrons are not the basic ones, they are secondary and resemble the Van der Waals and chemical forces between atoms.

Twelve fermions are subdivided into three generations – two quarks and two leptons in each:

		1st	2nd	3rd	Q
quarks		u	c	t	2/3
		d	s	b	-1/3
leptons		ν_e	ν_μ	ν_τ	0
		e	μ	τ	-1

Each electrically charged fermion has its antiparticle. It may be, that the same is true for neutrinos, but it is also possible that neutrinos, like photons, have no antiparticles: each neutrino is its own antiparticle. Another unsolved problem, whether neutrinos are massless or have nonvanishing masses.

What are the roles of the three fermionic generations? The atomic shells are made of electrons, the atomic nuclei are made of the u and d quarks held together by gluons inside protons and neutrons: $p = uud$, $n = ddu$. Electronic neutrinos are needed for weak reactions in the sun and stars. As a result

$$2e^- + 4p \to {}^4\text{He} + 2\nu_e + 27 \text{ MeV} .$$

Without electronic neutrinos there would be no sun and hence we would not exist. Thus, the first generation of basic fermions is absolutely necessary for the existence of our world.

The second and third generations seem, at first sight, to be absolutely useless. But, maybe, they were essential in the first nanoseconds of the Big Bang by preventing full annihilation of protons and electrons into neutrinos and photons. Maybe, they had (and have?) some other functions. They definitely played an important role in the history of physics. The study of strange particles (containing s-quark) lead to the discovery of quarks and to the discovery of violation of P, C, CP and T symmetries in weak interactions, which lead to unification of electromagnetic and weak interactions into one electroweak interaction. In accordance with this unified theory (in the Born approximation, i.e., neglecting electroweak radiative corrections):

$$\frac{m_W^2}{m_Z^2} = \frac{\alpha_W}{\alpha_Z} = 1 - \frac{\alpha}{\alpha_W} .$$

One of the key elements of the electroweak theory is the Z boson. Let us note that the experimental study of $2 \cdot 10^7$ Z boson events at LEP I collider (CERN) has proved that there are only three light (or massless?) neutrinos. Thus, new particles help to understand the old ones.

The last free box in the Table of basic fermions has been filled in only two years ago, when the heaviest quark t was discovered at the Tevatron collider (FNAL). The mass of this quark is 175 ± 15 GeV.

4 The higgs and the origin of mass

It might sound strange, but the value of the top mass is the most natural one of all leptons and quarks. In order to see this, let us consider the so-called Higgs mechanism, that is used in electroweak theory to generate masses of fundamental particles. At the basis of this mechanism lies the (still hypothetical) Higgs field, the quantum excitations of which are neutral scalar (spinless) bosons – higgses. The mass of the higgs is unknown at present. In the most popular scenario higgs is heavier than Z boson but lighter than top quark. The search for the higgs is the major priority of a new e^+e^- collider LEP II and of the future Large Hadron Collider (LHC) at CERN.

Higgs field is coupled to all massive particles, the value of the coupling constants being proportional to the particles masses. They are called Yukawa coupling constants.

The unique feature of the Higgs field is that it has a non-vanishing vacuum expectation value (VEV) $\eta = 250$ GeV throughout the world. Mass of a fermion is a product of its Yukawa coupling times η. Masses of W and Z bosons are $g_W \eta/2$ and $g_Z \eta/2$ respectively. The mass of the top quark is the most natural one in the sense that its Yukawa coupling is of the order of unity.

5 Running of α_s and confinement

For $\eta = 0$ all fundamental bosons and fermions would become massless. This however does not refer to hadrons. Most of them would remain massive even if the quarks were massless. For instance, the masses of the proton and neutron would be practically the same, as they are. This conclusion is deeply connected with the phenomenon of confinement and with the running of the coupling constant α_s.

According to quantum field theory the values of all charges, of all coupling constants depend on distance (or momentum, or energy). The constants are changing with these variables because of vacuum polarization. The famous

$$\alpha = 1/137.0359895(61)$$

is in fact the value of $\alpha(q^2)$ at a vanishing momentum transfer: $q^2 = 0$. In the interval from 0 to m_Z α increases from $1/137$ to $1/129$. The α_W and α_Z, in the same interval, change very little, they "crawl":

$$\alpha_W(0) = 1/29.01 \quad , \quad \alpha_W(m_Z) = 1/28.74$$
$$\alpha_Z(0) = 1/23.10 \quad , \quad \alpha_Z(m_Z) = 1/22.91$$

According to Quantum Chromodynamics (QCD) the behaviour of α_s is totally different; α_s runs in the opposite direction and runs fast:

$$\alpha_s(m_Z) \simeq 0.12 \quad , \quad \alpha_s(1\,\text{GeV}) \simeq 1 \quad ,$$

and it would "blow up" at smaller momentum transfers, or distances larger than the radius of confinement, if it were possible to separate unscreened colour charges by such distances. The non-perturbative strong self-interaction of gluons, and their interactions with quarks produces gluon and quark condensates with characteristic energy scale $\Lambda_{\text{QCD}} \approx 300$ MeV. It is Λ_{QCD} that sets the scale of masses of hadrons built from light quarks (u, d) and gluons.

6 Symmetries and grand unification

Up to this point I tried to avoid mentioning symmetries and groups, using physical, rather than mathematical, language. But in order to understand the essence of physics one has to appreciate its mathematical beauty, the beauty of symmetries. First of all, special relativity is represented by Poincaré group. Second, QCD is represented by a local SU(3) colour symmetry with gluons as quanta of gauge fields of this symmetry. Third, electroweak theory is described by SU(2)×U(1) gauge symmetry, which is spontaneously broken to U(1)$_{em}$ by the higgs VEV. Unification of all three types of interactions is expected to be based on a higher broken gauge symmetry described by such groups as unitary group SU(5), orthogonal group SO(10) or exceptional group E$_6$, which contain SU(3) and SU(2)×U(1) as their subgroups. This idea of grand unification finds strong support in the fact that the three gauge coupling constants α_s, α_W and α (the latter with a proper coefficient 8/3), being so different at low energies, tend to a single meeting point, at $E_{GU} \sim 10^{16}$ GeV, where all of them have the same value of the order of 1/30.

The fermionic multiplets of higher groups contain both leptons and quarks. For instance, in the case of SO(10) each generation of fermions (with account of antiparticles and of three colours of quarks) forms a 16-plet. Among the 45 vector bosons of SO(10) there are bosons with such couplings, that their exchange leads to the proton decay into a positron (or antineutrino) plus accompanying light hadrons (mesons). Another baryon number violating interaction produces decays of nuclei, in which two neutrons transform into mesons; it also transforms neutron into antineutron in vacuum.

The above decays of nuclei have lifetimes longer than 10^{32} years, because the corresponding bosons are very heavy: their masses are of the order of 10^{16} GeV. The search for such decays is one of the highest priorities of the new gigantic underground detector Super Kamiokande.

7 SUSY and superstrings

A symmetry, which might be broken not so badly as grand unification symmetry, is supersymmetry, or SUSY. According to SUSY, there exist at least one superpartner for each particle we already know. In this minimal case there exist bosonic analogues of leptons and quarks (sleptons and squarks with spin 0), and fermionic analogues of bosons (photino, gluino, zino, wino and higgsino with spin 1/2). The lighter of these superparticles may be discovered at LEP II and LHC. The lightest of them might be stable and constitute a substantial part of the so-called dark matter. It is interesting that Feynman loops of superpartners help to focus more accurately the three running gauge couplings at the grand unification point.

The energy of grand unification is only four orders lower than the Planck mass, m_P, introduced into physics by Planck when he discovered the quantum of action:

$$m_P = \left(\frac{\hbar c}{G}\right)^{1/2} = 1.2 \cdot 10^{19} \text{ GeV} \simeq 2.2 \cdot 10^{-5} \text{ grams} ,$$

where G is the gravitational (Newtonian) constant: $G = 6.6720(41) \cdot 10^{-8} \cdot \text{cm}^3 \cdot \text{g}^{-1} \cdot \text{sec}^{-2}$. The Planck length, l_P, and Planck time, t_P, were introduced in the same paper:

$$l_P = \frac{\hbar}{m_P c} = 10^{-33} \text{ cm.}$$

$$t_P = \frac{\hbar}{m_P c^2} = 3 \cdot 10^{-44} \text{ sec.}$$

At energies of the order of m_P, or distances as short as l_P, the energy of gravitational interaction becomes of the order of the total energy and quantum effects become important. This is the realm of quantum gravity.

The quantum of excitation of gravitational field is called graviton. It is massless, neutral and has spin 2. Its source is the energy-momentum tensor divided by m_P. Therefore at low energies ($E \ll m_P$) its coupling to matter is extremely weak. Therefore it has not been observed experimentally, and will not be observed in the foreseeable future. Even gravitational waves, classical ensembles of zillions of gravitons, have not been yet detected by specially built antennas. But for them prospects are quite realistic.

A consistent theory of quantum gravity has not been created yet. The most promising way to it is marked by the sign "superstrings". Superstrings are tiny one-dimensional objects of the characteristic Planck length l_P, with fermionic and bosonic excitations on them (therefore the prefix "super"). Most of these excitations are very heavy, of the order of m_P. But there are a few of them which remain massless. They look like pointlike particles, from distances much larger than Planckian. Some of the superstring models have patterns of massless degrees of freedom, which closely resemble some of the supersymmetric grand unification groups. Thus, superstrings, are believed not only to provide a self-consistent theory of quantum gravity, but to provide it in a broader framework of a unified theory of all interactions, a theory of everything (TOE). All values of known (and to be discovered) fundamental gauge and Yukawa coupling constants are expected to arise as dimensionless elements of the solution of the TOE equations. It was shown recently that various superstring models correspond in fact to perturbative expansions in vicinity of different points of the same theory.

If superstring ideas are correct, then nature is based on three fundamental dimensional constants: maximal velocity of particles c, quantum of action and of angular momentum $\hbar$, and Planck length l_P (or, what is equivalent in units of $\hbar, c$, Planck mass m_P, or Newton constant G). The dimensions of other physical quantities can be expressed in terms of dimensions of $c, \hbar, G$. In particular, the dimensions of length [L], time [T] and mass [M], with which elementary physics text-books usually start, are:

$$[L] = [l_P], \quad [T] = [t_P], \quad [M] = [m_P].$$

The $c, G, \hbar$ units has been considered as the most "natural units of nature" long before the superstrings (Eddington, Gamov, Ivanenko, Landau, Bronshtein, Zelmanov, Wheeler). From this point of view the program of Einstein to build a unified theory of gravity and electromagnetism, without using $\hbar$, was doomed from the beginning.

8 Anthropic universe

A remarkable feature of our world is how perfectly it is tuned to favour our existence. The anthropic properties of nature are discussed in many articles and books. Let me remind a few examples of such fine tuning in particle and nuclear physics. Start with proton and neutron. The mass difference $m_n - m_p$ is 1.3 MeV. Were it the case that this mass difference were 0.5 MeV or smaller, then the neutron would become stable, whilst the hydrogen atom would be unstable: $e^- + p \to n + \nu_e$. The most abundant element in the world would be helium, not hydrogen. The stars would explode at a rather young age. The genesis of life would become impossible for many reasons. Analogous dramatic changes are produced by making the electron 0.8 MeV heavier. Note that neutron-proton mass difference is determined essentially by the mass difference of d- and u-quarks ($m_d \sim 7$ MeV, $m_u \sim 5$ MeV). Note also that in two other generations the lower quarks (s, b) are not heavier, but substantially lighter than their upper partners (c, t). Compared to the Planck mass, the tuning of u-, d-, e-masses is of the order 10^{-22}!

Even more striking is the sensitivity of our world to much less fundamental quantities, such as the binding energy of the deuteron, $\varepsilon = 2.2$ MeV. Decreasing it by only 0.4 MeV would make impossible the main reaction of hydrogen burning in the sun, $pp \to de^+\nu_e$, so that only the much less effective reaction $ppe^- \to d\nu_e$ would survive.

Another example is given by energy levels of ^{12}C and ^{16}O. The famous carbon level at 7.65 MeV lies only 0.3 MeV higher than the sum of masses of three α-particles, and therefore resonantly enhances the cross-section of the reaction $3\alpha \to {}^{12}$C. The nucleus ^{8}Be being unstable, carbon cannot be produced in two body $\alpha + {}^8$Be collisions. Without the 7.65 MeV resonance the three-body formation would be not effective enough. As a result carbon would disappear in the reaction $\alpha + {}^{12}$C $\to {}^{16}$O much faster than it would be produced, and the universe would have not enough carbon to create life.

When looking at the diagram of ^{12}C levels (there are about 30 of them in the interval of 30 MeV) one cannot help admiring that the level 7.65 MeV does not lie 0.5 MeV lower. The list of such examples may go on and on. How thin is the margin of safety of everything which is so dear to our hearts! Most essential features of our world are determined by absolutely non-essential (from the point of view of fundamental constants) details of "hadronic chemistry", not speaking about ordinary chemistry and biochemistry.

The anthropic properties of the universe have led to formulation of a number of speculative principles.

The weak anthropic principle is based on the notion of an ensemble of an infinite number of universes with values of dimensionless fundamental constants, which have been fixed during their cosmological evolution. From the very fact of our existence it follows that we live in one of the best of the worlds.

The cosmological realization of the above statistical ensemble is an infinite network of universes each of which, at its early inflationary stage, produces innumerable daughter universes. They may have different symmetry breaking patterns, even different numbers of space-time dimensions, and unlimited variety of values of dimensionless fundamental constants. But here we arrive to the gates of Metaphysics.

9 Concluding remarks

Looking back at those who made great discoveries at the dawn of our century and at those who helped them, let us ask ourselves: Was it possible for any of these pioneers to predict the major steps in evolution of fundamental physics in the XX century, its impact on the life of mankind, and its present landscape? The negative answer seems to me obvious. It would be even more difficult for us to guess, what summits fundamental physics would reach in the next hundred years, unless external factors will terminate its development. Unfortunately, it would be very easy to predict the landscape of physics and of science in general, if the existing antiscientific trends would prevail. It will be devastation: intellectual, scientific, cultural, technological, environmental. The life on our planet, a unique phenomenon, based on a unique tuning of fundamental constants of nature, might be ruined. Our duty, as scientists, is to be unanimous and to do our best in defending and promoting fundamental science.

10 Bibliography

1. A. Pais, **Inward bound.** Clarendon Press. Oxford. 1986.

2. **Review of Particle Properties.** Phys. Rev. **D50** (1994) No. 3, part I (an updated edition will appear in 1996).

3. K.Gottfried and V.Weisskopf. **Concepts of Particle Physics.** v. I,II. Clarendon Press. Oxford. 1984.

4. L.B.Okun. **Particle Physics: The Quest for the Substance of Substance.** Harwood Academic Publishers. 1985.

5. S.Weinberg. **The First Three Minutes: A Modern View of the Origin of the Universe.** Basik Books, Inc. N.Y., 1977.

6. M.B.Green, J.Schwartz, E.Witten. **Superstring Theory.** Cambridge University Press.

7. A.D.Linde. **Particle Physics and Inflationary Cosmology.** Gordon and Breach. N.Y. 1990.

8. J.D.Barrow and F.J. Tipler. **The Anthropic Cosmological Principle.** Clarendon Press. Oxford. 1986.

RESPONSE OF ATOMS IN PHOTONIC LATTICES

D. VAN COEVORDEN
FOM Institute for Atomic and Molecular Physics,
Kruislaan 407, 1098 SJ Amsterdam, The Netherlands

R. SPRIK, P. DE VRIES AND A. LAGENDIJK
van der Waals-Zeeman Instituut, Universiteit van Amsterdam,
Valckenierstraat 65-67, 1018 XE Amsterdam, The Netherlands

It is rigorously shown that the rate of spontaneous emission of atoms placed in a dielectric is determined by a *part* of the local density of eigenmodes of the Maxwell equations: the local radiative density of states. Spontaneous emission is inhibited if the atom is located at a position where this local radiative density is vanishing, even when the total density of states is not small. The radiative density of states can be obtained from a purely classical calculation. We demonstrate this principle by a calculation of the optical bandstructure and density of states of a three-dimensional lattice of resonant two-level atoms in the dipole approximation. The formation of photonic bandgaps is exhibited. The bandstructure can be characterized by two dimensionless parameters. We find a longitudinal polarization mode as well as a class of vacuum modes that are unaltered by the interaction with matter.

1 Introduction

The rate of spontaneous emission of an atom inside a dielectric is a crucial property for applications and requires fundamental understanding of its quantum aspects.[1] Experimental and theoretical efforts focused up to now on the modification of the emission in (micro) cavities filled with a homogeneous dielectric.[2,3] Recently three-dimensional inhomogeneous dielectric structures with characteristic length scales matching the wavelength of the luminescence have been put forward as structures in which spontaneous emission can be modified.[4,5] In such 'photonic materials' one identifies for instance Bragg reflection, band gaps in the density of states, and many other phenomena in close analogy with the well-known wavelike propagation of electrons in a crystalline structure.[6,7] It has been shown that the emission of an atom with an emission frequency inside the photonic band gap of such a photonic crystal would be fully suppressed and may serve as the basis for designing a laser without threshold.[4] Moreover, from a fundamental point of view, such materials, after some randomization, are interesting for the observation of the localization of light.[8]

Another class of fascinating three-dimensional dielectric structures with strong dispersion at optical frequencies has been developed recently: laser-trapped atomic systems. The first steps towards full three-dimensional lattices of laser trapped atoms have been successfully taken using laser cooling techniques.[9] Propagation of light with wavelengths near the optical resonances of the constituting atoms is dominated by multiple scattering from occupied unit cells and, if all cells are occupied, may lead to the formation of well-defined optical band structures, and possibly to full band gaps in the density of states.

The spectacular suppression of spontaneous emission in a dielectric is predicted to occur under rather arduous conditions: a vanishing or almost vanishing of the *total* density of states; i.e. integrated over one unit cell. We will demonstrate, however, that a much weaker condition needs to be fulfilled: only a local property, that we shall call the local radiative density of states (LRDOS), is required to be small. This LRDOS is manifestly different from both total and local density of states and depends on the location of the radiating impurity atom in the unit cell of the photonic crystal. Due to strong interference the LRDOS changes rapidly as a function of place, and a judicious positioning of the impurity atom will give a much lower rate of spontaneous emission than the total density of states would indicate.

We will show that the density of states in a dielectric and its influence on the radiative life time of atoms may be obtained by a classical calculation of the eigenmodes of the system.[10] We demonstrate that the total density of states and the LRDOS deviate considerably for a lattice of resonating classical dipoles (or equivalently two-level atoms).

2 Density of States for Spontaneous Emission in a Dielectric

In this section we demonstrate that the LRDOS indeed arises in the expression for spontaneous emission and not the total density of states. We shall restrict ourselves to the transition from the excited atomic $|b\rangle$ –state to the $|a\rangle$ –groundstate . The initial state $|i\rangle = |b\rangle \otimes |\mathbf{E}_i\rangle$ and final state $|f\rangle = |a\rangle \otimes |\mathbf{E}_f\rangle$ are described by the state of the atom $|\sigma\rangle$ *and* the classical vector light field $|\mathbf{E}\rangle$. The coupling between the atom and the vacuum is given by $-\boldsymbol{\mu} \cdot \mathbf{E}(\mathbf{r})$, with $\boldsymbol{\mu}$ the dipole moment of the optical transition. With the use of Fermi's golden rule the transition rate is

$$\tau(\mathbf{r}) = A_{ab} \sum_{\{\mathbf{E}_f\}} |\langle \mathbf{E}_f | \mathbf{E}(\mathbf{r}) | \mathbf{E}_i \rangle|^2 \, \delta(\omega_f - \omega_i - \omega_{ba}) \equiv \frac{\hbar \omega_{ba}}{2\varepsilon_0} A_{ab} N_{\text{rad}}(\mathbf{r},\omega), \quad (1)$$

where $A_{ab} = 2\pi|\mu|^2/\hbar^2$ and ε_0 the permittivity in vacuum. The quantity $N_{\text{rad}}(\mathbf{r},\omega)$ is a fully classical object. Furthermore, it is *not* the local DOS obtained by counting the eigenvalues of the Hamiltonian: Two differences will be shown to exist; 1) The *local* nature of the density $N_{\text{rad}}(\mathbf{r},\omega)$, 2) N_{rad} is smaller than the total density of states.

The j-th mode of the Helmholtz wave equation obeys (in conventional bra-ket notation)

$$\left(\nabla \times \nabla \times - (\omega_j/c)^2 \varepsilon(\mathbf{r})\right)|\mathbf{E}_j\rangle = 0, \qquad (2)$$

where $\langle \mathbf{r}|\varepsilon|\mathbf{r}'\rangle \equiv \varepsilon(\mathbf{r})\delta(\mathbf{r}-\mathbf{r}')$ represents the dielectric matter. A density of states should be defined by counting eigenvalues. However, Eq. (2) is not in the form of a conventional eigenvalue equation if ε depends on $\mathbf{r}$. The traditional approach, to partition the dielectric constant into a constant and an inhomogeneous part, does not solve this problem. In that case the potential would become eigenvalue dependent. A correct way to manipulate Eq. (2) into a genuine eigenvalue equation is to divide by the operator ε and to symmetrize the non-Hermitian operator $\varepsilon^{-1}\nabla \times \nabla \times$. We get

$$\mathcal{L}|\mathbf{\Lambda}_j\rangle = (\omega_j/c)^2 |\mathbf{\Lambda}_j\rangle, \qquad (3)$$

where the Hermitian operator $\mathcal{L} \equiv \varepsilon^{-1/2}\nabla \times \nabla \varepsilon^{-1/2}\times$ and $|\varepsilon^{1/2}\mathbf{E}_j\rangle \equiv |\mathbf{\Lambda}_j\rangle$. As $\mathcal{L}$ is Hermitian its eigenvalues are real with orthogonal eigenfunctions and their number can be counted. The number of states with frequencies between ω^2 and $\omega^2 + d\omega^2$ is

$$N(\omega) = 2\omega N(\omega^2) = 2\omega \sum_j \delta(\omega^2 - \omega_j^2) = \frac{2\omega}{c^2}\mathrm{Tr}\left[\delta((\omega/c)^2 - \mathcal{L})\right]. \qquad (4)$$

It is convenient to decompose $N(\omega)$ into local contributions according to

$$N(\mathbf{r},\omega) = \frac{2\omega}{c^2}\mathrm{tr}\langle\mathbf{r}|\,\delta\left((\omega/c)^2 - \mathcal{L}\right)|\mathbf{r}\rangle, \qquad (5)$$

where tr denotes a trace over polarization states. We have obtained the total density of states $N(\omega)$, defined through (4) and the local density of states $N(\mathbf{r},\omega)$, defined through (5), both by diagonalizing a Hermitian operator and counting its eigenvalues. This genuine local density of states $N(\mathbf{r},\omega)$ is known to be important for the transport of radiative energy in dielectrics.[11] We emphasize that the local density of states $N(\mathbf{r},\omega)$ is not the local density $N_{\mathrm{rad}}(\mathbf{r},\omega)$ that features in the Einstein coefficient.

This can be ascertained by quantizing the radiation field in the emission rate (1) in a cavity of volume Ω according to

$$\varepsilon^{1/2}(\mathbf{r})\mathbf{E}(\mathbf{r}) = \sum_j\left\{\sqrt{\frac{\hbar\omega_j}{2\varepsilon_0\Omega}}ia_j^\dagger\mathbf{\Lambda}_j(\mathbf{r})\exp(i\omega_j t) + \mathrm{h.c.}\right\}, \qquad (6)$$

with $a_j^\dagger$ the creation operator of the mode $|\Lambda_j\rangle$. Only with this definition can the total Hamiltonian of matter and radiation be written as $\mathcal{H} = \sum_j \hbar\omega_j a_j^\dagger a_j$.[12] Eq. (6), when applied to a dielectric interface, agrees with the result of Kosravhi and Loudon.[13] Substitution of Eq. (6) in the definition of the field part of the Einstein coefficient (1) gives,

$$N_{\text{rad}}(\mathbf{r},\omega) = \frac{2\omega}{c^2}\varepsilon(\mathbf{r})^{-1}\text{tr}\langle\mathbf{r}|\delta\left((\omega/c)^2 - \mathcal{L}\right)|\mathbf{r}\rangle = \varepsilon(\mathbf{r})^{-1}N(\mathbf{r},\omega). \quad (7)$$

This demonstrates that the LRDOS $N_{\text{rad}}(\mathbf{r},\omega)$ that determines the Einstein coefficient differs from the conventional local density of states $N(\mathbf{r},\omega)$ by the factor $\varepsilon(\mathbf{r})$.[6] In random systems an average can be performed over disorder. In that case the separation of both densities of states is crucial since $\langle\langle N(\mathbf{r},\omega)\rangle\rangle \neq \langle\langle\varepsilon(\mathbf{r})\rangle\rangle\langle\langle N_{\text{rad}}(\mathbf{r},\omega)\rangle\rangle$.[10] Note finally that we can use the delta-function property $\delta(ax) = (1/a)\delta(x)$ to write:

$$\begin{aligned}N(\mathbf{r},\omega) &= \frac{2\omega}{c^2}\left[1 + (\varepsilon(\mathbf{r}) - 1)\right]\text{tr}\langle\mathbf{r}|\delta\left(\varepsilon(\mathbf{r})(\omega/c)^2 - \nabla\times\nabla\times\right)|\mathbf{r}\rangle \\ &\equiv N_{\text{rad}}(\mathbf{r},\omega) + N_{\text{mat}}(\mathbf{r},\omega),\end{aligned} \quad (8)$$

where the delta-function now features the wave operator $\varepsilon(\mathbf{r})(\omega/c)^2 - \nabla\times\nabla\times$ of the Helmholtz equation (2). The *material* DOS $N_{\text{mat}}(\mathbf{r},\omega)$ is manifestly zero outside the material, where $\varepsilon(\mathbf{r}) = 1$. It is closely connected with the stored energy in the material degrees of freedom.[11]

In section 4 we will investigate the various photonic properties for a lattice of resonant dipoles. The calculations illustrate photonic band gap formation, the concepts of the density of states and provide an easy accessible model system with frequency dependent and energy conserving dielectric properties. The point-like character of the resonant dipoles results in a closed form for the optical band structure and reduces numerical efforts required to calculate the density of states. In the next section the single dipole scatterer will be treated first.

3 The Resonant Point Dipole

The interaction of an atom with an optical field is treated in the dipole approximation with the linear polarizability $\alpha(\omega)$ described by

$$\alpha(\omega) = \alpha(0) \times \frac{\omega_0^2}{\omega_0^2 - \omega^2 - i\gamma\omega^3/\omega_0^2} \equiv -\frac{6\pi}{\omega^3}\sin(\eta)e^{i\eta}, \quad (9)$$

where $\tan(\eta) = \gamma\omega^3/\omega_0^2(\omega_0^2 - \omega^2)$. Eq. (9) is the simplest classical representation of the typical linear polarizability for two-level atoms and damped oscillators.[14] It exhibits a resonance at frequency ω_0, a (full) width γ and a static polarizability $\alpha(0)$. It is usually advantageous to formulate scattering problems for the corresponding t-matrix $t(\omega) = -(\omega/c)^2 \alpha(\omega)$. The t-matrix exhibits the ω^4 behavior of the scattering cross section $\sigma_{scat}(\omega) = |t(\omega)|^2/(6\pi)$ at low frequencies which is typical of Rayleigh scatterers. Atoms characterized by Eq. (9) scatter light elastically: its strength is fixed by the optical theorem:

$$-\frac{\text{Im}\, t(\omega)}{k} = \frac{|t(\omega)|^2}{6\pi}. \tag{10}$$

This relation describes that the extinction (l.h.s.) from the incoming light is transformed entirely into scattered intensity, given by the scattering cross-section σ_{scat}, i.e. there is no absorption. It also shows that Eq. (9) only features two independent quantities. In terms of the quality factor:

$$Q \equiv \frac{\omega_0}{\gamma} = \frac{6\pi}{\alpha(0)(\omega_0/c)^3}. \tag{11}$$

For atoms the quality factor is very high, $Q \gg 1$, so that Eq. (11) indicates that the ratio of the resonantly-enhanced scattering cross section and the geometrical cross section of the point scatterer (associated with the static polarizability $\alpha(0)$) is very large. Indeed, at resonance we have ($\lambda_0 \equiv 2\pi c/\omega_0$):

$$\sigma_{scat}(\omega_0) = \frac{3}{2}\frac{\lambda_0^2}{\pi}, \tag{12}$$

which is the expected maximum value for Rayleigh scatterers.[14]

It is instructive to plot the LRDOS for a single point dipole. In Fig. 1 we have plotted $N_{rad}(\mathbf{r},\omega)$ for three different values of ω/ω_0 and a fixed quality factor $Q = 100$. Near the scatterer itself we see the LRDOS acquires resonant enhancement, while it converges to the vacuum value for large enough r. Apparently the LRDOS has a considerable 'range' only when ω is close to the resonance frequency. This is in accordance with the maximal value of Eq. (12). In Fig. 2 we have fixed $\omega/\omega_0 = 0.95$ and varied the value of Q. (At resonance, the LRDOS is independent of Q.) The narrower the resonance (higher Q), the shorter the 'range', i.e. the LRDOS is pulled towards the scatterer.

4 Resonant Point Dipoles on a Lattice

4.1 Bandstructure

We next investigate an infinite lattice of point dipoles, so that $\varepsilon(\mathbf{x})$ in Eq. (2) becomes a periodic function. We decompose $\mathbf{E}$ in terms of the Bloch wave

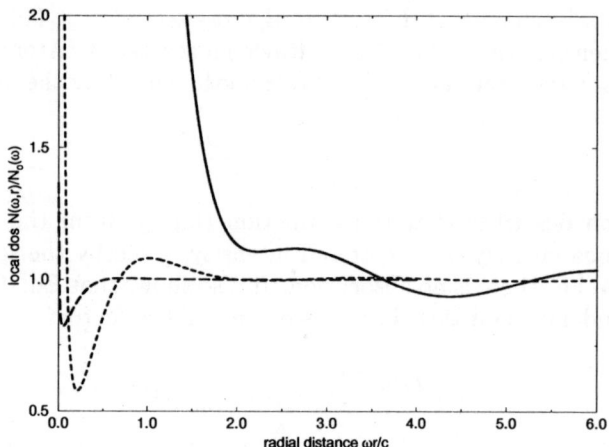

Figure 1: Local DOS for one point scatterer as a function of the scaled radial distance. The DOS is scaled by the "vacuum" value. Three values of ω/ω_0 are shown: 1.0 (solid line), 0.95 (dashed) and 0.5 (long-dashed).

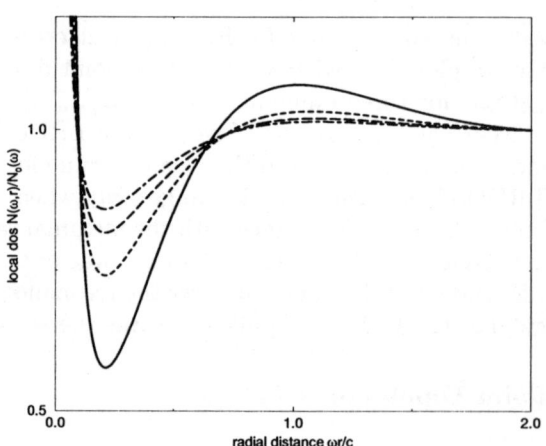

Figure 2: The LRDOS for various values of Q. We have taken $\omega = 0.95\omega_0$. Solid line: $Q = 100$; dashed: $Q = 200$; long-dashed: $Q = 300$ and dot-dashed: $Q = 400$.

vectors **k** in the first Brillouin zone and reciprocal lattice vectors **g** as follows[15]:

$$\mathbf{E_k}(\mathbf{x}) = \sum_{\mathbf{g}} \mathbf{E}(\mathbf{k} - \mathbf{g}) e^{i(\mathbf{k}-\mathbf{g})\cdot\mathbf{x}} . \tag{13}$$

Next the dielectric permeability $\varepsilon(\mathbf{x})$ is expanded in reciprocal lattice vectors **g**:

$$\varepsilon(\mathbf{x}) = \sum_{\mathbf{g}} \varepsilon(\mathbf{g}) e^{i\mathbf{g}\cdot\mathbf{x}}. \tag{14}$$

With the use of formalisms originally developed for band structure calculations of electrons in solid state[7,16], the photonic band structure $\omega(\mathbf{k})$ of the optical lattice can be calculated by diagonalizing the secular matrix for N plane wave states in the crystal coupled by a **k**-independent interaction $\Gamma(\omega) = -6\pi c \tan(\eta(\omega))/\omega V$, with V the volume of the unit cell. For a face-centered cubic (fcc) lattice: $V_{fcc} = a^3/4$, with a the length of a side of the full cubic cell. The secular equation for this interaction can be reduced to a 3×3 determinantal condition for $\omega(\mathbf{k})$:

$$\left\| \Gamma(\omega) \sum_{i=1}^{N} \left\{ \frac{1}{(\mathbf{k}-\mathbf{g}_i)^2 \Delta_{\mathbf{k}-\mathbf{g}_i} - \omega^2} + \frac{\mathbf{e}_{\mathbf{g}_i} \mathbf{e}_{\mathbf{g}_i}}{\omega^2} - \frac{\Delta_{\mathbf{k}-\mathbf{g}_i}}{\mathbf{g}_i^2} \right\} + \mathbf{I} \right\| = 0, \tag{15}$$

with $\{\mathbf{g}_i\}$ the set of reciprocal lattice vectors, **I** is the 3x3 identity matrix and $\Delta_\mathbf{k} = \mathbf{I} - \mathbf{e_k e_k}$ ($\mathbf{e_k} \equiv \mathbf{k}/k$).

We have solved $\omega(\mathbf{k})$ numerically from Eq. (15) for a fcc lattice. We define two dimensionless parameters that characterize the bandstructure picture completely. The first parameter is $P \equiv (\omega_{\mathrm{BZ}}/\omega_0)^3$, which we will call here the "polariton parameter". We introduce ω_{BZ}/c as the radius of the largest inscribing sphere of the Brillouin zone. It equals the modulus (in reciprocal space) of the L-point, multiplied by c. Thus $\omega_{\mathrm{BZ}} a/c \equiv \pi\sqrt{3} \approx 5.44$, where a is the lattice constant, and ω_{BZ} relates to the density of scatterers as $\rho \propto (\omega_{\mathrm{BZ}}/c)^3$. In terms of lengths we can say that P measures the ratio of the wavelength at resonance $\lambda_0 \equiv 2\pi c/\omega_0$ and a. The second parameter is the quality factor Q which was defined in Eq. (11).

As an illustration we consider two characteristic choices for the combined parameter set (P, Q). For parameter set I (Fig. 3), which represents the denser case, ω_0 is well below ω_{BZ}: $P = 1.29$ (corresponding with $\lambda_0 = 1.26a$) and $Q = 30$. For parameter set II (Fig. 4): $P = 0.47$ ($\lambda_0 = 0.9a$) and $Q = 21$. In both Figs. 3 and 4 we still see the linear dispersion law around the origin Γ of the Brillouin zone. There is a polarization degeneracy (for all modes but one) that is frequently lifted. Only one mode, the relatively straight dot-dashed line in the middle of Fig. 3 and upper part of Fig. 4, is not degenerate.

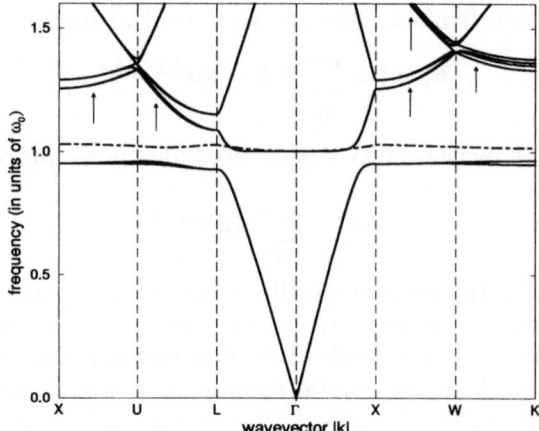

Figure 3: Photonic bandstructure for a fcc lattice of resonant dipoles. The resonance wavelength λ_0 for this case equals $\lambda_0 = 1.26a$ ($P = 1.29$ and $Q = 30$ in Fig. 8). Horizontally we have parametrized a path along the symmetry points X, U, L, Γ, X, W, K in the Brillouin zone. The dot-dashed line is a longitudinal polarization mode; the arrows indicate the vacuum modes that are unaltered by the interaction (see text).

Moreover inspection of its corresponding eigenfunction shows that it is mainly longitudinal (parallel to $\mathbf{k}$). It can therefore be interpreted as a coupling of light with the *longitudinal* polarization field. It originates from the dispersion law associated with Eq. (9) which exhibits negative values of the dielectric function. A simple estimate for the approximate position of the line at $\mathbf{k} = \mathbf{0}$ is given by the second zero of the dielectric function[15] $\varepsilon(\omega_{LO}) = 1 + 4\pi\rho\alpha(\omega_{LO}) \equiv 0$, from which follows $\omega_{LO} = 5.05c/a$; from Fig. 3 we have $\omega_{LO} = 5.02c/a$. Also in Fig. 3 the two branches in the lowest band level off at the boundary of the Brillouin zone, where they stay below the resonance frequency. In this flat region the dispersion is material-like, which indicates a polariton-type of propagation in the crystal.[15]

At the points U, L, X and W one observes the occurrence of avoided crossings; a phenomenon that in other physical situations may coincide with the presence of a Kronig-Penney type of bandgap.[7,17,18,19] A genuine bandgap however seems to exist only around the resonance frequency in Fig. 3. Whether or not a bandgap exists can be ascertained by considering every value of $\mathbf{k}$ in the Brillouin zone. This calls for a calculation of the (total) density of states (DOS) $N(\omega)$ of Eq. (4) in which the index j accounts for both the band and the polarization state.

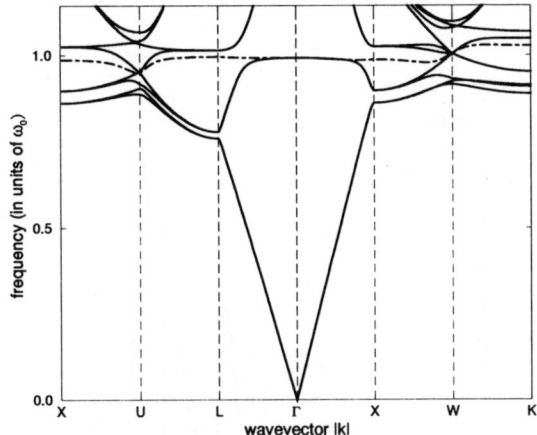

Figure 4: As in Fig. 3, but now for the more dilute case that $\lambda_0 = 0.9a$ ($P = 0.47$ and $Q = 21$ in Fig. 8).

4.2 The Density of States

In Fig. 5 $N(\omega)$ has been plotted for both parameter sets, scaled by the vacuum DOS $N_0(\omega)$. In the neighborhood of the resonance frequency, we see the corresponding resonant enhancement of the DOS. We see from Fig. 5 that for set I a gap is indeed present, whereas for set II no gap is found although a pronounced structure around the resonance frequency remains.

In Fig. 6 we have plotted the DOS for a random system which displays polariton behavior. Usually the hybridization of light with a resonant mode in the material is called a polariton. We see a similar decrease of the DOS for frequencies higher than the resonance frequency, which is a typical feature of resonant multiple scattering, sometimes called a pseudogap. Below ω_0 we have the corresponding enhancement of the DOS.

We have also calculated the radiative (local) DOS in the case of scalar waves. The Helmholtz equation (2) is frequently approximated by the scalar wave equation:

$$\left(-\nabla^2 - (\omega_j/c)^2 \, \varepsilon(\mathbf{r})\right) |\psi_j\rangle = 0 , \qquad (16)$$

which in photonic band gap materials is known to be inadequate in general. There are cases where bandgaps show up in the scalar wave approximation which are absent in a vector picture and vice versa.[7,17,18,19] In our case, however, the difference in eigenvalues is uniformly less than 1%. The longitudinal mode turns out to be absent here as it should for scalar waves. The total ra-

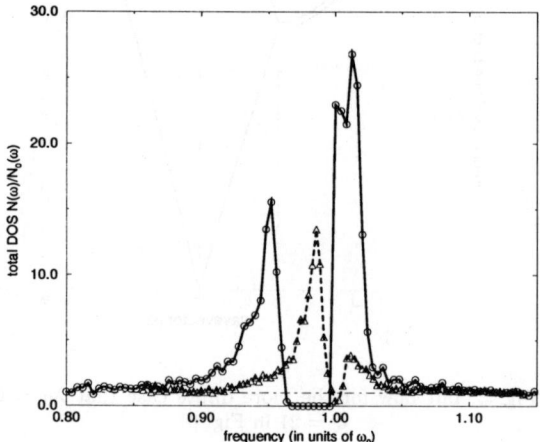

Figure 5: Scaled (total) density of states for the two special cases of Fig. 3 (solid line/circles) and Fig. 4 (dashed line/triangles).

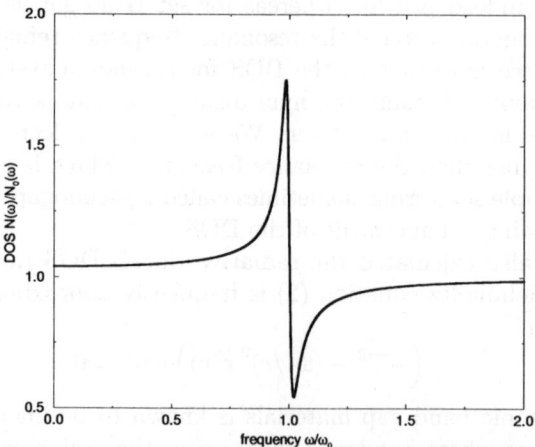

Figure 6: Polariton behavior in the DOS for a random system. We have chosen values of the density and resonance frequency which correspond with $Q = 30.0$ and $P = 1.29$ in the fcc lattice case, as in Fig. 3.

diation density of states for the resonant dipoles on a lattice can be expressed as:

$$N_{\text{rad}}(\omega) = 2\omega \sum_{\mathbf{k}} \frac{1}{F(\omega,\mathbf{k})+1} \delta(\omega^2 - \omega_{\mathbf{k}}^2),$$

$$F(\omega,\mathbf{k})^{-1} = \frac{4\pi\omega^4}{cVQ\omega_0^3} \sum_{\mathbf{g}} \frac{1}{\left[(\mathbf{k}-\mathbf{g})^2 - (\omega/c)^2\right]^2} \quad \text{(scalar)}. \quad (17)$$

The LRDOS can be expressed in terms of the functions $\psi_{\mathbf{k}}(\mathbf{r})$ that follow from the diagonalization as

$$N_{\text{rad}}(\mathbf{r},\omega) = 2\omega \sum_{\mathbf{k}} |\psi_{\mathbf{k}}(\mathbf{r})|^2 \frac{1}{F(\omega,\mathbf{k})+1} \delta(\omega^2 - \omega_{\mathbf{k}}^2) \quad \text{(scalar)}. \quad (18)$$

Note that due to the pointlike character of the resonant dipoles in our model the LRDOS and the local total density of states are identical for positions *between* atomic sites ($\mathbf{r} \neq 0$): only on atomic sites they differ (see end of Sec. 2). In Fig. 7 we have plotted the bandstructure (upper row), total radiative DOS (middle row) and LRDOS (lower row) for two parameter sets (with and without a gap present). In the bottom row of Fig. 7 the LRDOS is plotted as a function of position in the unit cell and frequency. The second parameter set (right figure) is very interesting. We see that the LRDOS depends strongly on position and at $\mathbf{r} = (a/5, 0, 0)$ the LRDOS is almost completely suppressed. An excited atom placed there with an emission frequency of about ω_0 would live considerably longer. This dramatic behavior could not be predicted on the basis of the density of states alone. Of course, also when there is a gap in the density of states, the spatial dependence of the LRDOS outside the gap is still very interesting. Efficient inhibition of spontaneous emission can even be obtained outside the gap, as is shown in the left-bottom part of Fig. 7.

4.3 Phase Diagram

We have investigated the possible presence of a gap for given values of P and Q : the resulting "phase diagram" is depicted in Fig. 8. Apparently a gap exists *always* for $P \geq 1.0$, independently of Q. In terms of the resonant wavelength, this condition translates to $\lambda_0 \geq 1.15a$. The line $P = 1.0$ constitutes for high values of Q an asymptotic boundary for the left region, where no gap is found. For smaller values of Q, say $Q < 50$ (corresponding with a broad resonance), the lower bound on P decreases. Note however that for atoms $Q \gg 50$. For reasons of graphical representation we have chosen low values of Q, since the

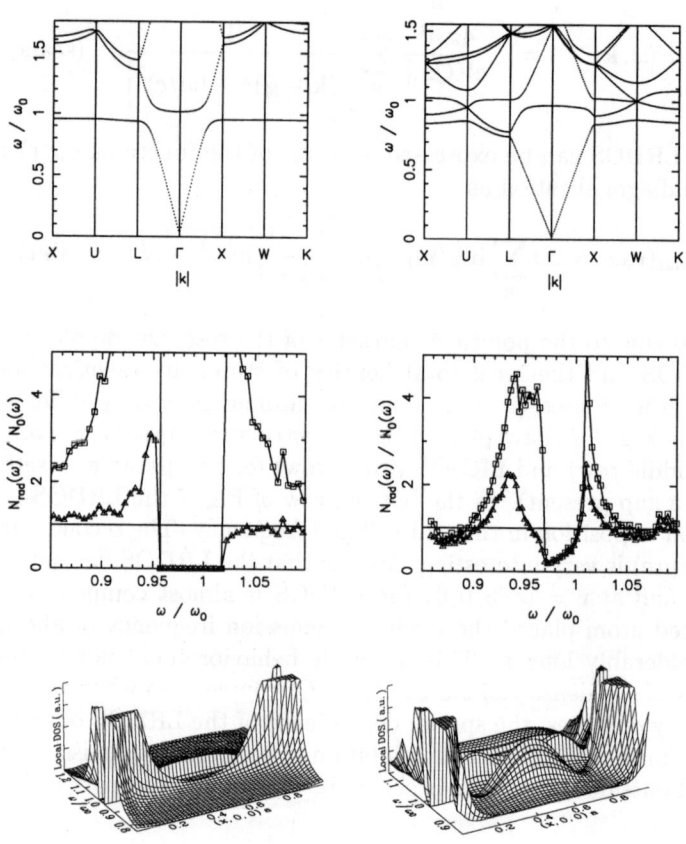

Figure 7: Photonic properties of a fcc lattice of resonant atoms in scalar approximation. In the left column $P = 2.52$, $Q = 6.25$, and in the right column $P = 0.47$, $Q = 3.6$. The top row: The photonic bandstructure. The middle row: The total density of states $N(\omega)$ (squares) and the total density of states for the radiation $N_{\rm rad}(\omega)$ (triangles) scaled to the density of states in vacuum $N_0(\omega)$. The bottom row: The local density of states for the radiating field $N_{\rm rad}(\mathbf{r}, \omega)$ on a traject moving from $\mathbf{r} = (0,0,0)$ to $\mathbf{r} = (a,0,0)$ with a the fcc lattice constant. For further dicussion see text.

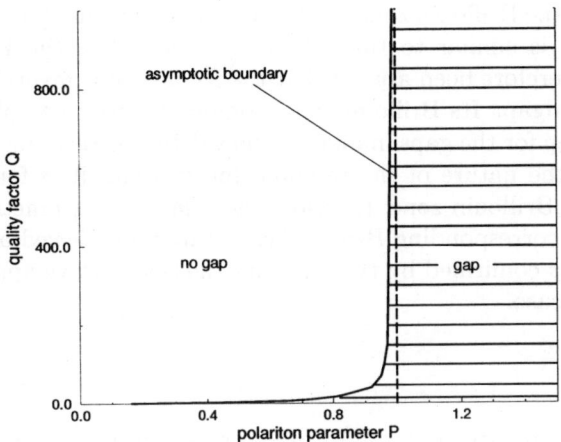

Figure 8: P, Q "phase diagram".

width of the gap $\Delta\omega \sim Q^{-1}$. From Fig. 5 we observe that $\Delta\omega/\omega_0 \approx 0.85 Q^{-1}$. The absence of a gap for $\omega_0 > \omega_{\text{BZ}}$ can be clarified by considering the lines in Fig. 3 that are indicated by arrows. These lines are identically present for all values of P and Q, and coincide with eigenvalues of the *empty lattice*. Thus free eigenvalues in some bands remain eigenvalues of the perturbed system (which is not true for the associated eigenfunctions). This feature was observed earlier in the Schrödinger case[20]. There it is shown that $\omega = c|\mathbf{k} - \mathbf{g}|$ is an eigenvalue of multiplicity $m \geq 1$ if and only if there are $m+1$ reciprocal lattice vectors $\mathbf{g}_0, \mathbf{g}_1, \ldots, \mathbf{g}_m$ such that:

$$|\mathbf{k} - \mathbf{g}_0| = |\mathbf{k} - \mathbf{g}_1| = \cdots = |\mathbf{k} - \mathbf{g}_m|. \tag{19}$$

This is precisely what we find in our numerical work, both for the scalar (see below) and Maxwell cases. Due to Eq. (19), this type of eigenvalue only occurs at $\mathbf{k}$-values of a certain symmetry. The rigidity of these lines disturbs the formation of a bandgap when ω_0 exceeds the minimum value ω_{BZ}.

Again a comparison with random systems can be made here. The Ioffe-Regel criterion[11] predicts that localizaton of light sets in when the wavenumber $k = 2\pi/\lambda$ matches the inverse scattering mean free path ℓ^{-1}, which is defined as $\ell = 1/\rho\sigma_{\text{scatt}}$. Evaluating ℓ at resonance using Eq. (12), with the density of the fcc lattice $\rho = 4/a^3$, we readily obtain: $k\ell = 1 \Leftrightarrow (\lambda_0/a)^3 = \pi^2/3$, which corresponds with $P = 2.1$. Thus, in these terms the random system would inhibit wave propagation at higher densities.

Usually for photonic crystals the frequency gaps are found near or above the boundary of the Brillouin zone[17,18,19] where the periodicity is nearly matched. Their formation is similar to that of the gaps found in the Kronig-Penney model. It has therefore been argued that the fcc lattice is favorable for finding an isotropic bandgap. Its Brillouin zone coming closest to a sphere implies a maximum overlap for the gaps in all directions.[17] In contrast our gap is a direct consequence of the nature of the resonant interaction. It is found *below* the boundary of the Brillouin zone; therefore the role of the geometry of the Brillouin zone (and corresponding Bravais lattice) may be viewed as less critical. This statement is confirmed by the fact that the scalar wave approximation is excellent in our case.

5 Discussion

In this work we described the properties of atomic lattices by applying an elastic t-matrix formalism. The polariton character of the resonance results in a gap in the frequency spectrum. We infer that the bandstructure (and existence of a gap) is determined entirely by two dimensionless parameters, which measure the scattering strength and the width of the atomic resonance, respectively. The emission properties inside the crystal are drastically modified at positions in between atomic sites and is proportional to the local density of states of the radiation, which is only a part of the total density of states. This should also hold for photonic band gap materials in general. Tailoring the positions of the active centers inside the unit cell relaxes the need for a full band gap to efficiently suppress spontaneous emission.

Acknowledgements

We acknowledge fruitful discussion and cooperation with B.A. van Tiggelen and A. Tip. Part of this work has been supported by the Stichting voor Fundamenteel Onderzoek der Materie (FOM), which is financially supported by the Nederlandse Organisatie voor Wetenschappelijk Onderzoek (NWO).

References

1. Part of this work has been published in: R. Sprik, B.A. van Tiggelen and A. Lagendijk, *Europhys. Lett.*, **35**, 265 (1996), and: D.V. van Coevorden, R. Sprik, A. Tip and A. Lagendijk, to appear in *Phys. Rev. Lett.*.
2. F. de Martini, M. Marrocco, P. Mataloni, D. Murra and R. Loudon, *J. Opt. Soc. Am. B* **10**, 360 (1993).

3. P. Meystre, M. Sargent III, in *Elements of Quantum Optics*, 2nd ed. (Springer, Berlin, 1991).
4. E. Yablonovitch, *Phys. Rev. Lett.* **58**, 2059 (1987).
5. S. John, *Phys. Rev. Lett.* **58**, 2486 (1987).
6. E. Yablonovitch and T.J. Gmitter , *Phys. Rev. Lett.* **63**, 1950 (1989).
7. see e.g. the NATO Workshop, *Photonic Band Gap Materials*, ed. Soukoulis C.M. (Kluwer, Dordrecht, 1996) and the topical issue of JOSA B **10** (1993).
8. S. John, *Phys. Rev. Lett.* **53**, 2169 (1983).
9. M. Weidemüller *et. al.*, *Phys. Rev. Lett.* **75**, 4583 (1995); G. Birkl *et. al.*, *Phys. Rev. Lett.* **75**, 2823 (1995); J.N. Tan *et. al.*, *Phys. Rev. Lett.* **75**, 4198 (1995).
10. B.A. van Tiggelen and E. Kogan, *Phys. Rev. A* **49**, 708 (1994).
11. A. Lagendijk and B.A. van Tiggelen, *Phys. Rep. C* **270**, 143 (1996).
12. R.J. Glauber and M. Lewenstein, *Phys. Rev. A* **43**, 467 (1991).
13. H. Khosravi and R. Loudon, *Proc. R. Soc. London A* **433**, 337 (1991).
14. R. Loudon, *The Quantum Theory of Light*, (2nd ed., Clarendon, Oxford , 1983).
15. N.W. Ashcroft and N.D. Mermin, *Solid State Physics* , (Saunders, 1976).
16. W. A. Harrison, *Solid State Theory*, (Dover, New York, 1979).
17. E. Yablonovitch, *J. Mod. Opt.* **41**, 173 (1994).
18. H.S. Sozuer and J.W. Haus, *J. Opt. Soc. Am. B* **10**, 296 (1993).
19. K.M. Ho, C.T. Chan and C.M. Soukoulis, *Phys. Rev. Lett.* **65**, 3152 (1990).
20. S. Albeverio, F. Gesztesy, R. Høegh-Krohn and H. Holden, *Solvable Models in Quantum Mechanics*, p.189 (Springer-Verlag, New York, 1988).

HYDROGEN-LIKE SYSTEMS AND QUANTUM ELECTRODYNAMICS

M.G. BOSHIER
*Sussex Centre for Optical and Atomic Physics, University of Sussex,
Falmer, Brighton BN1 9QH, United Kingdom*

1 Introduction

While much has changed in atomic physics since the first ICAP in 1968, a discussion of the impact of precise atomic spectroscopy on quantum electrodynamics (QED) has been a regular feature of this series of meetings. The reason for this longevity is of course the ongoing refinement of techniques for both measuring and calculating the energy levels of simple atoms which has continued since Lamb's famous 1947 experiment. One motivation for this long programme of work is illustrated by Figure 1, which shows how steady improvements in the precision of hydrogen spectroscopy over the last 150 years have probed new physical effects.

This paper will concentrate on the current status of tests of QED in light hydrogen-like atoms, where the best results are now mostly obtained by high resolution laser spectroscopy. In spite of its maturity, the field is at an exciting stage. Several theoretical groups have made substantial improvements in the calculation of QED corrections, and at the same time optical spectroscopy of hydrogen has reached the point where uncertainties due to strong interactions are an important ingredient in the interpretation of measurements (Sections 2 and 4). There is a very large and unexplained discrepancy in the He$^+$ Lamb shift (Section 3). Finally, new experiments on muonium complement this work on more conventional atoms (Section 5).

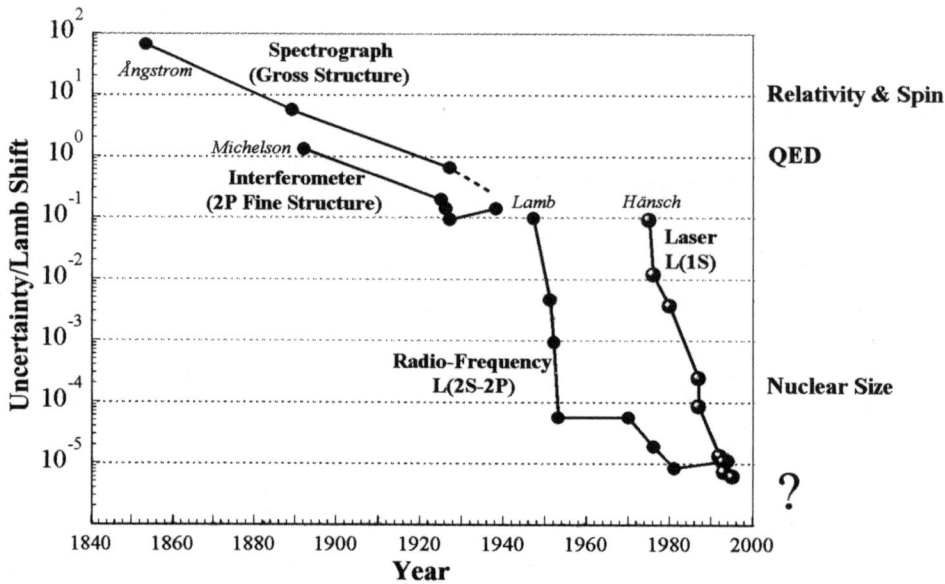

Figure 1: The progress made in 150 years of spectroscopy of the hydrogen atom.

2 Hydrogen

2.1 Measurement of the 1S Lamb Shift by Laser Spectroscopy

As can be seen from Figure 1, the most precise measurements of QED effects in hydrogen are now obtained by laser spectroscopy. As an example of these experiments, we consider a recent measurement carried out by our group at Yale University.[1] The experiment was based on Doppler-free two-photon laser spectroscopy of the hydrogen $1S$-$2S$ transition.[2-4] The use of this narrow line (natural width ~ 1 Hz) avoided the broad $2P$ state which has so far limited radio-frequency experiments, with the minor complication of requiring a comparison with a second hydrogen transition to separate the $1S$ Lamb shift from the much larger $1S$-$2S$ Dirac energy. An earlier version of this experiment[5] compared transitions using several intermediate frequency standards and the Rydberg constant. The direct comparison in the new experiment of the frequencies of the intervals H($1S$-$2S$) and H($2S$-$4P$) led to an improvement in precision relative to the earlier measurement of a factor of fourteen.

The apparatus (Figure 2) used two tunable ring dye lasers operating near 486 nm. The primary laser was scanned over both the single-photon $2S$-$4P$ transition and (after frequency-doubling) the two-photon $1S$-$2S$ transition, while the reference dye laser remained locked to a ^{130}Te$_2$ transition located between the two hydrogen lines. The frequency difference between the two lasers was measured by heterodyning them on a photodiode, leading to a direct measurement of the ~5 GHz difference frequency H($2S$-$4P$) - ¼H($1S$-$2S$). This quantity, which is of course zero in the non-relativistic Bohr theory, is due to relativistic corrections, hyperfine effects and the Lamb shifts. Since the first two contributions are well-understood, this measurement determined a combination of the Lamb shifts of the relevant levels.[4]

The $1S$-$2S$ spectrometer was an improved version of the system used in Ref. 5. The two-photon resonance was excited in a cell of flowing atomic hydrogen and detected through the 121 nm fluorescence produced by collisional quenching of atoms excited to the $2S$ level (Figure 3a). The $2S$-$4P$ transition was excited in a beam of metastable $2S$ atoms, produced by electron bombardment of ground state atoms. Most (88%) of the atoms excited to the $4P$ state decayed rapidly to the ground state, so the 2S-4P resonance (Figure 3b) could be observed as a decrease in the flux of metastable atoms reaching a

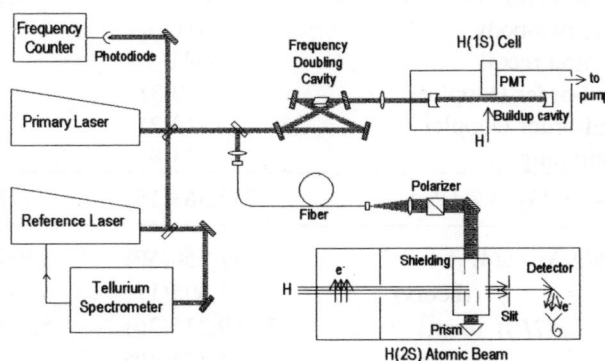

Figure 2: Schematic diagram of the apparatus used to measure the $1S$ Lamb shift in the Yale experiment.[1]

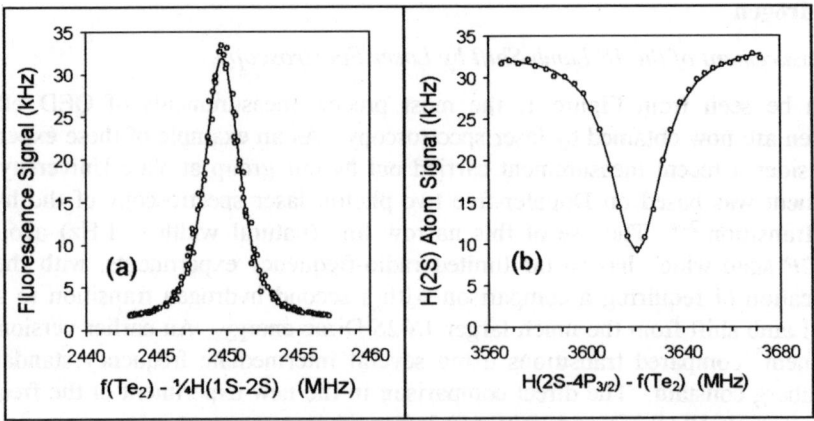

Figure 3: Typical spectra: (a) H(1S-2S) transition, (b) H(2S-4P$_{3/2}$) transition

surface ionization detector at the end of the beam machine. The laser beam was aligned at right angles to the collimated atomic beam and retroreflected by a 90° apex prism to reduce the first-order Doppler shift. The true perpendicular position was determined by taking spectra over a small range of atom-laser intersection angles around 90° and finding the angle giving minimum Doppler broadening.

Data were taken on the stronger $F=1$ to $F=1$ component of H(1S-2S) and both fine structure components of H(2S-4P). Combining the relevant pairs of spectra leads to the difference frequencies shown in the first line of Table 1. From these results are then subtracted the Dirac, relativistic two-body and hyperfine contributions to the interval, along with a correction to the 2S-4P frequency for photon recoil. Then systematic

Table 1: Reduction of the Yale data to obtain the 1S Lamb shift $\mathscr{L}(1S)$

	4P$_{1/2}$ (kHz)	4P$_{3/2}$ (kHz)
Raw H(2S-4P$_J$)-¼H(1S-2S)	4,698,337(12)	6,069,489(9)
H(2S-4P$_J$)-¼H(1S-2S) Dirac	-3,932,893	-5,300,854
hyperfine structure (1S,2S)	-33,291	-33,291
relativistic two-body	4,186	4,186
2S-4P photon recoil	-841	-841
2S-4P 2nd-order Doppler	50(3)	52(3)
1S-2S 2nd-order Doppler	-19(2)	-19(2)
optical pumping	33(8)	-17(4)
¼($\mathscr{L}(1S)-5\mathscr{L}(2S)+4\mathscr{L}(4P_J)$)	735,563(15)	738,705(10)
$\mathscr{L}(1S)-5\mathscr{L}(2S)+4\mathscr{L}(4P_J)$	2,942,250(59)	2,954,822(42)
$-4\mathscr{L}(4P_J)$ (theory)	5,605(1)	-7,069(1)
$+5(\mathscr{L}(2S) - \mathscr{L}(2P))$ (expt)	5,289,215(36)	5,289,215(36)
$+5\mathscr{L}(2P)$ (theory)	-64,175(10)	-64,175(10)
1S Lamb shift, $\mathscr{L}(1S)$	8,172,896(70)	8,172,792(56)
Combined value for $\mathscr{L}(1S)$	8,172,827(51) kHz	

corrections are applied for two effects which are not included in the theoretical lineshapes: the second-order Doppler shift of each transition and an optical pumping correction to the 2S-4P frequency, required because spontaneous decays from the 4P state to the 2S state do not repopulate the four 2S magnetic sublevels evenly. Finally, to obtain the ground state Lamb shift $\mathcal{L}(1S)$ the result is combined with the measured 2S-2P interval[6,7] and the theoretically well-understood P-state Lamb shifts.[8]

2.2 Calculation of the 1S Lamb Shift

The theoretical prediction for the 1S Lamb shift can be obtained by updating the expressions in Ref 8 to include new calculations of one-loop binding,[9] two-loop binding,[10,11] logarithmic two-loop binding,[12] and pure recoil[13] corrections. The reader is also referred to the recent review by Pachucki et al.[14] Numerical values for the contributions to the hydrogen 1S Lamb shift are given in Table 2. The $\alpha^2(Z\alpha)^5 m$ two loop binding correction is particularly important, since the two independent calculations by Pachuki[10] and by Eides and Shelyuto,[11] show that it contributes -292 kHz to the 1S energy, about thirty times more than expected from simple scaling arguments (as can be seen from Table 2.) The Lamb shift of an S state also includes a non-QED correction for the modification of the Coulomb potential inside the nucleus, shown at the bottom of Table 2. It follows simply from first-order perturbation theory that this correction is proportional to the mean square nuclear radius. It is unfortunate that the two most precise values for the proton size; 0.862(12) fm[15] and 0.805(11) fm,[16] disagree, so two calculated values for $\mathcal{L}(1S)$ are given in Table 2: 8,172,796(40) kHz for the "large" proton and 8,172,647(40) kHz for the "small" proton. The uncertainty in these values reflects similar contributions from the charge radius and from uncalculated or numerically evaluated QED terms.

Table 2: Contributions to the hydrogen 1S Lamb shift

Contribution	Frequency (MHz)	Error (MHz)
$\alpha(Z\alpha)^4 m$ one-loop	8,114.768	
$\alpha(Z\alpha)^5 m$ one-loop binding	56.950	
$\alpha(Z\alpha)^6 m$ one-loop binding	-3.193	
$\alpha^2(Z\alpha)^4 m$ two-loop	1.017	
$\alpha^2(Z\alpha)^5 m$ two-loop binding	-0.296	0.007
$\alpha^2(Z\alpha)^6 m \ln(Z\alpha)^n$ two-loop binding	-0.028	
$(Z\alpha)^5 m^2/M$ recoil	2.402	
$\alpha(Z\alpha)^5 m^2/M$ radiative recoil	-0.014	
$(Z\alpha)^6 m^2/M$ recoil	-0.007	
uncalculated higher order QED	0.000	0.015
proton size ($r_p = 0.862(12)$ fm)	1.164	0.032
1S Lamb Shift	**8,172.762**	**0.036**
(for $r_p = 0.805(11)$ fm:	8,172.613	0.032)

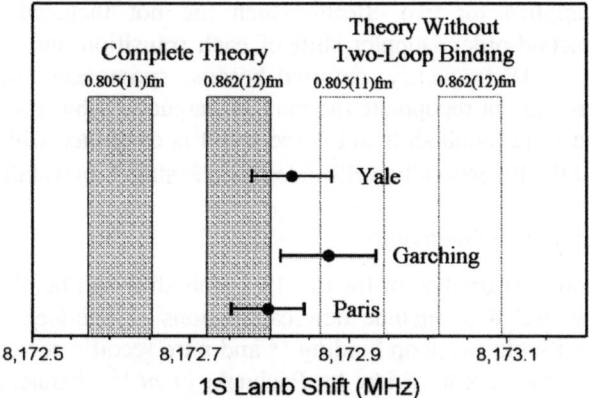

Figure 4: Comparison of recent measurements of the hydrogen 1S Lamb shift by groups at Yale,[1] Garching[17], and Paris[18] with theory for the two values of the proton charge radius (solid bars). Dashed lines indicate the position of the theory bars without the two-loop binding correction.

2.3 Comparison Between Experiment and Theory for ℒ(1S)

The final result for the 1S Lamb shift from this experiment is in excellent agreement with the most recent measurement of 8,172.874(60) MHz by Hänsch and co-workers,[17] which derives from a comparison of H(1S-2S) with the narrow but weak H(2S-4S, D) transitions, and with the value of 8,172.798(46) MHz obtained by the group of Biraben and Julien and colleagues[18] from a comparison of H(1S-3S) with H(2S-6S, D).

Figure 4 shows that these three recent measurements of the hydrogen 1S Lamb shift are in excellent agreement with each other, and that all are in good agreement with the "large proton" theory. The sensitivity of these measurements to the large two-loop binding correction can be assessed by subtracting it from the complete calculated Lamb shift. The results are indicated by the two dashed bars in Figure 4, where one sees that the ambiguity in the proton size prevents a clean test of the two-loop binding correction even though it is six times larger than the uncertainty in any one of the measurements.

2.4 The Proton Charge Radius

In view of this problem, we digress slightly to consider further these two values for the proton charge radius. Both measurements are derived not from a single experiment, but are instead drawn from compilations of data from several different experiments. The two determinations have no data in common. The relevant electric form factor data for the "Stanford"[16] and "Mainz"[15] values of the proton charge radius is shown in Figure 5. The lines represent least-squares quadratic fits to the two data sets. The proton charge radius is determined from the slope of the form factor at zero momentum transfer. It is clear from Figure 5 that the quality of the data which leads to the "Stanford" value is considerably worse that leading to the "Mainz" value, although the uncertainty quoted for the charge radius is slightly smaller than that assigned to the "Mainz" value. Also, an examination of the fit to the "Mainz" data shows that the χ^2 is smaller than expected for statistical uncertainties by about a factor of two, and it is also curious that Ref. 15 does not discuss the large discrepancy between the value reported there and the earlier

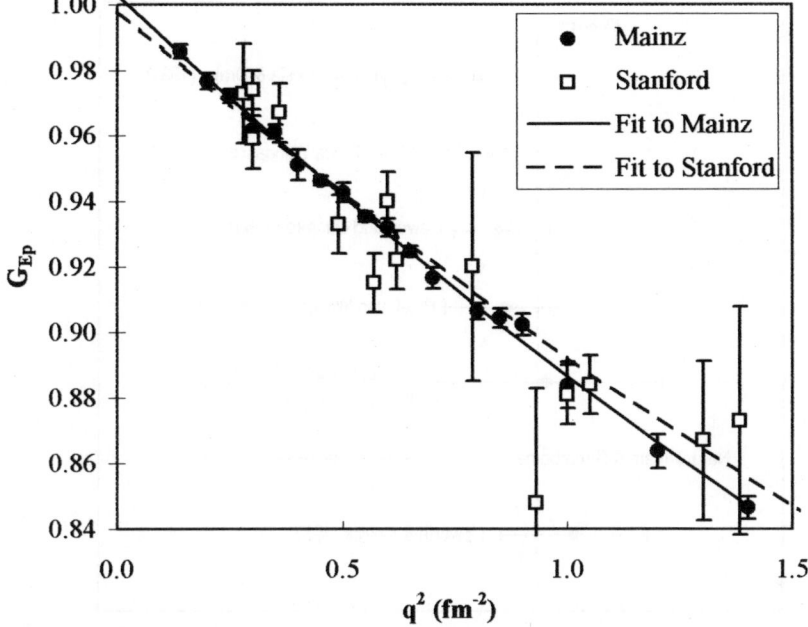

Figure 5: Measured values of the proton electric form factor G_{ep} as a function of momentum transfer q^2 for the data sets which lead to the "Stanford"[16] and "Mainz"[15] values of the proton charge radius. The lines are quadratic fits to the two data sets.

"Stanford" result, which had been the accepted value for the proton charge radius for many years. Finally, a recent analysis of world nucleon form factor data[19,20] concluded that the proton charge radius reported in Ref. 15 should be modified to 0.849(9) fm. It is clear that present knowledge of the proton charge radius is very unsatisfactory, and that a new measurement of this quantity is very desirable.

2.5 Future Prospects in Hydrogen

Although the proton size uncertainty presently imposes a barrier to improved tests of QED in hydrogen, one might hope that it will eventually be surmounted and so it is interesting to consider the prospects for improved measurements of Lamb shifts in hydrogen. The three experiments discussed above all involve measuring differences between pairs of hydrogen transitions chosen partly to facilitate the frequency comparison. In all three experiments, one of the two transitions has rather good statistics (H(1S-2S) and H(2S-6S,D)), while the others (H(2S-4S,D), H(2S-4P), H(1S-3S) respectively) have a larger statistical uncertainty and so limit the precision of the comparison. One might therefore hope to do better by comparing two narrow transitions via intermediate frequency standards. Steps in this direction have been made by the Garching and Paris groups, who have already measured the absolute frequencies of the H(1S-2S) and H(2S-8S,D) transitions respectively.[21,22] These two values can be combined to determine a value for the 1S Lamb shift with an uncertainty of 9 ppm, somewhat worse than the three recent values discussed above. Work on both transitions is continuing and it appears likely that this path will lead to an improved value for the

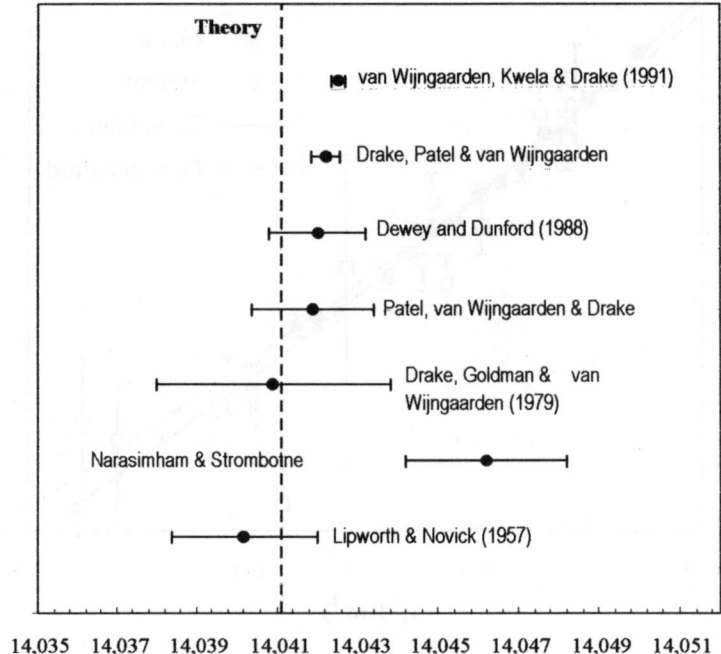

Figure 6: Measured values of the He$^+$(2S-2P) Lamb shift.[23-29] The dashed line shows the theoretical value.

1S Lamb shift. This potential for further improvement of course underlines the need for a better value for the proton charge radius.

3 Two-Photon Spectroscopy of Singly-Ionised Helium

A solution to the proton size problem is to compare a third transition in another system with the pair already compared in hydrogen, in order to separate the nuclear size and QED contributions to the Lamb shift. The two-photon 2S-nS,D transitions in singly-ionized helium are particularly suitable for this: the alpha particle size is well-known,[30] the 0.2 ms lifetime of the 2S state[31] permits beam experiments, the excited states are sufficiently narrow to reach an interesting level of precision, and the increase in Z emphasizes the interesting higher order corrections.

While bypassing the nuclear size uncertainty was the original reason for our interest in laser spectroscopy of He$^+$, another compelling motivation for the experiment has arisen out of the recent advances in the theory. As Figure 6 shows, with the new two-loop binding correction there is a large discrepancy between experiment and theory for the He$^+$(2S-2P) Lamb shift. The largest discrepancy, between theory and the best value,[29] is nine standard deviations! This is clearly a serious problem which must be resolved.

Since there is a large discrepancy for Z=2, it is natural to ask about the situation at still higher Z. Figure 7 summarises these measurements, and shows that only in He$^+$ do measurements test the two-loop binding correction. Measurements at higher Z do not

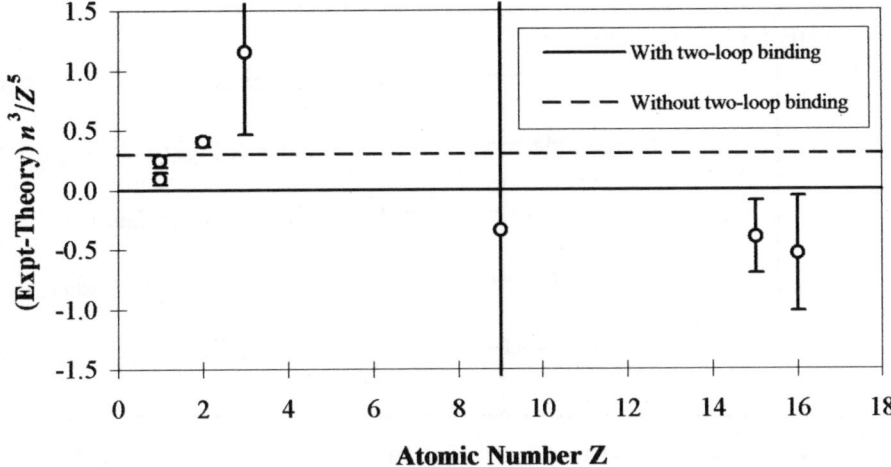

Figure 7: Differences between measured Lamb shifts and theory (normalised by the Z^5/n^3 scaling of the two-loop binding correction) versus atomic number Z. In addition to the hydrogen and He$^+$ measurements discussed in the text, the graph shows measurements on hydrogen-like lithium[32], fluorine[33], phosphorus[34] and sulphur.[35] The dashed line shows the position of the theory without the two-loop binding correction, indicating that only in He$^+$ are measurements sensitive to this large correction. Two points are shown for hydrogen because there are two theoretical values corresponding to the two experimental values for the proton charge radius.

have sufficient precision to test it (data exist for almost all Z < 20, but most of these measurements have such large error bars they do not fit on the graph).

Finally, it is interesting to note that to reach the level of the discrepancy one would need to split the 2S-4S resonance to only 15% of its 11 MHz natural width, making the experiment very attractive from the point of view of potential systematic errors. Work in this direction is now underway in our laboratory at Sussex.

4 The Proton Charge Radius

Since the usefulness of hydrogen for testing QED is at present limited by the uncertainty in the proton size, it is interesting to consider instead using the measurements of Lamb shifts in hydrogen to determine the proton charge radius (assuming of course that experiments in He$^+$ or some other system have resolved the discrepancies discussed above).

Figure 8 shows the differences between the best values for hydrogen Lamb shifts obtained by optical and radio-frequency spectroscopy, and the theoretical values corresponding to the two measurements of the proton charge radii. It is clear from this graph that if one had independent confirmation of the correctness of QED at this level, then the existing measurements would determine the proton charge radius to a considerably higher precision than has been possible from direction ep scattering. A new, more precise value for the proton charge radius is interesting for a number of reasons. It plays an important role in constraining models for the proton electromagnetic form factors,[19,20] and it also bears on attempts to calculate the hydrogen hyperfine structure. But the most intriguing possibility is that it may soon be possible to calculate the proton charge radius from first principles using lattice quantum chromodynamics. While initial results in this direction underestimate the charge radius

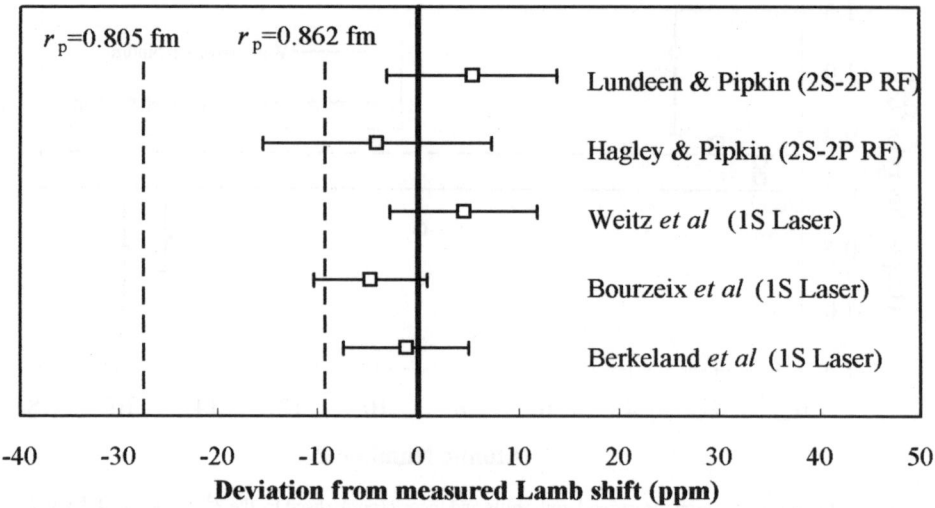

Figure 8: Comparison between measurements of hydrogen Lamb shifts[1,6,7,17,18] and the theoretical values corresponding to the two measurements of the proton charge radius. The solid line at the origin is the weighted average of the five experimental Lamb shift values.

by about 20%,[36,37] there is now some understanding of this problem,[38] and it appears that it might be possible to calculate the proton charge radius to a precision of a few percent within a few years.[39] Thus it may well turn out that after years of providing stringent tests of quantum electrodynamics, the role of hydrogen spectroscopy will shift to probing strong interactions and testing quantum chromodynamics.

5 Muonium

We turn now to another method of avoiding the complications of nuclear structure, namely the spectroscopy of the purely leptonic muonium atom μ^+e^-. Measurements on this system are also interesting because they test the assumption that the muon is just a heavy electron. The study of muonium has a long history, and the reader interested in more details is referred to the excellent review article by Hughes and zu Putlitz.[40] We will confine our attention here to two current experiments, laser spectroscopy of the 1S-2S transition, and microwave spectroscopy of the ground state hyperfine interval.

5.1 Laser Spectroscopy of the Muonium 1S-2S Transition

The energy levels of the muonium atom differ from those of the hydrogen atom by only 0.4% because the muon is 207 times heavier than the electron. As a result, the 1S-2S transition in muonium can be excited in much the same way as it is in hydrogen. The transition was first observed in 1988 at the KEK accelerator in Japan.[41] Subsequent work has taken place at the intense pulsed muon source provided by the ISIS facility at the Rutherford Appleton Laboratory in the United Kingdom.

We now consider some details of the ISIS experiment, which is a collaboration between Heidelberg, Oxford, Sussex, and Yale Universities, and the Rutherford Appleton Laboratory.[42] Thermal muonium atoms are formed in vacuum by stopping 27 MeV/c muons in a SiO_2 powder target. Most of the muons stop and form muonium in the target, and several percent of these atoms eventually diffuse to the surface and form a thermal cloud of atoms (average speed ~ 740 m/s). These atoms then interact with an intense pulsed standing wave of 244nm laser light in the region just above the target. The resulting excitation to the 2S level is detected by the photoionization of the excited atom with a third laser photon. This process produces a slow muon, which is accelerated by a pulsed electric field and transported through a sequence of energy- and momentum-selective electrostatic optics to a micro-channel plate detector (MCP). The signature of resonance on the 1S-2S transition consists of a particle arriving at the MCP in the appropriate time window, coupled with the detection of the subsequent decay positron in scintillator detectors surrounding the MCP.

The most recent published spectrum obtained in this way[42] is shown in Figure 9. The final value obtained from this experiment for the 1S-2S transition frequency is 2,455,529,002(33)(46) MHz, where the first uncertainty is statistical and the second systematic. This result is in reasonable agreement with the theoretical value of 2,455,528,934.0(3.6) MHz. The experiment determines the M(1S-2S) Lamb shift to be 6988(33)(46) MHz, in reasonable agreement with the theoretical value of 7056.1(1.0) MHz, and providing a test of QED at the level of 0.8%. This represents the best Lamb shift measurement in the muonium atom, improving slightly on the earlier microwave measurement of the 2S-2P Lamb shift.[43] The systematic error is dominated by effects associated with the intense pulsed laser needed to obtain an observable signal, in this case pulsed dye amplification of a 486 nm continuous-wave dye laser followed by frequency doubling to 243 nm. The most troublesome systematic error was a frequency chirp of the amplified pulse produced by the rapid time-variation of the refractive index of the dye in the three stages of the amplifier. A new version of this experiment is at present underway at ISIS in which it is expected that the chirp will be substantially reduced by the use of a new laser system based on an injection-seeded solid-state alexandrite laser, frequency-tripling to generate the required 244 nm light. In addition, any chirp will be monitored on a shot-by-shot basis by heterodyning the laser output pulse against a stable cw laser and digitising the beat signal.[44]

The 3.6 MHz uncertainty in the calculated value of the muonium 1S-2S interval is due almost entirely to the uncertainty in the muon/electron mass ratio, which enters

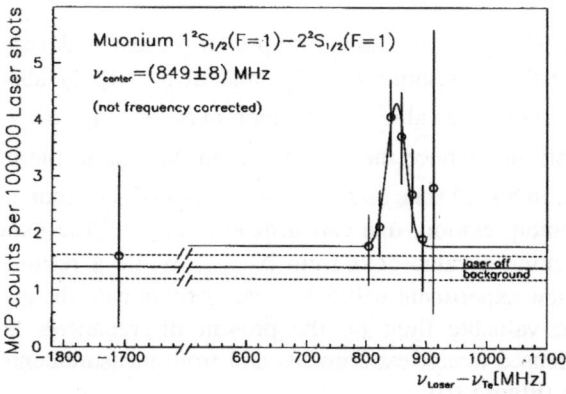

Figure 9: Spectrum of the muonium 1S-2S transition from Ref. 42. The laser frequency is measured at 486 nm relative to a calibration transition in $^{130}Te_2$.

the calculation via the simple reduced mass correction to the Dirac energy. Since most of the QED corrections in the H(1S-2S) - M(1S-2S) "isotope shift" cancel, a measurement of this quantity provides a very clean determination of the muon mass. The isotope shift measured in the first ISIS experiment[43] determined the muon mass to be 105.65880(29)(43) MeV/c^2, which is in good agreement with the accepted value.

The new experiment presently underway at the Rutherford Laboratory aims to reduce the uncertainty in the measurement of M(1S-2S) to the level of a few MHz, a further step along the path to obtaining tests of QED which are competitive in accuracy with those in hydrogen and He$^+$, but without the complications of nuclear structure.

5.2 Microwave Spectroscopy of the Muonium Ground State Hyperfine Interval

While the laser spectroscopy of muonium is a rather young field, radio-frequency spectroscopy has a rather long history.[40] In particular, some effort has been directed at precise measurement of the 4.46 GHz ground state hyperfine interval Δv because it is described almost completely by QED, with the exception of a small (15ppb) correction due to the weak interaction[45] (this is not at present accessible to experiment.) The current theoretical value for Δv is[46]

$$\Delta v_{\text{theory}} = 4\,463\,302.70(1.34)(0.06)(0.21) \text{ kHz}.$$

The principal uncertainty of 1.34 kHz arises from the uncertainty in μ_μ / μ_P, the ratio of the muon and proton magnetic moments. The second uncertainty arises from that in the values of α based on measurements of g-2 for the electron, and the third is an estimate of the uncertainty in a numerical QED calculation. It is known that some higher-order terms are probably significant at this level[46,47], and in addition there is a question about the $\alpha(Z\alpha)^2 E_F$ correction[48] which remains to be resolved. Nevertheless, one can still conclude that the best published measurement of Δv, made by a Yale-Heidelberg collaboration at Los Alamos Meson Physics Facility (LAMPF),[49]

$$\Delta v_{\text{expt}} = 4\,463\,302.88(16) \text{ kHz}.$$

is in good agreement with the theory. This same experiment also determined that

$$\mu_\mu / \mu_P = 3.183346(11).$$

A new experiment by the same group is at present underway at LAMPF, with the aim of reducing the uncertainty in Δv_{expt} and in μ_μ / μ_P by about a factor of five. This will lead to an improved value for the muon mass (via μ_μ / μ_P) and, since a better value of μ_μ / μ_P will also reduce the uncertainty in Δv_{theory}, to the most precise test of QED in the two lepton bound state and of the behaviour of the muon as a heavy electron. As an alternative interpretation, one can instead take the QED expression for Δv_{theory} and use it to determine a value of α from Δv_{expt}. While a result with the precision expected from the new experiment will not be competitive with the g-2 value of α, it will however cast some valuable light on the present discrepancies between the values obtained from condensed matter experiments and from measurements of the neutron de Broglie wavelength (Figure 10).

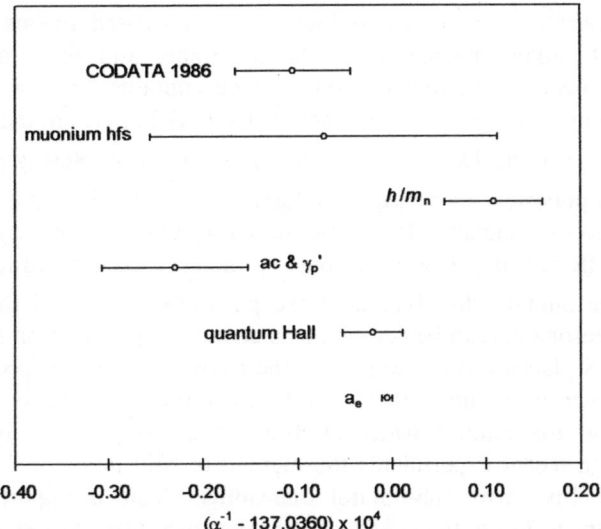

Figure 10: Recent determinations[50] of the fine structure constant α. The value labeled "CODATA 1986" is that given in the 1986 "Adjustment of Fundamental Constants".[51] The remaining values are determined from the muonium hyperfine structure (hfs),[50] the neutron de Broglie wavelength (h/m_n),[52] the combination of the ac Josephson effect and the gyromagnetic ratio of the proton in water (ac and γ_p'),[53] the quantized Hall effect (quantum Hall),[54] and from the anomalous magnetic moment of the electron (a_e).[50] A reduction in the uncertainty of the muonium hyperfine structure value by a factor five will enable it to discriminate between the condensed matter values.

We now briefly discuss some details of the new LAMPF experiment. The Breit-Rabi diagram for the relevant muonium ground-state energy levels is shown in Figure 11, which also includes the two microwave transitions ν_{12} and ν_{34} observed in the experiment at a magnetic field of 1.7 T. The hyperfine interval $\Delta\nu$ is determined from the sum $\nu_{12} + \nu_{34}$ and the magnetic moment ratio μ_μ/μ_p from the difference $\nu_{12} - \nu_{34}$.[40]

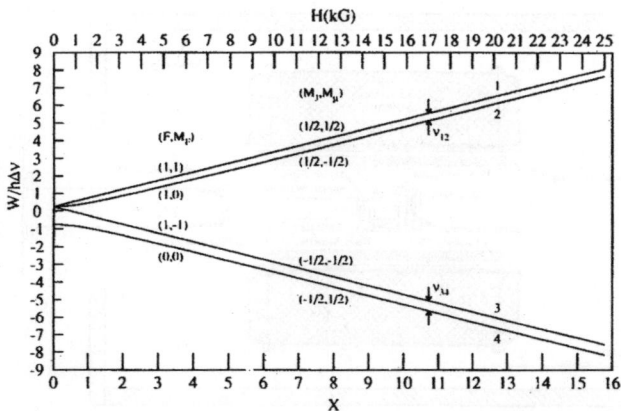

Figure 11: Breit-Rabi energy level diagram for the muonium $n = 1$ ground state in magnetic field H. $\Delta\nu$ is the hyperfine interval and X is a dimensionless parameter proportional to H.

The apparatus[55] is shown in Figure 12. Polarised muons produced in the LAMPF stopped muon channel enter the apparatus and form polarised thermal muonium atoms inside a microwave cavity which contains Kr at a pressure of 0.5 - 1.5 atm. The cavity is designed to resonate at 1.897 GHz (ν_{12}) in the TM_{110} mode and at 2.566 GHz (ν_{34}) in the TM_{210} mode. The cavity and its enclosing pressure vessel sit at the centre of a persistent-mode superconducting solenoid which provides the required stable, homogeneous magnetic field, monitored throughout the experiment with a sophisticated NMR system. The transitions ν_{12} and ν_{34} correspond to muon spin-flips and so, since the angular distribution of the positrons emitted in the muon decay is anisotropic, the resonance can be detected as a change in the positron rate observed in a set of scintillators placed downstream from the cavity. A typical spectrum obtained in this way is shown in Figure 13(a). This "conventional" line is power-broadened somewhat beyond the natural width of 145 kHz set by the 2.2 μs muon lifetime. However, in more recent experiments the method of "old muonium" has been used to obtain resonance lines with sub-natural linewidths. This is important because it is typically necessary to locate the centre of the resonance lines to less than 10^{-3} of their width, and so narrower lines help minimise many potential systematic errors. To obtain the "old muonium" lines, an electrostatic chopper just upstream of the solenoid divides the 650 μs LAMPF muon pulse into a series of 4μs muon pulses separated by 10μs periods without muons. The muonium resonance signal is then recorded as a function of time after the start of each muon pulse. Sub-natural linewidth resonances are then obtained from those atoms which have survived for several muon lifetimes (Figure 13(e)). There is of course a loss of statistical precision with this method, but this is partially compensated for by the narrower lines, so that the net loss of statistical power is only about a factor of two. So far data has been taken at several Kr pressures (to extrapolate the pressure shift of the resonance to zero), and using both magnetic field scans at fixed microwave frequency and frequency sweeps at fixed magnetic field. In addition to the line-narrowing obtained from "old muonium" and the two different methods of recording the resonance, the new experiment improves on the previous version in muon beam intensity and purity, and in the stability and homogeneity of the

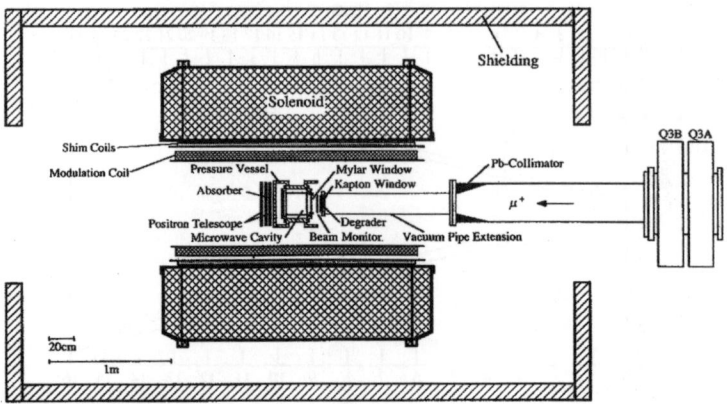

Figure 12: Apparatus used at LAMPF to measure the ground state hyperfine interval in muonium

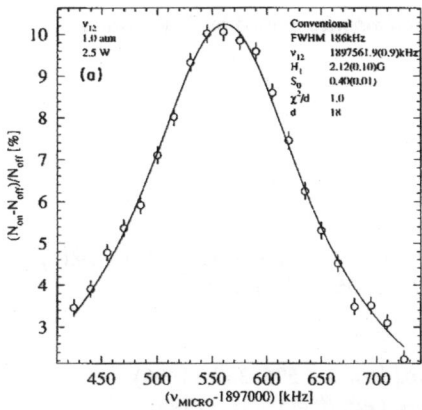 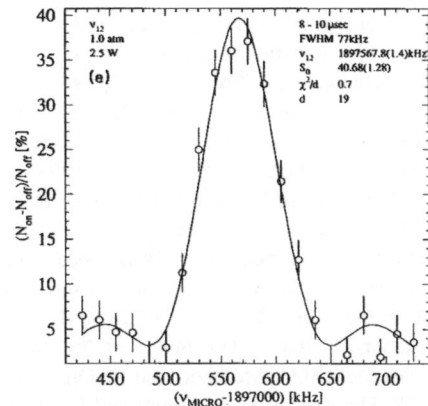

Figure 13: Muonium ν_{12} resonance lines obtained by a microwave frequency sweep with a 1.0 atm Kr target.[55] The left-hand spectrum (a) is a "conventional" line, whereas the right-hand spectrum (e) is an "old muonium" line (delay window = 8-10 μs), with a width of about half the 145 kHz natural width.

magnetic field. The analysis of the data is at present underway. It is expected to produce a substantial improvement on the previous result for $\Delta\nu$ and μ_μ / μ_p.

6 Summary

The current status of tests of QED in light hydrogenic systems can be summarised as follows. The highest precision (6 ppm) has been obtained in three recent measurements of the hydrogen 1S Lamb shift, although these results are limited as tests of QED by the uncertainty in the proton charge radius. In He^+ there is now a nine standard deviation discrepancy between theory and experiment for the 2S Lamb shift, and no other experiments are at present sensitive to the two-loop binding correction which is largely responsible for the problem. A new measurement of this Lamb shift by laser spectroscopy is underway in our laboratory at Sussex. New results are expected shortly from both laser and microwave spectroscopy of muonium, to complement work in hadronic atoms and also provide information about the muon mass. Finally, one other possibility for future tests of QED which has not been discussed here is laser spectroscopy of muonic atoms. More details can be found in Ref. 56.

7 Acknowledgements

The Yale hydrogen experiment was supported by Yale University, an NIST Precision Measurement Grant, and the NSF. I am happy to acknowledge many useful discussions with my colleagues from Yale, Heidelberg, Oxford and the Rutherford Laboratory.

References

1. D.J. Berkeland, E.A. Hinds, and M.G. Boshier, *Phys. Rev. Lett.* **75**, 2470, (1995).
2. B. Cagnac, G. Grynberg, and F. Biraben, *J. Phys. France* **34**, 845, (1973).
3. E.V. Baklanov and V.P. Chebotaev, *Opt. Comm.* **12**, 312, (1974).
4. T.W. Hänsch, S.A. Lee, R. Wallenstein, and C. Wieman, *Phys. Rev. Lett.* **34**, 307, (1975).
5. M.G. Boshier et al., *Phys. Rev. A* **40**, 6169, (1989).
6. S.R. Lundeen and F.M. Pipkin, *Phys. Rev. Lett.* **46**, 232, (1981).
7. E.W. Hagley and F.M. Pipkin, *Phys. Rev. Lett.* **72**, 1172, (1994).

8. J.R. Sapirstein and D.R. Yennie, in *Quantum Electrodynamics,* edited by T. Kinoshita, (World Scientific, Singapore, 1990).
9. K. Pachucki, *Ann. Phys.* **226**, 1, (1993).
10. K. Pachucki, *Phys. Rev. Lett.* **72**, 3154, (1994).
11. M.I. Eides and V.A. Shelyuto, *JETP Lett.* **61**, 478, (1995).
12. S.G. Karshenboim, *JETP* **76**, 541, (1993).
13. K. Pachucki and H. Grotch, *Phys. Rev. A* **51**, 1854, (1995).
14. K. Pachucki et al., *J. Phys. B* **29**, 177, (1996).
15. G.G. Simon, C. Schmitt, F. Borkowski, and V.H. Walther, *Nucl. Phys.* **A333**, 381, (1980).
16. L.N. Hand, D.G. Miller, and R. Wilson, *Rev. Mod. Phys.* **35**, 335, (1963).
17. M. Weitz et al., *Phys. Rev. A* **52**, 2664, (1995).
18. S. Bourzeix et al., *Phys. Rev. Lett.* **76**, 383, (1996).
19. P. Mergell, U.G. Meißner, and D. Drechsel, *Nuclear Physics A* **596**, 367, (1996).
20. H.W. Hammer, U.G. Meißner, and D. Drechsel, *Phys. Lett. B* **385**, 343, (1996).
21. T. Andrae et al., *Phys. Rev. Lett.* **69**, 1923, (1992).
22. F. Nez et al., *Europhys. Lett.* **24**, 635, (1993).
23. E. Lipworth and R. Novick, *Phys. Rev.* **108**, 1434, (1957).
24. M.A. Narasimham and R.L. Strombotne, *Phys. Rev. A* **4**, 14, (1971).
25. G.W.F. Drake, S.P. Goldman, and A. van Wijngaarden, *Phys. Rev. A* **20**, 1229, (1979).
26. J. Patel, A. van Wijngaarden, and G.W.F. Drake, *Phys. Rev. A* **36**, 5130, (1987).
27. G.W.F. Drake, J. Patel, and A. van Wijngaarden, *Phys. Rev. Lett.* **60**, 1002, (1988).
28. M.S. Dewey and R.W. Dunford, *Phys. Rev. Lett.* **60**, 2014, (1988).
29. A. van Wijngaarden, J. Kwela, and G.W.F. Drake, *Phys. Rev. A* **43**, 3325, (1991).
30. I. Sick, *Phys. Lett.* **116B**, 212, (1982).
31. E.A. Hinds, J.E. Clendenin, and R. Novick, *Phys. Rev. A* **17**, 670, (1978).
32. M. Leventhal, *Phys. Rev. A* **11**, 427, (1975).
33. H.W. Kugel et al., *Phys. Rev. Lett.* **35**, 647, (1975).
34. H.-J. Pross et al., *Phys. Rev. A* **48**, 1875, (1993).
35. A.P. Georgiadis et al., *Phys. Lett. A* **115**, 108, (1986).
36. T. Draper, R.M. Woloshyn, and H.-F. Liu, *Phys. Lett. B* **234**, 121, (1990).
37. D.B. Leinweber, R.M. Woloshyn, and T. Draper, *Phys. Rev. D* **43**, 1659, (1991).
38. D.B. Leinweber and T.D. Cohen, *Phys. Rev. D* **47**, 2147, (1993).
39. G.P. Lepage, (private communication, 1995)
40. V.W. Hughes and G. zu Putlitz, in *Quantum Electrodynamics,* edited by T. Kinoshita, (World Scientific, Singapore, 1990).
41. S. Chu et al., *Phys. Rev. Lett.* **60**, 101, (1988).
42. F. Maas et al., *Phys. Lett.* **A187**, 247, (1994).
43. K.A. Woodle et al., *Phys. Rev. A* **41**, 93, (1990).
44. I. Reinhard et al., *Appl. Phys. B* **63**, 467, (1996).
45. M.I. Eides, *Phys. Rev. A* **53**, 2953, (1996).
46. T. Kinoshita and M. Nio, *Phys. Rev. D* **53**, 4909, (1996).
47. S.G. Karshenboim, *Z. Phys. D* **36**, 11, (1996).
48. K. Pachucki, *Phys. Rev. A* **54**, 1994, (1996).
49. F.G. Mariam et al., *Phys. Rev. Lett.* **49**, 993, (1982).
50. T. Kinoshita, *Phys. Rev. Lett.* **75**, 4728, (1995).
51. E.R. Cohen and B.N. Taylor, *Rev. Mod. Phys.* **59**, 1121, (1987).
52. E. Kruger, W. Nistler, and W. Weirauch, *Metrologia* **32**, 117, (1995).
53. E.R. Williams et al., *IEEE Trans. Inst. Meas.* **38**, 233, (1989).
54. M.E. Cage et al., *IEEE Trans. Inst. Meas.* **38**, 284, (1989).
55. M.G. Boshier et al., *Phys. Rev. A* **52**, 1948, (1995).
56. M.G. Boshier, V.W. Hughes, K.J. Jungmann, and G. zu Putlitz, *Comments At. Mol. Phys.* **33**, 17, (1996).

NEW EXPERIMENTS WITH ATOMIC LATTICES BOUND BY LIGHT

A. GÖRLITZ, M. WEIDEMÜLLER*, T. HÄNSCH, AND A. HEMMERICH‡

*Sektion Physik, Universität München, Schellingstraße 4/III, D-80799 Munich, Germany
and Max-Planck-Institut für Quantenoptik, D-85748 Garching, Germany*

This paper summarizes our recent experimental work on two topics: the application of Bragg diffraction to optical lattices and the generation of a novel kind of optical lattice with extremely reduced fluorescence. We have used Bragg diffraction to observe the back action of the atoms on the lattice beams and to explore unusual vibrational modes with a position spread oscillating at the second harmonic frequency. We present a scheme for three-dimensional dark optical lattices and discuss their experimental realization.

1 Introduction

The refinement of laser cooling in the past decade has yielded the production of a novel type of cold matter: three-dimensional lattices of ultra cold atoms bound by standing light waves.[1] In 1992 two groups at NIST in Gaithersburg and at the ENS in Paris produced atoms confined by light into arrays of parallel planes.[2] Shortly after we were able to demonstrate first two- and three-dimensional atomic lattices in Munich by superposing four or six laser beams with controlled phases.[3] The ENS group also proceeded to realize three-dimensional lattices using fewer light beams.[4] In this case no phase control is needed, however, radiation pressure is not balanced on the microscopic scale.[5] Such *optical lattices* lend themselves as nearly ideal model systems in quantum optics because they uniquely combine the presence of a rich quantum structure easily accessible in experiments with the possibility of a theoretical understanding based on first principles. The trapping potential arises from the spatially varying light-shifts of the atomic ground state Zeeman levels and provides microscopic light traps at the intensity antinodes. Although these light traps are typically only a few hundred microkelvin deep their small size leads to vibrational energy levels quantized on the 100 kHz scale. Sub-Doppler cooling prepares a nearly thermal distribution of the atoms among these vibrational levels at temperatures of a few microkelvin where most atoms are trapped in the first few bound states.

The excitement about optical lattices initially arose because they appeared to be the first example of a neutral atom trap which operates in the Lamb-Dicke regime. In fact, the quantized atomic motion in the microscopic traps of optical lattices could be observed via probe transmission and fluorescence spectroscopy. In probe transmission spectra surprisingly sharp resonances were found originating from stimulated Raman transitions between different vibrational levels. A variety of other unexpected spectral features arising from multi-wave mixing processes as e.g. phase conjugation[6,7] or hyper-Raman transitions[8] were identified later on. Novel aspects of such spectra are still being discovered: for example the ENS group recently reported spectral features connected with propagating excitations which resemble stimulated Brillouin scattering in crystals.[9]

A novel kind of optical lattice with much reduced fluorescence has received considerable attention because it promises to be a good candidate for holding very dense atomic samples. Soon after a first proposal in one dimension by the ENS-group[10] we recently could demonstrate experimentally a three-dimensional version of such dark optical lattices[11].

Another class of recent experiments concentrates on aspects of optical lattices connected with their periodic order. For example, as reported by C. Salomon elsewhere in these Proceedings, Bloch oscillations could be observed not long ago.[12] Our group in Munich and the NIST group have applied the method of Bragg diffraction.[13,14] With this new tool both groups were able to explore exciting new aspects of optical lattices. One example is the observation of unusual vibrational modes with oscillating position spread.

2 Bragg diffraction in optical lattices

The situation we encounter with optical lattices differs from that of usual crystallography in some respects: the distance between adjacent lattice sites is typically several thousands of Ångstroms, i.e. three orders of magnitude larger than in crystalline solids. One can therefore employ optical frequencies for diffraction experiments. Optical lattices are very dilute, i.e. only a few percent of the lattice sites are typically populated. One could suspect that this should prohibit the observation of well defined diffraction maxima. But in contrast to crystals where empty lattice sites cause strong perturbations of the long range order, the periodicity in dilute optical lattices is basically determined by the lattice field and thus should be preserved over its entire extension. Only at higher filling rates we may expect distortions of the lattice due to back action of the atoms on the lattice beams. Consequently, Bragg diffraction is an interesting method to investigate this regime. Experimentally, we in fact find a contraction of the lattice with increasing atomic density.

The vibrational motion of the atoms inside the potential wells leads to an increase of the size of the atomic center-of-mass wave function, similarly as for the thermal motion of the ion cores in a crystal. This yields a reduction of the contrast of the Bragg maxima usually described by the Debye-Waller factor which can be easily observed experimentally. Thus, Bragg diffraction is a means to measure the degree of atomic localization inside the potential wells. By comparing the diffracted power of two Bragg maxima arising at different scattering angles we may even obtain the atomic localization without having to refer to parameters not easily accessed experimentally (atomic polarizability, density). We have used this new technique to observe the oscillating mean square extension of unusual vibrational states prepared by a non-adiabatic increase of the trapping potential depth.

2.1 Bragg diffraction and four-wave-mixing

Let us briefly discuss the connection between Bragg diffraction and four-wave mixing. If we irradiate the optical lattice by an additional probe beam, we find two different phase-matched four-wave mixing processes (fig. 1(a)). In the first process (left in fig. 1(a)) it is the interference between a lattice beam and the probe that introduces a spatial modulation of the atomic polarizability. A second lattice beam is then scattered off this polarizability grating onto the probe beam axis. This process can display an extremely sharp resonance behavior which is related to transport properties of the optical lattice.[7] However, it definitely cannot contain information on the atomic lattice geometry since the probe itself is contributing to the generation of the polarizability pattern.

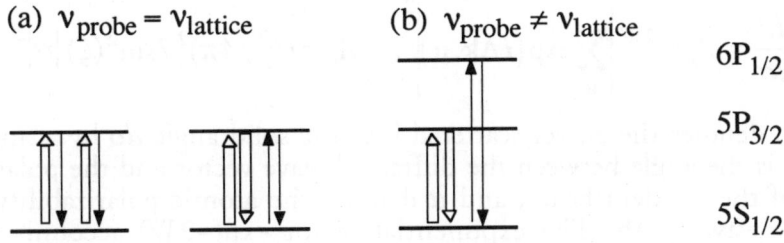

Figure 1: Possible phase-matched four-wave mixing processes arising in an optical lattice (unfilled arrows indicate lattice beams) in the presence of an additional probe beam (black arrows) when the probe frequency nearly equals that of the lattice (a) and when the two frequencies are different (b). In (a) only the second process gives rise to Bragg diffraction. In (b) an additional atomic level is used to resonantly enhance the Bragg process which then is by far dominant.

In the second four-wave mixing process (right in fig. 1(a)) it is the interference between two lattice beams yielding a polarizability distribution which leads to scattering of the probe beam. In this process the probe beam does not alter the spatial properties of the medium which are solely determined by the lattice beams. The first process is automatically phase-matched for all possible directions of incidence of the probe, whereas for the second process the phase-matching condition turns out to be equivalent to Bragg's law which imposes severe restrictions on the angle of incidence of the probe beam. Clearly, only the second process should be called Bragg diffraction, although in practice both processes may occur simultaneously and the first often is the dominating one, as e.g. in the experiments in refs. [4,6,7]. In particular, if the frequencies of the lattice beams and the probe beam are similar, both processes are near-resonant and can thus occur simultaneously (this is the case in fig. 1(a)). When the two kinds of beams have very different frequencies (fig. 1(b)), the Bragg process can be selectively chosen to be resonantly enhanced while the first process becomes negligible. This is the path we have followed in our experiments[14] while a second solution consists in quickly turning off the lattice beams during the Bragg scattering at the cost of destroying the lattice after a couple of microseconds. The latter was practiced in the experiments of the NIST group.[13]

2.2 Diffraction efficiency and atomic localization

The diffraction efficiency can be estimated by means of a simple model assuming each atom being bound to a harmonic potential centered at some lattice site u with root mean square deviations of the atomic positions from the trap centers ΔR_i, $i = 1,2,3$, along the three axes of a cartesian coordinate system. We consider a linearly polarized plane traveling wave with intensity I incident on the atomic lattice. The power dP diffracted into the solid angle $d\sigma$ is then given by the far-field expression[15]

$$\frac{dP}{d\sigma} = A_s e^{-2W} \left| \sum_{\mathbf{u}} \exp(i\Delta\mathbf{k}\cdot\mathbf{u}) \right|^2, \quad A_s \equiv (k_b^2/4\pi)^2 \, I \sin^2(\zeta) |\alpha|^2. \quad (1)$$

Here, A_s denotes the power scattered into the solid angle $d\sigma$ by a single free atom, ζ is the angle between the diffracted wave vector and the polarization vector of the incident beam, and α denotes the atomic polarizability at the incident wavelength. The exponential factor $\exp(-2W)$ accounts for the atomic localization in the potential well. The Debye-Waller factor W is given by

$$W = \frac{1}{2} \sum_{i=1,2,3} (\Delta k_i)^2 (\Delta R_i)^2 \quad (2)$$

where Δk_i denote the cartesian coordinates of $\Delta \mathbf{k}$. The sum in eq. (1) has to be taken over all lattice vectors u belonging to a populated potential well. This structure factor displays a resonance behavior for $\Delta \mathbf{k}$ satisfying the Bragg condition. In this case $\Delta\mathbf{k}\cdot\mathbf{u}$ is a multiple of 2π for any u and the sum becomes equal to the square of the number of irradiated atoms N. If N is sufficiently large, the divergence angle of the diffracted beam is limited by the diameter D of the irradiated atomic sample independent of the degree of occupancy, i.e. the angular width $\delta\theta$ of the diffracted beam is related to D by the approximate relation $\delta\theta \approx \lambda_r/D$. The same relation approximately holds for the acceptance angle of the incident beam.

2.3 Observation of Bragg diffraction

Fig. 2(a) shows a sketch of our Bragg diffraction experiment. The optical lattice is loaded by means of a magneto-optic trap (MOT) which collects Rubidium atoms from a hot dilute vapor at room temperature. After typically 100 ms of trapping time the MOT is turned off and the atoms are released into the optical lattice. The cubic body-centered lattice is formed within a few hundred microseconds by three mutually orthogonal optical standing waves at 780 nm tuned to the red side of the $5S_{1/2}(F=3) \rightarrow 5P_{3/2}(F'=4)$ transition of ^{85}Rb (for details see ref. [16]). The lattice typically has a lifetime of about 1 s limited by collisions with hot background atoms. For Bragg diffraction we use a blue light beam tuned near to the $5S_{1/2}(F=3) \rightarrow 6P_{1/2}(F'=2$ or $F'=3)$ transition at $\lambda_b = 421.7$ nm in order to enhance the scattering cross section. We use only a few ten microwatts in a collimated beam of 5 mm diameter (divergence 0.1 mrad) in order to avoid large radiation pressure on the trapped atoms.

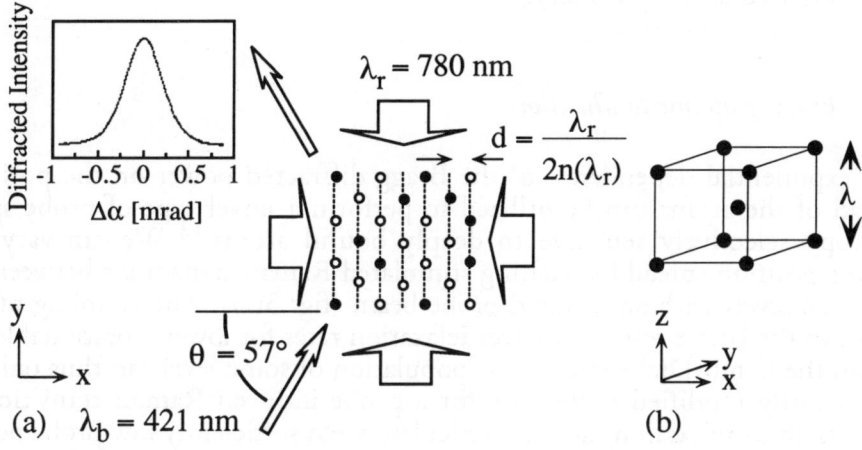

Figure 2: Sketch of our Bragg diffraction experiment (a). A violet laser beam at 421 nm is diffracted from the [201]-lattice planes of a cubic body-centered lattice of Rubidium atoms (b) bound by a infrared optical standing wave at 780 nm.

A primitive unit cell of the body centered cubic optical lattice sketched in fig. 2(b) is spanned by the vectors $\mathbf{a}_1 = \lambda_r \mathbf{x}$, $\mathbf{a}_2 = \lambda_r \mathbf{y}$, and $\mathbf{a}_3 = \lambda_r(\mathbf{x}+\mathbf{y}+\mathbf{z})/2$ where $\lambda_r = 780$ nm is the wavelength of the trapping light and x,y,z denote the cartesian unit vectors along the six lattice beams. A basis of the reciprocal lattice is then defined by the vectors: $\mathbf{b}_1 = k_r(\mathbf{x}-\mathbf{z})$, $\mathbf{b}_2 = k_r(\mathbf{y}-\mathbf{z})$, and $\mathbf{b}_3 = 2k_r\mathbf{z}$ where $k_r = 2\pi/\lambda_r$. According to the Bragg condition the difference $\Delta \mathbf{k}$ between the incident and the diffracted wave vector equals a reciprocal lattice vector, i.e. $\Delta \mathbf{k} = n_1\mathbf{b}_1 + n_2\mathbf{b}_2 + n_3\mathbf{b}_3$ where n_i are integers. The maximum possible $\Delta \mathbf{k}$ is given by the relation $|\Delta \mathbf{k}| \leq 2k_b$ where k_b is the wave number of the incident beam. If we restrict ourselves to an incident light beam traveling within the xy-plane (i.e., $2n_3 = n_1 + n_2$), the Bragg condition can be written $2\lambda_r \cos(\theta) = (n_1^2 + n_2^2)^{1/2} \lambda_b$, where θ denotes the angle between the incident wave vector and the normal to the set of lattice planes from which the incident beam is reflected. For most diffraction experiments we have chosen the [201]-lattice planes (i.e. $n_1=2$, $n_2=0$, $n_3=1$) for which the Bragg condition is satisfied for $\theta = 57.3°$.

When the Bragg condition is met to better than 0.5 mrad we observe a strong increase of the scattered intensity by at least three orders of magnitude as compared to the background of diffuse scattering (see upper left corner of fig. 1(a))[14]. The signal disappears when the time phase differences of the three one-dimensional standing waves forming the lattice are set to values for which no optical lattice is formed. We find that only ($F=3$, $|m|=3$)-atoms contribute to the Bragg diffraction by tuning the incident blue beam to the ($F=3$) → ($F'=2$) transition. If we now adjust the linear polarization along the z axis (π polarization), the Bragg signal disappears. This is only possible if all atoms are populating the outermost ground state Zeeman levels ($F=3$, $|m|=3$), since these levels do not couple to the excited state ($F'=2$) for π polarized light. Note that the population of the ($F=3$, $m=-3$)-level is negligible because of the spin-polarized lattice geometry.

2.4 Probing atomic localization

The exponential dependence of the Bragg diffracted power on the position spread of the atoms can be utilized to perform a novel type of probe spectroscopy selectively sensitive to deeply bound atoms.[14] We can vary the atomic position spread by exciting stimulated Raman transitions between vibrational levels with an infrared probe beam (fig. 3(c)). The coupling of the atoms to the lattice results in lower relaxation rates for lower vibrational levels due to the Lamb-Dicke effect. The population of some level can thus only be significantly modified if the rate for a probe-induced Raman transition is larger than its relaxation rate. In particular, when sufficiently low probe power is used only the population of the vibrational ground state is significantly altered. This probe-induced population transfer leads to an increase of the atomic position spread along the direction of the probe beam, e.g. ΔR_1 in eq. (2) increases when the probe travels along the x direction, and according to eq. (1) the diffracted power decreases.

This effect is demonstrated in fig. 3 where we have simultaneously recorded the diffracted power at 422 nm (a) and the probe transmission at 780 nm (b) versus the probe frequency detuning. The intensity of the probe beam was adjusted sufficiently low such that mainly the population of the vibrational ground state was modified. We clearly see that the resonances in the diffracted signal are narrower (particularly for the second Raman sideband near $\pm 2\nu_{vib}$) and exhibit practically no anharmonicity (i.e. the second pair of resonances appears exactly at $\pm 2\nu_{vib}$) in contrast to the probe transmission spectrum. This indicates that mainly the vibrational ground state contributes to the Bragg spectrum as expected for our experimental parameters. The different sensitivity of the Bragg spectrum and the probe transmission spectrum on the deeply bound atoms also becomes visible when we heat the lattice by tuning the lattice beams closer to resonance. In this case the Bragg signal completely vanishes while the transmission spectrum is still preserved showing significantly broadened Raman resonances.

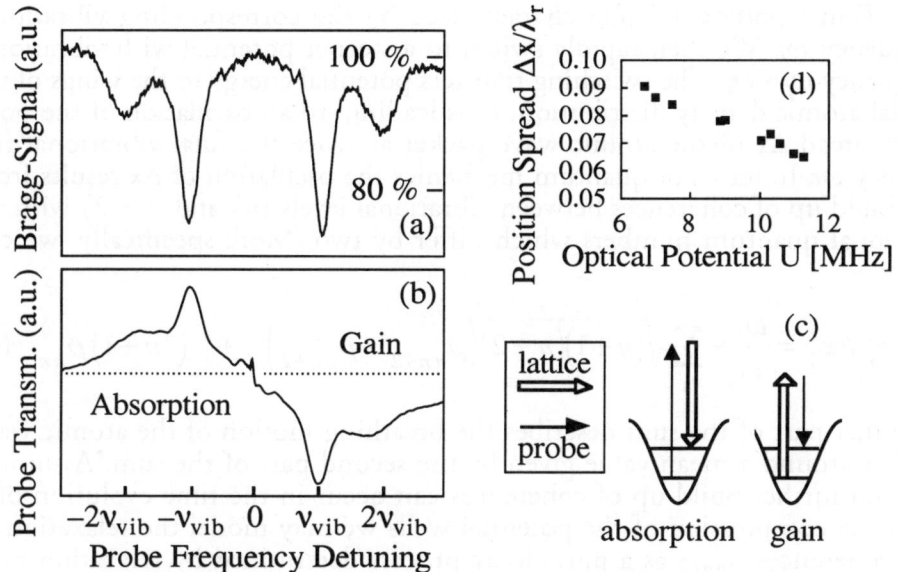

Figure 3: The Bragg diffracted power (a) and the transmission of a weak infrared probe beam (b) are simultaneously recorded versus the frequency detuning between the probe and the lattice field (In (a) 100% indicates the diffracted power without infrared probe). (d) Atomic Confinement plotted versus the depth of the potential well.

We can refine our observation technique if we compare the powers P_1 and P_2 diffracted from different lattice planes at different Bragg angles θ_1 and θ_2. For simplicity we assume isotropic potentials, i.e. $\Delta R^2 = \Delta R_1^2 + \Delta R_2^2 + \Delta R_3^2 = 3 \Delta R_i^2$ which is the case for our experiments. According to eq. (1) we then find

$$\Delta R^2 = \frac{3}{4 k_b^2} \frac{\ln(P_1) - \ln(P_2)}{\cos^2(\theta_2) - \cos^2(\theta_1)}. \quad (3)$$

Consequently, an absolute measurement of the atomic position spread is possible without having to refer to atomic parameters such as the density or the polarizability which are not easily accessible with high accuracy. In fig. 3(d) we show a measurement of the atomic position spread versus the potential depth, which was varied by changing the intensity of the lattice beams. The data were obtained by comparing the power diffracted from the [201] and the [312] lattice planes which yield Bragg maxima at $\theta_1 = 57.3°$ and $\theta_2 = 31.4°$, respectively.

Our method to perform direct measurements of the atomic position spread opens exciting possibilities such as, for example, the observation of unusual vibrational states for which this quantity oscillates in time.[17] Such states can be prepared by non-adiabatically switching the potential depth or alternatively by resonant parametric modulation of the potential. The basic physics of such states is understood in terms of a one-dimensional harmonic oscillator model. Our starting point is a thermal atomic state with temperature T_i in a potential $U(\omega_i)$ characterized by the corresponding vibrational frequency ω_i. We then rapidly switch to a steeper potential with vibrational frequency $\omega_f > \omega_i$. The switching transfers potential energy to the wings of the initial atomic density distribution, thus leading to an oscillation of the position spread Δx of the atomic wave packet at twice the final vibrational frequency ω_f. In terms of quantum mechanics the oscillation of Δx results from the build up of coherences between vibrational levels $|n\rangle$ and $|n+2\rangle$ with vibrational quantum numbers which differ by two. More specifically, we can write

$$k_r^2 \Delta x^2 = \frac{\omega_{rec}}{\omega_f} \sum_n \sqrt{(n+1)(n+2)} \left(\rho_{n\,n+2} - \rho^*_{n\,n+2} \right) + (2n+1) \rho_{nn} \quad (4)$$

The first part of the sum describes the breathing motion of the atomic wave packet around a mean value given by the second part of the sum. Assuming that no further build up of coherences can occur in the time evolution of ρ after the compression of the potential wells we may model the relaxation of the coherences ρ_{nn+2} as a pure decay process determined by relaxation rates Γ_{nn+2} (in contrast to the more complex relaxation of the populations). In our optical lattice we expect the coherences to decay due to the presence of spontaneous Raman transitions between vibrational levels which occur at a rate approximately given by $\Gamma_{nm} = \gamma(n+m+1)$ where the ground state relaxation rate γ equals the optical pumping rate Γ' for an unbound atom multiplied with the small Lamb-Dicke factor ω_{rec}/ω_f ($\hbar\omega_{rec}$ = photon recoil energy).[18] Since the coherences for high values of n may yield significant contributions to the oscillation of Δx^2 the decay rate of the breathing motion will be typically a few times γ, i.e. typically on the order of a few 10^5 s^{-1}. An other possible source of decoherence is the anharmonicity of the potential wells which yields slightly different oscillation frequencies ω_n for ρ_{nn+2} at different values of n. This leads to decays and subsequent revivals of the oscillating part of Δx on a time scale determined by the inverse of the frequency spread of the ω_n's. The time scale for the relaxation of the populations towards their steady state values in the compressed potential is closely connected with the rates for spontaneous Raman transitions which change the magnetic quantum number and which are responsible for the dissipation of kinetic energy in Sisyphus cooling mechanisms.[19] Such processes occur at rates which can be significantly smaller then γ and consequently cooling (heating) rates in optical lattices can be below 10^4 s^{-1}.

In our experiments we either raise the potential well rapidly within about 500 ns or alternatively we parametrically modulate the potential depth around some mean value at twice the normal vibrational frequency and record the Bragg diffracted power as a measure for ΔR^2. A typical observation is shown in fig. 4. After the potential well is raised at t = 12 μs we see a significant increase of the mean diffracted power accompanied by a rapid oscillation which decays in about 10 μs. We typically see three maxima of this oscillation, the third being slightly larger than the second. A second, much slower, decay is also observed which we attribute to the relaxation of the vibrational populations. We have checked that the observed oscillation frequency approximately equals $2\omega_f$ by recording a probe transmission spectrum (similar as that in fig. 3(b)) and directly extracting the value of ω_f.

When we model our observation within a purely harmonic model (using the experimental initial and final values ω_i and ω_f from fig. 4) the best assumptions are $T_i = 5$ µK and $T_f = 12$ µK for the initial and final temperatures respectively, a cooling rate $\gamma_f = 4 \times 10^3$ s^{-1} and a vibrational ground state relaxation rate $\gamma = 6 \times 10^4$ s^{-1}. However, the decoherence time scale due to anharmonicity is comparable with $1/\gamma$ in our experiment thus allowing a revival to occur within the relaxation time. This explains why the third oscillation maximum is not clearly below the second. In the parametric driving experiments we see a steady state oscillation in the Bragg signal at the driving frequency. We find a distinct resonance for the driving frequency at about $2\omega_f$ with a width of approximately 50 kHz limited by damping and the anharmonicity of the potential wells. Depending on the detuning with respect to this resonance the breathing oscillation is shifted in phase with respect to that of the modulated potential in accordance with our expectations for a driven oscillation.

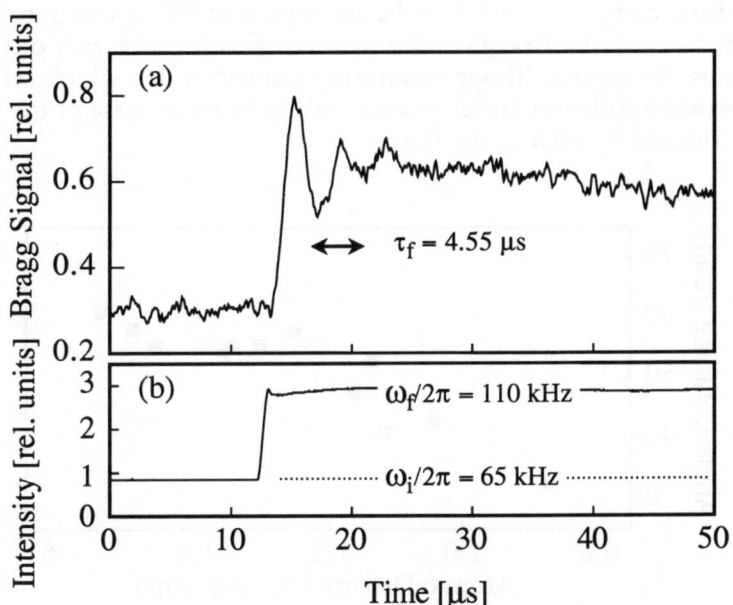

Figure 4: In (a) the diffracted power is plotted. At t = 12 µs the optical potential (with initial vibrational frequency ω_i) is raised non-adiabatically (b) (to the value ω_f) causing a transient oscillation in (a) at two times ω_f ($\tau_f = 2\pi/\omega_f$).

2.5 Back action of atoms on the lattice field

Despite the relatively low filling fraction in our lattices (few percent) we can observe back action of the atoms on the lattice beams. To lowest order a quite simple model applies which treats the atoms as an inhomogeneously distributed medium which gives rise to an index of refraction. In this picture, we account for first order scattering of the lattice beams from the periodically arranged atoms. Since the optical lattices treated in this section operate at negative detuning with respect to the atomic resonance frequency, the index of refraction is larger than one. This yields a reduced effective optical wavelength and thus a contraction of the lattice. The fact that the atoms are well confined enhances the refractive index as compared to that expected if the atoms were distributed homogeneously. This is because the atoms are trapped at high intensities where their interaction with the light field is larger as compared to the mean interaction strength in a homogeneously distributed atomic sample. In addition to forward scattering which leads to a refractive index in a homogeneous medium, the atomic localization in optical lattices yields additional scattering contributions. In case of simple lattice geometries (as that treated here) the wave vectors of the lattice beams represent Bravais vectors of the reciprocal lattice, i. e. the Bragg condition is satisfied for each pair of the lattice beams. Thus, we expect (Bragg) scattering contributions which redistribute photons between different lattice beams leading to an increase of the refractive index experienced by each of the beams.

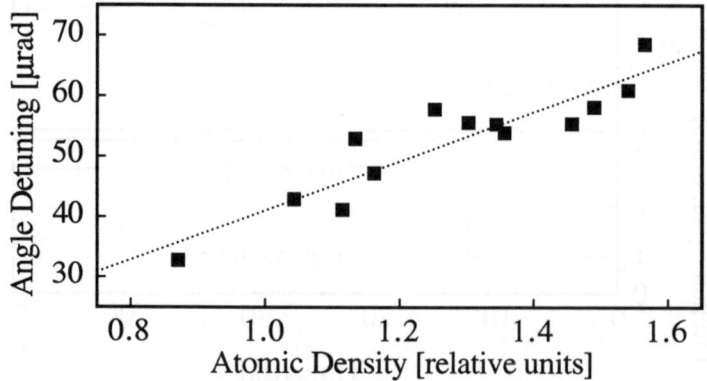

Figure 5: The angle satisfying the Bragg condition is plotted versus the density of the atomic sample loaded into the lattice.

In case of our cubic body-centered lattice the refractive index for each of the lattice beams is approximately given by the relation $n(\lambda_r) - 1 = (1 + \exp(-W_0) + 4 \exp(-W_{90})) \chi/4$ where χ is the steady-state atomic susceptibility for circular polarized light.[20] The right hand side consist of three terms: a "one" resulting from forward-scattering (which also occurs in homogeneous media), $\exp(-W_0)$ (with $W_0 = 2/3 \, k_r^2 \, \delta R^2$) due to backward scattering from the counter propagating beam, and 4 exp($-W_{90}$) (with $W_{90}=1/3\ k_r^2\delta R^2$) accounting for 90°-scattering from the remaining four lattice beams. The latter two contributions depend on the atomic position spread $\delta R^2 = \delta R_1^2+\delta R_2^2+\delta R_3^2$. The difference of the Debye-Waller factors for backward and 90° scattering results from the different separations between the corresponding lattice planes.

The effective lattice constant in fig. 2 is given by $d = \lambda_r/2n(\lambda_r)$. The corresponding angle for maximum Bragg diffraction deviates from its value θ if nearly no atoms are present by $\Delta\theta = -(\lambda_b/\lambda_r \sin\theta)(n(\lambda_r)-1)$ which is proportional to χ and thus to the atomic density. As is shown in fig. 5 we can in fact observe a change of $\Delta\theta$ when we vary the density in our lattice. This density variation is achieved by adjusting the efficiency of the MOT. The fluorescence of the lattice is monitored in order to serve as a relative measure for the number of loaded atoms. The density is then readily obtained because we know the sample volume through observation of the acceptance angle for Bragg diffraction.

3 Dark optical lattices

A particularly exciting perspective of optical lattices is the accomplishment of atomic densities where the number of trapped atoms exceeds the number of lattice sites. In this regime the lattice should acquire some solid state aspects and quantum statistics should play an important role for the system dynamics. Such hopes, however, are discouraged by the fact that the atomic density in near resonant optical fields is limited to the 10^{11} atoms/cm^3 level [21] due to light-induced interactions between the atoms. This has inspired research on so called *dark optical lattices* in which the elastic component of the fluorescence is strongly suppressed and the undesired light-induced interactions are much reduced. In combination with appropriate loading techniques this may allow one to reach the regime of high atomic densities. A scheme for a one-dimensional dark optical lattice has been proposed by the ENS group.[10] In Munich we have experimentally demonstrated a modified scheme which works in two and three dimensions.[11] The ENS group has proceeded to study three-dimensional weakly interacting (*grey*) optical molasses and measured very cold temperatures.[22]

3.1 Cooling scheme for dark optical lattices

The novel cooling schemes at the basis of dark lattices use polarization gradients which differ from those used in conventional Sisyphus-cooling and operate at blue detuning of the light field with respect to an atomic ($F\rightarrow F$)-transition or an ($F\rightarrow F$-1)-transition (F = total angular momentum). In addition a homogeneous magnetic field is employed. For simplicity, we will discuss our scheme for a ($F=1\rightarrow F'=1$)-transition. We assume a light field composed of two polarization components π and σ with field vectors parallel and orthogonal to the magnetic field, such that the nodes of π coincide with the antinodes of σ and vice versa. We operate at Larmor frequencies larger than the light shifts. At π antinodes all atoms are optically pumped to the ($m=0$)-level where they decouple from the light field whereas at σ antinodes the atoms are pumped to the ($m=\pm1$)-levels but remain strongly coupled (cf. fig. 6(a)). For blue detuning of the light field, the ground state light shifts are positive and thus the decoupling of the ($m=0$)-atoms occurs in a potential minimum. As illustrated in fig. 6(b), optical pumping processes (occurring predominantly as indicated by the arrows) provide efficient cooling of the atoms into the ($m=0$)-potential wells much in analogy to conventional Sisyphus-cooling.

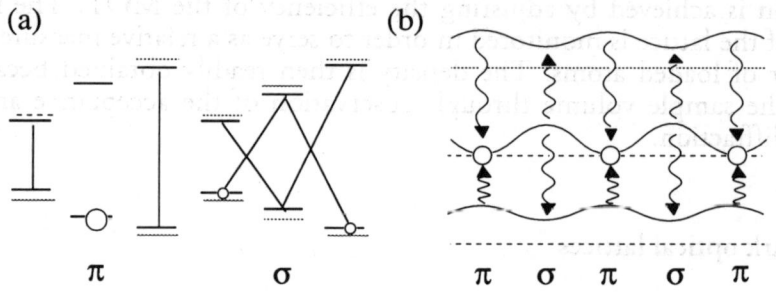

Figure 6: $F=1 \rightarrow F'=1$-level scheme in the presence of a homogeneous magnetic field (parallel to the axis of quantization) and π and σ light respectively (see text). The dashed lines depict the Zeeman-components when only the magnetic field is present. (b): Spatially modulated light-shifts of the ground state for alternating π and σ polarization. The arrows indicate the net direction and the locations of prevalence of optical pumping processes.

Using alternating π and σ components as in figure 6(b) we can also produce 2D and 3D potential wells. A 2D example is sketched in fig. 7. Four laser beams with polarizations as depicted in (a) are superposed.

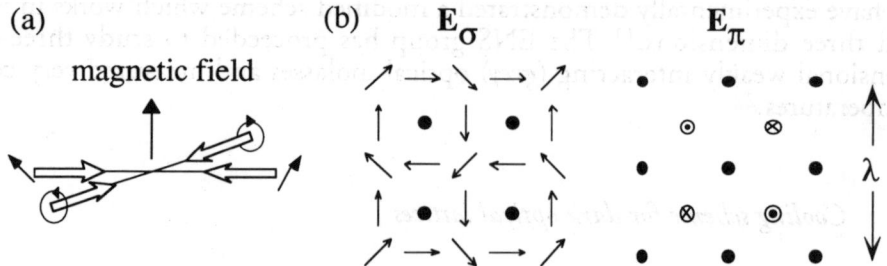

Figure 7: For laser beams are superimposed with polarizations and wave vectors as sketched in (a). For appropriate choice of the time-phase difference the resulting light field can be decomposed into π and σ components as illustrated in (b) (cf. text).

If the time-phase difference ϕ between the two standing waves produced by each pair of counter propagating beams is correctly adjusted (i.e. $\phi=90°$, this is the case in which the total energy density is spatially constant), the total light field can be decomposed into the π and σ components shown in fig. 7(b) which display the correct topography for 2D potential wells. In particular, the σ component gives rise to an optical potential acting on the $(m=0)$-atoms which provides dark potential minima. A 3D extension can be realized by superimposing a circularly polarized standing wave parallel to the magnetic field.

3.2 Observation of dark optical lattices

To observe dark optical lattices we proceed in a similar way as in case of conventional lattices: a cold sample of ^{85}Rb atoms is produced by a MOT. For a short time (10 ms) the trap is disabled and the dark lattice is active. In fig. 8(a) the fluorescence of the atoms during both periods is shown. We observe a dramatic decrease of the fluorescence during the lattice phase. When the MOT is reactivated, the fluorescence reappears at an intensity exceeding the steady state value in the MOT. When the lattice field is not present, the fluorescence also drops to zero because the atoms being no longer trapped disappear. In this case, however, the fluorescence takes more than ten milliseconds to recover when the MOT is switched on and never exceeds the dotted line in fig. 8(a).

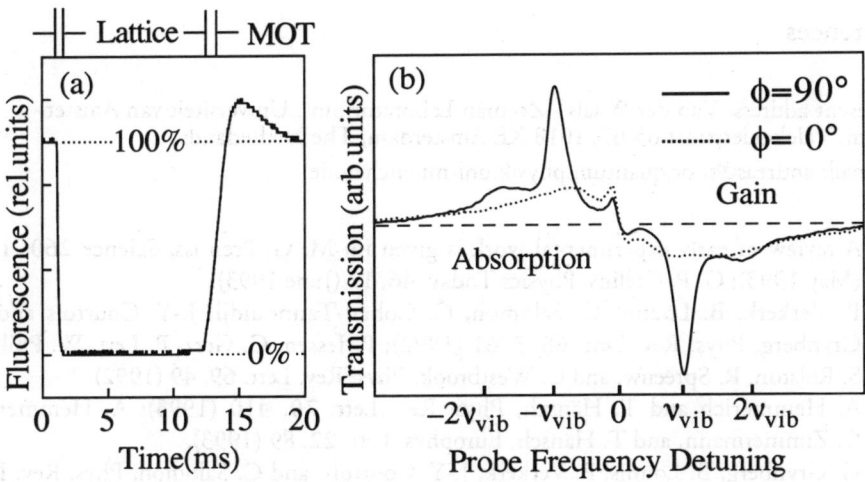

Figure 8: Comparison of the atomic fluorescence during the capture phase (MOT) and when the dark lattice is active. (b): probe transmission spectrum of 2D dark lattice. The linear probe polarization was oriented parallel with respect to the nearly copropagating pump beam (ν_{vib}=85 kHz).

Fig. 8(b) shows a probe transmission spectrum. The (F=3→F'=3)-component of the $D1$-transition of the ^{85}Rb isotope at 795 nm is used and the lattice light field has the topography shown in fig. 7. One clearly recognizes sharp resonances at $\pm v_{vib}$ due to Raman transitions between adjacent vibrational states of trapped atoms (solid trace). We can also identify small second order sidebands at around $\pm 2v_{vib}$. For comparison, when the time-phase difference is not correctly adjusted (i.e. when $\phi=0°$), the light field is not appropriate for 2D trapping and no vibrational resonances are observed (dotted trace).

One may be tempted to ask why Raman transitions can be observed at all despite of the fact that the atoms are trapped in the dark. The reason is that the atoms are trapped in quantum states which are perfectly dark only in the centers of the microtraps such that the lattice photons can still act on the atoms via the tails of the atomic wave functions which extend into bright regions. One can also argue that much in the same way as for conventional lattices the interference between the probe beam and the lattice field leads to an oscillating distortion of the optical potential which can excite vibrations if resonant.

Acknowledgments

Part of this work has been supported by the German BMBF under contract 13N6637/5.

References

[*] Present address: Van der Waals - Zeeman Laboratorium , Universiteit van Amsterdam, Valckenierstraat 65-67, 1018 XE Amsterdam, The Netherlands.

[‡] Email: andreas@thor.quantum.physik.uni-muenchen.de

1. A review of early experimental work is given in: M. G. Prentiss, Science **260**, 1078 (May 1993); G. P. Collins, Physics Today **46**, 17 (June 1993).
2. P. Verkerk, B. Lounis, C. Salomon, C. Cohen-Tannouidji, J.-Y. Courtois and G. Grynberg, Phys. Rev. Lett. **68**, 3861 (1992); P. Jessen, C. Gerz, P. Lett, W. Phillips, S. Rolston, R. Spreeuw, and C. Westbrook, Phys. Rev. Lett. **69**, 49 (1992).
3. A. Hemmerich and T. Hänsch, Phys. Rev. Lett. **70**, 410 (1993); A. Hemmerich, C. Zimmermann, and T. Hänsch, Europhys. Lett. **22**, 89 (1993).
4. G. Grynberg, B. Lounis, P. Verkerk, J.-Y. Courtois, and C. Salomon, Phys. Rev. Lett. **70**, 2249 (1993); K. Petsas, A. Coates, and G. Grynberg, Phys. Rev. A **50**, 5173 (1994).
5. A. Hemmerich and T. Hänsch, Phys. Rev. Lett. **68**, 1492 (1992).
6. B. Lounis, P. Verkerk, J.-Y. Courtois, C. Salomon, and G. Grynberg, Europhys. Lett. **21**, 13 (1993).
7. A. Hemmerich, M. Weidemüller, and T. Hänsch, Europhys. Lett. **27**, 427 (1994).
8. A. Hemmerich, C. Zimmermann, and T. Hänsch, Phys. Rev. Lett. **72**, 625 (1994).

9. J.-Y. Courtois, S. Guibal, D. Meacher, P. Verkerk and G. Grynberg, to be published in Phys. Rev. Lett. (1996).
10. G. Grynberg and J. Courtois, Europhys. Lett. 27, 41 (1994).
11. A. Hemmerich, M. Weidemüller, C. Zimmermann, T. Esslinger, and T. Hänsch, Phys. Rev. Lett. 75, 37 (1995).
12. M. Ben Dahan, E. Peik, J. Reichel, Y. Castin, and C. Salomon, Phys. Rev. Lett. 76, 4508 (1996).
13. G. Birkl, M. Gatzke, I.H. Deutsch, S.L. Rolston, and W.D. Phillips, Phys. Rev. Lett. 75, 2823 (1995).
14. M. Weidemüller, A. Hemmerich, A. Görlitz, T. Esslinger, and T. Hänsch, Phys. Rev. Lett. 75, 4583 (1995).
15. J. M. Cowley, "Diffraction Physics", 2nd ed., North Holland, Amsterdam (1981).
16. A. Hemmerich, M. Weidemüller, and T.W. Hänsch, Laser Phys. 4, 884 (1994);
17. S. Marksteiner, R. Walser, P. Marte, and P. Zoller, Appl. Phys. B 60, 145 (1995).
18. J. Courtois and G. Grynberg, Phys. Rev. A 46, 7060 (1992).
19. J. Dalibard and C. Cohen-Tannoudji, J. Opt. Soc. Am. B 6, 2023 (1989).
20. M. Weidemüller, A. Görlitz, T. Hänsch, and A. Hemmerich, to be published.
21. Examples of light-induced collective effects are considered in T. Walker, D. Sesko, and C. Wieman, Phys. Rev. Lett. 64, 408 (1990); M. Burns, J.-M- Fournier, and J. Golovchenko, Science 249, 749 (1990); A. Hemmerich, M. Weidemüller, T. Esslinger, and T.W. Hänsch, Europhys. Lett. 21, 445 (1993).
22. D. Boiron, C. Triche, D. Meacher, P. Verkerk, and G. Grynberg, Phys. Rev. A 52, R3425 (1995).

Bloch Oscillations of Atoms in an optical potential

Ekkehard Peik, Maxime Ben Dahan[(*)], Isabelle Bouchoule, Yvan Castin
and Christophe Salomon
*Laboratoire Kastler Brossel, Ecole Normale Supérieure, 24 rue Lhomond,
75231 PARIS CEDEX 05, FRANCE*

We describe the dynamics of ultracold atoms in a periodic optical potential submitted to a constant external force. Bloch oscillations in the fundamental and first excited energy bands of the potential are observed. We first give a solid-state interpretation of the observed effects and experimentally determine the first two bands. We then give a quantum optics interpretation in terms of photon exchanges between the atoms and the laser waves. Finally, efficient and dissipation-free acceleration of atoms by coherent transfer of a large number of photon momenta (≈ 100) is demonstrated. This technique produces atomic beams with a subrecoil longitudinal velocity spread that may find applications in atom optics, precision measurements and manipulation of Bose condensates.

1 Introduction

Manipulation of atoms with lasers has considerably progressed in the last decade[1]. In this paper, we analyze the behavior of ultra-cold atoms in the light field of two counterpropagating laser waves. By contrast to the studies devoted to the case of atoms trapped in dissipative optical lattices[2,3], we assume here that these waves are detuned far from any atomic resonance such that dissipation is negligible. The lightshift of the atomic ground state leads then to a conservative periodic potential with spatial period $\lambda/2$, half the laser wavelength. This configuration was first used in the context of atom diffraction[4] leading to the development of atom optics elements, interferometry[5,6] or studies of quantum chaos[7]. We use Raman subrecoil laser cooling techniques[8,9] to prepare the atoms with a momentum spread δp smaller than the photon momentum $\hbar k$ in the direction of the optical lattice. The corresponding coherence length $h/\delta p$ thus extends over several periods of the potential. As an example, cesium atoms have been cooled in one dimension to a temperature of 2.8 nanoKelvin corresponding to $\delta p = 0.12 \hbar k$ [9].

If the two counterpropagating waves have a time-dependent frequency difference they form a "standing wave" in a moving reference frame. If in addition the frequency difference varies linearly with time, the reference frame in which the optical potential is stationary is uniformly accelerated. In this frame the atoms experience a constant inertial force in addition to the force induced by the optical potential. As shown in recent papers[10,11,12], this simple system "atom+moving standing wave" models quite well the case of electrons in a perfect crystal under the influence of a constant electric field[13,14,15]. For instance Bloch oscillations (BO) of atoms in a light lattice[10,11] as well as Wannier-Stark ladders[12] have recently been observed.

These experiments have been made possible thanks to several differences that exist between electrons in solids and atoms prepared in an optical potential: (i) the initial momentum distribution of atoms is well defined, can be tailored at will and can be much narrower than h/a where a is the lattice period. By adiabatically switching on the optical potential, this narrow momentum distribution is turned into a statistical mixture of Bloch states in a given energy band with a quasimomentum width $\delta q = \delta p/\hbar$ much smaller than the width $2\pi/a$ of the Brillouin zone. (ii) the periodic potential, being created by light, can be easily turned on and off. With sudden switch-off we can directly measure the momentum distribution of Bloch states [10]. (iii) There is virtually no scattering from defects of the potential or from interactions between particles. (iv) Bloch oscillations in the time domain can have periods in the millisecond range, i.e., ten orders of magnitude longer than in semiconductors [16,17,18,19].

This paper is organized as follows: in section 2, we give a short description of the experimental setup. Then in section 3, we analyse our experiments in terms of atoms in a periodic potential induced by light under the influence of a constant force. We present results on Bloch oscillations in the fundamental energy band and in the first excited band. In section 4 we give a quantum optics description of Bloch oscillations using the concepts of adiabatic rapid passage and Landau-Zener transitions. Finally in section 5, we apply the previous analyses to the coherent acceleration of cold atoms.

2 Experimental setup

The experimental apparatus consists of a vapor-cell magneto-optical trap (MOT) for cesium atoms [20], a laser system for loading the trap and precooling the atoms in optical molasses [1,21] and a laser system for Raman cooling [8,9,22] and for generating the chirped standing wave. All lasers employed in the experiment are diode lasers, operating on wavelengths close to the cesium $6S_{1/2} \to 6P_{3/2}$ transition at 852 nm. After loading of the MOT, the magnetic field is switched off and allowed to decay in a time of 200 ms. During this period the atoms are cooled in an optical molasses to a momentum spread of $5\hbar k$, corresponding to a temperature of 6 μK. When Raman cooling is begun after the molasses period, the magnetic field has decreased to a value on the order of 100 μG. Raman cooling and interaction with a chirped standing wave are performed while the atoms are in free fall under gravity. The diameter of the horizontal laser beams limits our interaction time to 25 ms.

To drive the Raman transition between the two hyperfine ground states of the cesium atom, two grating stabilized extended cavity diode lasers are optically phase locked at a frequency offset of 9.192 GHz [23]. The detuning of these lasers from the $6S_{1/2}, F = 4 \to 6P_{3/2}, F = 5$ line is $\Delta = 30$ GHz, a value corresponding to about 5700 natural linewidths Γ of the cesium transition and larger than the hyperfine splittings of both the excited and the ground state. This detuning is sufficiently large to avoid excitation of the $6P_{3/2}$ level, which would be followed by the random momentum transfer due to the spontaneous emission of a photon. It also reduces the differential light shifts between the Zeeman sublevels of the ground state. This is essential for our method of Raman cooling using square pulses [9] and for the observation of BO, where all the atoms should experience the same light shift potential, irrespective of their magnetic quantum state.

Each of the two Raman beams is amplified by injection locking of a high power diode laser (200 mW). The two beams are superimposed, passed through an AOM that controls the envelope of the Raman pulses, a spatial filter, a Pockels cell and a polarizing beam splitter. The latter are used to exchange the directions of the two beams and hence the sign of the transfered momentum. The two beams finally pass the cesium cell with a diameter at $1/\sqrt{e}$ intensity of 6 mm, a peak intensity of 70 mW/cm^2 and having orthogonal linear polarizations. They are superimposed antiparallel to each other and aligned horizontally with a precision of about 5'. A DBR laser diode stabilized on the $6S_{1/2}, F = 4 \to 6P_{3/2}, F = 4$ transition is used to pump the atoms back to the $F = 3$ ground state after each Raman transfer. This beam is superimposed on one of the MOT axes and makes an angle of 3° with the Raman beams. With an intensity of 20 mW/cm^2 pulses with a duration of 10 μs were used to effectively repump all the atoms.

The chirped standing wave is generated from the Raman laser having the higher frequency. Its output is split into two beams that pass through two acousto optical modulators and spatial filters. One of these AOMs is driven by a quartz oscillator at a fixed frequency of 80 MHz and the other by a variable radiofrequency obtained by mixing a quartz oscillator at 60 MHz with the output of an arbitrary-function generator around 20 MHz. The system allows us to turn on the two beams with a variable rise time (on the order of 200 μs), to apply a frequency ramp with variable slope and to finally turn the beams off fast (within 1 μs). The frequency difference is controlled by monitoring the beat note between the two AOM drives. The two beams have a diameter at $1/\sqrt{e}$ intensity of 4.5 mm, a peak intensity of 40 mW/cm^2 and parallel linear polarizations. They are superimposed onto the horizontal optical axis of the Raman beams in counterpropagating directions.

Derived from the MOT laser is a vertical probe beam, composed of two counterpropagating waves having the same circular polarization. It is used to excite the $6S_{1/2}, F = 4 \to 6P_{3/2}, F = 5$ transition, so that the number of atoms in the $F = 4$ ground state hyperfine level can be measured via the detected fluorescence signal on a photodiode. The velocity distribution of the atoms is probed by velocity selective Raman transitions: the Raman beams are counterpropagating and a Raman pulse first transfers one velocity class of the $F = 3$ atoms to the otherwise empty $F = 4$ level. Then the probe beam measures the number of atoms in $F = 4$. This sequence (Raman pulse-probe beam) is repeated while scanning the detuning of the Raman lasers in order to record the atomic velocity distribution.

3 Bloch oscillations

In this section we analyse our experimental situation in terms of a particle in a one-dimensional periodic potential submitted to a constant force. This problem corresponds for instance to electrons in a crystal under the influence of a constant homogeneous electric field. This situation, originally studied by Bloch [13] in the late twenties led to the prediction of striking phenomena [14]. We recall here the main results.

3.1 Theory

Let us first consider the case where no external force is applied on the particle. The Hamiltonian of the particle is thus simply:

$$H = \frac{p^2}{2m} + V(x), \qquad (1)$$

where m is the mass of the particle and V satisfies $V(x + a) = V(x)$ (a is the period of the potential). The properties of the eigenstates and eigenenergies of this Hamiltonian are usually derived from the Bloch theorem [24]. This theorem states that the eigenenergies $E_n(q)$ and the eigenstates $|n,q\rangle$ of H are labelled by: (i) a discrete band index n, (ii) a continuous quasimomentum q. The eigenfunctions Ψ can be written:

$$\Psi_{n,q}(x) = \langle x|n,q\rangle = e^{iqx} u_{n,q}(x) \qquad (2)$$

where $|u_{n,q}\rangle$ is spatially periodic with the periodicity a of the lattice and satisfies the Schrödinger equation

$$H_q|u_{n,q}\rangle = E_n(q)|u_{n,q}\rangle \text{ with } H_q = \frac{(p+\hbar q)^2}{2m} + V(x). \qquad (3)$$

Furthermore $|n,q\rangle$ and $E_n(q)$ are periodic functions of the quasimomentum q with period $2\pi/a$ and q is thus conventionally reduced to the first Brillouin zone ($-\pi/a < q \le +\pi/a$). All these results can be collected in a band structure which is simply the energy spectrum of the particle (Fig. 1).

If a constant and spatially uniform force F is suddenly applied to the particle for $t > 0$, the initial Bloch state $|n,q\rangle$ is no longer an eigenstate of the resulting Hamiltonian:

$$H' = \frac{p^2}{2m} + V(x) - Fx. \qquad (4)$$

However the Bloch form (Eq.(2)) for the wave function is preserved:

$$\Psi(x,t) = e^{iq(t)x} u(x,t) \qquad (5)$$

with a time dependent quasimomentum $q(t)$ given by:

$$q(t) = q(0) + Ft/\hbar \qquad (6)$$

and a spatially periodic part evolving as:

$$i\hbar \frac{d}{dt}|u(t)\rangle = H_{q(t)}|u(t)\rangle. \qquad (7)$$

Furthermore when F is weak enough not to induce interband transitions the adiabatic approximation can be applied to Eq.(7). In this case $|u(t)\rangle$ is equal to $|u_{n,q(t)}\rangle$ up to the phase factor $\exp(-i\int_0^t d\tau\, E_n(q(\tau))/\hbar)$. Since the quasimomentum scans the reciprocal lattice with uniform speed (Eq.6), the wave function $\Psi(x,t)$ is periodic in time with a period

$$\tau_B = \frac{h}{|F|a}, \qquad (8)$$

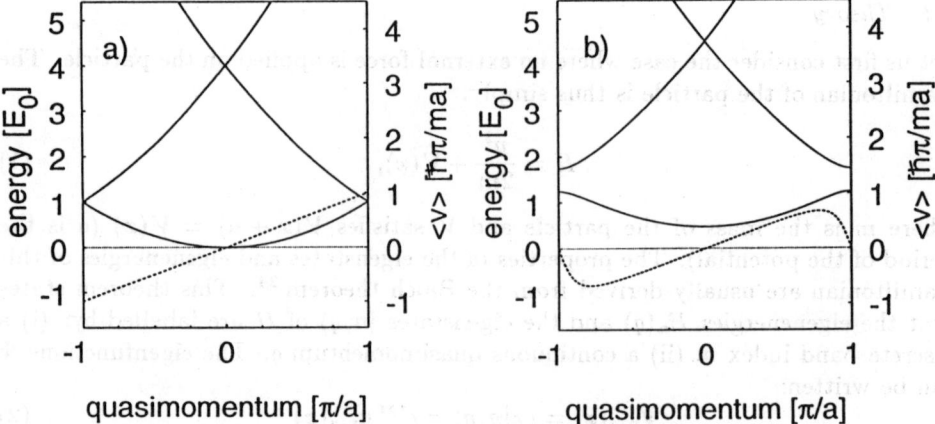

FIG. 1. Band structure $E_n(q)$ (solid line) for a particle in a periodic potential $U(x) = U_0 \sin^2 \pi x/a$ and mean velocity $\langle v \rangle_0(q)$ in the fundamental band (dashed line): a) free particle case, b) $U_0 = E_0 = \hbar^2 \pi^2 / 2ma^2$. A gap opens at $q = \pm \pi/a$. Under the influence of a weak uniform force, a particle prepared in the fundamental band remains in this band and performs a motion periodic in time called a Bloch oscillation.

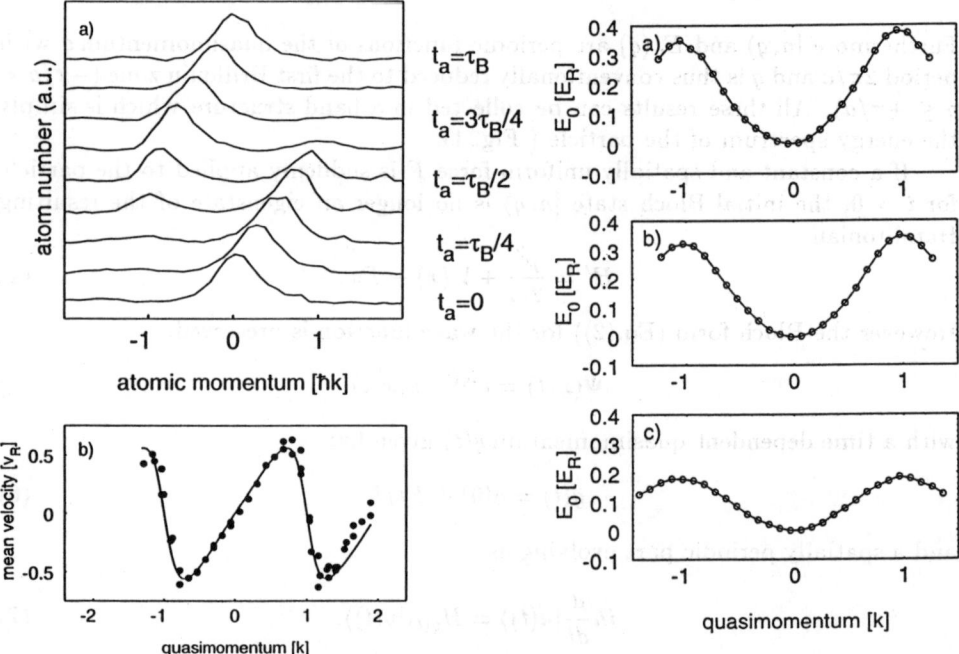

FIG. 2. Bloch oscillations of atoms in the fundamental band: (a) momentum distribution of Bloch states in the accelerated frame for equidistant values of the acceleration time t_a between $t_a = 0$ and $t_a = \tau_B = 8.2$ ms. (The small peak in the right wing of the first five spectra is an artifact created by a stray reflection of the Raman beams on the cell windows). (b) Mean atomic velocity in unit of the photon recoil velocity v_R. Solid line: theory. The light potential depth is $U_0 = 2.3 E_R$ and the acceleration is $a = -0.85$ m/s^2.

FIG. 3. Experimental determination of the fundamental energy band for: (a) $U_0 = 1.4~E_R$, (b) 2.3 E_R, (c) 4.4 E_R. This determination is obtained by integrating the mean velocity over quasimomentum (Eq. 9) corresponding to a $2\pi/a$ shift for the quasimomentum or equivalently to a full scan of the first Brillouin zone. The mean velocity of the particle in state $|n, q(t)\rangle$

$$\langle v \rangle_n(q) = \frac{1}{\hbar} \frac{dE_n(q)}{dq} \qquad (9)$$

is a periodic function of q as $E_n(q)$. Since q evolves linearly in time, $\langle v \rangle_n(t)$ is an oscillatory function with zero mean value. These so-called "Bloch oscillations" have never been observed in a natural lattice for electrons in a DC electric field because the coherent evolution is limited to a tiny fraction of the Brillouin zone by collisions with lattice defects or impurities [16]. However this effect has been recently observed in semiconductor superlattices [17,18,19].

As described in [10], the periodic optical potential results from the light-shift of the ground state of atoms in the light field of a one-dimensional linearly polarized standing wave provided by two counterpropagating beams with equal intensity. All Zeeman sublevels of $F = 3$ are equally shifted and the corresponding potential is equal to

$$U(z) = U_0 \sin^2(kx) = \frac{U_0}{2}(1 - \cos(2kx)) \qquad (10)$$

with

$$U_0 = (2/3)\hbar\Gamma(I/I_0)(\Gamma/\Delta) \qquad (11)$$

where I is the laser intensity in one beam and $I_0 = 2.2$ mW/cm^2 is the saturation intensity. This potential is periodic with spatial period $\lambda/2$ corresponding to a first Brillouin zone extending in the reciprocal lattice between $-\pi/(\lambda/2) = -k$ and $+\pi/(\lambda/2) = +k$. The natural energy scale is then the recoil energy $E_R = (\hbar k)^2/2m$. In our experiments, the potential depth is adjustable up to about 20 E_R. Resonant excitation to $6P_{3/2}$ and subsequent spontaneous emission depends on laser intensity and detuning and thus, for large enough detuning, dissipation can be made negligible. For typical conditions ($U_0 = 5E_R$ and $\Delta = 30$ GHz), the mean excitation rate per atom is 4 s^{-1}. Thus the excitation probability is very small during the 25 ms interaction time.

If we introduce a tunable frequency difference $\Delta\nu(t)$ between the two waves, the light field is no longer a standing wave in the laboratory frame. For a constant frequency difference, this corresponds to a uniform drift while for a difference linear in time the standing wave is stationary in an *accelerated* frame. In this frame, the atoms thus experience a constant inertial force in addition to the effect of the periodic potential:

$$F = -ma = -m\frac{\lambda}{2}\frac{d}{dt}\Delta\nu(t) = -\frac{m}{2k}\frac{d}{dt}\Delta\omega(t). \qquad (12)$$

3.2 Experimental sequence

In our experiments we first perform Raman subrecoil cooling which is essential to produce an initial momentum distribution with a width smaller than the extension of the first Brillouin zone [8,9,22]. Here we have used 1D Raman cooling with square pulses [9] to prepare atoms in 12 ms in a narrow momentum peak with a nearly Lorentzian lineshape of half-width at half maximum of $\sim \hbar k/4$.

We then adiabatically turn on the potential by linearly increasing the intensity of the standing wave with a rise time of about 200 μs. This duration is much longer than the adiabatic limit [11]. We thus prepare a statistical mixture of Bloch states around $q = 0$ with a half-width of $q/4$ in the fundamental band. The control of the initial state appears to be a great advantage when compared to solid-state experiments. We can actually prepare Bloch states with nearly any initial quasimomentum in any band by introducing a constant frequency shift $\Delta\nu$ between the two waves while we turn on the potential. In the standing wave frame the atoms are moving with velocity $v_0 = -(\Delta\nu)\lambda/2$. If the absolute value of mv_0 is between $n\hbar k$ ($n = 0, 1, 2, \ldots$) and $(n+1)\hbar k$ the corresponding Bloch state lies in the nth band. The quasimomentum is $q_0 = mv_0/\hbar \pm (n+1)k$ for odd n and $q_0 = mv_0/\hbar \pm nk$ for even n. $+$ and $-$ correspond respectively to negative and positive values of v_0. This is not valid for the zone boundaries ($mv_0 = \pm(n+1)\hbar k$) where levels are initially degenerate, making it impossible to prepare the atoms in a single band.

We apply a constant force for various times t_a limited to 8 ms and finally switch off the potential abruptly ($\simeq 1\mu s$). In this way the atoms keep the velocity distribution of the final Bloch states, which can then be measured using the Raman technique [8]. This is a second important difference with solid-state physics in which the effect of the periodic potential cannot be turned off. In our experiments, the Raman velocity selective transitions allow a resolution of about $v_R/18$, much smaller than the extension of the Brillouin zone and the initial momentum spread. This measurement provides the velocity distribution in the laboratory frame. The distribution in the accelerated frame is simply obtained from the distribution in the laboratory frame by a translation of $-mat_a$.

3.3 Results in the accelerated frame

Bloch oscillations in the fundamental energy band ($n = 0$) are presented in Fig. 2 [10]. It depicts the evolution of the Bloch states in the time domain $|n = 0, q(t)\rangle$ and the oscillation of the mean atomic velocity $\langle v \rangle_0(q)$. In these experiments, $U_0 = 2.3E_R$, the acceleration a is -0.85m/s^2 and $\tau_B = 2\hbar k/ma = 8.2$ms. Such value of the period is several orders of magnitude longer than in solid-state systems. The initial peak shifts linearly with time while its weight decreases. Simultaneously a second peak emerges at a distance $2\hbar k$; its increasing weight becomes equal to that of the first peak when $t_a = \tau_B/2$. It keeps growing until $t_a = \tau_B$ where the initial momentum distribution is recovered. The atoms have performed a full Bloch oscillation and further evolution reproduces this pattern periodically.

From these data it is possible to experimentally determine the shape of the fundamental band $E_0(q)$ by integrating the mean velocity over quasimomentum, as indicated by Eq.(9). The measured bands for different values of the potential depth are presented in Fig. 3. Note the very different radii of curvature of the band near $q = 0$ and $q = \pm k$ for weak potentials ((a) and (b)) and the flattening of the band as the potential depth U_0 increases.

BO can also occur in excited bands and we have observed these oscillations in the first excited band ($n = 1$) as follows: we prepare atoms in the state $|n = 1, q_0 \simeq -0.5k\rangle$ by introducing a constant frequency shift so that atoms have a velocity $1.5v_R$

in the standing wave frame. The subsequent acceleration is equal to -0.85m/s^2, as for the oscillations in the fundamental band. In Fig. 4 the full momentum distribution in the accelerated frame is displayed for various times between 0 and τ_B and a potential depth $U_0 = 9.5 E_R$. As in the fundamental band the time evolution is periodic but the Bloch states display a richer structure. The initial distribution is mostly a combination of two plane waves $|1.5\hbar k\rangle$ (dominant) and $|-0.5\hbar k\rangle$. At $t_a = \tau_B/4$ the quasimomentum arrives at the avoided crossing with the second excited band and the corresponding Bloch state $|n=1, q=0\rangle$ is predominantly a superposition of the two plane waves $|-2\hbar k\rangle$ and $|2\hbar k\rangle$ with equal weights. In the further evolution, the weight of the component near $2\hbar k$ vanishes while the component near $-2\hbar k$ grows. After $3/4\tau_B$ the zone boundary $q = k$ is reached and the atom is in a superposition of momentum states with $-\hbar k$ and $\hbar k$. Finally this latter momentum peak grows and after one Bloch period the initial distribution is recovered. Note in addition that, at $t_a = 0$, a third small peak is visible around $-2.1\hbar k$. We attribute the existence of this peak to the finite initial width of the momentum distribution ($0.6\hbar k$ FWHM in these data) which is not very small as compared to $\hbar k$. During the adiabatic turn on of the optical potential, atoms in the wing of the distribution, having an initial momentum p close to $2\hbar k$ couple to a Bloch state which is a linear superposition of two plane waves at $|p\rangle$ and $|p - 4\hbar k\rangle$. This finite width effect is also responsible for the slight curvature in the time evolution of the peaks in Fig. 4 a).

In Fig. 4 b), we also present the evolution of the mean velocity in the $n = 1$ band for a value of the potential depth $U_0 = 9.5 E_R$. This mean velocity is calculated in the same way as described in [10]. The amplitude of the velocity oscillations is greater than in the fundamental band and is $1.1\ v_R$ for $U_0 = 9.5 E_R$. The agreement with the theoretical curve obtained from a numerical calculation of the band structure (shown in solid line) is good. Note also that for an identical applied force, the slopes around $q = 0$ and $q = \pm k$ of the mean velocity in the excited and fundamental bands have opposite signs. This is expected from Fig. 1 (b): to a minimum of the fundamental energy band corresponds a maximum of the excited band and vice-versa. For values of $U_0 < 5 E_R$ – where we have observed BO in the fundamental band – the energy gap between the first and second excited bands is small and adiabatic evolution in the first band is difficult to obtain.

To conclude this section, we would like to underline that BO are a pure quantum effect which comes from the wave nature of the atoms. Here a matter wave is diffracted by a light structure, a situation entirely symetrical of the ordinary Bragg diffraction of light by matter lattice as first pointed out in [4]. As for light diffraction, the wave-like atoms strongly interact with the potential for special values of the wave vector. In the case of a weak potential where the Bloch states are very close to pure plane waves we can give a simple description of BO. An atomic plane wave, initially prepared with momentum $p \simeq 0$, is accelerated by an external force. The momentum p increases linearly according to Newton's law until it reaches a critical value satisfying the Bragg condition: $k_{at} = j\ k$ where $k_{at} = p/\hbar$ is the atomic wave vector and j is an integer. The atomic wave is then reflected and its momentum is reversed. The further evolution is nothing but the repetition of that process: an acceleration by the force followed by a Bragg reflection. This effect is illustrated in

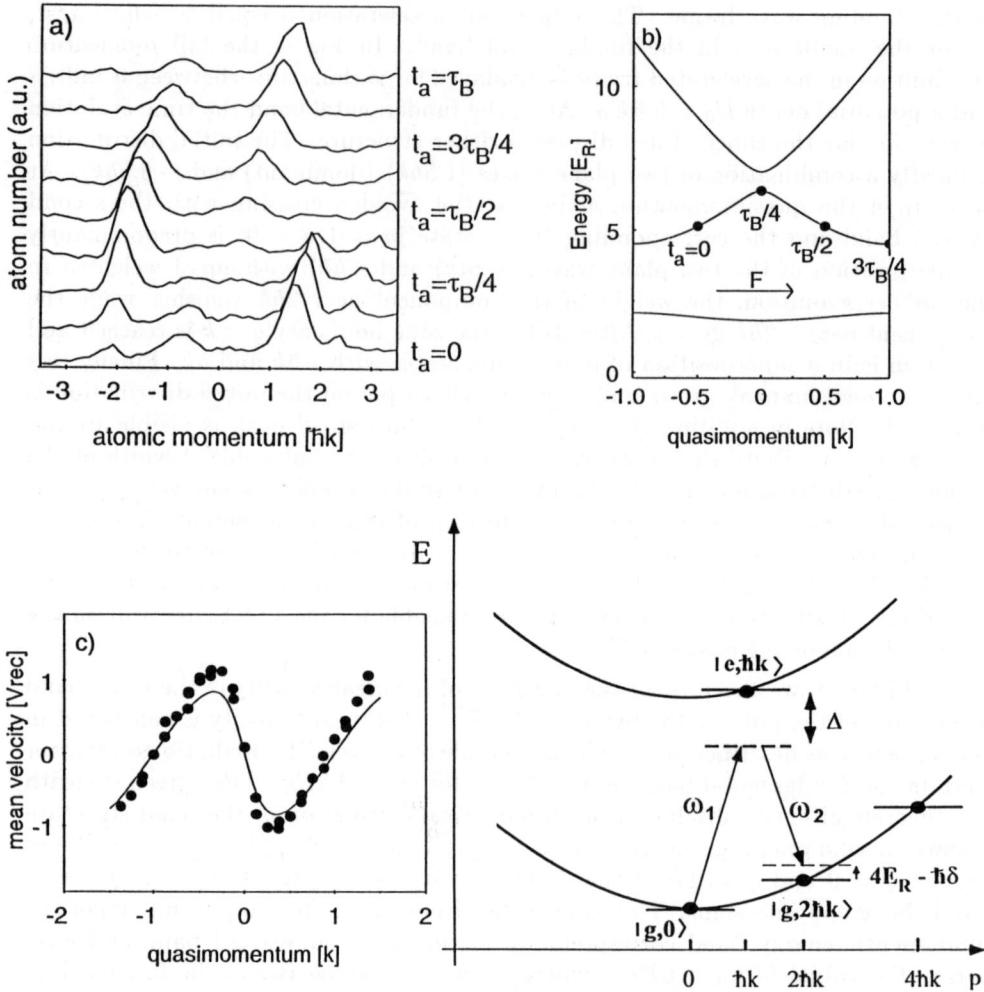

FIG. 4. Bloch oscillations of atoms in the first excited band. Before turning on the optical potential, the atoms are prepared with a momentum of $1.5\hbar k$. (a) momentum distribution of Bloch states in the accelerated frame for equidistant values of the acceleration time t_a between $t_a = 0$ and $t_a = \tau_B = 8.2$ ms. (The small peak between 2 and $3\hbar k$ is an artifact due to a stray reflection on a cell window). (b) Band structure and time evolution of the quasimomentum. (c) Mean atomic velocity. The light potential depth is $U_0 = 9.5\ E_R$ and the acceleration is $a = -0.85$ m/s^2. Solid line: theory with no adjustable parameter. Note the sign difference of the mean velocity for the oscillation in the first excited band and in the fundamental band (Fig. 2 (b)).

FIG. 5. Energy-momentum states in the laboratory frame. In the chirped standing wave, an initial state $|g, p\rangle$ is only coupled to $|g, p \pm 2j\hbar k\rangle$, where j is an integer, by stimulated two-photon Raman transitions.

Fig. 2 for the fundamental band. In the case of the first excited band (Fig. 4), the oscillations result from two successive Bragg reflections. The first one corresponds to a second order Bragg transfer of $-4\hbar k$ (from $|p = +2\hbar k\rangle$ to $|p = -2\hbar k\rangle$) while the second one leads to a change by $+2\hbar k$, a first order Bragg reflection.

4 A quantum optics approach

While our experimental results can be perfectly explained using the Bloch formalism in the accelerated frame to describe the motion of atoms in the optical lattice in presence of a constant inertial force, we can also give a different physical picture that is equally well suited to describe our experiment. We remain in the laboratory frame and we deal with free atoms interacting with two counterpropagating laser waves having a time dependent frequency difference.

In the absence of spontaneous emission the atoms momentum can change by units of $\hbar(k_1 - k_2) \approx 2\hbar k$ by absorbing a photon from one wave and emitting it into the other in a stimulated way as depicted in Fig 5. Because the atoms are initially prepared with a momentum spread much smaller than $2\hbar k$ and with a kinetic energy near zero, their possible states after interaction with the light fields are discrete points $|p = 2j\hbar k, E = 4j^2 E_R\rangle$ ($j = 0, 1, 2, 3\ldots$) on the momentum-energy-parabola of the free particle [25] (cf. Fig. 5). The gain in kinetic energy is provided by the frequency difference between the two laser waves: the atoms are accelerated in the direction of the beam with the higher frequency by absorbing photons from it and reemitting low frequency photons into the other. The transition $|p = 2j\hbar k, E = 4j^2 E_R\rangle \rightarrow |p = 2(j+1)\hbar k, E = 4(j+1)^2 E_R\rangle$ is resonant for an angular frequency difference $\Delta\omega = 4(2j+1)E_R/\hbar$. As we start with the atoms at rest ($j = 0$) and $\Delta\omega = 0$, these resonances are encountered sequentially and a gain of atomic momentum of $2\hbar k$ can be expected after each change in the frequency difference of $8E_R/\hbar$ as shown in Fig. 6. For a constant change in the angular frequency difference $\Delta\omega$ with the rate $\Delta\dot{\omega}$, the time required for this is

$$t = 8E_R/\hbar\Delta\dot{\omega} = 4E_R/\hbar k a = 2\hbar k/ma \qquad (13)$$

which is equal to the Bloch period for the inertial force $ma = m\Delta\dot{\omega}/2k$. Thus the mean atomic velocity increases by $2\hbar k/m$ during each Bloch period. As shown in Fig. 6 a), the Bloch oscillations in the laboratory frame appear as a periodic deviation of the mean velocity around the linear increase in time at. The method of exciting the transition between two energy levels with a electromagnetic wave of variable detuning that is scanned through resonance is well known under the term adiabatic rapid passage (ARP) [26]. For properly chosen parameters (i.e., a scan range that is greater than the peak Rabi frequency Ω and slow enough rate of change of the detuning $\Delta\dot{\omega} \ll \Omega^2$) the transfer between the states is complete and the method can be used to efficiently create an inversion between the levels. In our case a sequence of transfers between momentum states results in a coherent acceleration of the atoms in the laboratory frame.

Multiple ARP as a means of momentum transfer between light and atoms has already been proposed long ago, but considering the excitation and deexcitation of an internal state of the atom using a one-photon transition [27]. For instance, they

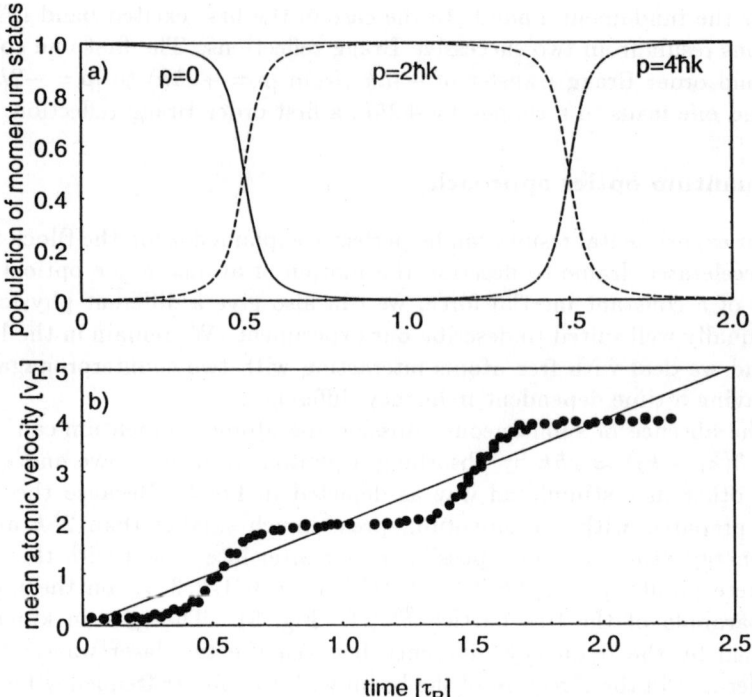

FIG. 6. (a) Population of momentum states $|p = 2j\hbar k\rangle$ as a function of time in the chirped standing wave (numerical simulation). (b) Experimental measurement of the mean atomic velocity in the laboratory frame as a function of time. Parameters are the same as in Fig. 2.

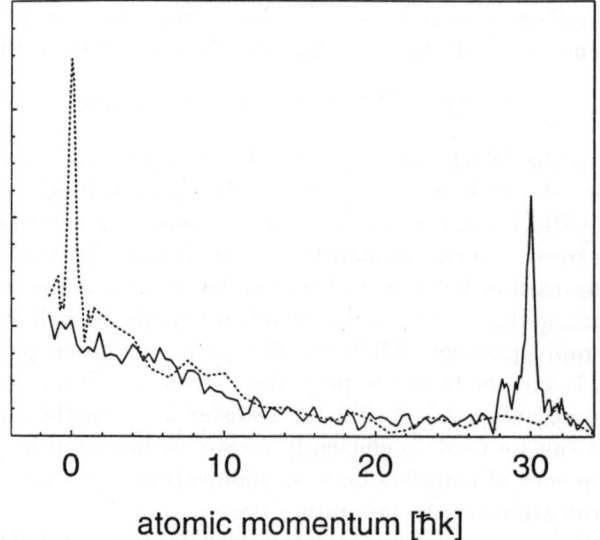

FIG. 7. Determination of the critical acceleration by coherent acceleration of atoms with a subrecoil r.m.s. momentum spread $(\hbar k/4)$. $U_0 = 9.5 E_R$ and the acceleration is 76m/s^2. Dotted line: initial distribution. Solid line, 30 photon recoils are transferred after 1.4 ms of acceleration. From this data we deduce a total transfer efficiency of 60% and a Landau-Zener loss rate per Bloch period of 0.033.

occur in saturation spectroscopy with curved wavefronts [28]. Our system has some peculiarities in comparison with previous studies of ARP: the states are linked by a two-photon transition; internal states of the atom are not excited, consequently there is no relaxation or dissipation; the sequence of levels is infinite, so that a large number of successive transfers can be made.

The two-photon Raman process can be characterized by an effective Rabi frequency

$$\Omega = \Omega_1 \Omega_2 / 2\Delta = U_0 / 2\hbar \qquad (14)$$

which is proportional to the depth of the light shift potential (Ω_1, Ω_2: Rabi frequencies of the two beams, Δ: detuning from the atomic resonance line). The two ARP conditions then read $\Delta\dot\omega \ll \Omega^2 \ll 64 E_R^2/\hbar^2$ and are well fulfilled for the conditions of our BO experiment in the fundamental energy band. The second condition, which is equivalent to the weak binding limit for the periodic potential, allows us to treat the transitions sequentially and to take only two momentum states at a time. The first condition, which ensures the adiabaticity of the process, is equivalent to avoiding interband transitions in the band structure model. It will be discussed quantitatively in the following paragraph.

The model of ARP between momentum states is also able to predict quantitatively the temporal behaviour of the BO, i.e., of the mean velocity in the laboratory frame $\langle v \rangle_{lab}(t)$. Focusing now on one of the momentum transfers displayed in Fig. 6, we consider only two states at a time and the mean velocity is simply given by

$$\langle v \rangle_{lab}(t) = \frac{2\hbar k}{m} \left(j P_{2j}(t) + (j+1) P_{2(j+1)}(t) \right), \qquad (15)$$

where P_{2j} is the time dependent probability of finding the atom in momentum state $|2j\hbar k\rangle$. These probabilities can be calculated from the vector model of the optical Bloch equations [29]:

$$P_{2(j+1)} = \frac{1}{2} + \frac{\delta}{2\sqrt{\delta^2 + \Omega^2}}, \quad P_{2j} = 1 - P_{2(j+1)} \qquad (16)$$

where

$$\delta = \Delta\omega - 4(2j+1) E_R/\hbar \qquad (17)$$

is the detuning from the two-photon resonance. For small values of Ω, $\langle v \rangle_{lab}(t)$ changes very little during the first half period ($-4 E_R/\hbar \leq \delta < 0$) and makes an abrupt jump of $2\hbar k/m$ as the resonance $\delta = 0$ is encountered. This corresponds to a BO with the maximal amplitude in velocity of $\hbar k/m$. For larger values of Ω, which corresponds to stronger binding of the atoms in the periodic potential, the increase in $\langle v \rangle_{lab}(t)$ approaches a linear behaviour in δ or in time, namely the amplitude of the BOs vanishes. Obviously this simple model cannot be extended to arbitrarily strong coupling, because the approximation of having only two momentum states involved at a time breaks down. In that case, we have solved numerically the Schrödinger equation involving a large set of momentum states. Note that for any value of the lattice potential these results are identical to those obtained with the Bloch approach presented in section 3. The full equivalence between the two descriptions is proven in [11]. A similar quantum optics approach has been independently developed for atomic three-level and two-level systems [30].

We have compared the experimental results to the simple two-state model (Eq. (16)) giving the height of the peak at $|2\hbar k\rangle$ (in the first Bloch period) as a function of the detuning and for different values of Ω, i.e., of the potential depth U_0. We find that the experimental data can be well fitted by Eq.16, but with Rabi frequencies about 12 percents smaller than those deduced from the measured intensities. This slight discrepancy may be attributed to an error in absolute power calibration as well as to the transverse spread of the atomic cloud in the Gaussian profile of the laser beams during the ~ 20 ms interaction time. This analysis can be pursued further using the dressed atom approach [31], which provides a simple physical picture of the Bloch oscillations in the dressed basis [11].

5 Preparation of ultracold atomic beams

Using the moving periodic potential of the detuned laser beams a large number of photon momenta can be transferred to the atoms coherently, accelerating them without any dissipation or heating. With this technique, we are able to produce an atomic beam with very small (i.e., subrecoil) momentum spread in the beam direction. For practical applications the most important figure of merit is the maximal acceleration under which the atoms stay bound to the potential. This determines the maximal velocity that can be reached in a given interaction time for atoms starting at rest.

Classically a periodic potential $U_0 \sin^2 kz$ would be tilted under the influence of the inertial force $-ma$. Acceleration is possible up to a critical value $a_{cl} = U_0 k/m$ from where on there exist no more local potential minima that can bind the particles. In a quantum treatment the requirement for acceleration is that the atoms stay in the fundamental band and do not perform transitions to higher energy bands. In the band structure of the potential $U_0 \sin^2 kz$, which is in our experiments smaller than $\sim 20 E_R$, the widths of the band gaps between the nth and the $(n+1)^{th}$ band diminish with n. Since the probability for interband transitions increases exponentially with diminishing width of the band gap, an atom that has passed the first and largest band gap and has made the transfer from the fundamental to the first excited band, will also make the transitions to higher bands and finally reach the continuum [32]. When the quasi momentum scans the Brillouin zone, the critical moment for interband transitions is at the zone boundary $q = k$ where the first excited band approaches the fundamental band the most. This is encountered once during each Bloch period. We suppose a transition rate per Bloch period given by a Landau-Zener formula [33]

$$r = \exp(-a_c/a) \quad (18)$$

where the critical acceleration a_c is proportional to the square of the band gap between the fundamental and the first excited band, which in the limit of weak binding is proportional to the potential depth U_0. Here we have

$$a_c/a = \pi \Omega^2 / 2\Delta\dot{\omega} \quad (19)$$

We determined these transition rates experimentally by accelerating atoms, initially cooled to a subrecoil momentum spread, for a fixed time $t_0 = n\tau_B$ and by measuring the fraction of atoms that reached the final velocity $v = at_0$ for different values of a and U_0. Typical transition rates were in the range 2–20% and followed well the exponential behaviour expected from the Landau-Zener formula. As an example of the data we show in Fig. 7 the initial momentum distribution and that obtained after a transfer of $30\hbar k$ in 1.4 ms ($a = 76$ m/s^2) in a potential $U_0 = 9.5 E_R$, leading to a total efficiency of 60% or a transition rate $r = 0.033$.

The critical acceleration is measured in units of $a_0 = \hbar^2 k^3/m^2$, which is equal to 91.6 m/s^2 for the cesium atom. For the range of potential depths investigated here ($U_0 < 10 E_R$) the relation between a_c and U_0 is well fitted by

$$\frac{a_c}{a_0} = 0.037 \left(\frac{U_0}{E_R}\right)^2 \qquad (20)$$

Our numerical value of 0.037 is in reasonable agreement with the value $\pi/64 \approx 0.049$ derived from Eq.19, considering our 30% laser intensity uncertainty.

The greatest momentum transfer we could measure in the experiment was $112\hbar k$, being limited by the laser power available for the optical potential and by the geometry of our detection system, which was designed to probe atoms nearly at rest and close to the position of the MOT. Numerical band structure calculations indicate that the proportionality between U_0 and the band gap holds up to $U_0 \approx 15 E_R$. At this value of U_0 for cesium an acceleration of 70.4 m/s^2 leads to a Bloch period of 0.1 ms and a transition rate $r = 2 \cdot 10^{-5}$ so that during a 10 ms interaction time $200\ \hbar k$ can be transferred to 99.8% of the atoms, leading to a final velocity of 0.7 m/s. Because of the scaling of a_0 with k^3/m^2 significantly higher velocities are possible with lighter atoms and shorter wavelengths. This simple acceleration method might also be an interesting alternative to the atom interferometer method for measuring the photon recoil and thus h/m in frequency units [34].

6 Aknowledgments

We are grateful to C. Cohen-Tannoudji, G. Grynberg, G. Bastard, J. Dalibard and M. Raizen for stimulating discussions. E.P. acknowledges support from the European Union through the HCM program. This work was supported in part by NEDO, CNES and Collège de France. Laboratoire Kastler Brossel is Unité de Recherche de l'Ecole Normale Supérieure et de l'Université Pierre et Marie Curie, associée au CNRS.

(*) Also member of the Direction des Recherches et Etudes Techniques (DRET).

References

1. See e.g. *Laser Manipulation of Atoms and Ions*, Proc. of the Int. School of Physics "Enrico Fermi", Eds. E. Arimondo, W. Phillips and F. Strumia, North Holland, Amsterdam (1992).
2. For a recent review see: M. G. Prentiss, Science **260**, 1078 (1993).

3. A. Hemmerich et al., these Proceedings
4. P. L. Gould, G. A. Ruff, and D. E. Pritchard, Phys. Rev. Lett. **56**, 827 (1986); P. J. Martin, B. G. Oldaker, A. H. Miklich and D. E. Pritchard, Phys. Rev. Lett. **60**, 515 (1988).
5. For a recent review see: C. S. Adams, M. Sigel and J. Mlynek, Phys. Rep. **240**, 143 (1994).
6. E. Rasel, M. Oberthaler, H. Batelaan, J. Schmiedmayer and A. Zeilinger, Phys. Rev. Lett. **75**, 2633 (1995); D. M. Giltner, R. W. McGowan and S. A. Lee, Phys. Rev. Lett. **75**, 2638 (1995).
7. F. L. Moore, J. C. Robinson, C. Bharucha, P. E. Williams and M. G. Raizen, Phys. Rev. Lett. **73**, 2974 (1994).
8. M. Kasevich and S. Chu, Phys. Rev. Lett. **69**, 1741 (1992).
9. J. Reichel, F. Bardou, M. Ben Dahan, E. Peik, S. Rand, C. Salomon and C. Cohen-Tannoudji, Phys. Rev. Lett. **75**, 4575 (1995).
10. M. Ben Dahan, E. Peik, J. Reichel, Y. Castin, and C. Salomon, Phys. Rev. lett. **76**, 4508 (1996).
11. E. Peik, M. Ben Dahan, I. Bouchoule, Y. Castin and C. Salomon, submitted to Phys. Rev. A. (July 1996).
12. S. Wilkinson, C. Bharucha, K. Madison, Qian Niu and M. Raizen, Phys. Rev. Lett. **76**, 4512 (1996) Qian Niu, Xian-Geng Zhao, G. Georgakis and M. Raizen, Phys. Rev. Lett. **76**, 4504 (1996). See also the contribution of M. Raizen in these Proccedings.
13. F. Bloch, Z. Phys. **52**, 555 (1929).
14. C. Zener, Proc. Roy. Soc. London Ser. A **145**, 523 (1934).
15. E. Mendez and G. Bastard, Physics Today, **46**, 34 (1993).
16. For a metal with $a = 0.1$ nm and an electric field of 10^{-2} V/cm the Bloch period is equal to 41 μs whereas a typical relaxation time is on the order of 10 fs. In a semiconductor superlattice the period is typically $a = 10$ nm and the field can be as high as 10 kV/cm ($F \approx 10^{-13}$ N), leading to a Bloch period in the picosecond range. In our atomic system the lattice period is $a = 426$ nm, the force typically on the order of 10^{-25} N and the corresponding Bloch period in the millisecond range.
17. J. Feldmann, K. Leo, J. Shah. D. A. B. Miller, J. E. Cunningham, T. Meier, G. v. Plessen, A. Schulze, P. Thomas and S. Schmitt-Rink, Phys. Rev. B **46**, 7252 (1992).
18. K. Leo, P. Haring Bolivar, F. Brüggemann, R. Schwedler and K. Köhler, Solid State Commun. **84**, 943 (1992).
19. C. Waschke, H. Roskos, R. Schwedler, K. Leo, H. Kurz and K. Köhler, Phys. Rev. Lett. **70**, 3319 (1993).
20. C. Monroe, W. Swann, H. Robinson and C. Wieman, Phys. Rev. Lett **65**, 1571 (1990).
21. C. Salomon, J. Dalibard, W. Phillips, A. Clairon and S. Guellati, Europhys. Lett. **12**, 683 (1990).
22. J. Reichel, O. Morice, G. M. Tino and C. Salomon, Europhys. Lett. **28**, 477 (1994).
23. G. Santarelli, A. Clairon, S. N. Lea and G. M. Tino, Opt. Commun. **104**,

339 (1994).
24. See e.g. N. W. Ashcroft and N. D. Mermin, *Solid state physics* (Saunders, Philadelphia, 1976).
25. C. Bord, in *Laser Spectroscopy 3*, ed. J. Hall and J. Carlsten, Springer Verlag, p. 121 (1977).
26. A. Abragam, *The principles of Nuclear Magnetism*, Oxford University Press, London, 1961.
27. I. Nebenzahl and A. Szöke, Appl. Phys. Lett. **25**, 327 (1974).
28. C. Bordé, in *Frequency Standards and Metrology*, Proc. of the fourth Symposium, ed. A. De Marchi, Springer, Berlin (1989).
29. L. Allen and J. Eberly, *Optical resonance and two-level atoms*, Wiley, New York (1975).
30. K. Marzlin and J. Audretsch, Phys. Rev. A, **53**, 4352 (1996) and preprint (May 1996).
31. See e.g. C. Cohen-Tannoudji, J. Dupont-Roc, G. Grynberg, *Atom-Photon Interactions*, p. 407, John Wiley, New-York (1992).
32. We neglect here possible returns from the second band to the first one which may lead to additional interference effects.
33. C. Zener, Proc. Roy. Soc. London Ser. A **137**, 696 (1932).
34. D. Weiss, B. Young and S. Chu, Phys. Rev. Lett.,**70**, 2706 (1993).

QUANTUM DECOHERENCE AND INERTIAL SENSING WITH ATOM INTERFEROMETERS

DAVID E. PRITCHARD, MICHAEL S.CHAPMAN, TROY D. HAMMOND, DAVID A. KOKOROWSKI, ALAN LENEF, RICHARD A. RUBENSTEIN, EDWARD T. SMITH

Department of Physics and Research Laboratory of Electronics, Massachusetts Institute of Technology, Cambridge, MA 02139, USA

JÖRG SCHMIEDMAYER

Institut für Experimentalphysik, Universität Innsbruck, A-6020 Innsbruck, Austria

Atom interferometers have now become powerful tools for the study of atomic properties and fundamental issues in quantum mechanics. In this paper we review two recent experiments. In the first of these, the MIT atom interferometer was employed to perform a version of Feynman's *gedanken* experiment, in which we scatter a single photon from each atom as it passes through the interferometer. The "which path" information that could in principal be gained by observing the scattered photon causes a loss of contrast in the atom interference fringes. We then regain the contrast by observing interference fringes formed only by those atoms which scatter a photon into a small subset of possible final directions. In a second experiment, we investigate inertial effects in our atom interferometer. We show that we can use our interferometer as a precision device for the measurement of small rotations, with a sensitivity of better than 50 milli-earthrate.

1 Introduction

This paper reviews some of the atom interferometry experiments that we have performed since our previous review at this conference in 1992 [1]. That earlier report concentrated on the application of interferometers to measurements of atomic and molecular properties, in particular to a 0.3% accurate determination of the polarizability of sodium atoms. Subsequent experiments using the technique of separated atomic de Broglie waves developed for those experiments included measurements of the index of refraction of various gases for de Broglie waves of Na atoms and Na_2 molecules [2]. The index of refraction is closely related to the phase shift in the collision, a quantity previously inaccessible to measurement in atomic physics.

Our general view, unchanged since that review is that atom interferometry is certain to have three major areas of application: studies of atomic and molecular properties, fundamental studies, and inertial measurements, and may also find application in atom lithography. This brief review discusses our application of atom interferometers to fundamental studies and to inertial measurements. More specifically it addresses the fundamental issue of the amount of loss of atomic coherence when a single photon is scattered from an atom passing through an interferometer. We will also demonstrate a method of recovery for this "lost" coherence. The particular inertial study which we discuss is a precise measurement of rotation, demonstrating reproducibility at the level of 50 milli-earthrates per $\sqrt{sec}$, performance typical of commercial laser gyros. More details of this work, as well as other work done with our interferometer are reviewed in Ref. [3], from which the following material is condensed. The complete review is available via the web at the URL http://coffee.mit.edu/.

2 Fundamental Studies of Coherence Loss

Interferometers of all types have had application to fundamental problems and precision tests of physical theories, especially quantum mechanics, and atom interferometers are sure to continue this tradition. In this section we focus our attention on the fundamental question, "what limits do the size and complexity of particles place on the ability of their center of mass motion to exhibit interference effects?" Quantum mechanical treatment of the center of mass motion of increasingly complex systems is an important theme in modern physics. This issue is manifest theoretically in studies of the transition from quantum through mesoscopic to classical regimes and experimentally in efforts to coherently control and manipulate the external spatial coordinates of complex systems, as exemplified by the wide interest in matter wave optics and interferometry. As demonstrated in our recent work [4-11], matter wave optics and interferometry have been extended to atoms and molecules – systems characterized by many degenerate and non-degenerate internal quantum states. In this section we will investigate if and where there might be limits – in theory or in practice – to coherent manipulation of the center of mass motion of larger and more complex particles due to the interaction of radiation with the particle as it is passing through the interferometer.

The key result here is our experimental realization of a *gedanken* experiment suggested by Feynman, in which one attempts, through the scattering of a single photon, to determine (i.e. localize) on which side of the interferometer an atom passes. We then consider the question of what happens to the coherence that is lost when the particle passing through the interferometer interacts with this radiation. We will demonstrate that the coherence becomes entangled with the scattered radiation – and show that this coherence can be regained by selectively detecting atoms that scatter this radiation into a restricted subspace of all possible final directions.

3 Coherence Loss due to Scattering a Single Photon – Discussion

The principle that a system can be in a coherent superposition of different states and exhibit interference effects is a fundamental element of quantum mechanics. Immediately, the question arises as to what happens to the interference if one tries to determine experimentally which state the system is in. This is the basis of the famous debate between Bohr and Einstein, in which they discussed *Welcher-Weg* ("which way") information in the Young's double slit experiment [12-15]. In a more recent *gedanken* experiment suggested by Feynman, a Heisenberg light microscope is used to provide *Welcher-Weg* information in a Young's double slit experiment with electrons [13] or atoms [16,17]. In this section, we will discuss our experimental realization of this *gedanken* experiment using our atom interferometer.

One starting point is the principle of complementarity. Since the contrast of the interference fringes we observe in our interferometer is a measure of the amount of the atomic coherence, complementarity demands that the fringes must disappear when the slit separation (more generally the path separation at the point of measurement) is large enough so that, in principle, one could detect through which slit the particle passed [18] using a Heisenberg microscope. Since the loss of contrast is related to the possibility of measuring the atom's position, for example by observing a photon scattered from the atom, it is necessary to consider a quantum treatment of this measurement process. The measurement interaction considered here is the elastic scattering of the photon by the atom, which causes their initially separable wave function to evolve into an entangled state [19] – a sum of separable wave functions, each of which conserves the total momentum and energy of the system, and which can no longer be written as a product of separate atom and photon wave functions. This entanglement can result in a loss of atomic coherence when information about the scattered photon is disregarded. The effects of such entanglement are relevant to important issues in contemporary quantum mechanics, including EPR type correlations, understanding of the measurement process, and the loss of coherence in the passage from quantum to classical mechanics. The experiment presented here reveals the details of the loss of coherence of one system due to entanglement with another. This is accomplished by scattering a probe particle off an interfering superposition of the observed system as it passes through an interferometer.

We now discuss experiments we performed to measure the loss of atomic coherence due to scattering single photons from sodium atoms inside our interferometer. Our experiments [20] demonstrate that the loss of coherence may be attributed to the random phase difference between the two arms of the interferometer which is imprinted on the atom during the scattering process. This random phase depends fundamentally on the spatial separation of the interfering waves at the point of scattering, relative to the wavelength of the scattering probe.

Our experiments also address the question: "where is the coherence lost to and how may it be regained?" Although the elastic scattering of a photon produces an entangled state, it is not *per se* a dissipative process and may be treated with Schrödinger's equation without any *ad hoc* dissipative term. Therefore the coherence is not truly lost, but rather becomes entangled with the scattered photon. Although the photon may be regarded as part of a measurement apparatus for determining the atom's position, it is here more naturally considered as a simple reservoir, consisting of the vacuum radiation modes accessible to it. We show the validity of this viewpoint by demonstrating that selective observation of atoms which scatter photons into a restricted part of the accessible phase space results in atomic fringes with regained contrast.

4 Coherence Loss due to Scattering a Single Photon – Experiment

To study the effects of photon scattering on the atomic coherence, single photons were scattered from the atoms passing through our three grating Mach-Zehnder interferometer (see Fig. 1). The contrast (which measures the remaining coherence) and the phase of the interference pattern were measured as a function of the separation of the atom paths at the point of scattering [20].

In the absence of scattering, the atom wavefunction at the third grating may be written

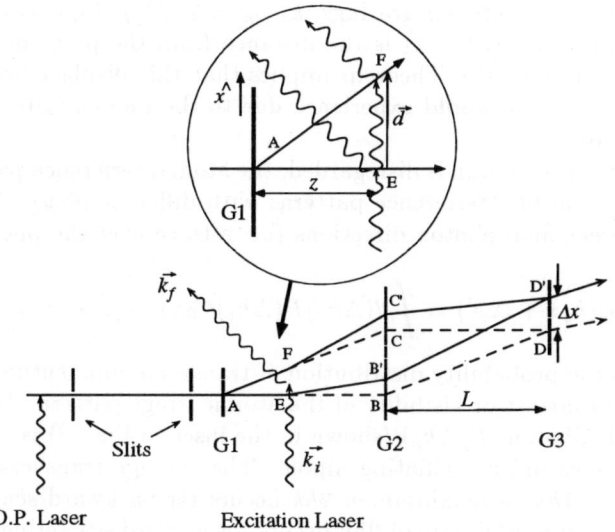

Figure 1: *A schematic, not to scale, of our atom interferometer. The original atom trajectories (dashed lines) are modified (solid lines) due to scattering a photon (wavy lines). The inset shows a detailed view of the scattering process.*

$$\Psi(x) = u_1(x) + e^{i\varphi} u_2(x) e^{i k_g x} \qquad (1)$$

where $u_{1,2}$ are (real) amplitudes of the upper and lower beams respectively, $k_g = 2\pi/d_g$ where d_g is the period of the gratings, φ is the phase difference between the two arms, and x is a coordinate perpendicular both to the atomic beam and to the grating bars. To describe the effects of scattering within the interferometer, we first consider an atom elastically scattering a photon with well-defined incident momentum $\vec{k_i}$ and final (measured) momentum $\vec{k_f}$, with $\left|\vec{k_i}\right| = \left|\vec{k_f}\right| = k_{photon}$. After this well-defined scattering event, the atomic wavefunction becomes

$$\Psi'(x) \propto u_1(x - \Delta x) + e^{i\varphi} u_2(x - \Delta x) e^{i k_g x + \Delta \varphi}. \qquad (2)$$

The resulting atomic interference pattern shows no loss in contrast but acquires a phase shift [21,22]

$$\Delta \varphi = \vec{\Delta k} \cdot \vec{d} = \Delta k_x d \qquad (3)$$

where $\vec{\Delta k} = \vec{k_f} - \vec{k_i}$, and $\vec{d}$ is the separation between the two arms of the interferometer at the point of scattering. Equation 2 implies that the photon recoil causes a spatial shift of the envelope of the atomic fringes by a distance

$$\Delta x = (2L - z) \Delta k_x / k_{atom}, \qquad (4)$$

where L is the distance between gratings, $k_{atom} = 2\pi/\lambda_{dB}$ (λ_{dB} is the atom's de Broglie wavelength), and $(2L - z)$ is the distance from the point of scattering to the third grating. Ehrenfest's Theorem implies that this displacement is just the shift that a classical atom would experience due to the momentum transferred by the scattered photon.

In the case that the photon is disregarded, the atom interference pattern is given by an incoherent sum of interference patterns with different phase shifts [23] corresponding to different final photon directions (i.e. a trace over the photon states),

$$C'cos(k_g x + \Delta\varphi') = \int d(\Delta k_x) P(\Delta k_x) C_0 cos(k_g x + \Delta k_x d) \qquad (5)$$

where $P(\Delta k_x)$ is the probability distribution of transverse momentum transfer and C_0 is the original contrast or visibility of the atomic fringe pattern. For the case of scattering a single photon, $P(\Delta k_x)$ (shown in the inset to Fig. 2) is determined by the radiation pattern of an oscillating dipole. The average transverse momentum transfer is $\hbar\overline{\Delta k_x} = 1\hbar k$ (a maximum of $2\hbar k$ occurs for backward scattering of the incoming photon and a minimum of $0\hbar k$ occurs for forward scattering). Due to the average over the angular distribution of the unobserved scattered photons, there will be a loss of contrast ($C' \leq C_0$) and a phase shift $\Delta\varphi'$ of the observed atomic interference pattern. It follows from Eq. 5 that the measured contrast (phase) of the interference pattern as a function of the separation d of the atom waves will vary as the magnitude (argument) of the Fourier transform of the probability distribution, $P(\Delta k_x)$. Equation 5 is equivalent to the theoretical results obtained for the two-slit *gedanken* experiment [16,17] (where d is identified with the slit separation), even though explicit which-path information is not necessarily available in our Mach-Zehnder interferometer because the atom wavefunctions can, and do, have a lateral extent (determined by the collimation of our atomic beam) much larger than their spatial separation at the point of scattering.

The atomic beam was first prepared in the $F = 2, m_F = 2$ state by optical pumping with a σ^+ polarized laser beam before the first collimating slit. We typically achieved $\sim 95\%$ optical pumping, as measured by a two-wire Stern-Gerlach magnet which caused state-dependent deflections of the atomic beam. We employed a short laser interaction region, and adjusted the excitation field strength to scatter, on average a single photon from each atom, approaching this ideal case as closely as possible (Fig. 1). The scattering cross section was maximized using σ^+ polarized laser light tuned to the D2 resonant line of Na ($\lambda_{photon} = 589nm$) connecting the $F = 2, m_F = 2$ ground state to only the $F' = 3, m'_F = 3$ excited state. This ensured that the scattering left the atom in the same hyperfine state. We typically achieved $\sim 95\%$ optical pumping, as measured by a two-wire Stern-Gerlach magnet which caused state-dependent deflections of the atomic beam.

The photons we wish to scatter from the atomic beam are provided by an excitation laser focused to a $\sim 15\mu m$ waist (FWHM of the field) along the atom propagation direction. A cylindrical lens was used to defocus the beam perpendicular to the atom propagation direction to ensure uniform illumination over the full $\sim 1mm$ height of the atomic beam. The transit time through the waist was roughly one third of the lifetime of the excited state. Hence, the probability for resonant

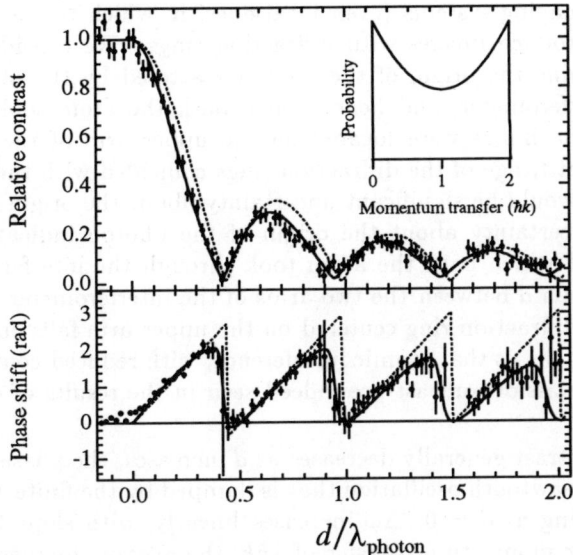

Figure 2: *Relative contrast and phase shift of the interference pattern as a function of the separation of the interferometer arms at the point of scattering. The inset shows the angular distribution of spontaneously emitted photons projected onto an axis perpendicular to the atomic beam. The dashed curve corresponds to purely single photon scattering, and the solid curve is a best fit that includes contributions from atoms that scattered zero photons (4 %) and two photons (14 %).*

scattering in the two-state system exhibited weakly damped Rabi oscillations, which we observed by measuring the deflection of the collimated atomic beam as a function of laser power. To achieve one photon scattering event per atom, we adjusted the laser power to the first maximum of these oscillations, closely approximating a π-pulse.

The contrast and phase of the measured atomic interference patterns are shown in Fig. 2 for different path separations. The contrast was high when the separation d at the point of scattering was much less than $\lambda_{photon}/2$ (corresponding to $\overline{\Delta k_x} d \ll \pi$), but fell smoothly to zero as the separation was increased to about half the photon wavelength, at which point $\overline{\Delta k_x} d \approx \pi$. (The zero would occur exactly at $d = \lambda_{photon}/2$ if the scattered photon angular distribution were isotropic.) As d was increased further, a periodic rephasing of the interference gave rise to significant revivals of the contrast and to a periodic phase modulation.

The observed behavior of the contrast is consistent with the complementarity principle. Considering the photon scattering as a position measurement of the atom, complementarity suggests that the fringe contrast must disappear when the path separation is approximately half the wavelength of the scattered light, since this is the smallest distance that can be resolved by a perfect optical microscope. A more careful consideration of the imaging process of this scattered photon, however, reveals still richer behavior.

All optical imaging systems (even ideal ones, in which the lenses can capture every photon), produce images with diffraction rings. We consider using such a system to determine the origin of the photon scattered by the atom as it passes through the interferometer, and hence which path the atom took. If the bright central spot of the image were located on the upper arm of the interferometer, and another bright fringe of the diffraction rings coincided with the position of the lower arm, there would be significant uncertainty about the origin of the scattered photon. This uncertainty about the origin of the photon indicates a significant uncertainty about which path the atom took through the interferometer. Hence, when the separation d between the two arms of the interferometer is such that the bright frnge of a diffraction ring centered on the upper arm falls on the lower arm, complementarity tells us that atomic interference, with reduced contrast, should reoccur. These revivals of contrast are indeed seen in the results of our experiment. (See Fig. 2.)

While the contrast generally decreases as d increases, the phase shift $\Delta\varphi$ of the fringes exhibits a sawtooth oscillation that is damped by the finite resolution of the apparatus. Starting at $d = 0$, $\Delta\varphi$ increases linearly, with slope 2π. This is the slope expected for momentum transfer of $1\hbar k$, the average momentum transfer of the symmetrical distribution (Fig. 2). After each zero of the contrast, the sign of the interference pattern is reversed, subtracting π from the phase and resulting in the observed sawtooth pattern.

In studying the decoherence and phase shift, we used a $50\mu m$ wide detector (rhenium hot wire), which is larger than the deflection Δx of the atom beam that results from the recoil momentum imparted by the scattered photon. The finite collimation of the atomic beam further degrades the overall momentum resolution of the apparatus. The result of this lack of resolution is that the measured interference patterns are averaged quite evenly over all values of Δx, which can be as large as $40\mu m$ in our experiment – corresponding to displacement of the envelope of the fringe pattern by ~ 200 fringes.

These numbers highlight the distinction between the expectation value of the atom's classical transverse position (the peak of the fringe envelope) and the phase of the fringes (which are never shifted by more than half a fringe). This distinction is emphasized by the fact that moving the point of scattering further downstream slightly reduces the displacement of the fringe envelope for a given $\vec{k_f}$, while monotonically increasing the corresponding phase shift. Therefore, the measured loss of fringe visibility cannot simply be understood as resulting from the transverse deflections of the atom at the detection screen (in our case the third grating) due to the photon momentum transfer, as it can be for the two-slit *gedanken* experiments [17].

The displacement of the envelope of the atom fringes, Δx, (or equivalently the x-component of the photon momentum transfer) is precisely what is measured in determining the transverse momentum distribution of an atomic beam after scattering a photon. These distributions have been measured for diffraction of an atomic beam passing through a standing light wave and undergoing a single [24] or many [8] spontaneous emissions, as well as for a simple collimated beam excited by a traveling light wave [7]. These results are usually discussed using a simple argument: the recoil momentum from spontaneous emission produces random angular displacements that smear the far-field pattern, a viewpoint also applicable to two-slit *gedanken* experiments. Clearly the quantum phase shift measured in our experiment is distinct from the "deflection" of the atom Δx due to the photon recoil. It reflects the phase difference of the photon wave function where it intersects the two arms of the interferometer.

The results here are also interesting as a contrast to the *gedanken* experiments recently proposed in which loss of contrast in an atom interferometer occurs after emission of a photon by the atom, even though there is insufficient momentum transfer to the wave function to explain this loss on the basis of dephasing [25]. In our experiment the opposite occurs: there is sufficient momentum transfer to the atom by the emitted photon to be easily detected, but the interference pattern is not destroyed for small separations. In both experiments the interaction with the radiation adds insignificant relative phase difference between the two arms of the interferometer. The crucial difference is that in the *gedanken* experiment of Ref. 25, the photon emitted by the atom is retained in one of two cavities located symmetrically on the two sides of the interferometer and can be used to determine which path the atom traversed (assuming the cavities were initially in number states), whereas in our experiment the scattered photon scatters without constraint and, for small d, no subsequent measurement can determine which path was traversed by the atom that scattered it. However, if a metal foil were interposed between the two sides of our interferometer and a beamsplitter and mirrors added so that the laser beam was split and excited both sides coherently, detection of the scattered photon would then determine which side of the foil the atom traversed, and would destroy the interference pattern even though the phase shift imparted to the atoms was negligible, just as in Ref. 25.

5 Regaining Entangled Coherence by Selective Observations

Returning to the loss of coherence induced by scattering of single photons from atoms in the interferometer, we now address the question: "where is the coherence lost to and how may it be regained?" We performed an experiment [20] to show that selective observation of atoms can result in fringes with regained contrast. This demonstrates that the coherence is not truly lost, but becomes entangled with the scattered photon which may be considered as a simple reservoir, consisting of the accesible vacuum radiation modes. When only atoms that scatter photons into a restricted part of this accessible phase space are observed, their distribution of possible phase shifts is narrower than without this selection, and they rephase more precisely and with greater coherence.

In this experiment, we observe atoms that are correlated with photons scattered into a narrow range of final directions. In principle, this could be achieved by detecting the photons scattered in a specific direction in coincidence with the detected atoms. With $\Delta k_x d$ now a known quantity, the predicted fringe shift would be the same for all the atoms (see Eq. 3); consequently the fringe patterns of this restricted set of atoms would line up and no coherence would be lost. Unfortunately, such an approach is not feasible in our experiment for a number of technical reasons – principally the slow response of our atom detector and the inefficiency of available photon detectors.

Fortunately an alternative experimental approach is made possible by the fact that the change of momentum of the photon, $\hbar\Delta k_x$, is imparted to the atom, and can be measured by the atom's deflection (see Eq. 4). Hence a measurement of an atom's Δx gives the Δk_x of the corresponding scattered photon. Furthermore, it is easily possible to measure Δx in our three grating interferometer because (since we scatter the photons close to the first grating), the deflection of the envelope of the atomic fringes for a particular Δk_x is 100 times larger than the associated fringe shift, $\Delta k_x d$. In practice this approach is superior to a correlation experiment because there are no inefficiencies or accidental coincidences introduced by the measurement of the scattered photon: the measurement of an atom's position reliably indicates the momentum transferred to that particular atom.

We have performed an experiment based on this technique to demonstrate the recovery of the entangled coherence. By using very narrow beam collimation in conjunction with a narrow detector, we can selectively detect only those atoms correlated with photons scattered within a limited range of Δk_x. This restricts the possible final photon states and results in a narrower distribution $P'(\Delta k_x)$ in Eq. 5.

We performed experiments with recoil distributions centered on three different photon momenta. Figure 3 shows three different realizations (referred to as Cases I-III) with the corresponding momentum transfer distributions, $P'_i(\Delta k_x)$, $i = I, II, III$. The contrast is plotted as a function of d for Cases I and III where we preferentially detect atoms that scattered photons in the forward and backward direction respectively. The contrast for Case II is similar to Case I and is not shown. The measured contrasts in this figure were normalized to the $d = 0$ (scattering laser on) values since a different number of atoms was detected with the laser off due to the absence of the deflection by the photons.

Our results show that the contrast falls off much more slowly than previously – indeed we have regained over 60% of the lost contrast at $d \approx \lambda/2$. The contrast falls off more rapidly for a faster beam velocity (Case III, $v_{beam} = 3200 m/s$) than for the slower beam velocities (Cases I and II, $v_{beam} = 1400 m/s$). This is because, for a given Δk_x, a higher velocity atom beam will result in a smaller displacement Δx at the third grating, resulting in diminished Δk_x resolution.

The phase shift is plotted as a function of d for the three cases in the lower half of Fig. 3. The slope of Case III is nearly 4π, indicating that the phase of the interference pattern is predominantly determined by the backward scattering events. Similarly, the slope of Case I asymptotically approaches a small constant value due to the predominance of forward scattering events. Case II is an intermediate case in which the slope of the curve, $\sim 3\pi$, is determined by the mean momentum transfer of $1.5\hbar k$. The lower inset shows the transverse momentum acceptance of the detector for each of the three cases (i.e. the functions $P'_i(\Delta k_x)$), which we determined using the known geometry and beam velocity. The fits for the data in Fig. 3 were calculated using Eq. 5 and the modified distributions $P'_i(\Delta k_x)$, and include effects of velocity averaging as well as the effects of those few atoms that scattered zero or two photons.

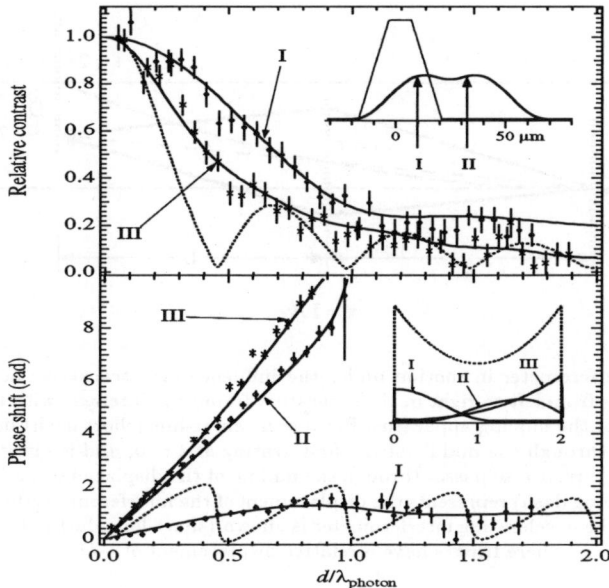

Figure 3: *Relative contrast and phase shift of the interference pattern as a function of d for the cases in which atoms are correlated with photons scattered into a limited range of directions. The dashed curve is for uncorrelated atoms. The upper inset shows atomic beam profiles at the detector when the scattering laser is off (thin line) and when the laser is on (thick line). The arrows indicate the detector positions for cases I and II. The lower inset shows the acceptance of the detector for each case compared to the original distribution (dotted line). Case I corresponds to predominantly forward scattered photons (minimal transfer of momentum), case III corresponds to backward scattered photons (transfer of two photon momenta), and case II lies in between.*

6 Inertial effects

Phase shifts caused by non-inertial motion of matter wave interferometers have been discussed by many authors in both non-relativistic and relativistic contexts (see for example Refs. [26-30]). Because such phase shifts increase with the mass of the interfering particle, atom interferometers are especially sensitive to inertial effects, and may be developed into rotation sensors, accelerometers, gravimeters, and gradiometers[30].

The inertial sensitivity of an atom interferometer arises because the freely propagating atoms form fringes with respect to an inertial reference frame. These fringes appear shifted if the interferometer moves with respect to this inertial frame while the atoms are in transit. To illustrate this, we now present a simple calculation of the fringe shift that results from acceleration a of a three grating interferometer in a direction perpendicular to both the grating bars and the atomic beam axis.

In a time $\tau = L/v$, where L is the distance between gratings and v is the velocity of the atom, the interferometer will move a distance $a\tau^2/2 \equiv D/2$ (Fig. 4). To simplify the calculation, we will make a convenient choice of initial transverse

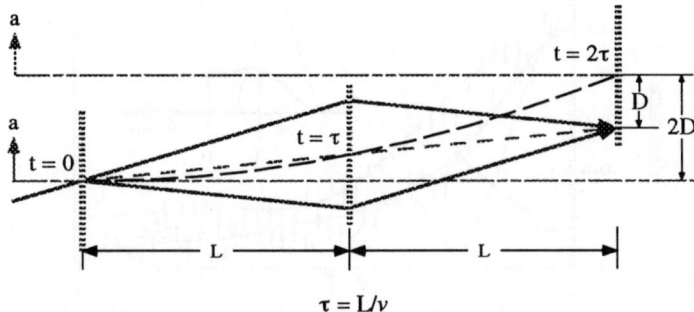

Figure 4: The interferometer in motion under the influence of a transverse acceleration. The atomic beam travels from left to right in the laboratory frame but interacts with the progressively displaced gratings of the moving apparatus. Because a center-line (short dash) between the atom beam paths passes through the middle of the first grating at $t = 0$, and is offset by a transverse velocity $v_{trans} = 1/2 a\tau$, it also passes through the middle of the displaced second grating at $t = \tau$. The dashed curve (long dash) represents the displacement of the interferometer due to acceleration. The center-line of the accelerating interferometer is shown (short-long dash) at $t = 0$, and $t = 2\tau$ where fringes have a relative displacement of $-D$.

velocity $v_{trans} = a\tau/2 = aL/2v$. Atoms in an accelerating interferometer which have this particular transverse velocity at the time they reach the first grating will pass through both the first and second gratings at exactly the same transverse positions as would atoms with *zero* transverse velocity moving in a *non-accelerating* interferometer. Because the atoms (or atom waves) in both cases intersect the first and second gratings at the same point, the relative phase shift between the two cases is given purely by the displacement of fringes at the third grating. Due to their transverse velocity, atoms form a fringe pattern at the third grating that is displaced by $v_{trans}\tau = a\tau^2 = D$ from the axis of the interferometer at $t = 0$. However, when these atoms reach the third grating, the interferometer has now moved $a(2\tau)^2/2 = 2D$ from its original position, resulting in an apparent fringe shift of $-D$. This corresponds to a phase shift

$$\varphi_{acceleration} = 2\pi \left(\frac{D}{d_g}\right) = 2\frac{2\pi}{d_g}\left(\frac{L}{v}\right)^2 a = -\frac{2\pi m^2 \lambda_{dB} A}{h^2} a \qquad (6)$$

where d_g is the period of the gratings, $\lambda_{dB} = h/mv$ is the de Broglie wavelength for an atom with mass m and velocity v, and $A = L^2(\lambda_{dB}/d_g)$ is the area enclosed by the paths of the interferometer. It should be noted that the phase shift in our three grating geometry is independent of the mass of the particle, and was derived using classical physics.

The phase shift due to rotation of the interferometer (called the Sagnac effect) follows by noting that rotation with angular rate Ω gives rise to a Coriolis acceleration $\vec{a} = 2\vec{v} \times \vec{\Omega}$, allowing one to use Eq. 6 to calculate the phase shift due to rotation about an axis parallel to the grating bars,

$$\varphi_{rotation} = \left[\frac{2\pi}{d_g}\left(\frac{L}{v}\right)^2 2v\right]\Omega = \left[4\pi\frac{mA}{h}\right]\Omega, \qquad (7)$$

where we call the bracketed factor the rotational response factor. This expression can also be directly obtained from considerations of the grating positons at the time of transit of the atoms through each grating[31].

The results of the simple derivations above agree with the non-relativistic phase contributions derived by various more sophisticated methods[28, 29,30]. The relativistic contributions to the phase shift caused by accelerations and rotations are of the order $E_{kinetic}/mc^2$ smaller than the non-relativistic terms[27] and are unresolvable in our experiments.

The Sagnac rotational response factor is independent of the velocity of the particle in an interferometer in which the area is constant (as it would be for an interferometer employing conventional beam-splitters). However, since all demonstrated atom interferometers employ diffractive beam splitters, their rotational response factors will exhibit $1/v$ dependence. This dependence arises from the variation of the enclosed area which in turn results from the variation of the diffraction angle with velocity.

In contrast to rotations, the phase shift due to linear accelerations (see Eq. 7) varies with velocity as $1/v^2$. Thus atom interferometers that use slow atoms will be relatively more sensitive to acceleration than to rotation.

Phase shifts due to rotation and acceleration, as well as shifts due to gravitational fields (which give the same response factor as acceleration due to the equivalence principle), have been observed in many kinds of matter wave interferometers. Accelerations were measured using neutron interferometers[26,29], and using atoms[32]. The Sagnac phase shift for matter waves has been verified with accuracy on the order of 1% for neutrons[29,33] and electrons[34], and to about 10% for atoms using both interferometers[21,35] and classical Moiré regime atom optics[36].

In view of the numerous demonstrations of the sensitivity of matter wave interferometers to non-inertial motion, the motivation for such experiments is principally technological: can such devices become the sensors of choice in practical applications or can they demonstrate such high sensitivity that they open up new scientific possibilities? With these considerations in mind, the observation that the rotation-induced phase shift in an atom interferometer exceeds the Sagnac phase for light of frequency ω by an amount $mc^2/\hbar\omega$, typically 10^{10}, suggests the tremendous potential of atom interferometer rotation sensors[30].

We now estimate the minimum angular velocities and accelerations detectable by our atom interferometer using the atomic velocity (1075 m/s) and signal intensity (a contrast of 12.9% and an rms count rate of 29 kcounts/sec) achieved in our apparatus. We assume that only Poissonian detection statistics degrade the signal-to-noise ratio, which is therefore proportional to $C\sqrt{N}$, where C is the fringe contrast and N is the total number of counts. The response factor for rotations is 1.86 rad/Ω_e, with a corresponding (purely statistical) rotational noise of $\sim 35m\Omega_e$ in one second of integration time. (Note: one earth rate (Ω_e) is $7.3\times 10^{-5} rad/sec$). For accelarations, the response factor is 116 rad/g, with a statistical noise of $\sim 5.5\times 10^{-4}$ g in one second of integration time.

We performed experiments to measure both the response factor for rotation and the rotational noise of our interferometer. Both measurements were made by suspending the interferometer by a cable from the ceiling and then driving it with a sinusoidally varying force applied at some distance from the center of mass, thereby giving the interferometer a rotation rate $\Omega(t) = \Omega_0 \sin(2\pi f t)$.

The rotation rate Ω_0 was typically several earth rates ($\Omega_e = 7.3 \times 10^{-5} rad/sec$) for the response factor measurement, and about $\Omega_e/10$ for the noise measurements. For the response measurements, f was chosen just over $1Hz$ in order to minimize deformations of our interferometer (which has several prominent mechanical resonances in the 10 to $30Hz$ frequency range). For the noise measurements, f was around $4.6Hz$, where the measured residual rotational noise spectrum of the apparatus had a broad minimum.

Our procedure was to measure the acceleration at the sites of the first and third gratings of the suspended interferometer using accelerometers. While modulating the grating phase, $\varphi_{grating}(t) = k_g(x_1(t) - 2x_2(t) + x_3(t))$, with a sawtooth pattern at a frequency just less than $1Hz$, we recorded accelerations from both accelerometers, $\varphi_{grating}$, and the atom counts each millisecond. Readings from the accelerometers allowed us to infer the atom phase expected from the acceleration and rotation rate of the interferometer using equations 6 and 7. We called this predicted inertial phase $\varphi_{predicted}$.

To study the magnitude and constancy of the response factor, we binned these data according to the $\varphi_{predicted}$ determined from the accelerometer readings after suitable correction for their known frequency response. Since the frequency of the sawtooth modulation of $\varphi_{grating}$ was chosen to be incommensurate with f, the data in a bin with a particular value of $\varphi_{predicted}$ had a variety of values $\varphi_{grating}$, allowing us to determine $\varphi_{rotation}$. A plot of $\varphi_{rotation}$ vs. $\varphi_{predicted}$ is shown in Fig. 5. The data reveal a linear response and an average response factor within error (0.8%) of that predicted from Eq. 7.

To study the reproducibility of our interferometer we employed a phase modulation technique to immediately convert atom counts into $\varphi_{rotation}(t)$. This was accomplished by scanning the second grating (and hence $\varphi_{grating}$) at $1Hz$ to produce a carrier modulation on the atom count rate. Since acceleration was negligible, the rotation of the interferometer introduced a phase modulation $\varphi_{rotation}(t)$ to the carrier which was demodulated by homodyne detection, using both the sine and cosine of $\varphi_{grating}(t)$ (measured by means of an optical interferometer with the same geometry as our atom interferometer) as the local oscillator.

¿From each of 21 data sets 32 seconds long, we analyzed samples of different sizes to find the rotation rate $\varphi_{measured}(t)$ determined from the rotation phase $\varphi_{rotation}(t)$ of the interferometer. Samples were taken from the middle of each data set and ranged in duration, T, from 0.66 to 10.66 seconds. Each sample was Fourier transformed and the magnitude of the amplitude of the rotation at the drive frequency f was found. The rms fluctuations in the amplitudes for given sample lengths were then determined for the various averaging times, T. In Fig. 6 they are plotted and compared to the shot noise limit.

We attribute the excess noise of the interferometer relative to shot noise seen in Fig. 6, for T greater than 2 seconds, to extraneous sources of rotational noise

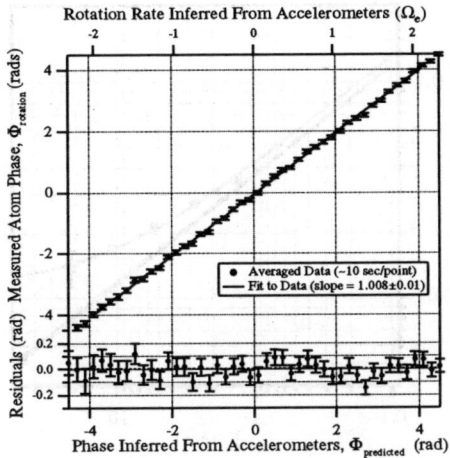

Figure 5: A plot of the measured interferometer phase, $\varphi_{rotation}$, verses the inferred phase from the accelerometer readings, $\varphi_{predicted}$, from a combination of 20 second runs totaling $\sim$ 400 seconds of data ($\sim$ 10 seconds of data per point). There is a 0.8% difference between these measurements with a total error of 1%.

rather than to any intrinsic failure of atom interferometers. The observed noise $\Delta\Omega$ can be fit as an uncorrelated sum of shot noise, SNL, and background rotational noise, B, times an overestimation factor, α,

$$\langle \Delta\Omega \rangle = \alpha \sqrt{(SNL)^2 + B^2}. \tag{8}$$

The over-estimation factor $\alpha = 1.09 \pm 0.02$, is close to unity, and is consistent with noise arising from imperfections in our modulation scheme together with the previous observation of super-poissonian noise from our detector[37]. The background noise determined from the fit is $B = 10m\Omega_e \pm 1m\Omega_e$.

We regard these results as highly encouraging for the future of inertial sensors using atom interferometers. Our interferometer was designed for separated beam interferometry, not inertial sensing. This resulted in restricting the usable area of our small $1mm \times 200\mu m$ gratings at both ends of the machine by a combined factor of 100. Furthermore, the vacuum envelope, with heavy diffusion pumps hung at odd angles, had numerous low frequency mechanical resonances. Despite these difficulties, we verified the rotational response factor to better than 1%, indicating that atom interferometric rotation sensors perform as predicted. Moreover, we achieved reproducibility at the $42m\Omega_e/\sqrt{sec}$ level. This is about three orders of magnitude more sensitive than previous rotation measurements using atom interferometry [35] and exceeds the sensitivities of much more difficult neutron interferometry measurements that required integration times of many minutes per point [29]. A dedicated rotation sensor using one cm^2 gratings and cesium atoms would perform many orders of magnitude better than ours, and should exceed the performance of laser gyroscopes.

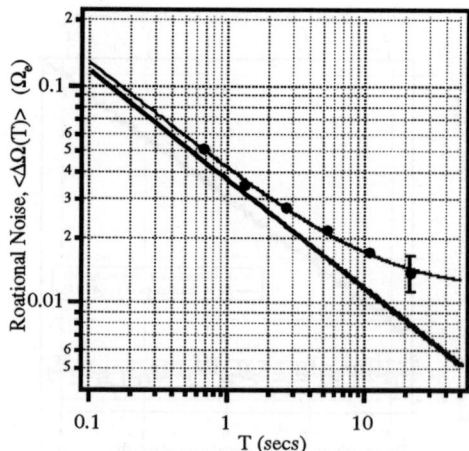

Figure 6: Reproducibility of the rotation rate measurements in the atom interferometer. Fluctuations in the spectral peak amplitude at the driving frequency, $f = 4.6 Hz$ (for 30 data sets), is compared to the predicted shot noise (dashed line) and plotted versus integration time, T. A fit to the data points with Eq. 8 yields an over-estimation factor $\alpha = 1.09 \pm 0.02$ and a background $B = 10 m\Omega_e \pm 1 m\Omega_e$.

7 Acknowledgments

The atom gratings used in this work were made in collaboration with Mike Rooks at the Cornell Nanofabrication Facility at Cornell University [38]. We are grateful for the existence and assistance of this facility, and also for help with nanofabrication from Hank Smith and Mark Schattenburg at M.I.T. The contributions of David Keith to these experiments are also gratefully acknowledged. This work was made possible by support from the Army Research Office contracts DAAL03-89-K-0082 and ASSERT 29970-PH-AAS, the Office of Naval Research contract N00014-89-J-1207, NSF contract 9222768-PHY, and the Joint Services Electronics Program contract DAAL03-89-C-0001, and the Charles Stark Draper Laboratory contract DL-H-484775 9. TDH and ETS acknowledge the support of National Science Foundation graduate fellowships. JS acknowledges the support of an Erwin Schrödinger Fellowship of the Fond zur Förderung der Wissenschaftilchen Forschung in Austria and an APART fellowship of the Austrian Academy of Sciences.

References

1. D.E. Pritchard, "Atom Interferometers" in *Atomic Physics 13 (conference held in Munich, Germany in August 1992)*, ed. H. Walther, T.W. Hänsch, and B. Neizert (AIP Press, New York NY, 1993).
2. J. Schmiedmayer, M.S. Chapman, C.R. Ekstrom, T.D. Hammond, S. Wehinger, and D.E. Pritchard *Phys. Rev. Lett.* **74**, 1043 (1995).
3. J. Schmiedmayer, M.S. Chapman, C.R. Ekstrom, T.D. Hammond, A. Lenef, R.A. Rubenstein, E.T. Smith, and D.E. Pritchard, "Optics and Interferometry

with Atoms and Molecules" to be published in *Atom Interferometry*, ed. P. Berman (Academic Press).
4. P.E. Moskowitz, P.L. Gould, S.R. Atlas, and D.E. Pritchard *Phys. Rev. Lett.* **51**, 370 (1983).
5. P.L. Gould, G.A. Ruff, and D.E. Pritchard *Phys. Rev. Lett.* **56**, 827 (1986).
6. P.L. Gould, G.A. Ruff, P.J. Martin, and D.E. Pritchard *Phys. Rev. A* **36**, 1478 (1987).
7. B.G. Oldaker, P.J. Martin, P.L. Gould, M. Xiao, and D.E. Pritchard *Phys. Rev. Lett.* **65**, 1555 (1990).
8. P.L. Gould, P.J. Martin, G.A. Ruff, R.E. Stoner, J.L. Picque, and D.E. Pritchard *Phys. Rev. A* **43**, 585 (1991).
9. D.W. Keith, C.R. Ekstrom, Q.A. Turchette, and D.E. Pritchard *Phys. Rev. Lett.* **66**, 2639 (1991).
10. M.S. Chapman, C.R. Ekstrom, T.D. Hammond, J. Schmiedmayer, B. Tannian, S. Wehinger, and D.E. Pritchard *Phys. Rev. A* **51**, R14 (1995).
11. M.S. Chapman, C.R. Ekstrom, T.D. Hammond, R.A. Rubenstein, J. Schmiedmayer, S. Wehinger , and D.E. Pritchard *Phys. Rev. Lett.* **74**, 4738 (1995).
12. N. Bohr in *A. Einstein: Philosopher - Scientist*, ed. P.A. Schilpp (Library of Living Philosphers, Evanston IL, 1949)
13. R. Feynman, R. Leighton, and M. Sands, *The Feynman Lectures on Physics* (Addison-Wesley, Reading MA, 1965)
14. W. Wooters and W. Zurek *Phys. Rev. D* **19**, 473 (1979)
15. A. Zeilinger in *New Techniques an Ideas in Quantum Mechanics* (New York Acad. Sci., New York NY, 1986)
16. T. Sleator, O. Carnal, T. Pfau, A. Faulstich, H. Takuma, and J. Mlynek in *Laser Spectroscopy X: Proc. 10th Int. Conf. on Laser Spectroscopy*, ed. M. Dulcoy, E. Giacobino, and G. Camy (World Scientific, Singapore, 1991)
17. S.M. Tan and D.F. Walls *Phys. Rev. A* **47**, 4663 (1993).
18. M.O. Scully, B.-G. Englert, and H. Walther *Nature* **351**, 111 (1991).
19. E. Schrödinger *Naturwissenschaften* **23**, 807; ibid. 824; ibid. 844 (1935).
20. M.S. Chapman, T.D. Hammond, A. Lenef, J. Schmiedmayer, R.A. Rubenstein, E.T. Smith, and D.E. Pritchard *Phys. Rev. Lett.* **75**, 3783 (1995).
21. C.J. Bordé *PLA* **140**, 10 (1989).
22. P. Storey and C. Cohen-Tannoudji *J. Phys. II France* **4**, 1999 (1994).
23. A. Stern, Y. Aharonov, and Y. Imry *Phys. Rev. A* **41**, 3436 (1990)
24. T. Pfau, S. Spälter, C. Kurstsiefer, C.R. Ekstrom, and J. Mlynek *Phys. Rev. Lett.* **29**, 1223 (1994)
25. B.-G. Englert, H. Fearn, M.O. Scully and H. Walther, "The Micromaser Welcher-Weg Detector Revisited" in *Proceedings of the Adriatico Workshop on Quantum Interferometry*, ed. F. De Martini, G. Denardo, and A. Zeilinger (World Scientific, Singapore, 1994)
26. R. Colella, A.W. Overhauser, and S.A. Werner *Phys. Rev. Lett.* **34**, 1472 (1975)
27. J. Anandan *Phys. Rev. D* **15**, 1448 (1977).
28. D.M. Greenberger and A.W. Overhauser *Rev. Mod. Phys.* **51**, 43 (1979).

29. S.A. Werner, J.L. Staudenmann, and R. Colella *Phys. Rev. Lett.* **42**, 1103 (1979).
30. J.F. Clauser *Physica B* **151**, 262 (1988).
31. A. Lenef, T.D. Hammond, E.T. Smith, M.S. Chapman, R.A. Rubenstein, and D.E. Pritchard to be published
32. M. Kasevich and S. Chu *Phys. Rev. Lett.* **67**, 181 (1991).
33. D.K. Atwood, M.A. Horne, C.G. Shull, and J. Arthur *Phys. Rev. Lett.* **52**, 1673 (1984)
34. F. Hasselbach, and M. Nicklaus *Phys. Rev. A* **48**, 143 (1993)
35. F. Riehle, T. Kisters, A. Witte, J. Helmcke, and C.J. Bordé *Phys. Rev. Lett.* **67**, 177 (1991)
36. M. Oberthaler, S. Barnet, E. Rasel, J. Schmiedmayer, and A. Zeilinger submitted to *Phys. Rev. A* (1996)
37. C.R. Ekstrom. Ph. D. Thesis, M.I.T., 1994
38. M.J. Rooks, R.C. Tiberio, M.S. Chapman, T.D. Hammond, E.T. Smith, A. Lenef, R.A. Rubenstein, D.E. Pritchard, and S. Adams *J. Vac. Sci. Technol. B* **13**, 2745 (1995)

QUANTUM EFFECTS IN HE CLUSTERS

J. Harms, M. Hartmann, W. Schöllkopf, J.P. Toennies and A.F. Vilesov
Max-Planck-Institut für Strömungsforschung, Bunsenstrasse 10, 37073 Göttingen, Germany

Several different molecular beam experiments are described for characterizing small clusters (N = 2-6) and large He droplets (N = 10^3-10^4). The former are analyzed non-destructively by diffraction from a transmission grating. The latter are studied by various scattering experiments and from the spectroscopy of embedded molecules. The spectroscopic experiments reveal sharp lines, indicating that the molecules rotate freely and provide information on the internal temperatures (0.37 K for ^{4}He droplets and 0.14 K for ^{3}He droplets) and that the droplets are superfluid.

1 Introduction

Although helium at ordinary temperatures is a rather "dull" inert gas, at low temperatures and especially below the superfluid transition at 2.2 K it exhibits a large number of very unusual and fascinating properties. These properties inspired Fritz London in 1954 to state that "superfluid helium also called liquid helium II, is the only representative of a particular *fourth* state of aggregation beside the solid, liquid and gaseous states." [1] London was also the first to call attention to the possible connection between the unusual behavior of He II and Bose-Einstein Condensation. [2] The remarkable feature of a superfluid with a significant condensate fraction is that its many particle wave function can be written in terms of a macroscopic superfluid wavefunction as first pointed out by Feynman [3]

$$\Psi_S(\vec{R}) = \left[\rho_S(\vec{R})\right]^{1/2} \exp(i\, S(\vec{R})) \quad , \tag{1}$$

where ρ_s is the density of the superfluid component and $S(\vec{R})$, the phase, is a real function of position $\vec{R}$. This unique property is explained in terms of off-diagonal long range order (ODLRO) as opposed to "ordinary" long range order exhibited by, for example, solids. Possibly related, large scale quantum coherent effects are also found in very small He clusters with N=2 and 3. Since as will be discussed below a similar macroscopic wave function is also expected for finite sized He clusters, these represent a unique opportunity to study an isolated macroscopic quantum system consisting of Bosons. Thus He clusters present a number of fundamental questions to theoreticians and experimentalists [4] such as: (1) What are the physical properties of He clusters, e.g. their temperatures, density distributions etc.? (2) Are the clusters superfluid and what is the critical size for superfluidity? (3) What is the condensate fraction? And finally (4), if superfluid, what new phenomena do they exhibit?

The question of their superfluidity has been rather extensively studied theoretically using several different approaches. In the first detailed calculations based on the variational Monte Carlo method Lewart, Pandharipande and Pieper [5] were able to calculate mean field potentials, radial density distributions and predict that even clusters with as few as 20 atoms should show a sizable condensate fraction. In the outer surface region of even small (N ~ 70 atoms) cold He clusters (T ≲ 1 K) where the particle density has fallen off to about 10% of that of the core which has the density of the bulk ($2.2 \cdot 10^{22}$ atoms/cm^3), they found a 100% Bose-Einstein condensed fraction compared to only about 10% condensate fraction in the interior.[6] Thus, in fact, at their surfaces He clusters provide a system with a very much higher particle density of Bose-condensed atoms than recently achieved in electromagnetic traps.[7] In 1989 Sindzingre, Klein and Ceperley[8] on the basis of finite temperature path-integral Monte Carlo calculations found manifestations of superfluid behavior for clusters of 64 atoms and larger for temperatures of $T_c \lesssim 1.5$ K which is about 0.7 K below the bulk T_c. Superfluidity in clusters has also been confirmed by calculations of the dispersion curves for elementary excitations which show manifestations of a sharp roton minimum for clusters of more than about 70 atoms.[9] The surprising observation of superfluid behavior in even such small droplets has also been confirmed by Chin in Monte Carlo calculations of the off-diagonal one body density matrix[10]

$$\rho_1(\vec{r}_1, \vec{r}\,'_1) = N \int d\vec{r}_2 \cdots d\vec{r}_N \, \psi_o^*(\vec{r}\,'_1, \vec{r}_2 \cdots \vec{r}_N) \psi_o(\vec{r}_1, \vec{r}_2 \cdots \vec{r}_N) \quad , \quad (2)$$

first introduced by Penrose and Onsager.[11] The convergence of $\rho_1(\vec{r}_1, \vec{r}\,'_1)$ to a constant value as $|\vec{r}_1 - \vec{r}\,'_1| = r \to \infty$ provides evidence of ODLRO which appears to be a sufficient condition for superfluidity. The value of the constant is a measure of the condensate fraction. Chin found that even for small N=40 clusters $\rho_1(r)$ has a plateau for $r \geq 4$ Å which remains roughly constant out to the rim of the cluster. This indicates that the size of the condensate region scales with the overall size of the cluster.

Very small clusters consisting of only a few He atoms have also attracted some theoretical attention ever since Lim[12] pointed out that the trimer may exhibit long range Efimov states. As predicted by Efimov in 1970[13] three identical spinless bosons are expected to have an infinite number of long range S-wave states if the binding energy of the two-body potential is exactly zero,

Table 1: Comparison of properties of typical superfluid He droplets with laser trapped Bose-Einstein condensed atoms.

	^{4}He Droplets	BEC in Laser Trap
Atoms	^{4}He ^{1}S	^{87}Rb ^{2}S, F=2 M$_F$=2
No. of atoms	10^4	10^6
Diameter D	100 Å	10^5 Å
Cooling mechanism	natural evaporation	rf-driven evaporation
Temperature T	0.37 K	$0.17 \cdot 10^{-6}$ K
De Broglie wavelength λ	30 Å	$6 \cdot 10^4$ Å
Density n	$2.2 \cdot 10^{22}$ atoms/cm^3	10^{15} atoms/cm^3
Av. distance $\bar{d}$	3.5 Å	10^3 Å
Av. pot. $\bar{v}$	-20 K	$2 \cdot 10^{-10}$ K
Two body scattering length a	100Å [14]	49 Å [Na] [15]

which corresponds to an infinite scattering length a. The scattering length of the He dimer is now known to be about 100 Å [14] so that there is good reason to expect that the trimer can support at least one Efimov state. [16] The unusual quantum properties of such small clusters have been discussed both in connection with the overall problem of superfluidity in bulk He [17] and in connection with Bose-Einstein condensation. [18]

Since He clusters have many conceptual similarities but also some significant differences compared to the recently much discussed Bose-Einstein condensed alkali atoms in electromagnetic traps some of the important properties of both are compared in Table 1. Whereas the numbers of atoms involved and the cooling mechanisms are not too different He clusters are much denser with a much greater average interaction potential and much warmer (0.4 K) than the ultra-low temperatures of alkali atoms in electromagnetic traps (10^{-7} K). The higher densities and much stronger interactions are, of course, the necessary requisites for superfluid behavior.

Several suggestions have been made for experiments to establish the superfluidity of finite sized He-droplets. Becker [19] and Gspann [20] have carried out a number of experiments with this aim but clear evidence for superfluidity was not found. Halley, Campbell, Giese and Goetz [21] have proposed a novel experiment for detecting ODLRO directly on large drops of liquid helium, which in principle could also be carried out on large clusters.[22]

The present brief review of recent experimental work on small clusters ($2 \leq N \leq 10$) and large clusters ($10^3 \leq N \leq 10^4$), which we call droplets, is organized as follows. The next section summarizes briefly how He clusters are produced in free jet expansions from both the gas and liquid phases and some of the properties of the cluster beams produced. Then we briefly describe the experimental technology used to select and identify very small He clusters and discuss some of their properties. In the subsequent sections we review recent spectroscopic studies in the infra-red and visible spectral regions of several different polyatomic and hydrocarbon molecules and of clusters of molecules within large He droplets ($N \approx 5000$ atoms). These experiments provide the first information on their low internal temperature (0.37 K) and first evidence for superfluidity of the ^{4}He droplets.

2 Production and Characterization of He clusters

He clusters and droplets with up to $N = 10^8$ atoms are readily produced by simply expanding helium, either as a gas to produce small clusters and droplets ($N \lesssim 10^4$), or as a fluid ($N \gtrsim 10^4$) through small apertures with diameters of typically $d = 5$ μm. The processes occuring in the expansion of a gas have been extensively studied and are fairly well understood.[23-25] Very low ambient temperatures of 10^{-3} K are achieved[25] and if the density is high enough extensive condensation sets in. Because of the low heat of evaporation of bulk helium (7.15 K per atom) the cluster temperatures have been predicted to be about 0.4 K.[20,26]

For the non-destruction detection and analysis of He droplets a special deflection technique shown schematically in Fig. 1 has been developed.[27] Initially mass spectrometers were used but because of the extensive amount of energy released by the strong electrostriction forces between the He atoms surrounding the nascent He$^+$ ion the fragile He clusters are extensively fragmented.[23] In the apparatus of Fig. 1 the cluster beam is crossed by a secondary beam of unclustered atoms or molecules expanded at room temperature. Since the He droplets are liquid the molecules can penetrate into the droplets and as determined experimentally a large fraction ($\approx 50\%$) are embedded and the remainder pass through after imparting most of their momentum.[28] Of course, in these collisions there is also extensive evaporation but since only weak van der Waals interactions are involved only typically about 300 He atoms are evaporated as is the case of a single embedded Xe atom.[28] For large droplets of say 5000 He atoms this is a relatively small loss. As a result of momentum conservation the cluster beam is deflected by angles of the order of $\Theta \approx \frac{m_{atom}}{N\, m_{He}}$

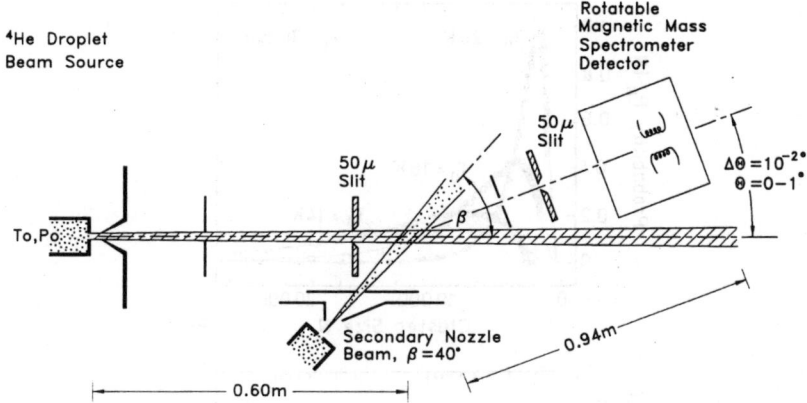

Figure 1: Schematic drawing of the apparatus used for detecting ^{4}He droplets nearly non-destructively and for measuring their size distributions.

or about 10^{-2}-10^{-3} radians. By narrowly collimating the incident beam with 50μ slits it has been possible to clearly resolve the distribution of deflected clusters with the mass spectrometer as a function of the deflection angle with a resolution of better than 1 mrad.[27] The measured deflection pattern can be directly converted into a size distribution of the incident neutral cluster beam. Some typical measurements obtained in this way are shown in Fig. 2a). The solid curves fitted to the data are log-normal distributions which have been found to fit all the data nicely. In this way the distributions for droplets in the range of $N = 10^3$-10^4 have been determined for a wide range of source conditions. The size of clusters and droplets in the free beams achieved for different combinations of source pressures P_o and temperatures T_o are indicated in the well known bulk phase diagram of He in Fig. 2b).

Related experiments have shown that it is possible to embed several atoms or molecules sequentially into He droplets.[28,29] The experiments indicate that they coagulate in the inside of the droplets where they form clusters within the droplets. The probability distributions for the clusters, the capture cross sections and the number of evaporated helium atoms following capture and recombination have also been investigated experimentally.[28]

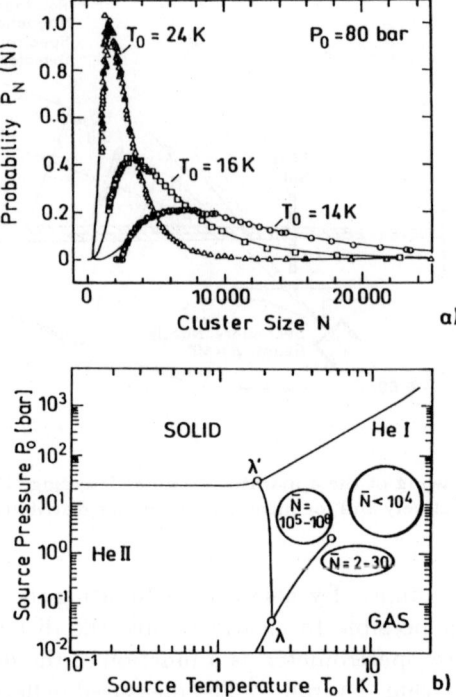

Figure 2: (a) Cluster size distributions at several source temperatures T_o ($P_o = 80$ bar) derived from the deflection resulting from a SF_6 secondary beam are compared with least-squares fits to a log-normal distribution model. Similar results are obtained using rare gas secondary beams. (b) The average clusters sizes $\bar{N}$ for different combinations of source pressures P_o and source temperatures T_o are indicated in the phase diagram of bulk helium.

3 Diffraction of Small He Clusters

For a long time it has been realized that the zero-point energy of the He dimer almost perfectly compensates the attractive negative part of the van der Waals potential and lies very close to zero, but the question of whether it is slightly positive (unbound) or slightly negative (bound) could only recently be clarified by several new theoretical approaches to the calculation of van der Waals potentials.[14] A related question concerns the number of Efimov states[13] of the He-trimer since if the He-dimer were critically bound (binding energy = zero) then according to Efimov the trimer should have an infinite number

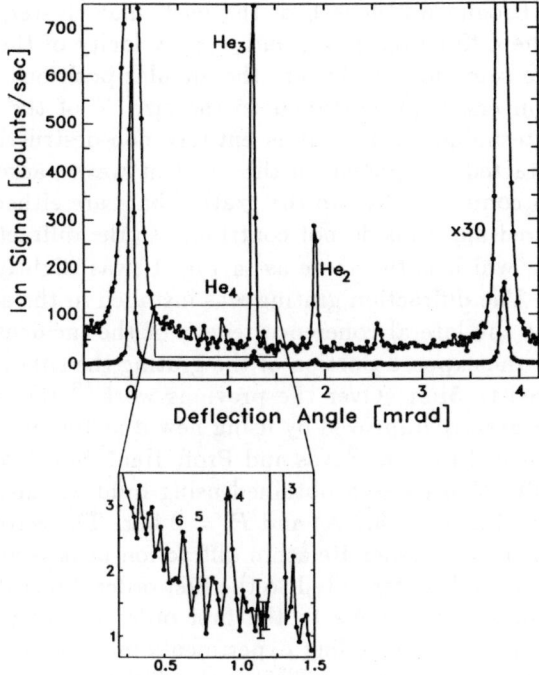

Figure 3: Angular distribution of a diffracted ^{4}He cluster beam measured at $T_o = 10$ K and $P_o = 2$ bar with the mass spectrometer set for m = 4 amu (He$^+$) for a grating with a period of $d = 100$ nm. The de Broglie wave length of the atoms is about 3.2 Å. First order diffraction peaks are seen for He, He$_2$, He$_3$, He$_4$, He$_5$ and He$_6$ (see inset).

of long range S-wave states. Similar long-range species are also expected to play a role in nuclear physics.[30] Obviously neither detection with a mass spectrometer[31,32] nor the deflection technique described in the previous section are ideally suited to select and identify these very fragile species. Thus we have developed another approach in which the cluster containing beam is diffracted with a transmission grating.[33] The deflection angles of the diffraction peaks are given by

$$sin\,\theta_n \;=\; n\frac{\lambda}{d} \;=\; n\frac{h}{N \cdot m_{He}\; v \cdot d} \quad , \qquad (3)$$

where λ is the de Broglie wavelength of the particular cluster, d the period of the grating, n the diffraction order, and v the velocity of the beam. Since v is very nearly the same for all clusters the angular positions for first order diffraction θ_1 are inversely proportional to the size N of the cluster. The advantage of this technique is that it is entirely non-destructive since only the coherently diffracted component of the incident wave packet is detected. Any clusters which come too close to the grating bars are either destroyed or incoherently scattered and thus do not contribute to the diffraction signal.

The apparatus, which is the same as in Fig. 1, was slightly modified for these experiments. The diffraction grating was installed in the scattering center. To assure that the lateral coherence length of the incident wave packet matches the size of the exposed portion of the grating the slits in the incident beam were narrowed to 5μm. Over the previous work [33] the resolution and dispersion could be greatly improved by using new $d = 100$ nm silicon-nitride gratings kindly provided by Tim Savas and Prof. Hank Smith at MIT. Fig. 3 shows a typical diffraction pattern obtained using mild expansion conditions (see Fig. 2) of $T_o = 10$ K ($\lambda \approx 3.2$ Å) and $P_o = 2$ bar. The zeroth-order peak is at the origin and the first-order He atom diffraction peak is at $\Theta \approx 4$ mrad. The sharp peak at 2 mrad is attributed to the first order dimer diffraction and the peak at 1.33 mrad corresponds to the first order trimer diffraction. As found in a large number of preceeding experiments made with a $d = 200$ nm grating for a wide range of source conditions the dimer and trimer diffraction peaks have by far the largest intensities. This is verified in these new diffraction patterns where for the first time the He_4, He_5 and He_6 clusters are clearly resolved. The large intensities of the dimer and trimer peaks suggest that their production may well be favored by their extreme quantum mechanical diffuseness resulting from their very small binding energies.

The diffraction technique described here is the first demonstration of a non-destructive mass spectrometer. Such a device opens up a number of new experimental possibilities and several studies have already been reported. [32]

4 Spectroscopy of Molecules Embedded in He Droplets

It has long been realized that spectroscopy of single atoms and molecules embedded in liquid helium provides a very promising technique for probing the quantum state of liquid He II. [34] Although liquid He I because of its weak interaction potential should be an ideal solvent it has been anticipated that superfluid He II may not be since the solvent molecules disrupt and obstruct the exchange of like particles. [35] Indeed it is well known that atoms and molecules tend to clump together and then either float to the surface or sink to the bottom of the container.[36,37] Using sophisticated laser techniques these problems could recently be circumvented.[38] For metal atoms the spectral absorption lines measured in this way have been found to be very broad and strongly shifted. Moreover in many cases the laser induced fluorescence signals were completely quenched.[38] These effects have been attributed to the strong repulsive interaction between the outer electrons of the metal atoms which repel the surrounding helium atoms.[39]

These problems could recently be circumvented by embedding non-metallic simple inorganic or organic polyatomic molecules in He droplets. Experimentally [28,40] and theoretically [41] it has been established that the most stable position for closed shell atoms and molecules is near the center of the cluster since the collective van der Waals interaction with the helium atoms is greatest there. In cluster beam experiments the absorption of even low energy infra-red radiation can be sensitively observed (see below).

In these experiments the ^{4}He droplets are passed through a scattering chamber where the gas of interest, in this case SF_6, is added at a known pressure, typically 10^{-5} millibar. The average number of molecules captured by the He droplet can be specified to a large extent by adjusting the pressure in the scattering chamber.[28] Upon capturing a room temperature molecule it is rapidly cooled to about 0.5 K within about 10^{-6} - 10^{-7} seconds by evaporation of helium atoms from the surface. Of the approximately 4000 helium atoms in a typical droplet about 600 evaporate during the process of capture of an SF_6 molecule.[28] The SF_6 containing droplets are then detected by electron impact ionization in a quadrupole mass spectrometer.

A semiconductor diode laser was directed antiparallel and coaxially to the cluster beam to excite the ν_3 vibrational mode (near 946.5 cm^{-1}) of the SF_6 molecules solvated in the droplets as they pass between the scattering chamber and the detector. The vibrationally excited molecules are rapidly quenched by interactions with the surrounding helium, resulting, in this case, in the evaporation of approximately 200 additional helium atoms. This laser induced decrease in the droplet size, as well as the corresponding recoil of the droplet, results in a depletion in the mass spectrometer signal of several percent, providing a sensitive signal change for monitoring the spectral absorption.

Fig. 4a) shows the measured spectrum for ^{4}He clusters with $\bar{N} = 4000$ atoms.[42] Surprisingly sharp spectral features are observed, which resemble those of the free SF_6 molecules except that the spacing of the rotational lines is only 1/3 of that of the free molecule. This sharpness and equal width of all the rotational lines indicates that the molecules rotate freely inside the droplets. To achieve the good fit shown by the weak continous lines in Fig. 4 it was necessary to fit not only the rotational constant but also the centrifugal

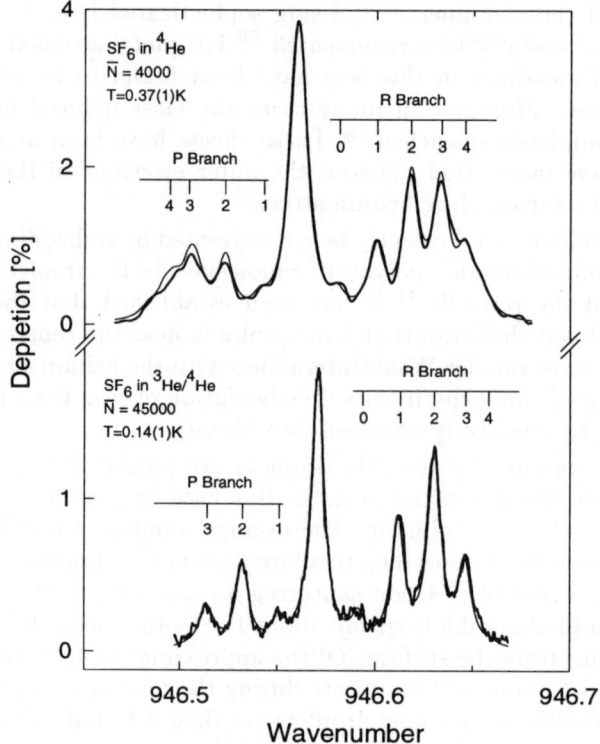

Figure 4: An expanded view of the measured infra-red spectrum for single SF$_6$ molecules (a) inside a pure ^{4}He cluster ($\bar{N}$ = 4000 atoms) and (b) inside a large ^{3}He cluster ($\bar{N}$ = 45000) with more than 45 ^{4}He atoms. From the relative line intensities of the P- and R-branches the rotational temperatures are found to be 0.37±0.01 K and 0.14±0.01 K.

coupling constant while keeping the Coriolis coupling constant the same as for the free molecule.[43] The observed large reduction in the rotational constant has been explained by assuming that effectively 8 He atoms are dragged by the rotating SF$_6$ molecule and rotate together with the molecule. The relative intensities of the resolved rotational lines in the $P-$ and $R-$ branches correspond to a rotational temperature of 0.37 ± 0.01 K. If we assume that the droplet has reached equilibrium after the capture process this rotational temperature can be taken as the first direct experimental determination of the temperature of the He droplets.[42] For the experimental cooling time of $0.8 \cdot 10^{-3}$ sec in the apparatus the theoretically predicted temperature is about 0.32 K[26] in good agreement with the measured rotational temperature.

Similar experiments have recently been carried out for SF_6-dimers, -trimers and -tetramers as well as for various mixed clusters such as SF_6-Rg, where Rg is a rare gas atom.[44] In all cases rotational side bands were observed indicating that these relatively bulky species also rotate freely within the liquid helium cages. This remarkable observation of free rotation and sharp rotational lines in SF_6 has led us to speculate that this might be a consequence of superfluidity.[44]

In an attempt to test for this possibility we have recently repeated these experiments with ^{3}He clusters, which are definitely not superfluid at these temperatures. Much to our surprise the measured spectrum showed no significant changes except that the rotational temperature is lowered to only 0.14 K as expected because of the lower heat of evaporation of ^{3}He of only 2.4 K. To explain the strong similarity of the two sets of rotational spectra the unavoidable small concentration of ^{4}He in the ^{3}He of about $1 \cdot 10^{-3}$ has to be taken into account. Since in the experiment of Fig. 4b) very large ^{3}He clusters were used with $\bar{N} = 45000$ they are expected to contain at least 45 ^{4}He atoms. The actual concentration of ^{4}He atoms may even be much greater since in the expansion the small amount of ^{4}He is expected to nucleate the condensation. Otherwise cluster formation is expected to be strongly inhibited by the fact that the small pure ^{3}He clusters with $N \leq 30$ are not stable.[45] Within the large ^{3}He droplets the ^{4}He atoms are expected to quickly replace all the weaker bound ^{3}He atoms next to the SF_6 molecule because of their lower zero-point energies. Thus in fact we observe the SF_6 molecules rotating within a cage of ^{4}He atoms within the large ^{3}He droplets.

These infra-red spectroscopy experiments are technically difficult and tedious since to achieve the high spectral resolution specially selected diodes each of which covers only a narrow range of frequencies (≈ 1 cm^{-1}) have to be used. Lasers in the visible are much more robust and can be easily tuned over a wide range of frequencies. For an initial study of the spectroscopy of electronic transitions in He droplets the organic molecule glyoxal ($C_2H_2O_2$) was chosen.[46] The reason for choosing this molecule instead of metal atoms, as used in earlier work, was the hope that the "cage" formed by the structural skeleton would provide for sufficient internal space for containing the intravalent excitation thereby reducing the interaction with the external helium environment. The depletion spectrum shown in Fig. 5 consists of four very sharp central lines each having a weak broader feature on its blue side which is ascribed to a phonon wing structure. The central lines have been identified as vibronic transitions between the ground state S_o and the first excited singlet state S_1 of glyoxal. These lines lie close to the corresponding lines observed in the laser induced fluorescence spectrum measured for free glyoxal molecules in

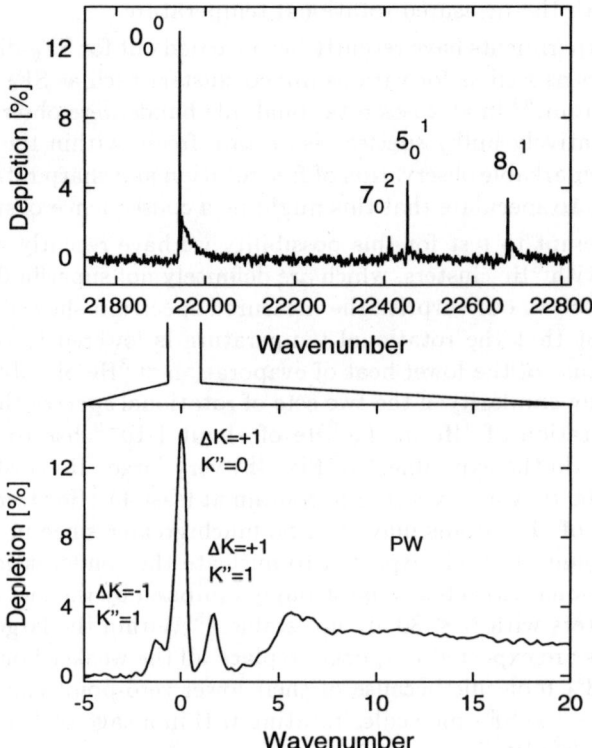

Figure 5: (a) Depletion spectrum of single glyoxal molecules inside ^{4}He droplets ($\bar{N} = 5500$ atoms) recorded at a laser output energy of 1.1 mJ/pulse. Four different vibronic lines of the $S_1 \leftarrow S_o$ transition are observed. (b) An expanded portion of the 0_o^o band region showing the rotational structure and the phonon wing side band (PW).

a molecular beam.[47] The position of the 0_o^o line corresponding to the ground vibrational state of both the ground and excited electronic states of glyoxal in a He droplet is found to be red shifted by 30.6 cm^{-1} relative to the corresponding line of the free molecule. From the relative frequencies of the other lines the vibrational frequencies of glyoxal in the S_1 state in the droplets were found to be slightly larger than in the free molecule by 4.6, 0.1 and 0.9 cm^{-1} for the ν_7 (torsion), ν_5 (CCO bend) and the ν_8 (CH wag) normal modes, respectively.

Of special interest are the phonon side bands since these contain information on the dynamic coupling of the molecule to the matrix. By using higher laser intensities it was possible to enhance the weak phonon wing intensities while the central line saturates. The results of such experiments are shown in Fig. 5b). Now the three branches of the rotational side bands of the nearly prolate symmetric top are clearly seen. Then separated off to the blue by a distinct gap a phonon wing is clearly apparent. Phonon wings, which are commonly seen in matrix spectroscopy experiments,[48] are usually attributed to an electron-phonon coupling of the excited molecule with the surrounding matrix environment. However up to now in most studies the phonon wing has always been structureless and sufficiently broad that it overlaps with the central line. This striking difference could be explained in terms of the special nature of the elementary excitations in liquid He II.

The Huang-Rhys theory developed to analyze the spectra of impurities in solid matrices[49] has been adopted to fit the data. The absorption cross section is given by

$$\sigma(w) = \frac{4\pi^2 w}{\hbar c} |M_{if}|^2 G(w - w_{if}) \quad , \quad (4)$$

where w is the transition frequency, M_{if} is the matrix element of the electronic transition and $G(w)$ is the spectral shape function for the phonon wing

$$G_n(w - w_{if}) = \sum_{\{n_Q\}} \left[\int d\vec{r}\, \Psi^f_{\{n_Q\}}(\vec{r})\, \Psi^i_S(\vec{r}) \right]^2 \delta(w - \sum_Q n_Q\, w_Q) \quad , \quad (5)$$

where Ψ^i_S and $\Psi^f_{\{n_Q\}}$ are the macroscopic cluster ground state and excited state wave functions, respectively. n_Q are the number of quanta in each normal mode of frequency w_Q with wavenumber Q. If Ψ^i_S is given by Eq. (1) then the single phonon excited cluster state can be expressed as $\Psi^f_{1Q} = f_{1Q}(R)\Psi_S(R)$ where $f_{1Q}(R)$ can be written as

$$f_{1Q}(R) = A\, \frac{\sin QR}{R} + B\, \frac{\cos QR}{R} \quad . \quad (6)$$

The final result for G_1 then becomes [46]

$$G_1(w) \propto \int_{w=const} dQ^3\, Q^2\, e^{-Q^2/2\beta^2}\, \frac{1}{|\frac{dw}{dQ}|} \quad , \quad (7)$$

where β is a range parameter which describes the fall-off in the helium density at the boundary to the molecule. The last term in Eq. (7) is equal to the density of phonon-roton states of liquid helium and in fact, largely determines $G_1(w)$. This is illustrated in Fig. 6 where at the top (a) the well-known curve

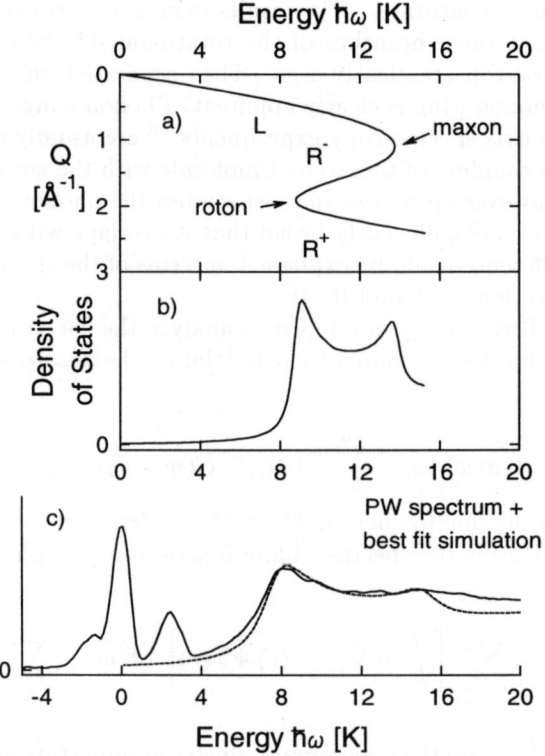

Figure 6: Composite diagram showing (a) the dispersion curve of sharp elementary excitations in superfluid ^{4}He II, (b) the corresponding density of states and (c) a comparison between a full calculation based on a slightly modified dispersion curve (- - -) and the experimental phonon wing similar to that of Fig. 5b) at 11 mJ/pulse.

of elementary excitations in bulk He II is shown. The nearly linear rise at low Q is attributed to longitudinal phonons, L. At $Q > 1$ Å beyond the maximum which is called a "maxon" the excitations are rotons (R$^-$ and R$^+$). The corresponding density of states, shown in Fig. 6b), has a strong similarity with the measured spectrum in Fig. 6c). The best fit of the spectra based on the complete expression, corresponding to Eq. (7), is shown in Fig. 6c). This fit was obtained for a roton energy of $E_{rot} = 7.8$ K (8.65 K) and a maxon energy of $E_{max} = 15.1$ K (13.7 K) where the values in parenthesis are the established values for the bulk. The deviations from the bulk liquid are expected in view of the sizable compression of the liquid He next to the molecule resulting from the van der Waals attraction. Thus the overall shape of the observed phonon side wing is entirely consistent with a single phonon dominated excitation of the surrounding superfluid medium. Of special significance in interpreting these results is the sharpness in the rise of the signal at about $\Delta E = 6$ K (4.2 cm^{-1}). As shown by numerous neutron studies in the bulk there is a dramatic narrowing in the energetic width of the roton by several orders of magnitude as the liquid passes from normal liquid helium into the superfluid state at T_c.[46] These observations of sharp phonon wings have since been confirmed by several other studies of the visible spectra of tetracence ($C_{18}H_{12}$) and pentacene ($C_{22}H_{14}$) molecules.[50]

5 Conclusions and Outlook

Many of the experimental challenges mentioned in the introduction concerning the physical properties and superfluidity of He clusters have recently been met and we now know the sizes, size distributions and the internal temperatures of ^{4}He droplets and of droplets consisting mostly of the ^{3}He isotope. Moreover it has recently been demonstrated that not only for the infra-red region but also in the visible region the spectral features are nearly as sharp as for free molecules. The observation of phonon wings with a distinct gap and a shape expected for the maxon-roton elementary excitation dispersion curve of superfluid helium first with glyoxal and more recently in several larger organic molecules provides the first evidence that the droplets are indeed superfluid.

These experiments raise a number of new questions. For one it is at present not clear whether the observation of free rotations, as in the case of SF$_6$, large (SF$_6$)$_2$ and SF$_6$-rare gas van der Waals dimers, is a manifestation of superfluidity or only reflects the weakness of the interaction with the surrounding He atoms. A related fundamental question concerns the dynamical nature of the interactions with the environment. The experiments with SF$_6$ indicate that effectively many He atoms are strongly coupled to the rotations, yet the preliminary analysis of the glyoxal rotational lines indicates that at most only one or two He atoms are firmly attached. This suggests that there may be different microphases of attached He atoms ranging from a solid or liquid phase to a loosely coupled gaseous-like phase. Certainly from these spectroscopic experiments discussed briefly here one can already conclude that the spectroscopy of molecular probes in ^{4}He droplets provides now detailed insight into the interaction of molecules with this special liquid environment, not available in other liquid matrices.

The experiments discussed here open up a vast new area of ultra-low temperature spectroscopy of single molecules and of van der Waals and chemically bound clusters. Preliminary experiments on several different metal clusters have confirmed the wide potential of the method.[51] Another advantage of this new method is, that it can easily be used to study complexes which can not be prepared by other methods. Vibrational band origins of the complexes in He can be obtained with high precision. This is progressively important for larger complexes which are difficult to cool by other means. In this case the analysis of the rotational structure is less important and the main structural information comes from the analysis of vibrational spectral features. In the study of large molecules with low volatility, the method has the additional advantage that the pressure in the capture cell needs only to be about 10^{-5} millibar. It is interesting to contemplate the possibility of studying large biological molecules at these low temperatures, both isolated in the droplet and interacting with different numbers of water molecules that have also been captured by the droplet.

Acknowledgements:

The authors are grateful to Prof. G. Benedek, F. Mielke, Prof. R.E. Miller and Prof. B. Sartakov for their contributions to some results of this work. M.H. and A.F.V. thank the Deutsche Forschungsgemeinschaft for financial support.

List of references

1. F. London, "Superfluids II, Macroscopic Theory of Superfluid Helium", John Wiley, New York, 1954, page xi.
2. F. London, Nature **141**, 643 (1938).
3. R.P. Feynmann, "Statistical Mechanics", Addison-Wesley, Redwood City, 1972, p. 312 ff.
4. K.B. Whaley, Int. Rev. Phys. Chem. **13**, 41 (1994).
5. D.S. Lewart, V.R. Pandharipande, and S.C. Pieper, Phys. Rev. B **37**, 4950 (1988).
6. A. Griffin and S. Stringari, Phys. Rev. Lett. **76**, 259 (1996).
7. M.H. Anderson, J.R. Ensher, M.R. Matthews, C.E. Wieman, and E.A. Cornell, Science **269**, 198 (1995); K.B. Davis, M.-O. Mewes, M.R. Andrews, N.J. van Druten, D.S. Durfee, D.M. Kurn, and W. Ketterle, Phys. Rev. Lett. **75**, 3969 (1995).
8. P. Sindzingre, M.L. Klein, and D.M. Ceperley, Phys. Rev. Lett. **63**, 1601 (1989).
9. M.V. Raman Krishna and K.B. Whaley, Phys. Rev. Lett. **64**, 1126 (1990), J. Chem. Phys. **93**, 6738 (1996).

10. S.A. Chin, J. Low Temp. Phys. **93**, 921 (1993).
11. O. Penrose and L. Onsager, Phys. Rev. **104**, 576 (1936).
12. T.K. Lim, S.K. Duffy, and W.K. Lambert, Phys. Rev. Lett. **38**, 341 (1977).
13. V. Efimov, Phys. Lett. **33B**, 563 (1970).
14. K.T. Tang, J.P. Toennies, and C.L. Yiu, Phys. Rev. Lett. **74**, 1546 (1995) and references therein.
15. K.B. Davis, M.-O. Mewes, M.A. Joffe, M.R. Andrews, and W. Ketterle, Phys. Rev. Lett. **74**, 5202 (1995).
16. B.D. Esry, C.D. Liu, and C.H. Greene, Phys. Rev., in press.
17. V. Efimov, Comments Nucl. Part. Phys. **19**, 271 (1990).
18. B.D. Esry, C.H. Greene, Y. Zhon, and C.D. Lin, J. Phys. B: At. Mol. Opt. Phys. **29**, L51 (1996).
19. R. Becker, Z. Phys. D **3**, 101 (1986).
20. J. Gspann, in Physics of Electronic and Atomic Collisions ed. by S. Datz North Holland, Amsterdam, 1982, p. 79ff.
21. J.W. Halley, C.E. Campbell, C.F. Giese, and K. Goetz, Phys. Rev. Lett. **71**, 2429 (1993).
22. C.E. Campbell, J. Low Temp. Phys. **93**, 907 (1993).
23. H. Buchenau, E.L. Knuth, J. Northby, J.P. Toennies, and C. Winkler, J. Chem. Phys. **92**, 6875 (1990).
24. M. Farnik, U. Henne, B. Samelin, and J.P. Toennies, Z. Phys. D, in press.
25. J.P. Toennies and K. Winkelmann, J. Chem. Phys. **66**, 3965 (1977).
26. D.M. Brink and S. Stringari, Z. Phys. D **15**, 257 (1990).
27. M. Lewerenz, B. Schilling, and J.P. Toennies, Chem. Phys. Lett. **206**, 381 (1993).
28. M. Lewerenz, B. Schilling, and J.P. Toennies, J. Chem. Phys. **102**, 8191 (1995).
29. A. Scheidemann, J.P. Toennies, and J.A. Northby, Phys. Rev. Lett. **64**, 1899 (1990).
30. J. Al-Khalili, Physics World, June 1996, p. 33.
31. F. Luo, G.C. McBane, G. Kim, C.F. Giese, and W.R. Gentry, J. Chem. Phys. **98**, 3564 (1993).
32. W. Schöllkopf and J.P. Toennies, J. Chem. Phys. **104**, 1155 (1996).
33. W. Schöllkopf and J.P. Toennies, Science **266**, 1345 (1994).
34. C.M. Surko and F. Reif, Phys. Rev. **175**, 229 (1968).
35. D.M. Ceperley, Rev. Mod. Phys. **67**, 279 (1995).
36. I.F. Silvera, Phys. Rev. B **29**, 3899 (1984).

37. E.B. Gordon, V.V. Khmelenko, A.A. Pelmenev, E.A. Popov, O.F. Pugachev, and A.F. Shestakov, Chem. Phys. **170**, 411 (1993).
38. T. Yabuzaki, A. Fujisaki, K. Sano, T. Kinoshita, and Y. Takahashi, Proc. of the 13th Int. Conf. on Atomic Physics, 1992, p. 337.
39. B. Tabbert, M. Beau, H. Günther, W. Häussler, C. Hönninger, K. Meyer, B. Plagemann, and G. zu Putlitz, Z. Phys. B **97**, 425 (1995).
40. A. Scheidemann, B. Schilling, and J.P. Toennies, J. Phys. Chem. **97**, 2128 (1993).
41. M.A. McMahon, R.N. Barnett, and K.B. Whaley, J. Chem. Phys., in press.
42. M. Hartmann, R.E. Miller, J.P. Toennies, and A.F. Vilesov, Phys. Rev. Lett. **75**, 1566 (1995).
43. J. Harms, M. Hartmann, J.P. Toennies, A.F. Vilesov, and B. Sartakov, to be submitted to J. Mol. Spectr.
44. M. Hartmann, R.E. Miller, J.P. Toennies, and A.F. Vilesov, Science **272**, 1631 (1996).
45. S. Stringari and J. Treiner **87**, 5021 (1987).
46. M. Hartmann, F. Mielke, J.P. Toennies, A.F. Vilesov, and G. Benedek, Phys. Rev. Lett. **76**, 4560 (1996) and references therein.
47. E.P. Peyroula and R. Jost, J. Mol. Spectr. **121**, 177 (1987).
48. V.E. Bondybey, A.M. Smith, and J. Agreiter, Chem. Rev., in press.
49. M.H.L. Pryce, "Interaction of Lattice Vibrations at Point Defects", in: Phonons in Perfect Lattices and in Lattices with Point Imperfections, Oliver and Boyd, Edinburgh and London, 1966, p. 403.
50. M. Hartmann, J.P. Toennies, and A.F. Vilesov, in preparation.
51. A. Bartelt, J.D. Close, F. Federmann, N. Quass, and J.P. Toennies, submitted.

ATOMS IN SUPER-INTENSE RADIATION FIELDS

H.G. MULLER
FOM-Institute for Atomic and Molecular Physics,
Kruislaan 407, 1098 SJ Amsterdam

The behaviour of atoms in radiation fields that are strong enough to pull the electron directly out of the atom is discussed. Unexpectedly, there still exist atomic states under such conditions, that can have arbitrarily long lifetimes. In fact the lifetime of such states usually increases with intensity. This super-intense world of atomic spectroscopy features atomic states that look very different from the atomic structure we know and love. This can lead to very interesting dynamical processes.

1 Through the looking glass

Current laser technology makes it possible to realise intensities above *10^{20} W/cm^2*. At such enormous intensities, the electric field component of the electromagnetic wave completely overwhelms the field that binds a typical valence electron to an atom. One might expect that as a result atoms (or at least their outer shells) are completely blown apart, and that atomic physics no longer is a meaningful enterprise at these intensities. One of the surprises of the past decade was that this expectation is completely false, and that well defined atomic structure does exist at such super-high intensities. The explanation of this paradox can be found in the fact that the electromagnetic field is an ac field, which vanishes in the time average. Thus, on the *average*, the force binding the electron to the atom will always dominate the laser field, no matter how much the *instantaneous* laser field dominates the Coulomb force.

As might be expected, the atomic structure that results from this curious situation of mutual domination is completely different from ordinary atomic structure. It has been shown first theoretically and later experimentally that this structure actually gets more stable (in the sense of having longer lifetimes with respect to photo-ionisation) at higher intensity. This behaviour is quite opposite to the usual course of affairs at low intensities, where perturbation theory and Fermi's Golden Rule apply.

It thus seems that atomic structure in a laser field has two stable limiting cases, the ordinary structure at zero intensity, and a completely different one at infinite intensity. These two worlds have little in common, and if the intensity is sufficiently low or high, the atom looks very much like one of these limiting cases, except that is will not be entirely stable and will slowly photoionise. Both of these limits have their own spectroscopy. In the border between the ordinary and 'super-intense' world, neither limit is valid and life gets really complicated. This regime is characterised by very interesting atomic dynamics.

2 The meaning of Super Intense

To specify more precisely what is meant by 'super-intense', let us take the example of a hydrogen atom in its ground state (Fig. 1). The *1s*-wave function has maximum probability for the electron to be at a distance of *1 Bohr* from the nucleus, and the electric field it feels at this distance is *1 au* (atomic units with $\hbar = e = m_e = 1$ are used throughout this paper). Such a field amplitude is reached by the laser at an intensity of *$3.5 \cdot 10^{16}$ W/cm^2*. However, this strongly overestimates the electric field required to directly pull the electron from the atom. Even in the unperturbed atom the outer turning point of the *1s*-orbit (i.e. the border of the classically allowed region in which the electron can move)

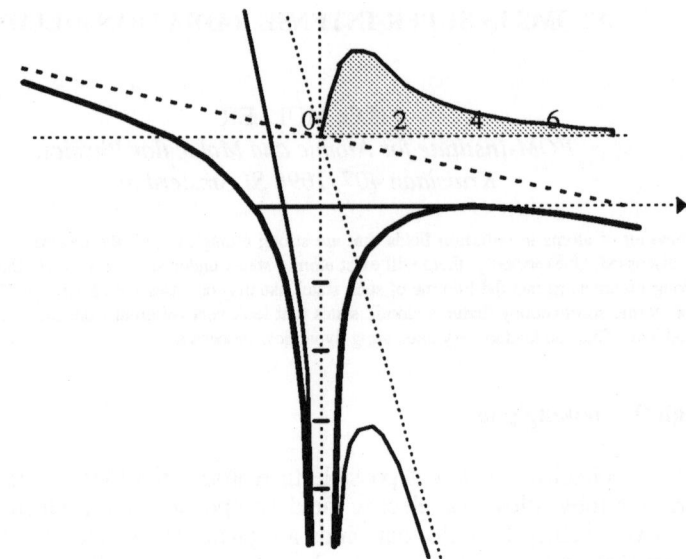

Figure 1: The potential of a hydrogen atom in homogeneous electric fields of *1 au* (thin line) and *1/16 au* (fat line). In the latter case the potential barrier is suppressed enough for the *1s*-state to move over it and escape.

is at *2 Bohr*, and in the presence of a field the electron gains energy moving downstream through the field, shifting the turning point even farther out. The Coulomb force diminishes as the inverse square of the distance to the nucleus, making the field required for immediate ionisation significantly smaller than *1 au*.

The exact number is easily obtained from analysing the potential generated by an ion of charge Z in the presence of a homogeneous field $\mathcal{E}\hat{z}$, $V(r) = -Z/r + \mathcal{E} z$. This potential has a saddle point on the z-axis, at a distance $(\mathcal{E}/Z)^{-1/2}$ from the nucleus, where the energy is $-2(\mathcal{E} Z)^{1/2}$. The electron has to overcome this barrier in order to escape the nucleus, and if its binding energy E_o is less than $2(\mathcal{E} Z)^{1/2}$ it can move straight over it. The condition for this 'barrier-suppression' ionisation (BSI) is thus $I = \mathcal{E}^2 = (E_o^2/4Z)^2$. For hydrogen ($Z=1$, $E_o = 0.5$) the BSI intensity is thus only *1/256*, or $1.4 \cdot 10^{14}$ *W/cm²*. Even below this intensity, quantum mechanics allows some ionisation due to tunnelling through the barrier.

The analysis above assumed the field was static, i.e. that the electron would always have time to move to the saddle point and extract the energy from the field $\mathcal{E}$ while doing so. This is only true if the time scale on which the field changes is slow compared to the natural motion of the electron in the atom, i.e. if $\hbar \omega < E_o$, at low frequencies. For atomic ground states and optical frequencies this is usually the case, especially for noble-gas atoms. For high frequencies, the potential picture with the saddle point has little merit, because the electron will not be able to move over a significant distance before the field reverses. In such a case the electron will react to the time-average potential. Even for fields very much above the barrier-suppression field, over-the-barrier or tunnelling ionisation is then impossible. In this case ionisation has to occur through an entirely different physical mechanism, known as multiphoton ionisation.

The situation is in many ways similar to emptying a bowl of water. One way is to gently tilt the bowl (or alternatively, smoothly accelerate it in the horizontal direction) so that the water surface, while remaining flat, rises to above the rim of the bowl. Alternatively, shaking the bowl at a high frequency creates waves where the water mass is driven against the moving wall of the bowl, and the waves might get high enough to splash over the rim. The latter process is most effective where the walls are steep. In all cases it helps if the water mass as a whole can be set in a sloshing motion by resonantly shaking the bowl.

This analogy carries over quite well to atoms in laser fields. Tunnelling ionisation happens near the outer turning point of the electron, because the electron can extract enough energy from the field by moving there to escape the atom. Hardly any energy absorption from a high-frequency field can take place there, since the potential gradients at this outer turning point are much too low, making the effect of the laser vanishes in the time average. Multiphoton ionisation therefore takes place mainly where the electron encounters the Coulomb singularity of the potential near the nucleus. There the time scale of the orbital motion is such that the work performed by the field on the electron (proportional to $\int \mathcal{E} \cdot v \, dt$) does not average out (can not be approximated by $v \cdot \int \mathcal{E} \, dt$). The electron can then derive enough energy from a single or repeated collisions with the nucleus to escape the atom without having to pay any attention to barriers.

Although tunnelling and multiphoton ionisation are thus physically distinct processes, the names are a little misleading, because it suggests that photons are not involved in the tunnelling process. In a quantised description of the field, the tunnelling is just as much accompanied by absorption of photons as ionisation in the multiphoton regime. The electron spectrum will show the effects of light quantisation both in the tunnelling regime and the multiphoton regime, and the photoelectrons will have to come out with energies U that are an integer number of photons above the (possibly Stark-shifted) initial state, described by Einstein's (generalised) law for the photo-electric effect,

$$U = N\hbar \omega - IP(I). \tag{1}$$

The intensity dependence of the ionisation potential IP through Stark shift is indicated explicitly in Eq. (1). Since the number of absorbed photons N is arbitrary (as long as it is large enough to overcome IP), even ionisation of a single atomic initial state to a single final ionic state can produce several peaks in the photo-electron spectrum. In the strongly non-perturbative situation of super-intense fields, there is no particular preference for lower-order processes as compared to higher-order ones, and a large number of similarly sized peaks separated by the photon energy results for each (ionising) transition. Such Excess-Photon Absorption is quite a common feature of many transitions in the super-intense regime.

The quantisation of final-state energies even persists when the field is treated semi-classically (and thus shows up in experiments with lasers), due to the periodic repetition of events on each optical cycle. As a practical matter the peaks might not be resolvable in an experiment if the photons are too small in the face of inhomogeneous Stark shifts and space-charge broadening, but in principle they are always there if it takes more than one optical cycle to completely ionise the state. For historical reasons the absorption of excess photons during photo-ionisation is known as above-threshold ionisation (ATI).

3 Quiver motion

The problem of a *free* electron in an oscillating electromagnetic field is easily solved. In the non-relativistic case (which is the only one discussed in this paper) the electron will be driven into a harmonic motion by the field, quivering about an 'average' position. This average position just moves in the way a free electron would move in the absence of any fields. In the super-intense world the instantaneous force due to the laser dominates everything else, and as a result the electron can not avoid being driven in such a quiver motion. Most of the physics that goes on in high-intensity laser-atom interactions involves this quiver motion in one way or an other.

As is often the case in physics, a judiciously chosen frame of reference can cause an enormous simplification of the description of the problem[1]. For atoms in super-intense radiation fields the proper choice is not the stationary (lab-)frame, but the (accelerated) co-ordinate frame that follows the quiver motion of a free electron. In this frame, known as the Kramers-Henneberger (KH) frame, the *inertial* forces exactly cancel the *electro-magnetic* force on the electron. As a result, free electrons move on straight-line trajectories at constant speed, and the only source of acceleration is their mutual repulsion and the attraction to the nucleus. The nucleus, which was nearly stationary in the lab frame, now reflects the quiver motion, and all effects due to the laser have to be mediated by this non-stationary nucleus (Fig. 2).

For any atomic system at any intensity a description in the KH-frame is possible, but a simplification of the system is only achieved for those electrons that indeed perform the quiver motion. An atomic state that is too tightly bound to the nucleus to be set quivering will follow the nuclear motion in the KH-frame, thus changing its position and momentum strongly during the course of a single optical cycle. Such a state would be more conveniently described in the length gauge in lab co-ordinates.

The different co-ordinate frames reflect the division between the ordinary and super-intense world. Described in the appropriate co-ordinates each world looks simple, and it is only states that belong to the other world that look complicated in that same description. The binding force to the nucleus determines to which world a state belongs. Thus the two worlds can coexist in the sense that under the same physical conditions, states belonging to either world can be present.

Processes that involve states of both worlds, e.g. transitions between such states, of necessity look complicated in either description, and might be considered inherently complicated. This is usually reflected in the observed phenomena accompanying to such a process. One of the first examples of this to be discovered was ATI with low-frequency light[2], where a transition between an 'ordinary' tightly bound ground-state, and a wildly quivering 'super-intense' continuum state causes the photoelectron to exhibit a rich energy spectrum (Figure 3), with many peaks corresponding to absorption of large numbers of photons, which feature complicated angular distributions and intensity dependences.

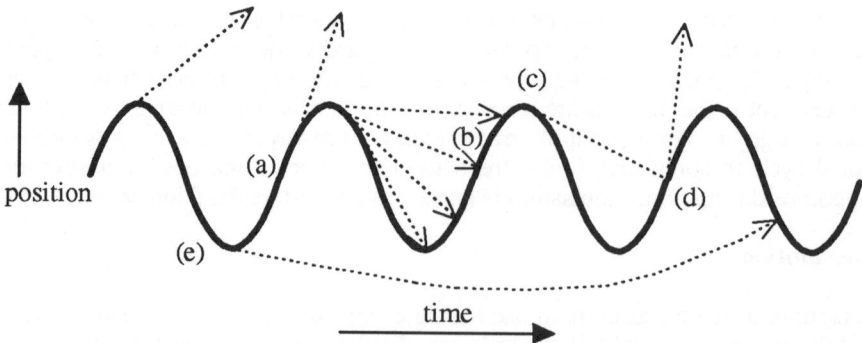

Figure 2: In the KH-frame, the nucleus performs the quiver motion (fat line), that is in phase with the electric field. Electrons behave as if no laser was present, and move in straight lines tangent to the nuclear trajectory once the atom has shaken them off (dashed lines). Electrons shaken off during a zero crossing (a) have the largest drift velocities. Electrons emitted during the quarter cycle after the peak of the field will re-encounter the nucleus (b). The maximum impact energy is reached for those electrons emitted only slightly after the peak (c). If such electrons backscatter elastically, quite high energies are achieved (d). Due to the Coulomb attraction, a slow electron emitted before the peak field (e) might actually be bend back and suffer a later collision after spending some time in orbit.

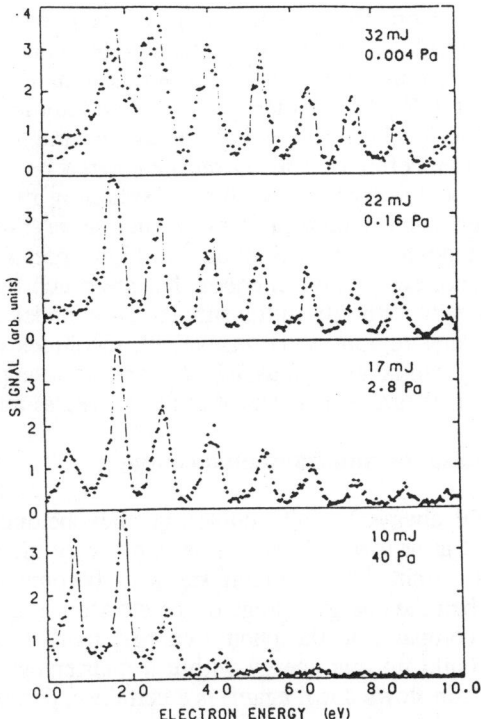

Figure 3: The photoelectron spectrum of xenon ionised with *1.06μm* light. The various peaks are due to ionisation with different numbers of photons, illustrating that different order processes have similar probabilities. Even the lowest-energy peak is due to an eleventh-order process. Such equality of orders is typical for transitions between the ordinary and super-intense world.

On the other hand, processes involving only states belonging to the same world can be described entirely in one type of co-ordinates, often have only a simple phenomenology. Ionisation of a Rydberg state with high angular momentum is such a process in the super-intense world, because both the Rydberg and the continuum electron fully experience the quiver motion. In such a process usually only one or a few photons are involved. Deep in the super-intense world, when all atomic states (including inner shells) are quivering, the only process that is inherently complicated is a hard collision with the nucleus, since this nucleus, due to its large mass, will be the only stationary object left at super intensities.

A very useful way to think about processes involving states from both worlds, is to split up phase-space in two parts. The region within and below the barrier would contain the tightly bound states, that are nearly stationary with sharply defined energies, and subject to multiphoton transitions. Outside this region the electron is practically free, and can be considered as a moving particle driven by the laser field. In the simplest approximation (known as the simpleman's model[3]) the atomic forces are neglected in this region, and the motion is treated classically. Most of the interesting dynamics is then explained by the behaviour of the electron in this second region.

To return to the concept of super-intense fields, it is useful to get a feeling for the typical value of some physical parameters at which these occur. It was derived above that the barrier-suppression field in hydrogen is *1/16 = 0.06 au*, and a typical photon energy in the optical range happens to be *1.6eV = 0.06 Hartree*. The velocity change associated with the quiver motion is then $\delta\omega = 1$ *au*, The displacement $\delta\omega^2 = 16$ *Bohr* = *8Å*, and the quiver energy $U_p = \mathcal{E}^2/4\omega^2 = 0.25$ *Hartree = 7eV*. The barrier is located at a distance of *4 Bohr* from the nucleus. In practical experiments, the peak field strength only increases by a few percent from optical cycle to optical cycle, because even the fastest pulses available contain tens of cycles. So the maximum ionisation rate that can be achieved is about *10%* of ionisation per cycle: an attempt to reach higher rates would lead to a rate at this level on many preceding cycles, and thus to total depletion of the ground-state population before this higher rate can be reached.

The Simpleman's model is quite successful in explaining many of the high-energy processes that can occur in an intense laser field, such as high-harmonic generation[4], enhancements[5] and strange angular distributions[6] in high-order ATI, and direct double ionisation[7]. All these processes involve energies that are much larger than U_p, and the required energy occurs quite naturally in the simpleman's model when the electron, being accelerated in the outer region by the laser field, recollides with the nucleus. The maximum impact energy can be as high as *3.17 U_p*, and various processes then employ this energy in radiative recapture, elastic rescattering and (*e,2e*)-ionisation, respectively.

4 Above-Threshold Ionisation and continuum coupling

Since free electrons will always be fully quivering, they belong to the super-intense world at any intensity. But at low intensities this is not very interesting, since the two worlds coincide in this limit. The situation starts to become non-perturbative, and therefore interesting, when the energy change of the electron in the course of the quiver motion becomes large compared to the photon energy. In that case absorption of the number of photons to build up this energy within a quarter cycle (and their later re-emission when the electron slows down again) is a certainty, demonstrating that there is no rate-penalty associated at all with absorption of these photons. When $U_p/\omega = I/4\omega^3 >> 1$ this will definitely be the case, but in fact a simple classical argument shows that high-energy electrons become non-perturbative much earlier than that.

The instantaneous velocity of an electron in an oscillating field is given by

$$v_t = v_{avg} + \mathcal{E}/\omega \sin \omega t, \qquad (2)$$

and the associated instantaneous kinetic energy is proportional to its square,

$$U_t = \tfrac{1}{2}mv_{avg}^2 + v_{avg}\mathcal{E}/\omega \sin \omega t + \mathcal{E}^2/2m\omega^2 \sin^2 \omega t = U_o + (8U_oU_p)^{1/2} \sin \omega t + 2U_p \sin^2 \omega t. \qquad (3)$$

For large energy U_o the energy modulation is almost exclusively due to the cross product in Eq. (3). In ordinary ATI the in the tunnelling regime the electron emerges with near zero energy, and U_o is typically negligible. This means that in that case only the last term of Eq. (3) is important. If the original ionisation process would produce high-energy electrons, the number of contributing orders would be typically enhanced compared to that in ATI at a similar intensity by a factor $(2U_o/U_p)^{1/2}$, which can be large. One could say that the non-perturbative continuum coupling gets larger at higher energy, but the basic physics is just that a similar velocity modulation produces a much larger energy modulation at high energy.

To test this classical model of continuum coupling we performed an experiment where U_o is quite large[8]. To this end, we made use of Auger decay as a source of energetic ($\approx 200eV$) electrons. By exposing the decaying atom to a laser field, the ionisation continua to which the atom can decay get coupled amongst each other, and the decay can go accompanied by absorption or emission of optical photons. Thus an isolated Auger line would acquire 'sidebands' spaced from the original line by the photon energy, just as the ATI peaks are spaced in the electron spectrum due to photoionisation. We named the process resulting in these sidebands Laser-Assisted Auger Decay, because the laser is not instrumental to cause the decay, but helps in setting the final energy of the electron.

If the decay itself would not be possible in the absence of the laser, the process would be called Laser-Induced, rather than Laser-Assisted. Such a situation could for instance occur in cases where regular Auger decay is parity forbidden, such as with C^+ $(1s)(2s)^2(2p)^2$ ^3P. Decay to the C^{2+} $(1s)^2(2s)^2$ ^1S state would then require ejection of an electron into a p-continuum in order to conserve total angular momentum, but this final state would have odd overall parity. The spherically symmetric Coulomb interaction between electrons has to conserve both parity and angular momentum, and therefore does not couple this final state to the even-parity initial state. The $KL_{2,3}L_{2,3}$ process thus can occur only in the presence of an electric field that mixes some S or D character into the initial state. The decay rate would then be directly modulated by the electric field due to the laser, leading to a correlation between the instant of decay and the phase of the light field. This would strongly affect the side-band structure of the Laser-Induced Auger line, the absence of the central 'ordinary' line being the most conspicuous.

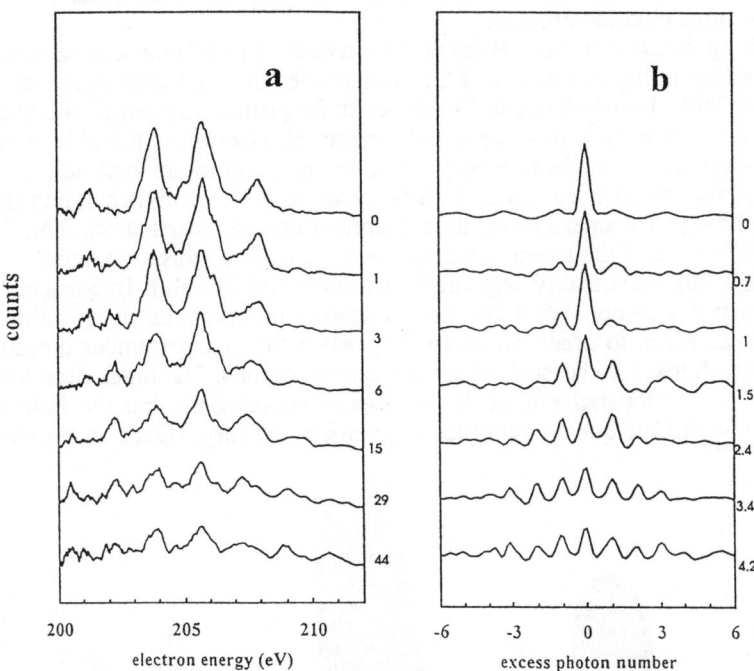

Figure 4: Electron spectra of Laser-Assisted Auger Decay. On the right are the deconvoluted line shapes corresponding to the raw line shapes on the left. The top trace is the unperturbed Auger line. The lower traces are taken at successively higher laser intensity, dressing the line with more and more sidebands, spaced by the photon energy, as can be seen on the right.

No experiments on LIAD have been reported yet, but the LAAD experiment showed that the continuum coupling can be well described by the simpleman's theory. Figure 4 shows the LAAD spectra taken at various laser intensities. Due to the fine structure of the involved states, the line itself shows already appreciable structure, that makes the formation and growth of the sidebands almost impossible to discern. Deconvoluting all experimental spectra with the unperturbed Auger spectrum (shown in the upper trace) recovers the modification a single narrow Auger line would undergo, and clearly shows that such a line acquires of a number of sidebands spaced by the photon energy. This number grows proportional to the square root of intensity, just as the classical model predicts.

5 Atomic Structure at Super Intensities

Apart from the rich *dynamics* that occurs from transitions between states from the ordinary and super-intense world, the *static structure* of states from the super-intense world is interesting in itself. It is quite different from the structure of ordinary atoms, because the electrons are bound to (in the KH-description) a quivering nucleus, that acts as an extended structure, rather than as a point charge. The super-intense condition states that the inertia of the electron is too large for it to follow this quivering nucleus, and as a result it can only experience the force that attracts it to this nucleus in some average sense. For linear polarisation the nucleus moves in a straight line, and spends comparatively large amounts of time at the turning points of its quiver excursion. This makes an atom in the super-intense regime resemble a diatomic molecule much more than an ordinary atom. For circular polarisation the situation is even more exotic, since the nucleus turns into a circular structure.

Describing the atom in the KH-frame does reveal an important scaling, since systems subject to different light intensities I and frequencies ω, but identical quiver amplitude $\alpha_o = I^{1/2}/\omega^2$, look almost identical. The different frequency with which the nucleus traverses its path is of minor importance only, since the electron is not able to follow this motion anyway. In the high-frequency limit, where I and ω are both taken to infinity, keeping α_o constant, the nuclear motion becomes completely invisible, and the nucleus acts only through its smeared-out time-averaged charge distribution. This leads to a strong modification of the atomic structure, with vanishing ionisation rates.

Although this universality was originally predicted for high frequencies[9], so high that the electron can not follow the nuclear quiver motion even in its ground state, it applies just as much to other situations, e.g. when the electron under consideration is unable to attach itself to the nucleus for dynamical reasons. The latter situation occurs if an atom is in a comparatively weak (not super-intense) field, but the field has a frequency so low that the quiver amplitude and velocity are large. Once an electron subject

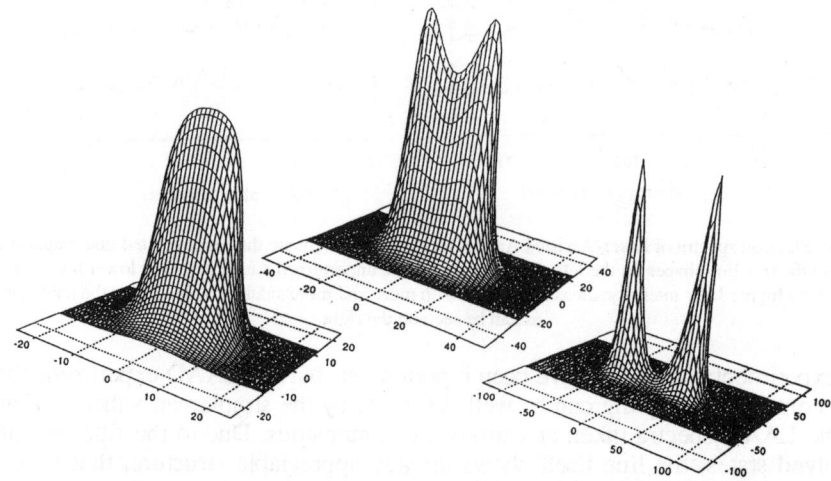

Figure 5: Ground-state wave function of a hydrogen atom in a field with $\alpha_o = $ *5, 20 and 50au*, showing the development of dichotomy in the wave function. In the lab frame the entire wave function quivers with the laser frequency, bringing the dichotomous lobes alternately near the nucleus (adapted from ref. 10).

to such a field has severed its ties with the nucleus, it will be driven in a large quiver motion, and encounter the nucleus henceforth at velocities too large to experience much momentum transfer per encounter.

In Fig. 5 the similarity between super-intense states of the hydrogen atom and a molecule is apparent: the wave function splits into two lobes (an effect known as dichotomy), each residing at opposite ends of the nuclear charge line. Fig. 5 also reveals an important characteristic of the super-intense world: although binding energies in general go down because of the smearing of the nuclear charge, the binding power of such a smeared distribution can still go up, because the electrons can move in larger volumes without losing proximity to the charge, which means that their kinetic energy can be a lot lower. Excitation energies thus typically decrease faster than the binding energies, and this can lead to an increase in the number of bound states.

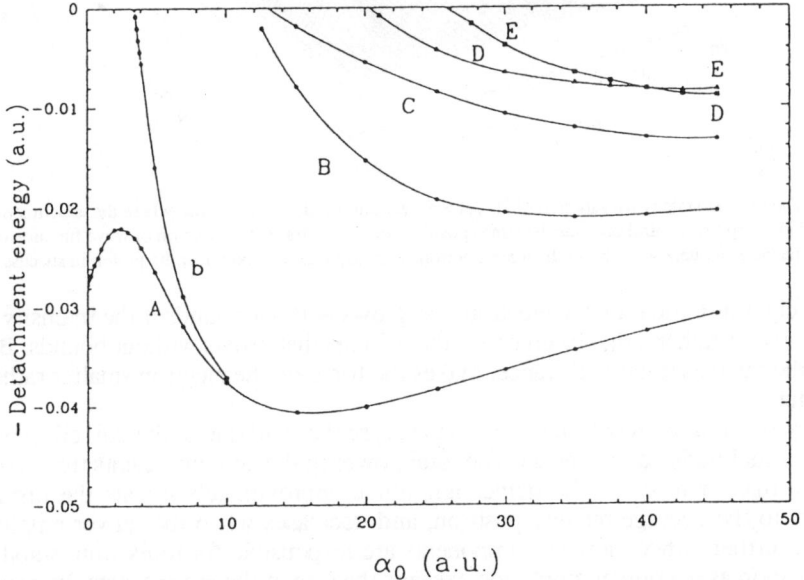

Figure 6: The detachment energy of H⁻ in high frequency fields, as a function of the quiver amplitude $\alpha_o = I^{1/2}/\omega^2$. The ground-state energy goes through a maximum of $1eV$ before decreasing. Many excited states enter the picture as the scaled intensity increases. A (hardly) avoided crossing between a state in which one of the electrons is excited radially, and a state in which both electrons are excited axially occurs in the picture.

Such an increase in binding power is very clearly manifested in negative ions. In the super-intense regime, even H⁻ acquires a number of bound states[10], amongst which even doubly excited states below the first ionisation threshold (Fig. 6). It becomes even possible for a single proton to bind more than two electrons, because the dichotomy of atomic states leaves positive charge exposed between the end-points, to which a third, or even more electrons could bind. Fig. 7 shows the results from a calculation[11] of ground-state wave function of H^{2-}, where the three non-overlapping electron clouds can be clearly seen.

6 Stabilisation

One of the most remarkable effects in the super-intense world is that the lifetime of atomic states actually increases with intensity, an effect known as stabilisation. Again the KH-frame is the key to understanding this phenomenon. In this frame the laser field

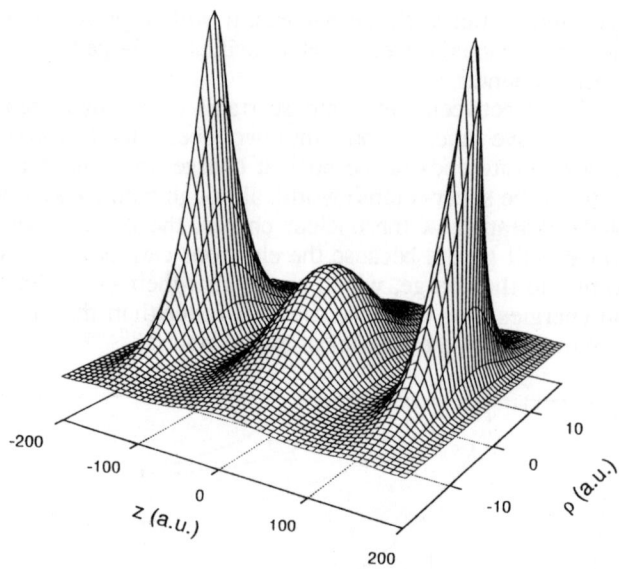

Figure 7: The ground-state wave function of H^{2-}, at a quiver amplitude $\alpha_o = 155\ au$, where the attachment energy of the third electron (the central one) just becomes positive. In the lab frame the position of wave function oscillates with the laser frequency, enabling the three electrons to occupy the position near the proton in succession.

does not enter as a force on the electron that grows without bounds if the intensity is cranked up, but rather a displacement of the nucleus that grows without bounds. But moving the nucleus to large distance, makes the force on the electron smaller rather than larger.

At any point in the KH-frame, one can analyse the field due to the quivering nucleus in terms of its Fourier components. The total power in the ac-components turns out to go through a maximum when the quiver amplitude approximately equals the distance of that point to the average nuclear position, and decreases when the quiver amplitude is increased further. Since these ac components are responsible for ionisation, stabilisation results as soon as the quiver amplitude exceeds the size of the atomic state. In practice, it might occur even at lower intensities in highly excited states, because the outskirts of such states usually contribute little to ionisation due to the absence of strong potential gradients there. The fact that the dc component of the force on the electron also decreases with intensity makes the electronic wave function expand towards these regions, and thus enhances the stabilisation effect.

Stabilisation is a quite general phenomenon, and at sufficiently high intensity, every state will exhibit it. But it is also a phenomenon that is exclusive to the super-intense world, and there might not always be an obvious connection between stabilised states and ordinary states of the atom, like for instance in the case of light-induced states. Even if there is a continuous evolution of an ordinary atomic state into a stabilised one, it might be difficult to lead the atom along this path into the stabilised regime. On the one hand the danger exists that the atom completely decays when conditions are close to that of its lifetime minimum if the passage is too slow, and non-adiabatic effects might prevent the stabilised state to be reach if the passage is too sudden. This "Death-Valley problem" is less severe at higher frequencies, or for high Rydberg states, where lifetimes are comparatively large anyway.

We have demonstrated the occurrence of stabilisation in an experiment on the $5g$-state of neon[12]. Although stabilisation is predicted to be ubiquitous, it was actually hard to find a system where it occurs unobscured by other processes, such as ionisation of the ion core of the Rydberg state. The experiment itself is a nice example of the state of the art in this field, proving that elementary quantitative information can be obtained in super-intense fields if the experiment is set up with sufficient care.

The usual complication of averaging over focal intensities was solved by preparing the target states in only a small part of the laser focus, by using another, even more tightly focused laser to excite neon atoms to the $5g$-state. Therefore all Rydberg atoms in the sample experience essentially the same laser pulse. In addition, by moving the laser foci with respect to each other, and recording the magnitude of a signal due to both colours, *in situ* measurement of the focal size is possible. Sub-picosecond timing of the pulses with respect to each other, combined with Larmor precession of the atom around a strong magnetic field (that was also used to efficiently collect the photo electrons), enabled control over the magnetic quantum number of the Rydberg state. A third (nanosecond) laser pulse was used to empty all Rydberg populations afterwards by photoionisation, revealing also the part of the final population in bound states. Because of the relatively low intensity such a long pulse can have, problems like Stark shifts and non-linear processes can be avoided, making identification of the final states unambiguous.

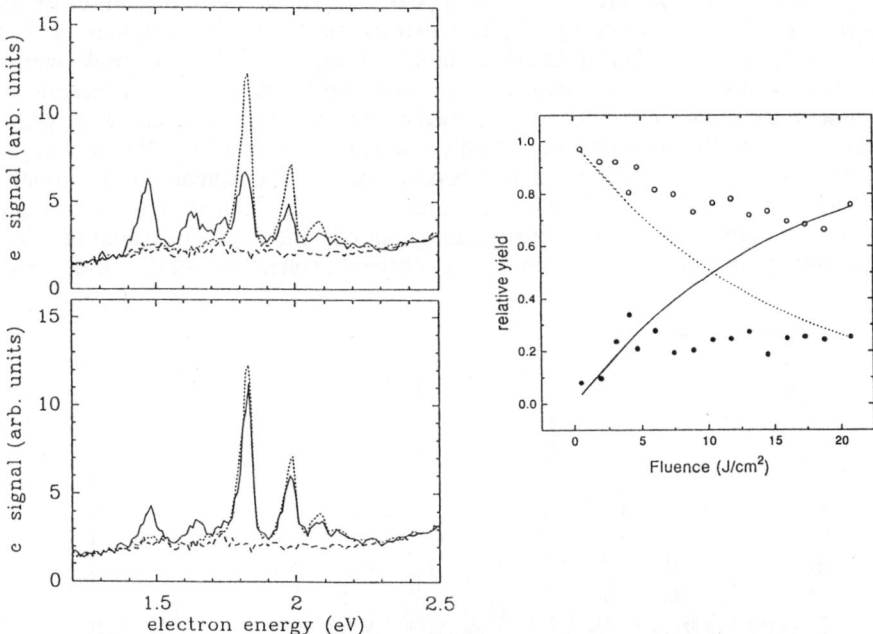

Figure 8: Experimental measurement of adiabatic stabilisation, the light-induced resistance towards photo-ionisation. On the left the measured electron spectra are shown. In each spectrum the ng-Rydberg series is visible twice, once by ionisation with an intense 2 *eV* femtosecond laser pulse (present only in the solid spectra), once by post-ionisation with a 2.33 *eV* nanosecond pulse (in both the solid and dotted spectra). The first signal (black dots in rightmost graph) grows at the expense of the second (open circles), and saturates at about 25% ionisation when the intensity of the femtosecond pulse is increased. The 75% remaining in $5g$ refuses to ionise.

The results of this experiment are displayed in Fig. 8, and clearly show that around a peak intensity of $3 \cdot 10^{14}$ *W/cm²* the ionisation yield reaches its maximum. Increasing the peak intensity by a factor of three does not provide measurable additional ionisation. In fact the experimental result is compatible with an abrupt decrease of the ionisation *rate* at intensities above the critical one. But an experiment of this type can not prevent the ionisation *yield* of the entire pulse to stay at the level of the maximum: the passage through Death Valley will have to take place somewhere in the temporal wings of the pulse, and provide a constant level of ionisation no matter how high the intensity will grow later when the laser pulse approaches its peak.

7 Spectroscopy of Super-Intense states

A very recent result[13] on the straightforward photoionisation of argon by *800 nm* radiation at around $7 \cdot 10^{13}$ *W/cm²* shows an example of the rich structure one encounters in the super-intense world (Fig. 9). The magnitude of this field might not sound very impressive, and the argon ground state is indeed only perturbed a little by it, with ionisation rates of 10^{-4} per cycle or less. Nevertheless, the corresponding field suppresses the barrier (located at $r = 5$ *Bohr*) to a level of *-11.5 eV*. All excited states live well above and outside this barrier, and thus are in the super-intense regime. Indeed, extrapolation of perturbative lifetimes of excited states to this intensity would produce lifetimes of less than a femtosecond even for the lower f- and g-states.

Despite all this, the high-energy part of the photo-electron (ATI) spectrum shows very narrow peaks, that don't seem to be broadened by the large Stark shift of the ionisation potential (≈ 4 *eV*). Three sets of such peaks are visible (within each set the peaks are spaced by the photon energy; the sets extrapolate to binding energies of *1.25eV*, *0.85eV* and *0.55eV*), and their relative magnitude changes with the laser peak intensity.

In an experiment like this, with all intensities up to the maximum present at the same time in the laser focus, such well defined energy structures can only arise when the process producing the peak operates at only one particular intensity. Production over a range of intensities would lead to broadening, due to the spread in the (intensity-dependent) ionisation potential this would cause. Thus all electrons falling in a certain peak must be born at the same intensity, an apparently the atom is capable of selecting this intensity quite precisely. The commonly accepted explanation for this selectivity is

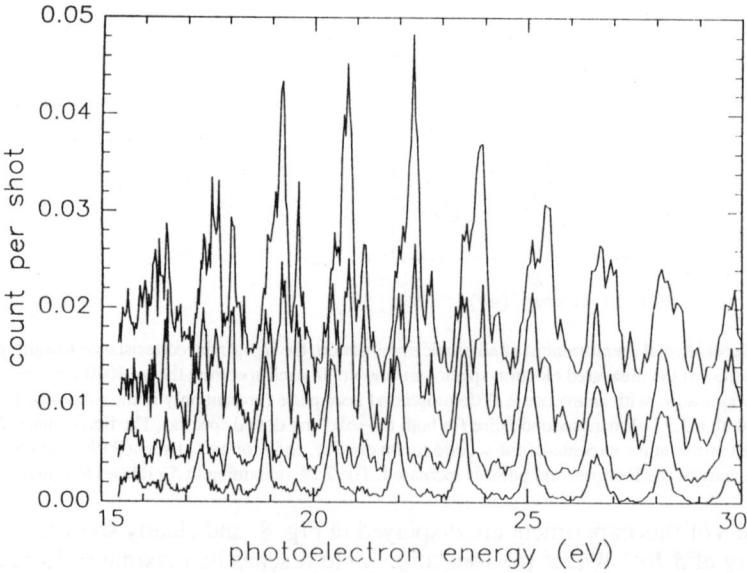

Figure 9: High-energy part of the electron spectrum of argon ionised with *800 nm* light (*1.56 eV*) around *50 TW/cm²*. Different traces (offset vertically for clarity) represent slightly different intensities, and show quite different ratios between the three ATI series, as the responsible states Stark shift into resonance one by one.

the occurrence of resonances, where ionisation proceeds by transferring population to an excited state at the intensity where Stark shifts make the transition energy coincide with a laser harmonic, followed by ionisation of the much more weakly bound excited state.

The observed sensitivity to laser intensity (or equivalently, through Stark shift, to level energy) requires atomic states beside the ground state that live for tens of optical cycles in the field. In view of the perturbative extrapolation, a lifetime this large can only be explained for states of the observed binding energy if some kind of stabilisation is going on, and the stabilised states apparently play a key role in the production of high-energy electrons. How exactly they do this is one of the still unresolved mysteries.

8 Conclusions

All results presented above show that research on atoms in super-intense radiation fields is a very interesting business, holding many surprises in store. A whole new world of atomic spectroscopy is there, waiting to be discovered, with sometimes surprising dynamics and qualitatively completely different from the world of atoms that we are used to. The super-intense world has only been probed at its fringes, since at optical frequencies the direct path to super-intense conditions leads through the Death-Valley of lifetime minima, A novelty in this kind of spectroscopy is that some of the parameters governing the structure are tuneable, so that the creation of 'custom-built' atomic states is one of the possibilities.

Acknowledgements

This work is part of the research program of FOM, which is funded by NWO. Work on Laser-Assisted Auger Decay was supported by the EC (contracts Sci-0103-C, ERB4050PL921025 and ERB4001GT921553), and the high-order ATI work by the Ultra-Fast Center of the University of Michigan.

References

1. Muller and H.B. van Linden van den Heuvell, *Laser Physics* **3**, 694 (1993)
2. P. Kruit, J. Kimman, H.G. Muller and M.J. van der Wiel, *Phys. Rev.* A **28**, 248 (1983)
3. van Linden van den Heuvell and H.G. Muller in *Multiphoton Processes*, eds. S.J. Smith and P.L. Knight (Cambridge University Press, Cambridge, 1988) p.25;
4. Corkum, N.H. Burnett and F. Brunel, *Phys. Rev. Lett.* **62**, 1259 (1988)
5. Kulander, this volume
6. Paulus, W. Nicklich, H. Xu, P. Lambropoulos, H. Walther, *Phys. Rev. Lett.* **72**, 2851 (1994)
7. Yang, K.J. Schafer, B. Walker, K.C. Kulander, P. Agostini and L.F. DiMauro, *Phys. Rev. Lett.* **71**, 3770 (1993)
8. Walker, B. Sheehy, L.F. DiMauro, P. Agostini, K.J. Schafer and K.C. Kulander, *Phys. Rev. Lett.* **73**, 1227 (1994)
9. Schins, P. Breger, P. Agostini, R.C. Constantinescu, H.G. Muller, G. Grillon, A. Antonetti and A. Mysyrowicz, *Phys. Rev. Lett.* **73**, 2180 (1994)
10. Pont, N. Walet and M. Gavrila, *Phys. Rev.* A **41**, 477 (1990)
11. Muller and M. Gavrila, *Phys. Rev. Lett.* **71**, 1693 (1993)
12. E. van Duijn, H.G. Muller and M. Gavrila *Phys. Rev. Lett.* **77**, 3759 (1996)
13. N.J. van Druten, R.C. Constantinescu, J.M. Schins, H. Nieuwenhuize and H.G. Muller, *Phys. Rev.* A, accepted (1996)
14. M. Hertlein, P.H. Bucksbaum and H.G. Muller, *J. Phys.* B, to be published (1996)

WAVE PACKET DYNAMICS OF EXCITED ATOMIC ELECTRONS IN INTENSE LASER FIELDS

K. C. KULANDER

TAMP Group, Physics Directorate, Lawrence Livermore National Laboratory, Livermore CA 94551 USA

K. J. SCHAFER

Department of Physics and Astronomy, Louisiana State University, Baton Rouge LA 70803 USA

The dynamics of multiphoton ionization in the tunneling (long wavelength, high intensity) regime is described. Photoemission by tunnel ionized atoms is dominated by the odd harmonics of the driving laser field. Excitation by ultra short ($\sim$ 10–20 fs) pulses produces high harmonics with characteristics which will allow them to be compressed to give coherent sources of VUV and possibly XUV radiation with pulse lengths near or below 1 fs.

1 Introduction

The ionization of rare gas atoms in strong ($I > 10^{14}$ W/cm^2) visible and IR laser fields is dominated by the sequential stripping of one electron at a time. In this regime the ionized electron is freed from its ion core by tunneling through a potential barrier comprised of the Coulomb attraction of the screened nuclear charge and the instantaneous electric field of the laser.[1,2,3] Tunneling will occur more rapidly near the two maxima in the electric field during each optical cycle. After escaping through the barrier the electron oscillates in response to the laser field as it drifts away from its ion core. The energy of oscillation can be quite large in comparison with the drift energy so that at least half the time the electron revisits the vicinity of the ion core before becoming completely free. During these collisions the electron may gain a substantial amount of drift energy, becoming a high energy photoelectron, or the electron may be recaptured into a bound state of the atom while simultaneously emitting a high energy photon. The resultant photoemission spectrum is dominated by frequencies which are integer multiples of the driving frequency, harmonic conversion. Because of the inversion symmetry of an atom, a linearly polarized laser field generates only odd harmonics. Thus a simple picture of ionization emerges in the strong–field, long wavelength regime: each half cycle a wave packet of electron probability escapes into the continuum via a tunneling process, then the wave packet drifts relatively slowly away from the ion, with a significant fraction of the density recolliding with ion core because of the large amplitude, oscillatory motion imposed by the driving field.[4,5,6] In this paper we will show how the observed characteristics of the excitation and emission processes can be easily interpreted in terms of this simple picture.

A major topic of current research focuses on the effects of using increasingly shorter, high intensity pulses on excitation and ionization processes. Pulse lengths as short as ten optical cycles (10s of fs) are becoming widely available.[7,8] Changes in the observed harmonic emission spectrum generated using these ultra short pulses reflect modifications in the ionization dynamics that can be explained within the tunneling model. Here we will first present a discussion of the high order harmonic generation process including results obtained in longer pulse (> 1 ps) experiments. We then describe calculations of the emission spectrum for ultra short pulses and show that we can expect to exploit the short pulse harmonics to create VUV and XUV wavelength light sources with pulse lengths on the order of and possibly shorter than 1 fs.

2 Harmonic Generation

Because high order harmonic generation in a gas is a coherent process its efficiency depends on both the emission from each individual atom and the propagation of the emitted fields through the excited volume as well, *i.e.*, good phase matching is essential.[9] Almost all experiments to date have been performed with tightly focused lasers to achieve high pump intensity. This focusing introduces a coherence length for the higher order harmonics which can be much shorter than the interaction length. This is alleviated to some degree in the high intensity regime by the non–perturbative nature of the single atom emission. It turns out that high order harmonics are produced over a much larger volume of the gas than would be expected based on perturbative calculations.[10,11] This allows harmonic generation of very short wavelengths to be efficient enough to be a usable, coherent light source. In this paper we ignore phase matching effects to concentrate on the single atom emission for two reasons: (1) Most of the extremal properties of the full (single atom + phase matching) problem are constrained by the single atom behavior (in particular the minimum pulse length and maximum photon energy of the harmonics are so constrained), and (2) The advent of multi terrawatt lasers means that harmonic generation experiments with unfocused lasers are now possible, leading to simplified phase matching, at least for intensities below saturation (complete ionization) of the neutral gas atoms.

We calculate the single atom harmonic generation spectrum using the single active electron (SAE) approximation which has been extensively discussed in previous publications.[12] Briefly, we solve the time–dependent Schrödinger equation (TDSE) for an atom in a linearly polarized laser pulse assuming that only the outermost valence electron responds to the field. Comparing our predictions using the SAE approximation to many different multiphoton experimental results has shown that for rare gas atoms and optical wavelengths this is generally a very good approximation. The SAE pseudopotential used in our calculations is obtained from a series of Hatree–Slater calculations for different angular momentum channels and results in very accurate excited state energies for rare gas atoms. We can numerically integrate the TDSE for the SAE to obtain its time–dependent wave function, ψ, which provides all the information about the various electron or photon emission processes. After choosing a laser wavelength, peak intensity and pulse envelope we calculate the emitted harmonic radiation by first calculating the dipole acceleration, $a(t)$, given by

$$a(t) = \frac{d^2 \langle z(t) \rangle}{dt^2} = -\langle \psi | H, [H, z] | \psi \rangle \qquad (1)$$

for each time step during the pulse. Here H is the full (atomic + laser interaction) Hamiltonian. For the SAE pseudopotential, the commutator must be evaluated numerically. This form has been found to be computationally more stable than the usual dipole expression when significant ionization is present.[13] The emitted radiation spectrum is proportional to the Fourier transform of the acceleration

$$A(\omega) = \int dt e^{i\omega t} a(t) \equiv \mathcal{A}(\omega) e^{i\phi(\omega)} \qquad (2)$$

where $\mathcal{A}(\omega)$ is the (real) spectral envelope function. The acceleration spectrum $A(\omega)$ is equal to $\omega^2 D(\omega)$, where $D(\omega)$ is defined to be the "dipole spectrum" obtained by Fourier transforming $\langle z(t) \rangle$. We can also calculate an approximate $D(\omega)$ as outlined in Ref. 13 by taking into account only transitions back to the ground state and find that it agrees with $A(\omega)/\omega^2$ for all of the harmonics in the plateau and cutoff. This clearly demonstrates that the harmonics are produced only by excited electrons which return to the immediate vicinity of the ion core.

We next present some of the basic features of high order harmonics, including the effects of using shorter pulses on the spectral structure and maximum harmonic order. Following that we demonstrate that the harmonics created at the peak of a short laser pulse can be compressed using a VUV grating pair arranged to provide positive group velocity dispersion to give pulses approaching the attosecond regime.

2.1 Single-Atom Spectra

At long wavelength and high intensity harmonic generation involves the excitation of electrons into the continuum and rescattering from the parent ion.[4,5,6] This implies a close relationship between harmonic generation and ionization. In particular the highest order harmonic that is emitted is limited by the highest intensity that any atom experiences before ionizing in accord with the cutoff rule $E_{max} \sim IP + 3U_p$ where IP is the ionization potential and U_p the intensity-dependent ponderomotive energy.[14] In Fig. 1 we show the spectrum generated by neon when excited by a "long" pulse 1064 nm laser at an intensity of 6×10^{14} W/cm^2. By long pulse we mean one in which the envelope of the pulse varies slowly compared to the laser frequency (> 1 ps.) The spectrum consists of narrow peaks at odd multiples of the fundamental which have approximately constant strength (the plateau) up to a maximum harmonic order predicted by the cutoff rule (arrow). The width of the individual harmonic lines is determined by the shorter of the pulse width or the ionization time. As shown in Eq. (2), these harmonics will have a phase, $\phi(\omega)$, relative to the driving field. In the long pulse case these phases vary rapidly with intensity and almost randomly, modulo 2π, from one harmonic to the next. We note that the SAE emission spectra agree well with those observed in the many recent experimental studies of high order harmonic generation.

By contrast, the emission spectrum from an atom in an ultra short pulse is very different. This is demonstrated in Fig. 2 where the high end of the "harmonic" spectrum from argon in a 27 fs, 810 nm pulse is displayed in the upper plot and the corresponding relative frequency-dependent phase is shown in the lower plot. First we note that the cutoff rule still holds, predicting the end of the plateau

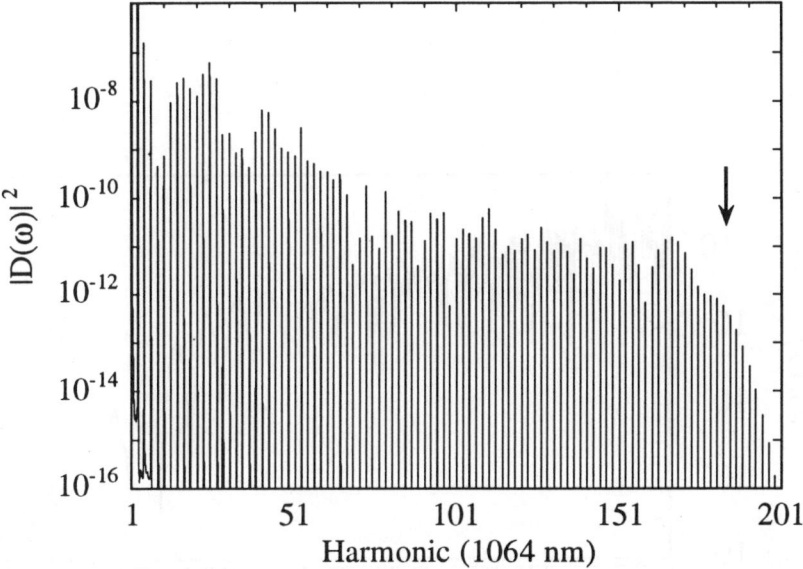

Figure 1: Harmonic emission spectrum from neon excited by a 1064 nm long pulse laser at an intensity of 6×10^{14} W/cm^2.

to be around the 47th harmonic. We find that only the last few harmonics are distinct and they are very broad compared to the long pulse case. In fact they are even broader than one would expect from this very short pump pulse, *i.e.*, $k^2 \Delta \lambda_k \gg \Delta \lambda_1$. Here k is the order of the harmonic and λ_1 is the wavelength of the pump. Surprising also is that the phases of these highest harmonics are well behaved. The spectrum below the vicinity of the cutoff, where the plateau harmonics are expected, is highly structured with the odd harmonic peaks being hard to distinguish and their frequency–dependent relative phases varying rapidly. We find this very different spectrum to be typical in the ultra short pulse regime.

The mechanism for producing spectra such as that shown in Fig. 2 becomes clear when we consider the ionization dynamics of the tunneling model described above. The 27 fs pulse has a half–width of ten optical cycles at this frequency. This means that during both the rising and falling parts of the pulse, the laser intensity is changing significantly on the time scale of a single cycle. Consider a tunneling wave packet created in one half cycle during the pulse rise. The field which turns the escaping wave packet around to rescatter from the ion core one half cycle later is stronger than it was when the wave packet was created. Therefore the recollision velocity is higher than the one encountered in a constant intensity field. Also the time between when the wave packet is created and its return is shorter when the intensity is rapidly rising. Both these effects cause a blue shift in the "harmonic" frequency. Similarly, on the falling edge of the pulse, a red shift occurs. Only those harmonics produced within a short time interval near the peak of the pulse will be

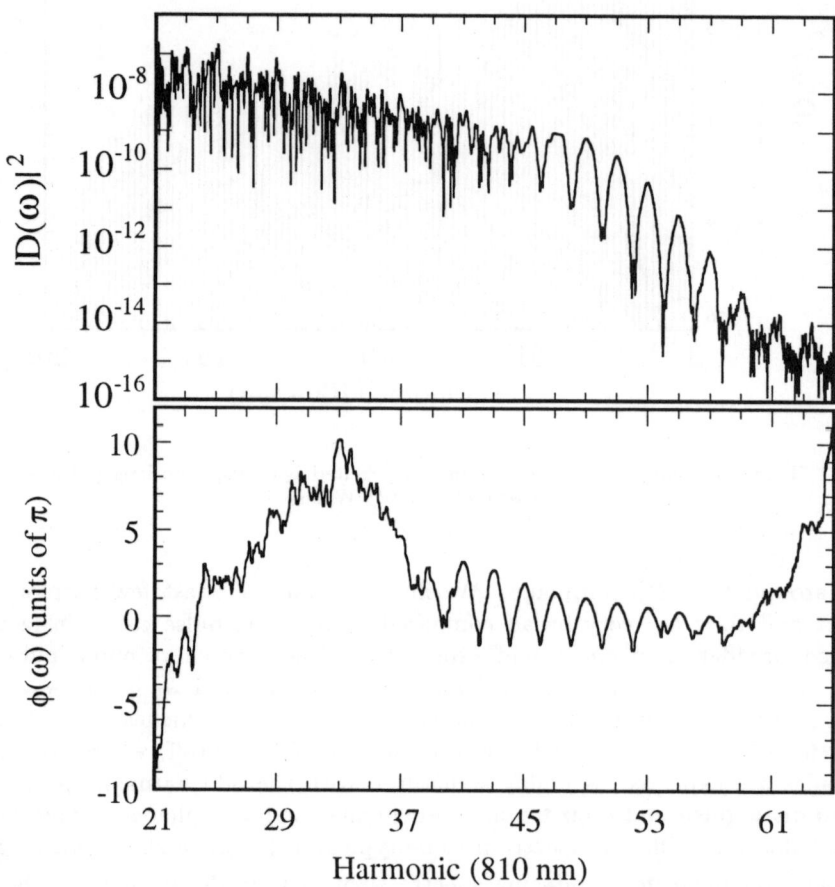

Figure 2: Harmonic emission spectrum (upper plot) from argon excited by a 810 nm, 27 fs pulse at an intensity of 3×10^{14} W/cm^2. The lower plot shows the phases of the emitted radiation, $\phi(\omega)$, defined in Eq. (2).

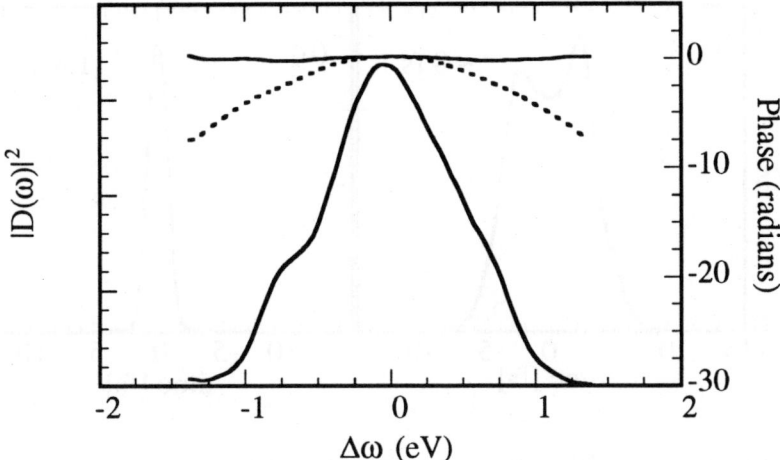

Figure 3: Emission spectrum in the vicinity of the 49th harmonic from argon excited by a 810 nm, 27 fs pulse at an intensity of 3×10^{14} W/cm^2. The dotted line shows the phases of the emitted radiation and the upper solid line shows the phase with its quadratic component removed.

unshifted. According to the cutoff rule, there is a well defined relationship between the laser intensity and the maximum harmonic produced. For the harmonics within the plateau, the intensity is high enough for them to be generated during a large fraction of the pulse. These are emitted, with a blue shift, as the pulse rises and with a red shift as the intensity declines. The resulting spectrum at these mid-plateau harmonic wavelengths is broadened to the extent that the peaks have disappeared and the emission strength shows substantial interference. On the other hand the rule indicates that those harmonics at the end of the plateau are produced only at the peak of the pulse. The short time interval during which they are emitted minimizes any interference. From the discussion above we would still expect these cutoff harmonics to have a linear chirp, which is equivalent to a quadratic phase change, if they are produced just as the pulse passes through its maximum. This is exactly what is found as indicated in Fig. 2. We show this in more detail in Fig. 3 where the emission strength and the phase for the 49th harmonic are plotted. The phase shift, shown by the dotted line, is clearly parabolic. The amplitude of the harmonic is peaked almost exactly at the expected, unshifted line center. These unexpected properties of the cutoff harmonics turn out to be ideal for producing pulses of VUV and XUV radiation which are more than an order of magnitude shorter than the driving pulse width, and, in fact, even shorter than an optical cycle of the incident field.

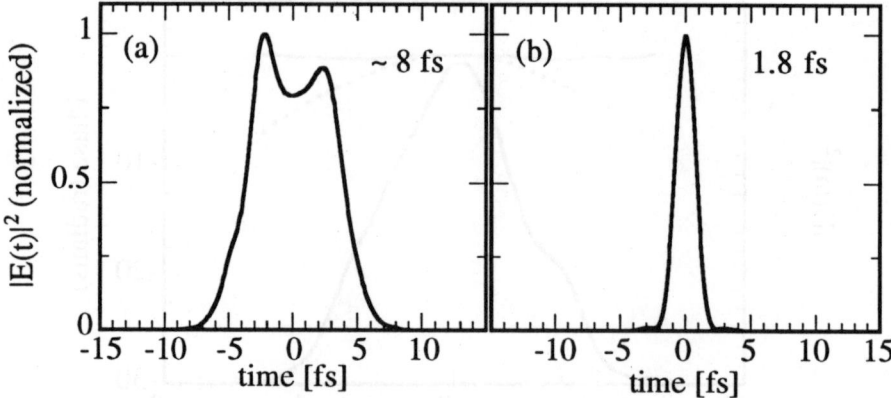

Figure 4: The (a) raw and (b) compressed envelope of the 49th harmonic from argon excited by an 810 nm, 27 fs pulse at an intensity of 3×10^{14} W/cm^2.

2.2 Compressing The Cutoff Harmonics

To calculate the pulse width of the kth harmonic we examine the emitted electric field intensity envelope. This is obtained by back transforming the acceleration spectrum after applying a filter function around the frequency ω_k:

$$\mathcal{E}(t) = e^{-i\omega_k t} \int d\omega e^{-i(\omega-\omega_k)t} \left[\mathcal{A}(\omega)e^{i\phi(\omega)}F((\omega-\omega_k))\right] \equiv E_k(t)e^{-i[\omega_k t - \phi_k(t)]}. \quad (3)$$

$F(\omega - \omega_k)$ is a square filter that roughly duplicates the action of a grating in picking out a specific harmonic. As an example we again focus on the 49th harmonic (~75 eV) in Fig. 3. Since the energy falls by several orders of magnitude before and after the 49th harmonic the shape of the filter function used is not crucial. The raw phase of this harmonic (dotted line in Fig. 3) is almost purely quadratic in the detuning $\Delta\omega = \omega_{49} - \omega$. The phase minus its quadratic component

$$\tilde{\phi}(\omega) = \phi(\omega) - 0.5 \times (\omega_{49} - \omega)^2 \frac{d^2\phi}{d\omega^2}\bigg|_{\omega=\omega_k} \quad (4)$$

is shown as a solid line above the envelope in Fig. 3. The calculated time history is displayed in Fig. 4a which clearly shows that this harmonic is emitted only near the peak of the pulse. A similar calculation for a mid–plateau harmonic would show a much longer time of emission and considerable structure because of interference. The emission envelope for the 49th harmonic shows a small asymmetry with the first peak being slightly stronger than the peak which appears just after the maximum in the laser pulse. Since the early time contributions to the harmonic are blue shifted, we expect and see that the blue wing of the harmonic (Fig. 3) is slightly stronger.

The linear chirp in this harmonic can be entirely removed using a grating pair by introducing a positive group velocity dispersion which exactly compensates for the quadratic component of the phase.[15] We can mimic this by replacing $\phi(\omega)$ by $\tilde{\phi}(\omega)$ in Eq. (3) and recalculating the pulse envelope. The compressed spectrum is shown in Fig. 4b. Its width, 1.8 fs, is within a few percent of the transform limit for the spectral envelope function. This is because the phase becomes almost constant when the quadratic term is removed as can be seen in Fig. 3. This pulse width is a factor of 15 smaller than the incident laser pulse and shorter than the 2.7 fs period of the driving field. The peak heights in Fig. 4 have been normalized to one for the purposes of plotting. In reality the energy of the pulse is conserved, leading to a much higher peak field in the compressed case.

We have repeated the calculations illustrated in Figs. 2–4 for a range of peak intensities in argon and neon at 810 nm. Qualitatively the results are the same. The harmonics at the end of the plateau exhibit a simple linear chirp which can be removed to provide < 2 fs pulses from a single harmonic. The highest efficiency is obtained by compressing harmonics that are at the end of the plateau for peak intensities just below the saturation intensity. Above that intensity interferences destroy the simple quadratic phase dependence. Using neon, with its higher ionization potential, instead of argon allows us to reach higher photon energies.

In addition, as is clear from Fig. 2, several of the harmonics near the cutoff have almost the same group delay ($d\phi/d\omega$). This is a general feature in all of the short pulse harmonics calculations that we have carried out. In principle the same compressor that can produce femtosecond pulses from a single harmonic can be used to produce sub–femtosecond pulses by combining several harmonics. Removing the average group delay of three cutoff harmonics and back transforming yields an intensity profile similar to that shown in Fig. 4b.[15] However, in this case the width of the central feature is found to be as short as $\sim$ 400 attoseconds. This seems to be one of the most promising ways to reach the sub–femtosecond regime.

3 Conclusion

We have presented the simple picture which has recently emerged of the dynamics of strong field multiphoton ionization of atoms. The electrons escape from the atom by tunneling through an instantaneous suppressed barrier. Each half cycle of the field a wave packet is promoted into the continuum and its subsequent evolution is determined predominantly by its interaction with the oscillating electric field of the laser. The driven oscillations of the wave packet cause additional encounters with the parent ion core resulting in the surprisingly efficient generation high order harmonics. If the driving field is an ultra short pulse, the wave packet dynamics favors the production of very well defined but broad harmonic peaks for the highest energy members of the plateau. These harmonics are created with a linear chirp which can be removed to yield pulses as short as $\sim$ 1–2 fs. If harmonics from several neighboring peaks are combined, the pulse lengths can be pushed down into the attosecond regime.

Acknowledgments

This work has been carried out in part under the auspices of the U. S. Department of Energy at the Lawrence Livermore National Laboratory under contract No. W-7405-ENG-48.

References

1. H. B. van Linden van den Heuvell and H. G. Muller, in *Multiphoton Processes*, eds. S. Smith and P. L. Knight (Cambridge University Press, Cambridge, 1988).
2. T. F. Gallagher, *Phys. Rev. Lett.* **61**, 2304 (1988).
3. P. B. Corkum, N. H. Burnett and F. Brunel *Phys. Rev. Lett.* **62**, 1259 (1989).
4. K. J. Schafer, B. Yang, L. F. DiMauro and K. C. Kulander, *Phys. Rev. Lett.* **70**, 1599 (1993).
5. K. C. Kulander, K. J. Schafer and J. L. Krause, in *Super–Intense Laser–Atom Physics*, eds. B. Pireaux, A. L'Huillier and K. Rzazewski (Plenum, New York, 1993).
6. P. B. Corkum, *Phys. Rev. Lett.* **71**, 1994 (1993).
7. J. Zhou, J. Peatross, M. M. Murnane, H. C. Kapteyn and I. P. Christof, *Phys. Rev. Lett.* **761**, 752 (1996).
8. C. P. J. Barty *et al.*, *Opt. Lett.* **21**, 668 (1996).
9. A. L'Huillier, K. J. Schafer and K. C. Kulander, *J. Phys. B* **24**, 315 (1991).
10. A. L'Huillier, K. J. Schafer and K. C. Kulander, *Phys. Rev. Lett.* **66**, 2200 (1991).
11. A. L'Huillier, P. Balcou, S. Candel, K. J. Schafer and K. C. Kulander, *Phys. Rev. A* **46**, 2778 (1992).
12. K. C. Kulander, K. J. Schafer and J. L. Krause, in *Atoms in Intense Laser Fields*, ed. M. Gavrila (Academic Press, New York, 1992).
13. J. L. Krause, K. J. Schafer and K. C. Kulander, *Phys. Rev. A* **45**, 4998 (1992).
14. J. L. Krause, K. J. Schafer and K. C. Kulander, *Phys. Rev. Lett.* **68**, 3535 (1992).
15. K. J. Schafer, K. C. Kulander, J. A. Squire and C. P. J. Barty, in *Generation, amplification and measurement of ultra short pulses, III*, *SPIE* **2701**, (1996) in press.

NONLINEAR LASER-ELECTRON SCATTERING

D.D. MEYERHOFER

Dept. of Mechanical Engineering[*]
220 Hopeman Bldg
University of Rochester
Rochester, NY 14623

Abstract

In the field of an intense laser, photon-electron scattering becomes nonlinear when the oscillatory energy of the electron approaches its rest mass. The electron wave function is dressed by the field with a concomitant increase in the effective electron mass. When the photon energy in the electron rest frame is comparable to the electron rest mass, multiphoton Compton scattering occurs. When the photon energy is significantly lower than the electron rest mass, the electron acquires momentum from the photon field and emits harmonics. Nonlinear photon-electron scattering processes and results from two recent experiments which have observed them are reviewed.

I. Introduction

The interaction (scattering) of light with free electrons is a well-known phenomenon. Thomson scattering, the scattering of an electromagnetic wave from an electron under the condition that the photon energy is much less than the electron rest mass, $\hbar\omega \ll m_e c^2$, was discovered in the late 19th century.[1] Compton scattering, the scattering of a photon and electron under the conditions where the photon energy in the electron rest frame is large enough that electron recoil effects must be included was discovered in 1923,[2] and the kinematics was described in 1927.[3] The scattering of two photons to produce an electron-positron pair was calculated by Breit and Wheeler in 1934.[4] All of these processes are linear. In Thomson scattering, the electron response to the field is harmonic. In Compton and Breit-Wheeler processes, the scattering rates are linearly dependent on the photon fluxes. Recent advances in high intensity laser systems[5, 6] have allowed the study of the nonlinear analogues to the scattering processes described above. Nonlinear Thomson[7] and nonlinear Compton[8] scattering have been observed and an experiment is underway to study nonlinear Breit-Wheeler scattering.[9] In Sec. II nonlinear laser-electron scattering is described, both in the Thomson and Compton regimes. Nonlinear photon-photon scattering is described in Sec. III. Experimental measurements of nonlinear laser-electron scattering are reviewed briefly in Secs. IV and V for the Thomson and Compton scattering respectively. The conclusions are presented in Sec. VI.

II. Nonlinear laser-electron scattering

The interaction of an electron with an intense laser field is characterized by[10]

$$\eta^2 = \frac{e^2 E_{rms}^2}{m_e^2 \omega_L^2 c^2} = -\frac{e^2}{m^2 c^4}\langle A_\mu A^\mu \rangle \quad (1)$$

where E_{rms} is the rms. electric field strength, ω_L is the laser frequency, and A_μ is the four vector potential. As defined in Eq. 1, η^2 has the same value for linear and circular polarization at a given laser intensity. For counter propagating electron and photon beams, η^2 is an invariant, independent of the electron beam energy in the laboratory.

In 1929, Volkov calculated the electron wave function in an electromagnetic wave.[11] One can interpret these wave functions as dressed states of an electron in an electromagnetic field. The electron's rest mass is dressed by the field giving an effective mass, $m^{*2} = m_0^2(1+\eta^2)$. At low intensities $\eta^2 \ll 1$, η is equal to the ratio of the quiver velocity in the field to the speed of light and $\eta^2/2$ is the ratio of the average oscillatory energy to the rest mass. η^2 is directly related to the ponderomotive potential,[12, 13]

$$\Phi_{pond} = \frac{e^2 E_{rms}^2}{2m\omega^2} = \frac{\eta^2}{2} m_e c^2 \quad (2)$$

and has the numerical value, $\eta^2 = 3.6\times 10^{-19} I(W/cm^2) \lambda_L^2(\mu m)$. For 1 μm wavelength light, η^2 approaches 1 as the intensity approaches 10^{18} W/cm^2.

In a quantum mechanical picture, multiphoton-electron scattering processes can be calculated as transitions between dressed states.[14-16] Many other quantum mechanical calculations have been carried out. Classically, intense field interactions with electron can be calculated as anharmonic corrections to the electron trajectory, which become prominent as η^2 approaches 1.[17-20]

An n photon-electron scattering process can represent either a single interaction of n photons with an electron (multiphoton or nonlinear scattering), or a series of single photon-electron scattering events (plural scattering), or a combination thereof.[8] The electron kinematics of these two process are almost identical. Fig. 1 shows a representative diagram of the two scattering processes.

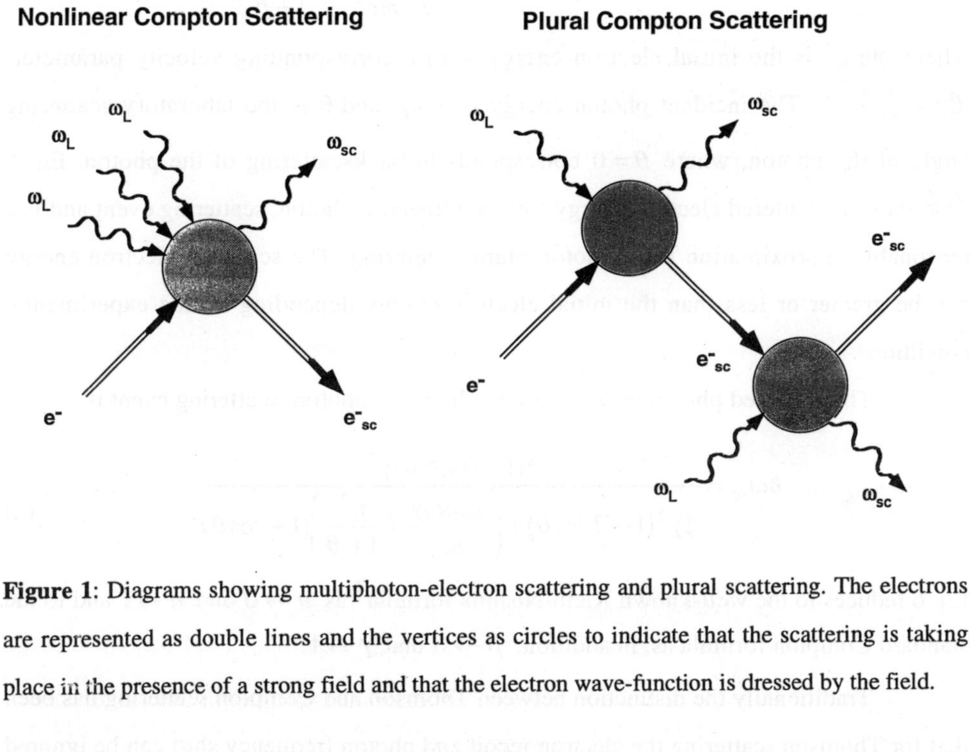

Figure 1: Diagrams showing multiphoton-electron scattering and plural scattering. The electrons are represented as double lines and the vertices as circles to indicate that the scattering is taking place in the presence of a strong field and that the electron wave-function is dressed by the field.

In the figure, the electrons are represented as double lines and the vertices as circles to indicate that the scattering takes place in the presence of a strong field and that the electron wave-function is dressed by the field. A nonlinear, n photon-electron scattering event can be described as a single interaction,

$$e^- + n\omega_L \to e'^- + \omega_{sc} \tag{3}$$

while a plural scattering event can be described as a sequence of interactions

$$e^- + \omega_L \to e'^- + \omega_{sc}$$
$$e'^- + \omega_L \to e''^- + \omega'_{sc} \quad (4)$$
$$e''^- + \omega_L \to e'''^- + \omega''_{sc}.$$

The kinematics of the two processes are almost identical and the scattered electron energies for counter-propagating electron and photon beams in the laboratory frame can be written

$$E_{sc} = \gamma mc^2 + \hbar\omega_L - \frac{2(1+\beta)n\hbar\omega_L\gamma^2}{2\gamma^2(1-\beta\cos\theta) + \left(\frac{2n\hbar\omega\gamma}{mc^2} + \frac{\eta^2}{1+\beta}\right)(1+\cos\theta)}, \quad (5)$$

where $\gamma m_e c^2$ is the initial electron energy with a corresponding velocity parameter, $\beta\gamma = \sqrt{\gamma^2 - 1}$. The incident photon energy is $\hbar\omega_L$, and θ is the laboratory scattering angle of the photon, where $\theta = 0$ corresponds to backscattering of the photon. Eq. 5 describes the scattered electron energy for a nonlinear, n photon, scattering event and is a reasonable approximation for n photon plural scattering. The scattered electron energy can be greater or less than the initial electron energy depending on the experimental conditions.

The scattered photon energy for a nonlinear, n-photon, scattering event is

$$\hbar\omega_{sc} = \frac{2(1+\beta)n\hbar\omega_L\gamma^2}{2\gamma^2(1-\beta\cos\theta) + \left(\frac{2n\hbar\omega\gamma}{mc^2} + \frac{\eta^2}{1+\beta}\right)(1+\cos\theta)}. \quad (6)$$

Eq. 6 reduces to the well-known Klein-Nishina formula[3] as $\eta \to 0$ and $n \to 1$ and to the standard Compton formula as, in addition, $\beta \to 0$ and $\gamma \to 1$.

Traditionally the distinction between Thomson and Compton scattering has been that for Thomson scattering the electron recoil and photon frequency shift can be ignored because $\gamma\hbar\omega/m_e c^2 \ll 1$, while for Compton scattering, the photon energy in the rest frame of the electron, $\gamma\hbar\omega/m_e c^2$, is large enough that the electron recoil and photon frequency shift are non-negligible. While this distinction is reasonable when single photon scattering is considered, it becomes less clear when multiple photon-electron scattering occurs. In this case, the more relevant parameter is $\gamma n\hbar\omega/m_e c^2$. In the Thomson regime, $\gamma\hbar\omega/m_e c^2 \ll 1$, sufficiently large numbers of photons can participate in the interaction to make the electron recoil and photon frequency shift non-negligible. For an electron initially at rest, the scattered electron energy for an n-photon scattering is approximately,

$$\Delta E_e \approx \frac{n\hbar\omega_L}{m_e c^2} \frac{n\hbar\omega_L}{1 + \frac{n\hbar\omega_L}{m_e c^2}}. \tag{7}$$

Thus, if sufficiently large numbers of photons are scattered, the electron can gain an energy comparable to its rest mass. This occurs as η^2 approaches 1. Equation 8 is equally valid for nonlinear and plural scattering.

In the traditionally Compton regime, the probability of multiphoton, or nonlinear, Compton scattering,

$$e^- + n\omega \rightarrow e'^- + \omega_{sc}, \tag{8}$$

scales approximately as $P_n \propto \eta^{2n}$.[16] As η^2 becomes large enough to cause significant scattering, the dressing (shift) of the electron rest mass by the electromagnetic field, $m^* = m_0\sqrt{1+\eta^2}$, cannot be ignored. The scattered photon energy can be related to the incident photon energy for photons colliding with electrons with energy $\gamma m_0 c^2$. The maximum backscattered photon energy for head-on collisions is[10]

$$\hbar\omega_{sc} = \frac{4n\hbar\omega_L \gamma^2}{1 + \frac{4n\gamma\hbar\omega_L}{m_e c^2} + \eta^2}, \tag{9}$$

which corresponds to a minimum scattered electron energy,

$$E_{sc} = \frac{\gamma m_e c^2 (1+\eta^2)}{1 + \frac{4n\gamma\hbar\omega_L}{m_e c^2} + \eta^2}, \tag{10}$$

when the photon energy in the laboratory frame is much less than the electron rest energy. Thus, η^2 enters both in the rate of multiphoton emission and in causing a shift in the scattered photon frequency due to the mass shift. The Thomson scattering process, both cross-section and kinematics, is recovered as the low energy limit of the Compton scattering cross-section.[10]

The transition to multiphoton electron scattering in a high-intensity laser focus occurs when the oscillatory energy of a free electron in the laser field approaches the electron rest energy. Under these conditions, the electron recoil due to the laser field momentum can no longer be ignored, and free electrons acquire momentum in the direction of the wave vector of the laser. This momentum was predicted from both quantum mechanical[21-25] and classical considerations[19, 20] in the 1960's.

In the Thomson regime ($\hbar\omega \ll m_e c^2$), where the photon energy is much less than the electron rest energy, harmonic emission and a mass shift of the electrons are also governed by η^2.[21-25] Under these conditions the mass shift can be thought of as a Doppler shift associated with the forward momentum of the electrons in the field.[24] The forward momentum arises from conservation of energy and momentum during the dressing of the electron.

The electron is born at rest in a plane wave and is accelerated by the Lorentz force

$$\frac{d\vec{p}}{dt} = -e\left[\frac{1}{c}\frac{\partial \vec{A}}{\partial t} + \frac{\vec{p} \times (\vec{\nabla} \times \vec{A})}{\gamma mc}\right]. \tag{11}$$

The Lorentz force equation can be solved by separating it into an equation in the plane of polarization and the $\vec{k}(\hat{z})$-direction. The solution of these equations, for an electron initially at rest, is[26]

$$\vec{p}_\perp = \gamma mc \vec{\beta}_\perp = \frac{e}{c}\left[\vec{A}(t) - \vec{A}(0)\right],$$

$$p_{\vec{k}} = \gamma mc \beta_{\vec{k}} = mc(\gamma - 1), \tag{12}$$

$$\gamma mc^2 = mc^2 + \frac{1}{2}\frac{e^2}{mc^2}\left[\vec{A}(t) - \vec{A}(0)\right]^2.$$

In this case, the parallel direction is that of the k-vector of the laser and the perpendicular direction is in the plane of polarization. The mass shift of the electron in the field is given by the time average of the equation for γmc^2. For linearly polarized laser light with, for example, $\vec{A}(t) = A_0 \sin \omega t \, \hat{x}$, Eqs. 13 yield a trajectory which includes anharmonic motion of the electron in the field and a forward drift. The anharmonic motion is a figure-8 motion in the plane made by the laser polarization and k-vector for linear polarization.[18, 20, 26] The nonlinear motion leads to the radiation of harmonics from the laser,[17, 19, 20] while the forward motion leads to a wavelength shift of the harmonics as shown in the quantum mechanical calculations.[24] This wavelength shift is equivalent to the quantum mechanical mass shift in the field.[24] The rate of harmonic emission scales approximately as $R_n \propto \eta^{2n}$.[17]

Conservation of momentum and energy during the photon scattering process can be used to derive the relationship between the perpendicular and parallel momenta in Eq. (13).[27-29] If the electron gains an energy ΔE_e from the photon field, it also gains a longitudinal ($\vec{k}$ - direction) momentum $cp_{\vec{k}} = \Delta E_e$. Thus, if an electron, initially at rest, gains an energy $\Delta E_e = (\gamma - 1)m_e c^2$ its final parallel momentum will be related to its perpendicular momentum by[27-29]

$$p_{\vec{k}} = \frac{p_\perp^2}{2m_0 c}. \tag{13}$$

When the electron is in the laser field, the perpendicular momentum is due oscillatory motion in the field.

In a laser focus, the electron feels a ponderomotive force, $\vec{F}_p = -\vec{\nabla}\Phi_{pond}$, which causes the electron to be ejected from the laser focus with a directed kinetic energy equal to its oscillatory energy in the field.[12] The angle of the ejected electron relative to the wave vector of the laser (θ) depends on the final electron energy γmc^2 as[27-29]

$$\tan\theta = \sqrt{\frac{2}{\gamma - 1}}. \tag{14}$$

Thus, for low intensities, the electron is ejected with an energy much less than its rest mass nearly perpendicular to the $\vec{k}$ - direction. In ultra-relativistic conditions $\gamma \gg 1$, the electron leaves the laser focus almost parallel to the $\vec{k}$ - direction.

In an intense field, the interaction of the laser pulse with free electrons yields shifted harmonics of the laser frequency. This is true in both the Compton and Thomson regimes, though in the Compton regime, the scattered photon energy can be much higher in the laboratory frame than in the rest frame of the electron.[10]

In 1983 Englert and Reinhart[30] reported a preliminary observation of second harmonic emission from the scattering of a moderate intensity, multi-mode, laser from a low energy electron beam. They observed second harmonic photons with the expected Doppler shift associated with the electron beam. The observed rates were extremely low but appeared to be consistent with predictions.[17, 20]

Results from two recent multiphoton-electron scattering experiments ($\eta^2 \sim 1$) have been reported. In one experiment, the electrons are initially at rest in the laboratory frame,[7] while in the other, 50 GeV electrons produced at the Stanford Linear Accelerator Center collide with a counter propagating laser beam.[8] In both cases, nonlinear photon-electron scattering has been observed. These experiments are briefly described in Sec. IV and V

III. Nonlinear photon-photon scattering

The rate of pair production by photon-photon scattering was calculated by Breit and Wheeler in 1934.[4] For pair production with counter propagating photons, the product of the photon energies must be greater than the pair mass squared, $\hbar\omega_1 \hbar\omega_2 \geq m_e^2 c^4$. When the laser field, E_L, approaches the QED critical field,[31] E_{crit}, nonlinear photon-photon scattering,

$$\omega_1 + n\omega_2 \rightarrow e^- + e^+ \quad (15)$$

becomes possible.[10, 15, 16, 32] The photon energy requirement for nonlinear pair production becomes

$$\hbar\omega_1 n\hbar\omega_2 \geq m_e^2 c^4 \quad (16)$$

The QED critical field[31], corresponds the electric field strength where an electron gains an energy equal to its rest mass in a Compton wavelength,

$$eE_{cr}\lambda_C = m_e c^2 \quad (17)$$

and is equal to $E_{cr} = m_e^2 c^3 / e\hbar = 1.3 \times 10^{16}$ V/cm. The QED critical field also corresponds to binding electric field of the ground state of a Hydrogenic ion with nuclear charge equal to 137.

The QED critical field corresponds to a laser intensity of 4×10^{29} W/cm^2, well beyond the reach of current laser technology.[6] This field strength can be reached, however, in the rest frame of an energetic electron because the laser field transforms as $E_{rest} = \gamma(1+\beta)E_{lab}$. The QED critical field strength can also be reached in the center of momentum frame associated with the collisions of photons of different energies. For a collision of a high energy photon, ω_1 with an intense laser, the field strength in the center of momentum frame is

$$E_{com} = \sqrt{\frac{\omega_1}{\omega_L}} E_L \quad (18)$$

Thus nonlinear photon-photon scattering may be observed during the scattering of an intense laser with energetic electrons in two-step process.[9,32] High energy gammas produced by Compton scattering can subsequently interact with the laser field, leading to the production of electron positron pairs. The ratio of the laser field in the rest frame of the electron is to the QED critical field has been given the symbol Upsilon, $Y = E_{rest}/E_{cr}$.

IV. Nonlinear laser-electron scattering in the Thomson regime

When electrons born at rest through field ionization in a high intensity laser focus they acquire energy and momentum from the laser field through their quiver energy in the field and conservation of momentum. As the electron's kinetic energy approaches its rest energy, the absorption of momentum from the field cannot be ignored and the electrons gain forward momentum. In Ref. 7 the experimental observation of electron recoil during high intensity laser-electron scattering (Compton scattering) in the Thomson regime with photon energy much less than the electron rest mass, $\hbar\omega \ll m_e c^2$, was described. A particularly simple formulation in terms of conservation of energy and momentum from the field was shown in Eq. 14. In this low electron energy experiment, electrons were injected into the field at rest during the ionization of Ne and Kr atoms and ions by a high-intensity laser.[7,33] The electrons gained both energy and longitudinal momentum from the field. They were subsequently accelerated out of the focus by the ponderomotive force, retaining their longitudinal momentum. A schematic of the experimental setup is shown in Fig. A magnetic spectrometer was used to measure the energy and angular (relative to $\vec{k}$) distributions of electrons emitted from a high-intensity laser focus.[34] The spectrometer consists of an energy-resolving magnet and a detector consisting of a scintillator coupled to a photo-multiplier tube (PMT). The energy resolution was $\Delta E/E \sim 0.3$ FWHM.. The angular distribution of electrons in θ (relative to $\vec{k}$) is measured by rotating the entire spectrometer about the cylindrical axis that passes through the laser focus at 90° to the laser axis. The experiments were conducted with a 1.053-μm, 1-ps laser using chirped-pulse amplification (CPA).[35] The laser is focused with f/3 optics producing a 5-μm ($1/e^{-2}$ radius) focal spot and a peak laser intensity of approximately 10^{18} W/cm^2 into neon and krypton at a pressure of 10^{-3} Torr with circular polarization.

In experiments by Moore et al.[7] the angular and energy distributions of electrons injected into the field during the field ionization[36] of Ne and Kr were measured. Electrons associated with Ne^{8+} emerged from the circularly polarized focus with an energy of 80±5 keV and at an angle of 75±1.5° from the laser axis ($\vec{k}$-direction). These results were extended to the observation of electrons injected during the production of Kr^{10+} and Kr^{11+}.[33] The energy and angular spectra are in good agreement with the predictions of calculations which include the relativistic mass shift.

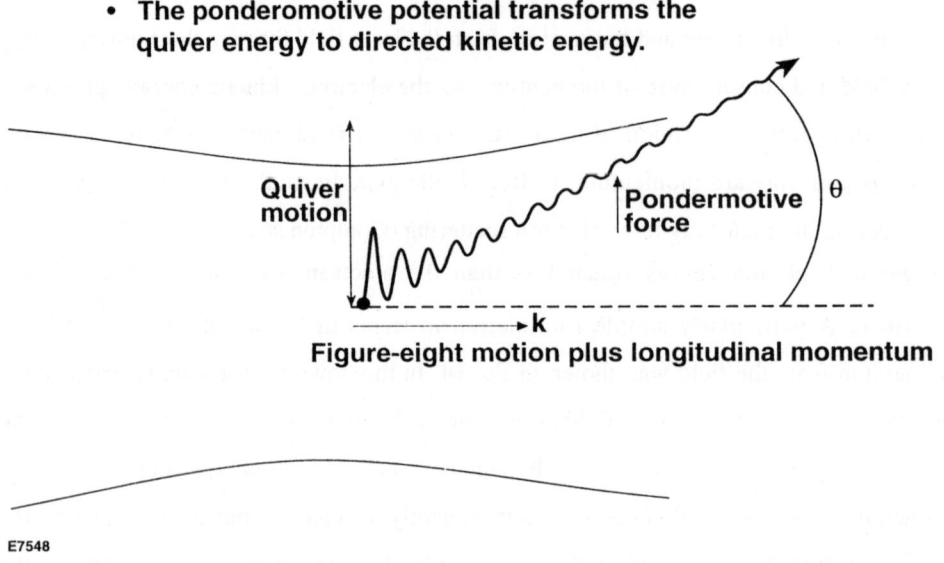

Figure 2: Schematic figure of the experimental setup of the scattering of low energy electrons born by ionization in a laser focus by an intense laser. Electrons are observed with forward momentum due to conservation of photon momentum.

Figure 3 shows the observed energy and angular distribution for $Ne^{3+} - Ne^{8+}$ and Kr^{10+}, and Kr^{11+} (from Refs. [7, 33]). In all cases, the observations are in good agreement with the relativistic calculations, with the energies predicted by classical field ionization[36] and subsequent ejection from the focus by ponderomotive and canonical momentum effects. The solid curve is that predicted by Eq. 15. The energy and angular spectra were in good agreement with the predictions of calculations which include the relativistic mass shift.[7, 33] For electrons from the highest charge states from Kr, differences in the energy and angular distribution due to the electron mass shift in the field were observed.[33]

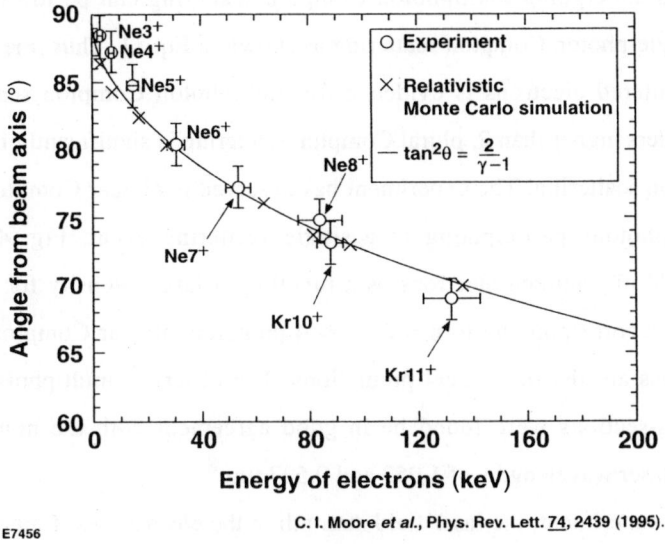

Figure 3: Comparison of observed (with errors) and relativistic Monte Carlo calculations (X) of the energy and ejection angle for electrons born during the production of various charge states of Ne and Kr. Equation (15) is plotted for comparison. (From Ref. [33])

V. Nonlinear laser-electron scattering in the Compton limit

Nonlinear Compton scattering of 2-4 laser photons has been reported during the interaction of a high intensity laser ($I \sim 10^{18}$ W/cm^2) counter-propagating with 47 GeV electrons at the Stanford Linear Accelerator Center.[8] In this case, the photon energy in the electron rest frame is comparable to the electron rest mass ($\gamma \hbar \omega_L / m_e c^2 \sim 1$).

The electron beam interacts with a 0.5 Hz repetition rate, 1 µm, 1 ps, terawatt, chirped pulse amplification (CPA) laser system using a flashlamp-pumped, Nd:glass, zig-zag slab amplifier.[37] The laser produces infrared pulses with energies greater than 2 Joules with an average power in excess of 1 W and 1.4 times diffraction limited focusing. Frequency doubling of these pulses is accomplished with up to 55% efficiency. The laser has been synchronized with a 47 GeV electron beam at the Stanford Linear Accelerator. with a temporal jitter of $\sigma = 2$ ps between the two beams.[38]

In the experiment the scattered electrons are observed by passing them through a magnetic spectrometer and detecting them with segmented silicon calorimeters.[8,9] The number of scattered electrons as a function of scattered energy is measured. The energy of the electrons undergoing multiphoton Compton scattering can be lower than those allowed for single-photon Compton scattering as shown in Eq. 11. Thus, the existence of low energy scattered electrons is evidence for multiphoton Compton scattering. For multiphoton orders higher than 2, plural Compton scattering is significantly less probable than multiphoton scattering. The experiment has observed nonlinear Compton scattering with up to 4 photons participating in a single scattering event. Fig. 4 shows the normalized yield of scattered electrons as a function of laser intensity for a variety of electron energies corresponding to n = 2, 3, & 4 photon nonlinear Compton scattering. The shaded areas are the theoretical predictions. The observed multiphoton Compton scattering cross-sections were found be in good agreement with the nonlinear QED predictions for laser wavelengths of 1.053 and 0.527 μm.[8]

In this experiment, the laser field strength in the electron rest frame approaches the QED critical field corresponding to a laser intensity of approximately 10^{29} W/cm^2. Multiphoton-photon scattering processes are currently under investigation.[9]

VI. Conclusions

In the focus of a high intensity laser, photon-electron and photon-photon scattering processes become nonlinear. The electron's oscillatory energy in the field can exceed its rest mass. Recent experiments have observed nonlinear scattering processes during laser - electron scattering with low energy electrons, born at rest[7] and with 46.6 GeV electrons.[8] Studies of nonlinear photon-photon scattering are underway.[9] In the past few years advances in laser technology have begun to open up a new area of high field studies where the traditional boundaries between Thomson and Compton scattering become less clear. For example, nonlinear Compton scattering of n photons and an electron, takes place in the presence of an intense laser field, mixing quantum mechanical and classical pictures.

Acknowledgments

I thank J.H. Eberly, S. Goreslavskii, J.P. Knauer, K.T. McDonald, A.C. Melissinos, and N. Narozhny for many stimulating discussions. This work was supported by the National Science Foundation. The experiments on multiphoton-electron scattering with low energy electrons were carried out in collaboration with C.I. Moore, J.P. Knauer, and S.J. McNaught. The nonlinear QED experiments at SLAC are carried out by a collaboration between the University of Rochester, Princeton University, the University of Tennessee, and the Stanford Linear Accelerator Center.

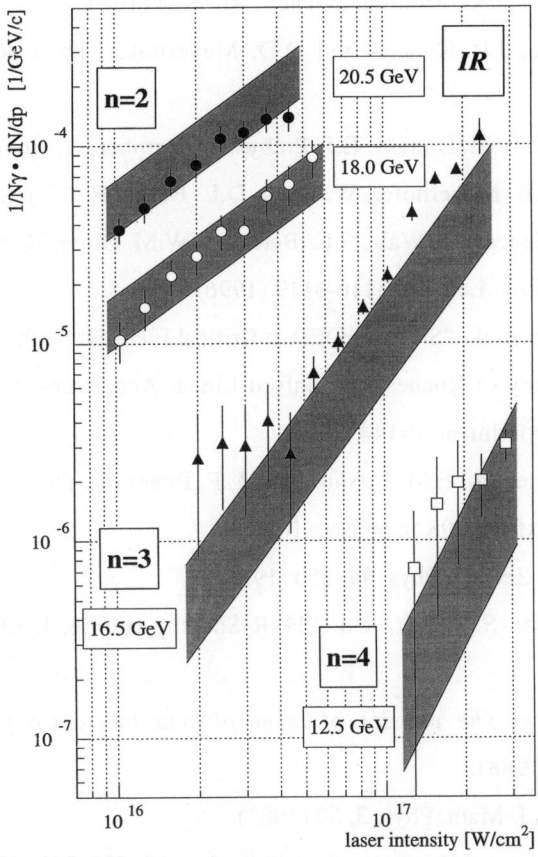

Figure 4: Fig. 4 shows the normalized yield of scattered electrons as a function of laser intensity for a variety of electron energies corresponding to n = 2, 3, & 4 photon nonlinear Compton scattering. The shaded areas are the theoretical predictions. (From Ref. 8)

References

* also Dept. of Physics and Astronomy and Laboratory for Laser Energetics

1. J.J. Thomson, *Recent Researches in Electricity and Magnetism* (Clarendon Press, Oxford, 1893).

2. A.H. Compton, Phys. Rev. **22**, 411 (1923).

3. O. Klein and Y. Nishina, Zeits. f. Physik **52**, 853 (1929).

4. G. Breit and J.A. Wheeler, Phys. Rev. **46**, 1087 (1934).

5. P. Maine, D. Strickland, P. Bado, M. Pessot, and G. Mourou, IEEE J. Quant. Elect. **QE-24**, 398 (1988).

6. M.D. Perry and G. Mourou, Science , 917-924 (1994).

7. C.I. Moore, J.P. Knauer, and D.D. Meyerhofer, Phys. Rev. Lett. **74**, 2439 (1995).

8. C. Bula, K.T. McDonald, E.J. Prebys, C. Bamber, S. Boege, T. Kotseroglou, A.C. Melissinos, D.D. Meyerhofer, W. Ragg, D.L. Burke, R.C. Field, G. Horton-Smith, A.C. Odian, J.E. Spencer, D. Walz, S.C. Berridge, W.M. Bugg, K. Shmakov, and A.W. Weidemann, Phys. Rev. Lett. **76**, 3116-3119 (1996).

9. C. Bula and et al., "Study of QED at Critical Field Strength at SLAC," Princeton University, University of Rochester, Stanford Linear Accelerator Center, University of Tennessee SLAC experiment, E-144 (1992).

10. V.B. Berestetskii, E.M. Lifshitz, and L.P. Pitaevski, *Quantum Electrodynamics* (Pergamon Press, Oxford, 1982), pp. sec. 101.

11. D. Volkov, Zeits. f. Phys. **94**, 250 (1929).

12. H.A.H. Boot, S.A. Self, and R.H. R-Shersby-Harvie, J. Elect. Control **4**, 434 (1958).

13. W.L. Kruer, *The physics of laser plasma interactions* (Addison-Wesley, Redwood City, Ca, 1988).

14. H.R. Reiss, J. Math. Phys. **3**, 59 (1962).

15. A.I. Nikishov and V.I. Ritus, Sov. Phys. Jetp **19**, 529-541 (1964).

16. N.B Narozhny, A.I. Nikishov, and V.I. Ritus, Sov. Phys. JETP **20**, 622-629 (1965).

17. Vachaspati, Phys. Rev. **128**, 664 (1962).

18. Vachaspati, Proc. Nat. Inst. Sci. (India) **29**, 138-142 (1962).

19. J.H. Eberly and A. Sleeper, Phys. Rev. **176**, 1570-1573 (1968).

20. E.S. Sarachik and G.T. Schappert, Phys. Rev. D **1**, 2738-2753 (1970).

21. L.S. Brown and T.W.B. Kibble, Phys. Rev. **133**, A705 (1964).

22. O. von Roos, Phys. Rev. **135**, A43-A50 (1964).

23. Z. Fried and J.H. Eberly, Phys. Rev. **136**, B871 (1964).

24. T.W.B. Kibble, Phys. Rev. **150**, 1060 (1966).

25. J.H. Eberly, "Interaction of very intense light with free electrons," in *Progress in Optics,* , edited by E. Wolf (North-Holland, Amsterdam, 1969), Vol. 7.

26. L.D. Landau and E.M. Lifshitz, *The Classical Theory of Fields*, Course of Theoretical Physics, Vol. 2 (Pergamon Press, Oxford, 1971).

27. J.N. Bardsley, B.M. Penetrante, and M.H. Mittleman, Phys. Rev. A **40**, 3823-3835 (1989).

28. H.R. Reiss, J. Opt. Soc. Am. B **7**, 574-586 (1990).

29. P.B. Corkum, N.H. Burnett, and F. Brunel, "Multiphoton Ionization in Large Ponderomotive Potentials," in *Atoms in Intense Fields,* , edited by M. Gavrila (Academic Press, New York, 1992), pp. 109-137.

30. T.J. Englert and E.A. Rinehart, Phys. Rev. A **28**, 1539 (1983).

31. J. Schwinger, Proc. Nat. Acad. Sci. **40**, 132 (1954).

32. H.R. Reiss, Phys. Rev. Lett. **26**, 1072-1075 (1971).

33. D.D. Meyerhofer, J.P. Knauer, S.J. McNaught, and C.I. Moore, J. Opt. Soc. Am. B **13**, 113 (1996).

34. C.I. Moore, "Observation of the transition from Thomson to Compton scattering in optical multiphoton interactions with electrons," PhD thesis, University of Rochester (1995).

35. Y.-H. Chuang, D.D. Meyerhofer, S. Augst, H. Chen, J. Peatross, and S. Uchida, J. Opt. Soc. Am. B **8**, 1226-1235 (1991).

36. S. Augst, D. Strickland, D. D. Meyerhofer, S. L. Chin, and J. H. Eberly, Phys. Rev. Lett. **63**, 2212 (1989).

37. C. Bamber, T. Blalock, S. Boege, J. Kelly, T. Kotseroglou, A.C. Melissinos, D.D. Meyerhofer, W. Ragg, and M. Shoup III, Opt. Lett. , submitted (1996).

38. T. Kotseroglou and et al., Nucl. Instrum. & Methods A , tobe published (1996).

COMPARING THE ANTIPROTON AND PROTON AND PROGRESS TOWARD COLD ANTIHYDROGEN

G. GABRIELSE, D.S. HALL, A. KHABBAZ, T. ROACH, P. YESLEY
Department of Physics, Harvard University
Cambridge, MA 02138, USA

C. HEIMANN, H. KALINOWSKY
Institut für Strahlen- and Kernphysik, University of Bonn
53115 Bonn, Germany

W. JHE
Department of Physics, Seoul National University 151-742 Seoul, Korea

B. BROWN
Department of Physics, Mount Holyoke College
Mount Holyoke, MA

The most stringent test of CPT invariance with a baryon system is a 1 part in 10^9 comparison of the charge-to-mass ratios of the antiproton and proton. An improved Q/M comparison for antiproton and protons at the 10^{-10} level of accuracy is underway, utilizing an antiproton and an H^- ion confined at the same time in a Penning trap. Precise spectroscopic comparisons of cold antihydrogen and hydrogen atoms should produce much more accurate CPT tests. The recent experimental progress toward cold antihydrogen includes the accumulation of more than a million 4.2 K positrons in ultrahigh vacuum, and a demonstrated low temperature interaction of cold electrons and protons in a nested Penning trap. The first attempt at producing cold antihydrogen is scheduled for late 1996. Collaborations and facilities for antihydrogen studies after 1996 will be mentioned.

1 Introduction and Motivation

Our TRAP Collaboration originally estimated that the charge-to-mass ratios of the antiproton and proton could be compared to 1 part in 10^9 (1 ppb), an improvement by a factor of 45,000 over earlier measurements. Techniques to slow, confine and cool antiprotons by 10^{10} in energy had to be developed, along with techniques for eventually isolating a single antiproton. Recently this 1 ppb goal was achieved[1] by comparing the cyclotron frequencies of a single, trapped antiproton with a similarly confined proton. A radio detector damps the cyclotron motion, causing the cylotron frequency to shift because of special relativity. This is the most stringent test of CPT invariance with a baryon system by many orders of magnitude. An additional improvement by as much as a factor of 10 is expected soon by comparing the cyclotron frequencies of a H^- ion and an antiproton which are trapped together. The history of comparisons of antiproton and proton is shown in Fig. 1. The goal for 1996 is below the graph.

The next experimental goal is a greater challenge for which we have been preparing as we developed the techniques to cool and accumulate low energy antiprotons: comparing hydrogen and antihydrogen to high accuracy. The 9 antihydrogen atoms that were recently observed[2] add to the interest and excitement, even though their

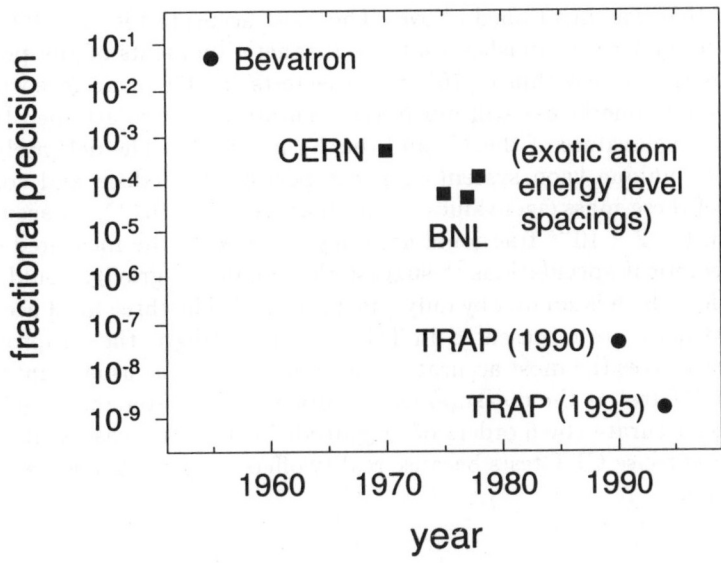

Figure 1: The accuracy at which antiprotons and protons have been compared.

small number and extremely high energy make it impossible to make any accurate comparisons of antihydrogen and hydrogen. Cold trapped antiprotons merged with cold trapped positrons should produced cold antihydrogen [3] of low eneough energy to be confined for precise comparisons with hydrogen [4]. At the end of 1996, the we will make the first attempt to produce cold antihydrogen. The best that could be hoped for this one week experiment would be to simply observe cold antihydrogen. A more realistic expectation would be to observe positron cooling of antiprotons. There will certainly not be time to attempt any precise studies.

Unfortunately the Low Energy Antiproton Storage Ring (LEAR) at the CERN laboratory in Geneva closes at the end of 1996. Insofar as the TRAP collaboaration has demonstrated the possibility to accumulate cooled antiprotons in a trap [5,6], rather than at high energies in a much more expensive storage ring, the possibility to make a simpler antiproton source available at CERN (one ring instead of three) is being seriously considered. Scientific approval has been attained at CERN, but the needed financial contributions needed from outside CERN are still being negotiated. In anticipation of precise studies of cold antihydrogen after 1996, the TRAP collaboration is expanding into a larger ATRAP collaboration [7] which includes groups with complementary expertise. A competing collaboration has also been formed [8], with 47 members signing the initial statement of intent.

An improved CPT test is one important motivation for experiments which compare antihydrogen and hydrogen. A reasonable requirement on any new CPT test is that it eventually be carried out with an accuracy that exceeds the accuracy of the best tests with baryons and leptons. Right now, the most accurate test of CPT invariance with a baryon system is the 1 ppb comparison of Q/M for the antiproton and proton that was mentioned above. The most accurate test of CPT invariance with a lepton system establishes that the magnetic moments of the positron and electron[9] are the same within 2×10^{-12}. These tests of CPT invariance with baryons and leptons are nonetheless still much less accurate than was attained by comparing the mass eigenvalues of the K_0 and the $\bar{K}_0$ mesons[10]. The delicately balanced nature of the unique kaon system makes it possible to deduce and compare the differences of these mass eigenvalues to an accuracy of 2×10^{-18}, an accuracy much greater than the 2×10^{-3} fractional accuracy required in the measured quantities. (Recent theoretical speculations[11] suggest that quantum gravity could produce a CPT violation which is smaller by only a factor of 10.) The three most accurate tests of CPT invariance are represented in Table 1. Interestingly, the antiproton-proton comparison involves the most accurate measurement, and (as usually interpreted) it provides a CPT test at the measurement accuracy. The lepton and meson measurement are less accurate (by 6 orders of magnitude in the latter case) but nonetheless provide more precise CPT tests because reality offers substantial and free sensitivity enhancements.

Table 1: Comparing the CPT Tests

	CPT Test Accuracy	Measurement Accuracy	Free Gift
Mesons ($K_0 \bar{K}_0$)	2×10^{-18}	2×10^{-3}	10^{15}
Leptons ($e^+ e^-$)	2×10^{-12}	2×10^{-9}	10^3
Baryons ($p\bar{p}$) (goal in 1996)	1×10^{-9} (1×10^{-10})	1×10^{-9} (1×10^{-10})	1 1

The narrow resonance lines realized in the laser spectroscopy of the hydrogen 1s - 2s transition[12] illustrate the possibility to use laser spectroscopy to greatly improve the accuracy of CPT tests involving baryons and leptons. In a very encouraging development, this transition was recently observed for the first time with cold, trapped hydrogen atoms[13]. If such a line were available for antihydrogen as well as hydrogen, the signal-to-noise ratio would be sufficient to allow the frequencies to be compared to at least 1 part in 10^{13}, a large increase in accuracy over the current tests. The line was only slightly narrower than had been observed before, but the linewidth seems to be entirely limited by laser stability which could be substantially improved. It may be possible to carry out precise laser spectroscopy of as few as 100 trapped antihydrogen atoms[14], but this estimate remains to be demonstrated. The antihydrogen dream is to attain the 5×10^{-14} natural linewidth of the 1s - 2s transition, with a signal-to-noise ratio that allows splitting the line by another factor of 250. A comparison at this level of accuracy is extremely unlikely in the foreseeable future, but would equal the accuracy in the meson CPT test.

A second motivation for experiments which compare cold antihydrogen and hydrogen is the possibility to search for differences in the force of gravity upon antimatter and matter [15]. Making gravitational measurements with neutral antihydrogen atoms certainly seems much more feasible than using charged antiprotons, for which the much stronger Coulomb force masks the weak gravitational force. The possibility of gravitational measurements with trapped antihydrogen has been considered [16] and the free fall of cold atoms released from a trap [17] is now observed routinely by many groups. We are intrigued by the possibility of experimental comparisons of the force of gravity upon antihydrogen and hydrogen, but are not yet ready to make specific estimates of attainable measurement accuracies.

The recombination of antiprotons and positrons to form cold antihydrogen is discussed in Sec. 2. Considerable progress toward the production and study of cold antihydrogen has already been made insofar as the raw materials for cold antihydrogen are now available. Our TRAP Collaboration developed the techniques whereby antiprotons from LEAR are now routinely slowed in matter, trapped [18], and then electron-cooled to 4 K [5,6]. The surrounding vacuum is so good that antiprotons have been stored for months at an energy 10^{10} times below the energy of antiprotons in LEAR [6]. These techniques were demonstrated years ago (some have recently been confirmed by others [19]), so nothing more will be said about them here. Instead, in Sec. 3 we will focus upon techniques to accumulate many 4K positrons in a similarly good vacuum. More than a million such positron have been confined in recent weeks. With the same radioactive source, the accumulation rate is now about 50 times higher than was realized when we reported accumulating 3×10^4 positrons one year ago [20]. Moreover, the positrons are now within the nested Penning trap which will be used for initial attempts to observe cold antihydrogen, rather than in a hyperbolic Penning trap with small aperatures. In Sec. 4, we summarize a recent demonstration of the electron cooling of protons in a nested Penning trap [21]. The protons and electrons were shown to interact at low relative velocity where recombination rates are expected to be largest.

2 Recombination Processes

Before and during the time that the TRAP Collaboration was developing the techniques to cool and accumulate 4 K antiprotons, we were also considering the production of cold antihydrogen from cold antiprotons and positrons accumulated in separate traps [4,3]. To recombine, a positron and antiproton must have sufficient kinetic energy to approach each other. This energy must be removed to form an atomic bound state. Energy and momentum cannot be conserved unless a third particle is involved. Different processes have been considered, including radiative recombination, stimulated radiative recombination, a three body recombination, and a recombination process involving positronium. We look briefly at each of these in turn.

Radiative recombination can be thought of as producing an excited hydrogen atom which radiates a photon to conserve energy and momentum.

$$\bar{p} + e^+ \rightarrow \bar{H} + h\nu \tag{1}$$

This process suffers from a low rate because it takes approximately a ns to radiate a photon, much longer than the duration of the interaction of the $\bar{p}$ and an e^+, even if these particles have an energy corresponding to 4 K. For 10^6 antiprotons at 4 K within a 4 K positron plasma of density $n_e = 10^8/cm^3$, the estimated recombination rate [3] is $3 \times 10^3/s$. The radiative recombination process has the attractive feature that most of the antihydrogen formed would be in the ground state, but the anticipated rate is lower than for other processes.

The radiative recombination rate can be considerably increased by stimulating the process with a laser.

$$\bar{p} + e^+ + h\nu \rightarrow \bar{H} + 2h\nu \qquad (2)$$

Laser-stimulated, radiative recombination has been observed in merged beams [22,23], but has not yet been observed in a trap. Under realistic conditions in a trap [3], the laser should increase the radiative recombination rate for the production of antihydrogen in the ground state by up to a factor of 10^2. The laser intensity cannot be increased arbitrarily to speed the recombination because the cross section for field ionization increases sharply at high intensities. There is also the suggestion [3] to use a CO_2 laser to stimulate recombination to principal quantum numbers $n = 10$ (or to nearby levels with various diode lasers), rather than to the ground state. The advantage is an additional increase in rate over unstimulated radiative recombination by a huge factor of 10^5 for the antiproton number and positron density mentioned above, provided that sufficient laser power is available and can be managed within the cryogenic environment. We intend to pursue the stimulated radiative recombination process.

The three body recombination

$$\bar{p} + e^+ + e^+ \rightarrow \bar{H} + e^+ \qquad (3)$$

is attractive because the recombination rate promises to be much higher than for any other processes [3]. However, deexcitation to lower states takes some time, though a recent calculation [24] exagerated this difficulty by overlooking replacement collisions [25]. For 10^6 antiprotons at 4 K, submerged within an extended plasma of 4 K positrons at density $10^7/cm^3$, the recombination rate is an astounding $10^9/s$ [3]. A strong magnetic field (for the trap containing antiprotons and positrons) would reduce this high rate [26] by approximately a factor of 10, and an electric field (also part of the traps) has some effect [24], but the rate still seems higher than for other recombination processes. The related process

$$\bar{p} + e^+ + e^- \rightarrow \bar{H} + e^- \qquad (4)$$

has also been mentioned [3], but this seems more difficult to arrange.

Radiative recombination with and without laser stimulation will be attempted in both a nested Penning trap and in a combined Penning-Paul trap (discussed in the next section). The nested Penning trap also offers some control over the three body recombination process. An antihydrogen atom leaving the trapped positron plasma will experience an electric field whose magnitude can be adjusted. This field will ionize atoms which are highly excited. For example, an electric field of 7V/cm will ionize all the antihydrogen produced with principal quantum number $n > 100$, returning the positron and antiproton to their respective wells. With a strong electric field the radiative recombination can thus be "switched off" entirely (to investigate radiative recombination, for example). With a carefully adjusted field, three body recombination to the lowest possible states can be selected. There is also the possibility to analyze the excited state distribution in the hope of designing an experimental procedure to deexcite these states as rapidly as possible.

The last recombination process uses positronium, the bound state of an electron and a positron, with the electron carrying off the excess energy [27].

$$\bar{p} + e^+e^- \to \bar{H} + e^-. \tag{5}$$

One advantage of of this process is that the antihydrogen is produced preferentially in the lowest states. Also when antiprotons are not available, the charge-conjugate process can be studied (recombining protons and positronium to form hydrogen and positrons). This was recently observed [28] using beams of protons and electrons. Unfortunately, comparable quantities of antiprotons and positrons are very difficult to arrange. It should be possible to increase the recombination rate by initially exciting the positronium atom with a laser [29]. Nonetheless, the disadvantages of the recombination using positronium is that the projected rate is extremely low for realizable numbers of antiprotons, and existing postronium beams. It may be possible to make a miniature positron storage ring to greatly increase the available positronium. However, the apparatus required is somewhat different than is required for the other processes and we thus do not plan to initially pursue this approach.

Much experimental investigation is still required to determine the most efficient route to low energy antihydrogen. We will initially investigate radiative recombination (Eq. 1), stimulate this process with a laser (Eq. 2), and attempt to characterize and control the three body recombination (Eq. 3).

3 Great Increase in 4.2 K Positrons Accumulated in High Vacuum

In two steps, we have recently found ways to greatly increase the number of positrons which can be accumulated in an extremely good vacuum sufficient to avoid the annihilation of antihydrogen. In the first step [20], a hyperbolic Penning trap was used to accumulate positrons. High energy positrons from a ^{22}Na source passed through the trap (following the magnetic field lines) to strike and enter a single crystal of tungsten. This crystal was suspended by extremely thin wires to allow it to be cleaned and annealed by briefly heating it white hot while its surroundings are at 4 K. Positrons diffuse in the crystal without being trapped at defects, slowing via collisions. Most of the positrons thermalize in the crystal and some diffuse back to the surface of the crystal where they are ejected at low energies by the work function [30].

The cooled positrons, with energy spread less than 1 eV, reenter the trap, again traveling along the (vertical) magnetic field direction. The trap is carefully biased to keep some of the positrons inside for one "magnetron" revolution period during which they oscillate harmonically along the direction of the magnetic field. These positrons would naturally leave the trap after this $10\mu s$ revolution, exiting the trap through their entrance hole. During their time in the trap, however, the rapid vertical oscillations induce a current in a resonant tuned circuit attached to the electrodes. The positrons lose the energy that is dissipated in this circuit and remain in the trap indefinitely if the energy loss is sufficient to prevent their leaving through the entrance hole.

The number of positrons is determined from the frequency spectrum of the voltage across the resonant RLC circuit. Without positrons in the trap, a resonance is observed at the resonant frequency of the circuit, due to the random thermal (near 4 K) motions of the electrons in the circuit. When positrons are added, they effectively "short" the noise signal, producing a pronounced dip as illustrated in Fig. 2a. The number of trapped positrons is determined by fitting this dip and comparing to electrical signals observed with one through five trapped positrons.

In a second step that is actually still underway, we have accumulated positrons directly into the nested Penning trap. Both a reflection moderator and a transmission moderator have been used. Fig. 2b shows how the positrons accumulate linearly in time. The peak observed accumulation rate is 11 per second, almost 50 times higher than the rate discussed above. More than 10^6 positrons have been accumulated in slightly more than one day, but no serious attempt to accumulate more positrons has yet been carried out. The rush to prepare apparatus which will allow antiprotons and positrons to interact later this year is making it difficult to study and optimize a new and extremely promising approach to positron accumulation.

A high positron density is desired since the antihydrogen recombination rate depends either linearly or quadratically upon the positron density, depending on the recombination process. The observed density is nearly $10^8/cm^3$. The density depends primarily upon the size of the trap and upon the potential difference across the trap electrodes, making it more difficult to attain the desired high densities in a large trap. The achieved density is more than adequate for antihydrogen production. Nonetheless, experiments are underway to obtain more trapped positrons in order to fill a larger volume with positrons at the observed density. It would also be convenient if the positrons could be loaded more rapidly.

The most straightforward way to increase the number of trapped positrons is to increase the size of the radioactive source used to produce them. This approach, limited primarily by safety concerns, is being pursued by the TRAP Collaboration with attempts now being made to fabricate a $1\,Ci$ ^{58}Co source for the experiments planned to take place at LEAR at the end of 1996. Such a source should improve the accumulation rate by up to a factor of 30.

A more efficient positron accumulation would allow more positrons to be accumulated more rapidly with less concern about safety. One alternative is to capture positrons which collide with laser-cooled ions in a a trap [31,32]. There are at least two other approaches to positron loading that we will not pursue initially. The first is to capture positrons that cool via collisions with a background pressure of neutral atoms introduced into the trap. This scheme is attractive because of the large number of positrons which have been trapped in this way [33,34]. The challenge is to introduce enough differential pumping to get the pressure to the extremely low value needed to avoid the annihilation of antihydrogen. A minimum requirement thus seems to be a substantial, room temperature apparatus outside the cryogenic

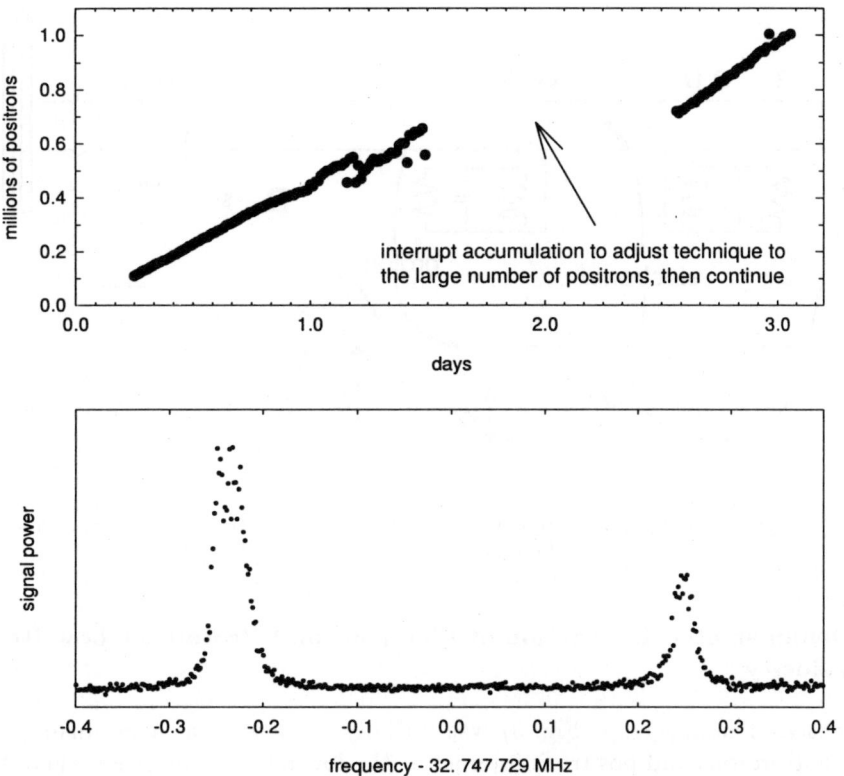

Figure 2: Millions of trapped positrons accumulated vs. accumulation time (above). The measured noise spectrum across the RLC damping circuit (below) is fit to determine that 10^6 positrons have been accumulated.

environment needed to allow the confinement and study of cold antihydrogen. We prefer to avoid this if possible, but will reconsider if we are dissatisfied with the number of positrons which we accumulate by the methods described above. The second alternative is to slow and capture positrons from electron-positron pairs produced at an accelerator. (We understand that such studies are being planned at the Aarhus microtron.) The attractive feature of this scheme is the possibility to trap large numbers of positrons quickly. The disadvantage is the need to add a very substantial and expensive facility to an already complicated apparatus. Especially given the big steps demonstrated with simpler and cheaper methods, it seems prudent to investigate the simpler positron accumulation schemes while awaiting a demonstration of the accumulation of microtron positrons into a cryogenic vacuum.

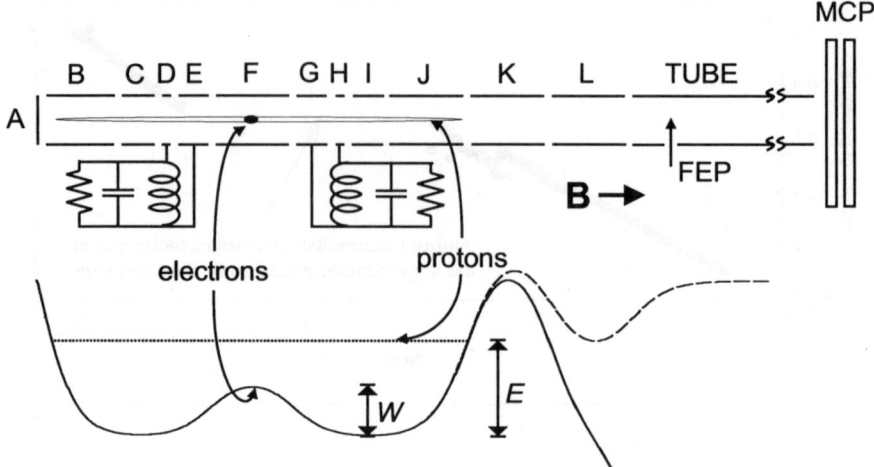

Figure 3: Nested Penning trap should allow antiprotons and positrons to interact at low relative velocity, where the recombination rate is largest.

4 Demonstrated Interaction of Electrons and Protons at Low Relative Velocity

The nested Penning trap (Fig. 3) was initially proposed [3] as a promising way to make antiprotons and positrons interact at the low relative velocities where the recombination rates are highest. While recombination has not yet been observed in a nested Penning trap, the interaction of cold electrons and protons with extremely low relative velocities has now been clearly demonstrated by members of our collaboration [21]. Fig. 4a shows the energy spectrum of hot protons when there are no electrons in the inner well. (Highest energies are to the left because the highest energy particles are detected first.) When electrons are introduced into the inner well, they rapidly cool via synchrotron radiation to come into thermal equilibrium with their 4K surroundings. The protons slow as they climb the potential hill to enter the cloud of electrons, then lose energy via collisions with the electrons. This process continues until the protons lose just enough energy to keep them from entering the electron cloud again. The cooled protons thus reside in the two side wells, decoupled from the electrons in the center well.

An example of the resulting energy spectrum of the cooled protons residing in the side well to the right is shown in Fig. 4b. The measured energy is with respect to the bottom of the side wells. It should be stressed, however, that the relative velocity of the trapped protons and electrons is what is important, not the energy of the cooled protons with respect to the bottom of the side well. This relative velocity of protons and electrons is extremely low when the cooling stops. We would thus expect recombination to be triggered when the depth of the electron well is reduced slightly to make the protons and electrons interact again at low relative velocity. Fig. 5 illustrates that the protons always cool to have the lowest possible velocity with respect to the cold electrons. The measured proton energy with respect to the bottom of the side wells, corrected for adiabatic cooling, is shown as a function of the depth of the central electron well. The straight line represents the protons cooling to the energy of the electrons (very near to the bottom of their well), where the relative velocities of the protons and electrons are very low.

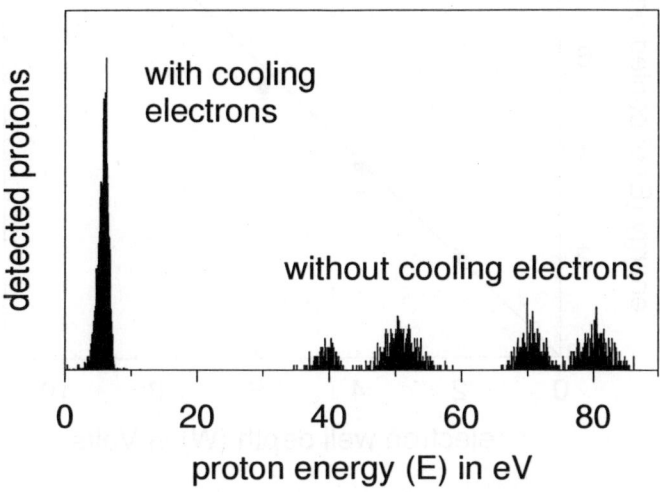

Figure 4: Number versus energy of hot protons (above) and cooled protons (below) within a nested Penning trap.

The demonstration of the nested Penning trap, in which protons are cooled to low velocities relative to the cooling electrons, took place in the extremely good cyrogenic vacuum required to avoid annihilating antihydrogen. In fact, the 4K apparatus used was virtually identical with apparatus used by the TRAP Collaboration (PS196) for the low energy antiproton experiments.

An alternative approach to getting protons and electrons to interact is to store protons in a Penning trap, and to store electrons in a superimposed Paul trap. Ions have been confined in Paul traps that were run in such a combined mode since the early days of particle trapping, with an application to protons and electrons analyzed more recently [35]. The simultaneous confinement of electrons and ions (1/8 were protons) in such a combined trap was recently demonstrated [36]. Several thousand electrons and ions were confined at the same time. The interaction of the electrons and ions was observed insofar as the presence of electrons reduced the number of ions which could be stored in the trap, presumably due to the heating of the electrons by the microwave driving field used to confine them. This heating is expected to increase with increasing electron number, so the full effect of the heating will now be studied to see how it can be managed. The hope is to soon repeat the experiment with only protons and electrons, after heating out the heavier ions. When electrons and protons are confined simultaneously, laser-stimulated, radiative recombination to $n = 10$ will be attempted with a CO_2 laser [3]. The demonstration apparatus had a pressure of 10^{-8} Torr, but there are plans to develop

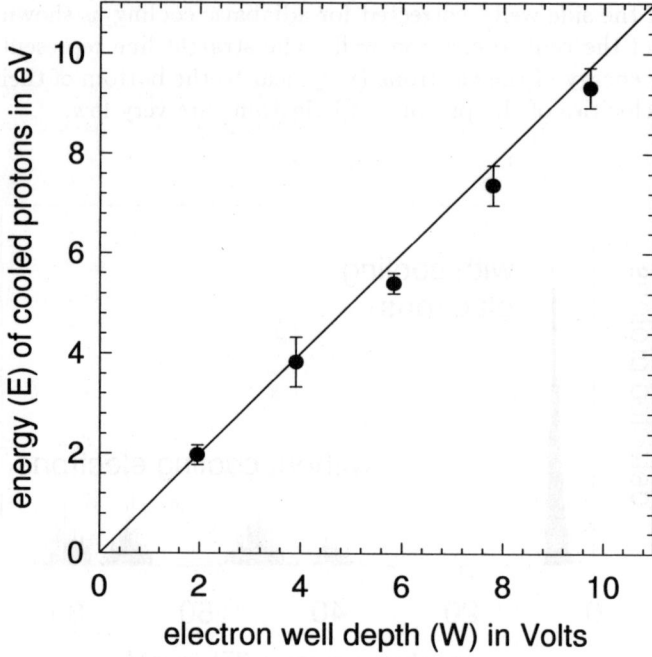

Figure 5: Measured proton energy above the bottom of trap as a function of the depth of the central electron well in a nested Penning trap. A straight line corresponds to the protons cooling to the level of the highest energy electrons. That is, the protons cool so that they have the lowest possible velocity relative to the cold electrons.

and test a cryogenic apparatus which has the extremely low pressure required to avoid annihilating antihydrogen atoms.

5 Conclusion

In conclusion, the goal for 1996 is to improve the comparison of the charge-to-mass ratios of the antiproton and proton to 1 part in 10^{10}, thereby making this comparison 450,000 times more accurate than previous comparisons with other techniques. The ingredients for low energy antihydrogen are now available, and protons and electrons in a nested Penning trap have been used to demonstrate that positrons and antiprotons would interact with low relative velocities in this environment. An initial attempt to synthesize and observe cold antihydrogen is scheduled to take place later in 1996. If modifications to a CERN acelerator ring take place as planned so that low energy antiprotons wil again become available in two years, recently formed collaborations will seriously pursue precise measurements which compare cold, trapped antihydrogen and hydrogen.

Acknowledgments

We thank J.W. Kim for experimental assistance. We also thank members of the ATRAP Collaboaration for helpful discussions, including T. Hänsch, C. Zimmermann, J. Walraven, T. Hijmans, W. Oelert, W.D. Phillips, S.R. Rolston and D. Wineland.

References

1. G. Gabrielse, D. Phillips, W. Quint, H. Kalinowsky, G. Rouleau, and W. Jhe. Special relativity and the single antiproton: Forty-fold improved comparison of $\bar{P}$ and P charge-to-mass ratios. *Phys. Rev. Lett.*, 74:3544, 1995.
2. G. Baur, G. Boero, S. Brauksiepe, A. Buzzo, W. Eyrich R. Geyer, et al. *Phys. Lett. B*, 368:251, 1996.
3. G. Gabrielse, S.L. Rolston, L. Haarsma, and W. Kells. Antihydrogren production using trapped plasmas. *Phys. Lett.*, A129:38, 1988.
4. G. Gabrielse. *Fundamental Symmetries*, page 59. Plenum, New York, 1987.
5. G. Gabrielse, X. Fei, L.A. Orozco, R.L. Tjoelker, J. Haas, H. Kalinowsky, T.A. Trainor, and W. Kells. Cooling and slowing of trapped antiprotns below 100 mev. *Phys. Rev. Lett.*, 63:1360, 1989.
6. G. Gabrielse, X. Fei, L.A. Orozco, R.L. Tjoelker, J. Haas, H. Kalinowsky, T.A. Trainor, and W. Kells. Thousand-fold improvement in the measured antiproton mass. *Phys. Rev. Lett.*, 65:1317, 1990.
7. G. Gabrielse, H. Kalinowsky, W. Jhe, T. Hänsch, C. Zimmermann, J. Walraven, T. Hijmans, W. Oelert, W. Phillips, S. Rolston, and W. Wineland, 1996.
8. M.H. Holzscheiter, G. Bendiscioli, A. Bertin, G. Bollen, M. Bruschi, M. Charlton, M. Corradini, and et al D. DePedis. Antihydrogen production and spectroscopy, 1996.
9. R.S. Van Dyck, Jr., P.B. Schwinberg, and H.G. Dehmelt. New high-precision comparison of electron and positron g factors. *Phys. Rev. Lett.*, 59:26, 1987.
10. R. Carosi, et al. A measurement of the phases of the cp-violating amplitudes in k^0 to 2 pi decays and a test of cpt invariance. *Phys. Lett.*, B237:303, 1990.
11. J. Ellis, N. E. Mavromatos, and D. V. Nanopoulos. Testing quantum mechanics in the neutral kaon system. *Phys. Lett.*, B293:142, 1992.
12. F. Schmidt-Kaler, C. Zimmermann, D. Leibfried, W. Künig, M. Weitz, and T.W. Hänsch. High resolution spectroscopy of the 1s-2s transition of atomic hydrogen and deuterium. *Phys. Rev. A*, 51:2789, 1995.
13. C.L. Cesar, D.G. Fried, T.C. Killian, A.D. Polcyn, J.C. Sandberg, I.A. Yu, T.J. Greytak, D. Kleppner, and J. Doyle. Two-photon spectroscopy of trapped atomic hydrogen. *Phys. Rev. Lett.*, 77:255, 1996.
14. T. Hänsch and C. Zimmerman. Laser spectroscopy of hydrogen and antihydrogen. *Hyper. Int.*, 76:47, 1993.
15. R.J. Hughes. Antihydrogen and fundamental symmetries. *Hyper. Int.*, 76:3, 1993.

16. G. Gabrielse. Trapped antihydrogen for spectroscopy and gravitation studies: Is it possible? *Hyper. Int.*, 44:349, 1988.
17. P.D Lett, R.N. Watts, C.I. Westbrook, W.D. Phillips, P.L. Gould, and H.J. Metcalf. Observation of atoms laser cooled below the doppler limit. *Phys. Rev. Lett.*, 61:169, 1988.
18. G. Gabrielse, X. Fei, K. Helmerson, S.L. Rolston, R.L. Tjoelker, T.A. Trainor, H. Kalinowsky, J. Haas, and W. Kells. First capture of antiprotons in a penning trap: A kiloelectronvolt source. *Phys. Rev. Lett.*, 57:2504, 1986.
19. M.H. Holzscheiter, X. Feng, T. Goldman, N.S.P King, M.M. Nieto, and G.A. Smith. Are antiprotons forever? *Phys. Lett.*, A214:279, 1966.
20. L.H. Haarsma, K. Abdullah, and G. Gabrielse. Extremely cold positrons accumulated electronically in ultrahigh vacuum. *Phys. Rev. Lett.*, 75:806, 1995.
21. D. Hall and G. Gabrielse. Electron cooling of protons in a nested penning trap. *Phys. Rev. Lett.*, 1996. (in press).
22. U. Schramm and J. Berger and M. Grieser and D. Habs and E. Jaeschke and G. Kilgus and D. Schwalm and A. Wolf and R. Neumann and R. Schuch. Observation of laser-induced recombination in merged electron and proton beams. *Phys. Rev. Lett.*, 67:22, 1991.
23. F.B. Yousif and P. Van der Donk and Z. Kucherovsky and J. Reis and E. Brennen and J.B.A. Mitchell and T.J. Morgan. Experimental observation of laser-stimulated radiative and recombination. *Phys. Rev. Lett.*, 67:26, 1991.
24. L.I. Men'shikov and P.O. Fedichev. Theory of elementary atomic processes in an ultracold plasma. *Zh. Éksp. Teor. Fiz.}*, 108:144, 1995. (JETP **81**, 78).
25. P.O. Fedichev. private communication.
26. M. Glinsky and T. O'Neil. *Phys. Fluids*, B3:1279, 1991.
27. J.W. Humberston, M. Charlton, F.M. Jacobsen, and B.I. Deutch. On antihydrogen formation in collisions of antiprotons with positronium. J. Phys. B, 20:L25, 1987.
28. J.P. Merrison, M. Charlton, B.I. Deutch, and L.V. Jorgensen. Hydrogen formation using positronium, 1993.
29. M. Charlton. Antihydrogen production in collisions of antiprotons with excited states of positronium. Phys. Lett., A143:143, 1990.
30. P.J. Schultz and K.G. Lynn. Interaction of positron beams with surfaces, thin films, and interfaces. Rev. Mod. Phys., 60:701, 1988.
31. D.J. Wineland, C.S. Weimer, and J.J. Bollinger. Laser-cooled positron source, 1993.
32. J.J. Bollinger, D.J. Heinzen, W.M. Itano, S.L. Gilbert, and D.J. Wineland. Xxxxx. IEEE Trans. on Instrum. and Meas.}, 40:126, 1991.
33. C.M. Surko, M. Leventhal, and A. Passner. Positron plasma in the laboratory. *Phys. Rev. Lett.*, 62:901, 1989.
34. T.J. Murphy and C.M. Surko. Positron trapping in an electrostatic well by inelastic collisions with nitrogen molecules. *Phys. Rev.*, A46:5696, 1992.

35. G.-Z. Li and G. Werth. The combined trap and some possible applications. *Phys. Scr.}*, *46:587, 1992*.
36. J. Walz, S.B. Ross, C. Zimmermann, L. Ricci, M. Prevedelli, and T.W. Hänsch. Combined trap with the potential for antihydrogen production. *Phys. Rev. Lett.*, *75:3257, 1995*.

Author Index

Andrews, M.R., 192

Bakarezos, M., 82
Bergquist, J.C., 31
Berkeland, D., 31
Bharucha, C.F., 62
Bollinger, J.J., 31
Boshier, M.G., 328
Bouchoule, I., 358
Brown, B., 446
Brune, M., 1
Burnett, K., 82
Burt, E., 132

Castin, Y., 358
Cesar, C.L., 158
Chambers, D., 82
Chapman, M.S., 374
Cirac, J.I., 16
Cohen-Tannoudji, C., 237
Cornell, E.A., 132

Dahan, M.B., 358
Dalibard, J., 180
Dangor, A.E., 82
de Vries, P., 313
Delos, J.B., 253
Dreyer, J., 1
Durfee, D.S., 192
Dwivedi, L., 82
Dyson, A., 82

Eikema, K.S.E., 169
Ensher, J., 132

Fews, P., 82
Fried, D.G., 158

Gabrielse, G., 446
Gardiner, S., 16
Ghrist, R., 132
Görlitz, A., 343
Greytak, T.J., 158

Hagley, E., 1
Hall, D.S., 446
Hammond, T.D., 374
Hänsch, T., 343
Harms, J., 391
Haroche, S., 1
Hartmann, M., 391
Heimann, C., 446
Hemmerich, A., 343
Hogervorst, W., 169
Holden, M., 82
Holden, P.B., 82

Itano, W.M., 31

Jhe, W., 446
Jin, D., 132

Kagan, Yu., 145
Kalinowsky, H., 446
Ketterle, W., 192
Key, M.H., 82
Khabbaz, A., 446
Killian, T.C., 158
King, B.E., 31
Kleppner, D., 158
Klinkenberg, P.F.A., 221
Kokorowski, D.A., 374
Kox, A.J., 212
Kulander, K.C., 422
Kurn, D.M., 192

Lagendijk, A., 313
Lee, P., 82
Leibfried, D., 31
Lenef, A., 374
Lewis, C.L.S., 82
Lindgren, I., 271
Loukakos, P., 82

Maali, A., 1
Madison, K.W., 62
Maitre, X., 1

Matthews, M., 132
McPhee, A.G., 82
Meekhof, D.M., 31
Mewes, M.-O., 192
Meyerhofer, D.D., 431
Miller, J., 31
Monroe, C., 31
Moore, F.L., 62
Morgenstern, R., 97
Moustazis, S., 82
Muller, H.G., 409
Myatt, C., 132

Neely, D., 82
Norreys, P.A., 82

Okun, L.B., 305
Otten, E.W., 113

Peik, E., 358
Pellizzari, T., 16
Persson, H., 271
Pert, G.J., 82
Ploues, J.A., 82
Polcyn, A.D., 158
Poyatos, J., 16
Preston, S.G., 82
Pritchard, D.E., 374

Raimond, J.M., 1
Raizen, M.G., 62
Roach, T., 446
Robinson, J.C., 62
Rubenstein, R.A., 374

Salomon, Ch., 358
Salomonson, S., 271
Sanpera, A., 82
Schafer, K.J., 422
Schmiedmayer, J., 374
Schöllkopf, W., 391
Shlyapnikov, G.V., 145
Smith, C.G., 82
Smith, E.T., 374
Sprik, R., 313
Stöhlker, Th., 289
Sundaram, B., 62

Sunnergren, P., 271
Surkov, E.L., 145

Tallents, G.J., 82
Toennies, J.P., 391
Townsend, C.G., 192

Ubachs, W., 169

van Coevorden, D., 313
van Druten, N.J., 192
Vassen, W., 169
Vilesov, A.F., 391

Wark, J.S., 82
Watson, J.B., 82
Weidemüller, M., 343
Wieman, C.E., 132
Wilkinson, S.R., 62
Wineland, D.J., 31
Wunderlich, C., 1

Yesley, P., 446

Zeilinger, A., 47
Zepf, M., 82
Zhang, J., 82
Zoller, P., 16